Racecar Technology Full Course

Includes Level One, Two and Three

L1 - Learning All About The Parts and Pieces Of A Race Car
An Entry Level Racers Guide To Race Cars

L2 - To The Race Car – How To Setup For Competition

L3 - Advanced Race Car Engineering and Applications

Bob Bolles

CRD Publishing

CRD Publications

Published by Chassis Research and Development Corporation, aka/Chassis R&D

P.O. Box 730542, Ormond Beach, FL 32173-730542. USA

Email: chassisrd@aol.com

Race Car Technology – Full Course

All photos by Bob Bolles unless otherwise noted.

PRINTING HISTORY

First CRD edition / June 2019

ISBN: 978-1-7324884-6-5

CREATED/PRINTED IN THE UNITED STATES OF AMERICA

Cover design by: LS Bulow, Graphic Designer

Racecar Technology – Level One

Table Of Contents – RT Level One

Race Car Technology – Level Two

Table Of Contents – RCT Level Two

Race Car Technology – Level Three

Table Of Contents – RCT Level Three

INTRODUCTION

Around 1992 I decided to change careers and become a race car engineer. My early experience in racing as a kid was spending countless hours at Daytona International Speedway in the pits, in the stands and around the mechanics, drivers and owners listening and hearing about how the cars handled. I was always fascinated by the design and setup of race cars, be they stock cars, road racing cars, formula or weekend SCCA cars.

I attended many years of races at New Smyrna Speedway and Barberville, now known as Volusia Speedway Park. I was interested in the "race", but always fascinated by the way the cars handled and how all of that was accomplished. I watched Dick Trickle prepare his car in 1975 at a friend's garage in preparation for the Speed Weeks show at New Smyrna and thought, I would like to be able to do that.

I am an engineer by education, degree and by nature and I knew someday I would have to get involved with racing. When that day came, I threw myself into the task of learning and inventing with more energy and determination than at any point in my life with anything I had ever done. It was a passion combined with a purpose. I was determined to find the truth about chassis dynamics and race car setup.

My work, and indeed my racing business, was born out of frustration and failure in trying to find really helpful information that I could use to set up a racecar. So, I set out on a journey that followed in no one's footsteps. Instead, I used one of my greatest personal assets, a profound and acutely developed ability to apply a common sense approach to problem solving. That is exactly what you will find in this book, a common sense approach to chassis setup, vehicle dynamics and race car design, together with solid engineering theory.

This is not a controversial race car setup book and agrees in principle with technology and theory taught in major college motorsports programs. To many, according to what I hear, the books of mine that have preceded this one have become their bible of racing knowledge. Much of the technology presented here has been more recently developed over just the last five years or so. For those who believe that we had already pushed the envelope of vehicle dynamics about as far as it could go by the early 2,000's, this book does not follow that line of thinking.

Regardless of what is on the pages, the proof is on the racetrack, and the methods in this book have been tested and proven to improve performance in race cars. They increase speed, improve basic stability and have already been used to win many races and championships in many classes of racing.

How then did these RCT books come about? When I began my career working with race cars, I found plenty of information on chassis theory, but I couldn't find conclusive information that would tell me how to set up my race car in the shop the right way the first time. I had to read between the lines and keep trying different setups, working by trial and error. I personally don't like trial and error. I want to be able to know exactly how to set up my race car and know not only how something works, but why.

This book will help you avoid the trial-and-error approach to chassis setup. It will teach you sound, proven technology that is both easy to understand and easy to use, so you can set up your race car in the shop and see the positive results on the track immediately, with very little tweaking. What follows is a common-sense approach to chassis setup, vehicle dynamics and race-car design, founded on solid engineering theory. However, you will need to have an open mind, and be willing to accept new ideas that may go against previous chassis setup thinking.

Just to make it clear, the technology presented here applies to all race cars, from quarter midgets to Formula One and everything in between. This book tends to lean towards stock car racing because it represents most of the world's automobile racing. But know that not only will it be useful for all forms of circle track racing from asphalt types to dirt cars, a great deal of the technology applies to all race cars.

Success comes at all levels of endeavor, and we can't all be champions. But we can all get better at what we do. The goal of this book is to give good, solid information that has been tested and evaluated and found to be the truth. It is not, and will never be, complete as long as we continue to push the envelope in the search for better performance, but it will lay the foundation upon which future race engineers can build their programs.

Race Car Technology – Level One

Welcome To Race Car Technology - Level One

Thank you for enrolling in the RCT Level One course. You have taken the first step in furthering your knowledge and understanding of the systems and parts of a dedicated race car. This course will explain all of the relevant systems race teams and race engineers work with on a daily basis. This course will teach you how to speak the language of race car technology.

Level One provides the foundation for the next three Levels in the Race Car Technology school. There is information contained in this Level that has never been presented before in any format. For that very reason, even seasoned veteran racers will come to appreciate the presentation of this unique knowledge.

The information presented is in some ways about specific race car types, and then many ways, much of what you will learn is relevant to all race car types. The basic technology is common among race cars from simple street stock circle track cars to sophisticated road racing and formula cars.

The understanding of, and the setup for, any race car depends on certain common knowledge of all of the parts of the race car, and how all of those parts interact. Whether it involves a late model circle track car, or a world class formula car, what you will learn here will be relevant to any and all race cars.

A very sophisticated formula car shares many common traits with less sophisticated race cars. The part and pieces are constructed differently and the costs are much different, but all in all, we are talking about a race car. With this Course you will be able to piece together a race car part by part so that you will come to fully understand how they all work.

Please know that you can proceed at your own pace. This is not a race and you will get no better grade for finishing quickly. In fact, it is much better to take your time and try to fully understand the concepts and information contained in each of the Lessons presented here.

The Lessons are arranged in a predetermined order so that you can piece this all together at the end. In fact, the order in some cases is critical so that the information builds on what has previously been discussed.

There will be a test at the end of each Lesson and a Final Exam at the end of the course. These exams are necessary to ensure that you have grasped and learned all of the information presented here that will allow you to earn your Certificate of Completion upon completion of the three Levels for Race Car Technology. By passing these exams, you will have proven to yourself and to others that you have retained important knowledge about race car technology.

So, let's get started. As you go to Lesson One, again remember to take your time and repeat reading each section as many times as it takes to understand the concepts and information presented. Good luck and let's get started.

Race Car Technology – Level One

Lesson One – Goals: What We Trying To Achieve

The Race Car Technology – Level One Course is designed to introduce the student to the parts and pieces of a race car. We will explain what the systems are, what they are used for, and how these systems work with the other components on the race car. Many of these systems are common among various race car types and uses. Once you have completed Level One, you will know what the systems are, and what to call the parts used to make up the systems. This is an important and necessary course leading into our RCT Level Two Course, which takes us to the actual setting up of the race car.

Before we get into the nuts and bolts of race car technology, we need to fully understand what our goals are for this school and exactly what we are trying to achieve in RCT Level One. For every racer, no matter what class, type, sanction or level of expertise, the goal is simply to win races. To win races you need to be faster than any of the other competitors. To be faster, you need knowledge. We are going to give you some important knowledge.

The beginning of knowledge, someone once said, is the admitting that you don't know. Not knowing says you do know there is knowledge out there that you have not yet been introduced to. Being open to this philosophy helps you to seek and gain knowledge throughout your journey in life.

For any student to proceed through a course, you need to be aware of and understand the words that are going to be used in the process, and the meaning of those terms. You can always refer to the Glossary if you forget or need to remind yourself.

The race engineer's goals are very simple in concept, but more complicated to carry out. On any race course, be it with circle track or road racing, on a dirt or asphalt surface, there are key areas of performance where your car needs to be made better than any of your competition and those are the following;

The goals for designing our race cars are basically the same be it a formula road racing car or a late model circle track car. Performance gains in the slower portions of the track cause all of the speeds to increase around the track including on the straights.

Maximizing Grip - When non-chassis elements like powerplants are more or less equal, Grip is what will help you win races and championships. Motors accelerate you, brakes slow you, but Grip makes you faster through the slower turn portions of the track and relatively small gains in lateral Grip will produce huge gains in speed and performance.

What are the Key elements of Grip? The tire contact patch is where Grip is produced. Causing the tire contact patch to be larger increases Grip and causing the most vertical loading on that tire contact patch are the two key ingredients for maximizing Grip.

Maximizing the amount of traction that is available from the four tires on a race car, any race car, will make you as fast as you can be, all other things being equal. Everything we present in RCT L1 will ultimately lead to optimization of the race cars Grip and the use of that Grip to go faster.

What are the basic parts that make up Grip?

• Loading On The Tire – The more load we can put on a tire, the more Grip that tire will have, period. But, this is gain in grip is not linear as we will explain later on. Load can come from the weight of the car, mechanical downforce from banking, and aerodynamic downforce, all of which will be explained in detail later on.

• Contact Patch Area – The greater the size of the contact patch, the more Grip. If we can find ways to make the tire contact patch larger, then that tire will produce more Grip.

• Tire Compound – The physical and chemical makeup of the tire can provide more Grip. The softer the material, the more Grip within limits. We have rules that govern the softness of the tires, but we need to stay very close to those limits.

• Load Distribution – A pair of tires on the same end of the car, or same axle, will produce more Grip when they are more equally loaded. The most Grip from a pair of opposing tires will come when they are equally loaded. There is a variation to this concept for dirt racing that will be addressed later on.

• Angle of Attack – What is called Angle of Attack, or Slip Angle, is when a tire is pointed slightly to the inside of the arc it is following through a turn. If a tire were to follow the exact tangent line around a curve or arc of the turn, it would not generate any side force to counter the centrifugal force.

So, in consideration of the other items that make up Grip, it is fair to say that none of those would be useful if it weren't for the creation of Angle of Attack. No matter what amount of Loading or size of the Contact Patch area is, or how soft the tire compound is, or how equal the load distribution between opposing tires, the car would not stay on the course without the tires developing an Angle of Attack.

So, there you have it. Those five things represent the parts that help make the Grip we seek to make us faster so we can win races. As we go through each part of the race car in the other Lessons, we will explain how to optimize of those parts to enhance our Grip and make us faster through the slow speed turns. And we will understand how that will in-turn make our whole lap faster at every point around the course.

Why Does More Grip Make Us Faster?

When a race car turns, a lateral force called Centrifugal Force tries to push the car to the outside of the turn. The tire contact patches resist this force. The speed we can drive through the turn is limited by the amount of Grip we have in our tires. The more Grip, the faster we can drive through the turns.

The one often overlooked benefit of achieving faster turn speeds is this: the faster you exit a slow speed turn, the higher the speed at which you will start accelerating down the straight part of the track in-between the turns. So, speed gained in the slow speed turns will be carried down the straights too. It's not just that we gain speed in the slow speed turns with more Grip, we gain everywhere around a race track.

To give you an example, on a typical or average length Formula One track, one tenth of a second is about 17 feet on the race track. Some teams are a full second slower per lap than the fastest teams. That is 170 feet per lap that the faster cars move ahead of the slower cars each lap. In a fifty lap race, that equals 1.6 miles, or 2.5 kilometers.

We often think of the formula one cars as being the ultimate race cars and perfect in every way. If that were so, then there wouldn't be a two second gap between the fastest and slowest cars. There must be a difference in designs that causes that huge gap in speeds. We like to think it is lack of mechanical grip that makes not only the slow turns faster, but the entire lap.

We already know that some teams get lapped in a F1 race, and the average length of those tracks is around 3 miles. So, those lapped cars are on average 2.0 seconds per lap slower than the winning team. And they all run with the same choice of tires. That is hard to fathom.

Is the gain realized by the winning F1 teams all Grip? No, as we have read, some engine packages are down on horsepower and that is a factor in the difference in lap times. But what about the teams who have the same engine package as the winners? Why are they so slow? It could be that they lack mechanical grip in the slower portions of the race track.

Building mechanical grip is one of the most important areas of race car engineering, even over and above aerodynamic grip. The reason is this; To gain maximum aero grip, you must have high speeds and the most gains from mechanical grip happen in the slower turns where the speeds and the aero downforce is lower for all of the teams. So, the only thing we can point to as the reason for the gain in performance for the winning teams is Mechanical Grip.

The Concept Of Mechanical Balance - If your car has more overall Grip than other cars early in the race, that Grip might not stay superior and you might end up being a slower car through the turns later in the race. This is due to the Balance Factor. There are two definitions of balance. They are not the same and they cannot be confused. One is handling balance which is defined as a car that is neither pushing (understeering) or loose (oversteering). Pushing is when the front set of tires has less Grip than the rear set of tires. Loose is the opposite.

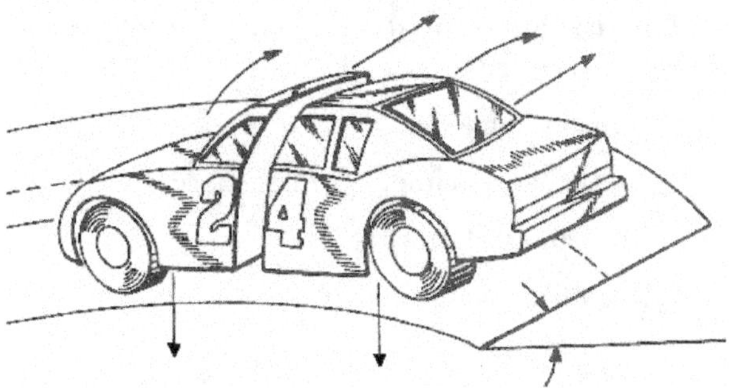

The Mechanical Balance concept is simple. We want both ends of the car to work in sync so that one does not try to cause the other to do what it does not want to do. When the two ends of the car are working in unison, we can achieve the goal of mechanical balance.

A team can attain handling balance fairly easily, but just the fact that the car is neutral in handling doesn't translate to speed or consistency. A car can have less overall Grip than other cars and still be neutral in handling. Although a neutral handling car is a goal, it is not the first and primary goal and not what we are discussing here.

The other balance is Mechanical Balance. Simply put, this is when the two suspension systems, front and rear, are working in sync and where the load transfer is predictable and maximized. This is what teams need to search for. The end result of all we do with chassis engineering must be towards the goal of Mechanical Balance.

Circle track race cars are complicated, but much easier to setup that a road racing car. This is because the setup in a circle track car can be asymmetrical, meaning the spring rates, cambers, etc. from side to side can be different. The circle track car is only turning in one direction and so the setup need only be correct for that direction of turning.

The Search and Desire for Mechanical Balance should be the foremost goal for every race engineer. A lot of good things come with achieving MB. The tires do not work as hard and will last longer. The car is much easier to drive. The car maintains higher turn speeds as the tires wear later in the race, or later in each stint for cars that are allowed to pit and take on new tires.

So, to reiterate what we have covered so far, our goals for this course are to gain knowledge, understand the terminology, and to learn how to setup a race car to maximize Grip in a way that utilizes a Mechanical Balanced state. In the Lessons to come, we will be presenting information you will need to understand and apply principles that will help achieve these two important goals.

And we are not leaving aerodynamic downforce out of the discussion, we just need to separate the mechanical Grip from Aerodynamic enhanced Grip. We need to do this because, many times in the search for the maximum aero properties, a team might deviate from the goal to achieve a Mechanical Balance. We think you can have the best of both worlds.

Exam In The Context Of This Lesson:

What is the beginning of knowledge?

1) Reading books
2) Working on a race team
3) Asking Questions
4) Admitting You Don't Know Everything

What wins you races and championships?

1) Superior Grip
2) Handling Balance
3) Mechanical Balance
4) All of the above

Which of these are basic Elements of Grip?

1) Tire Pressures
2) Tire Temperatures
3) Contact Patch Area
4) Loading
5) 3 and 4

What is the name of the force that tries to push the car off the turns?

1) Downforce
2) Drag
3) Centripetal
4) Centrifugal
5) Gravity

Low speed corner gains from Mechanical Grip increases performance where?

1) Down the straights
2) In the high-speed turns
3) At the start of the race
4) At the end of the race
5) All of the above

The first and primary goal for performance in a race car is?

1) Equal weights on all four tires
2) A car that is neither tight nor loose
3) Mechanical Balance
4) Aerodynamic Downforce

Lesson Two - Why Do Race Cars Turn?

• ***Tangent Line of An Arc*** – The tangent of an arc is a line that is at right angles to a line from a point on the arc to the radius. This tangent forms a "T".

• ***Angle of Attack*** – A direction of a tire that is pointed towards the inside of the turn from the tangent line.

What we are about to explain is a concept that has never been fully explained as far as we know. It is a simple, but effective, description of why a car can drive through a turn at high speeds. It is one of those items of new knowledge that we promised the veteran racers they would get when considering enrolling in RCT Courses.

In Lesson One, we talked about how a tire develops Grip. One of the key ways is from Angle of Attack or Slip Angle. A tire running parallel to the path of the arc of the curve cannot generate the Grip it needs to maintain its direction along the arc. It would "slip" off the arc.

Every race car in the world, or 99.999 percent of them, is a front-end steering car. That is, the driver turns the car using a steering mechanism that is used to change the direction and angle of the front wheels.

In this way, the driver creates an angle of attack for the front tires so they can develop the side force necessary to counter the centrifugal force. Without this Angle of Attack, the car would not turn no matter how much tire loading, or size of the tire contact patch there was.

But what about the rear tires? We cannot steer them, setting aside the effects of mechanical rear steer that will be covered later on. How do they develop the angle of attack needed to stay on the arc of the turn? The answer is: they already have their angle of attack, and here's why.

In Figure 2-1 you will see how each tire follows its own arc. The arc radius of the inside (inside of the turn) tires is less than that of the outside (outside of the turn) tires. In Figure 2-2, each tire is placed over an exaggeration of the arc lines with a radius equal to the distance from each set of tires to the common radius point of the arcs.

In the Figure 2-2 sketch, all four tires are pointed straight ahead to show their relationship to the direction of the tangent line. One arc is for the inside, or left side tires and one is for the outside, or right side tires in this example.

As you can clearly see, at the front, the tires are pointed away from the tangent line. For those tires to turn the car, not only do they need to turn left to the direction of the tangent line, they need to go a little bit further to create the angle of attack needed to develop the amount of Grip needed to stay on course.

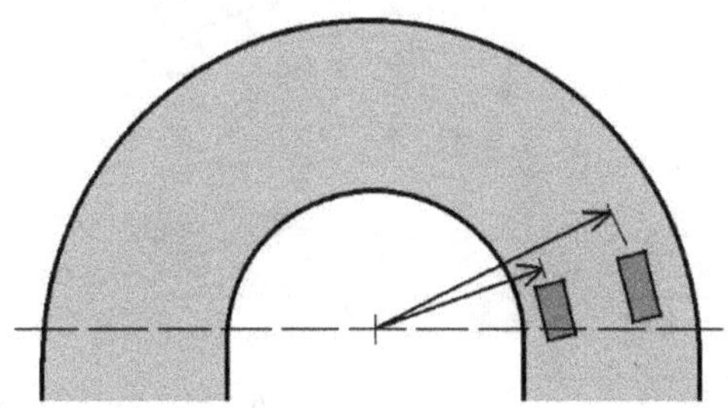

Figure 2-1

In this illustration, we show how the inside and outside tires have different radii and must follow different arcs when driving through the turn. A line perpendicular to the line extending to the radius point is called a Tangent. To develop the Grip needed to counteract Centrifugal force, all of the tires must be pointed to the left of the Tangent line.

For wheels that can be steered, this is easy. The driver just need to turn the steering wheel in that direction. And the driver will do that to the extent and to the angle needed to turn the car and follow the line they think they need to run through the turns. Sounds pretty simple, right?

Now let's look at the back tires. The direction they are pointed is to the left of the tangent line. Unlike the front tires, they already have an angle of attack, whether or not what it has is exactly what is needed to stay on track. Interesting, right? It gets better.

Now let's say there's not enough angle of attack in the rear tires to keep the tires on course and the rear tires slip a bit. Guess what, as the rear tires slip, the angle of attack increases. In reality, for a race car, the rear will slip if it doesn't have quite the Grip it needs until the needed angle of attack is reached and then it will stop slipping. It is therefore self-compensating.

The front tires react the opposite if they slip. The angle of attack decreases and the steering must be increased in order for the car to stay on the course. We will tell you more about steering angles and angles of attack later on. There are limits to what angle the front tires can be steered to.

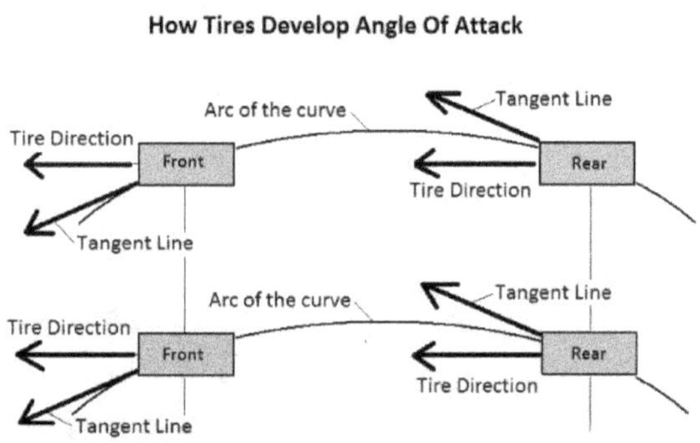

Figure 2-2

The front and rear tires are pointed in different directions in relation to the Tangent of the arc they are following. In the front, the tires are pointed away from the Tangent and in the wrong direction needed to turn the car.

We must steer those tires left in this left turn in order to not only track parallel to the Tangent line, but a bit more to develop an Angle of Attack needed to generate the amount of Grip necessary to keep the car on the course.

At the rear, we have a different situation. With the rear tires positioned straight ahead, they are already pointed left of the Tangent line and have some Angle of Attack.

If the angle is less than needed, the rear of the car will slip out until the tire angle is enough to provide sufficient Angle of Attack needed to stay on the course.

When you come to understand this concept of how the car is able to makes it around the turns, it will answer a lot of questions people might ask about vehicle dynamics. I have often wondered, given that we have a lot of setup tools to make the front ends turn, why and how does the rear turn with the front? Now we know.

Here we can actually see the lines of the arc of the turns. The front wheels are steered to create the Angle of Attack needed to develop the force to counter the G-lateral forces. The rear tires are already pointed to the inside of the turn and have their own Angle of Attack. This is why race cars turn the corner.

Know that every race car, no matter the design or classification, is subject to the concept explained above. Farther along in this Course, we will get into various design effects, goals and geometry for the rear suspension whereby we can change and fine tune the angle of attack for the rear tires.

Although this Lesson is relatively short, it is vital to understanding much of the coming presentations about vehicle setup and dynamics and the parts and pieces of the race car that make all of that happen. We'll keep referring back to this concept in Lessons to come.

Exam - In The Context Of This Lesson:

Which Element of Grip Is Essential Above All Others?

1) Tire Loading

2) Tire Compound

3) Angle of Attack

4) Contact Patch Area

How Do We Make The Front Tires Develop Angle of Attack?

1) Add More Camber

2) Add Tire Pressure

3) Drive Slower

4) Turn the Steering Wheel

How Do We Make The Rear Tires Develop Angle of Attack?

1) Add More Camber

2) Add Tire Pressure

3) Drive Slower

4) We don't need to, they already have Angle of Attack

How Do We Adjust The Angle of Attack For The Front Wheels?

1) Lengthen the Tie Rods

2) Change The Speed

3) Change The Tire Cambers

4) Turn the Steering Wheel Either Way

How Do We Adjust The Angle of Attack For The Rear Wheels?

1) Change The Tire Stagger

2) Change The Tire Pressure

3) Adjust The Speed

4) We Don't, The Rear Is Self-adjusting

Race Car Technology – Level One

Lesson Three – Cambers: Definition and Effect
The Front Geometry Group

When we start to develop our Grip, we start with the tires. There are several components related to tire optimization, or making the tire happy. Camber is one of those components.

Camber Is - Simply put, Camber is the tilting of the tire/wheel combination either with the top of the tire in towards the center line of the chassis, or out away from the centerline of the chassis. It is measured in degrees of tilt from vertical relative to the racing surface.

When the top of the tire is tilted in, that is called Negative Camber. When the top is tilted out, that is called Positive Camber. Wheel Cambers are adjustable in most race cars. And wheel cambers can change with chassis motion.

For this discussion, we will be talking only about cambers in a double A-arm suspension. Stock cars mostly have only front AA-arm suspensions and a solid axle rear suspension. Many prototype racing cars and formula cars have independent AA-arm front and rear suspensions.

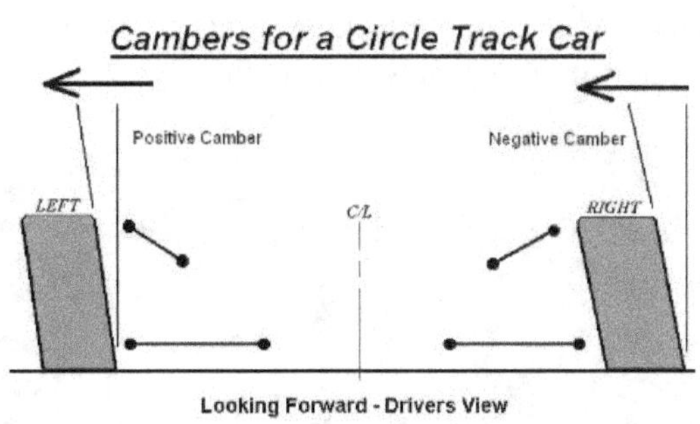

The reason why we want and need camber is because the tire distorts as lateral loading is applied to it when cornering. This distortion changes the contact patch shape and may cause the tire to have less Grip. The outside tire (meaning outside of the turns away from the radius) carries more load in a turn and therefore needs more camber than the inside tire.

So, we add negative camber (top tilts in) to the outside front wheel. For a circle track car turning in only one direction, we add negative camber to the outside wheel and positive camber to the inside wheel.

For road racing cars, we need to set negative camber to both front wheels since we will be turning left and right and in a right hand turn the left front tire will need negative camber just as in a left hand turn the right front tire will need negative camber.

The amount of camber that a particular tire needs is dependent on three factors. These are: 1) The tire composition, mainly the side wall stiffness (softer sidewalls require more camber), 2) the amount of lateral loading (the more loading the more the tire will roll over and then the more camber we will need), and 3) the camber change characteristics of the suspension.

So, let's recap. A tire needs camber to correct what the tire contact patch does when cornering with a high lateral load. Since the tire distorts, the camber serves to correct the situation and provide a more ideal contact patch shape.

A tire will distort to the inside of the turn since the rest of the tire is trying to move out away from the turn. In circle track racing, we apply positive camber to the inside tire and negative camber to the outside tire.

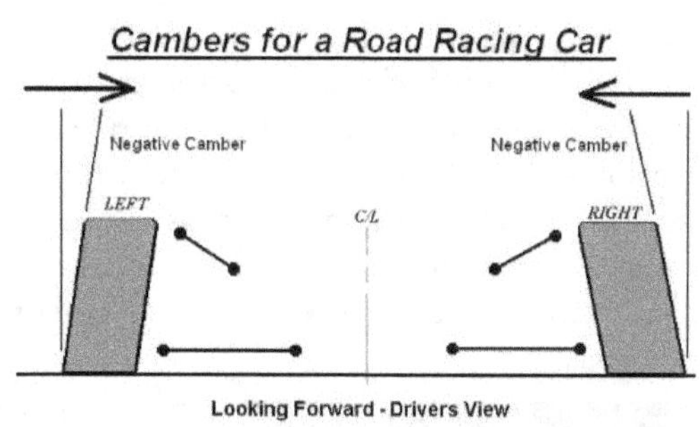

In road racing, both tires will be the "outside" tire depending on which way we are turning, so we cannot put positive camber in either wheel.

Let's say we took a circle track car, that only turns left, to a road course. It would turn very well in the left turns, but in the right turns, both of the tires would have the opposite of the direction of camber it needs to correct the tire distortion.

Again, in road racing we sacrifice the tire contact patch of the inside tire to benefit the outside tire that will be carrying the most loading through the turns. And there may be ways to minimize that sacrifice that will be explained later.

Camber Change – In a modern AA-arm suspension, as the chassis dives and rolls through the turn, because of the design of most AA-arm suspensions, the wheel camber will change relative to the racing surface. Camber Change can be designed differently to optimize or minimize the camber change for a particular chassis design.

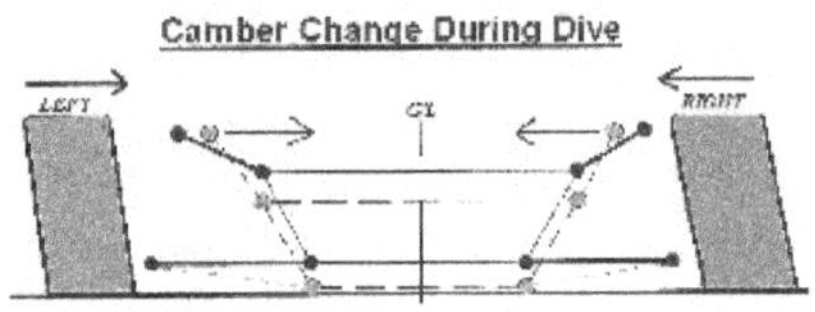

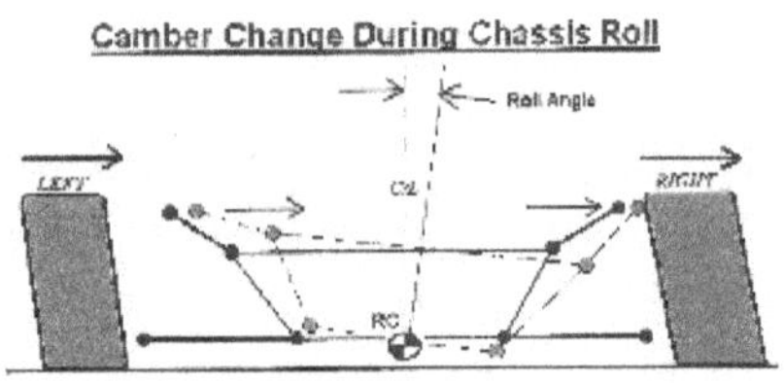

Camber Change is related to the combination of motions of the chassis. In a turn, the chassis moves. Most of the time, in all race cars, it gets lower and it rolls.

It has been common in the past to talk about camber change only related to vertical chassis movement. A team might set the car at ride height on supports and then move the wheel up to see how much the camber changes in degrees per inch of movement. This is camber change, but only from one single part of the movement of the chassis.

We can measure the static camber in our wheels by using a bubble level gauge and directly read the camber amount.

We can also measure the static camber in our wheels by using a digital gauge that reads the camber amount in degrees and decimals.

For the purpose of this Course, and to look at camber change in a more realistic way, we will only be thinking about camber change when it is related to both the vertical chassis movement which is combined with chassis roll. Here is why.

If we roll the chassis without moving it vertically (i.e. the center remains at the same height), the front wheel cambers will change the same amount in degrees as the chassis rolls. If we move the wheels vertically without rolling the chassis, the wheel cambers will change some amount, but in the opposite direction. So, if we combine the movements, we can then determine what the net camber change is relative to the racing surface. That is what matters.

In extreme cases, with very soft sidewalls, it might be an advantage to run very high amounts of negative camber in the outside wheel in order to generate the greatest tire contact patch amount.

Designing Camber Change – We can change the amount of camber change in some race cars. If you can make changes to the arm lengths and/or arm angles, then you can change the amount of camber change.

Installing shorter or longer upper control arms, or making changes to the angles of the upper control arms will change the amount of camber change.

In most cases, race tires do not like camber change. When entering a turn, the tire takes a set, or rolls over and begins to develop Grip. If the cambers do not remain stable, the tire will take more time to adjust, and frankly, we don't have a lot of time for that.

The result is loss of Grip and loss speed as the tires struggles to find Grip. So, in modern race car design, we need to plan out our AA-arm geometry so that there will be minimum camber change. How do we do that? There are several ways to accomplish the goal of minimum camber change.

Camber is easily adjusted for this circle track car by changing the spacers between the frame mounting plate and the upper control arm shaft. For formula cars, the cambers are changed in a similar way, by moving the inner mounts of the upper control arms or extending the length of the upper control arms.

First, we can arrange the control arms with angles so that with the dive and roll we expect going into and through the turns, will result in minimum camber change. This has been done with great success in what we call conventional setups with both stock cars and some formula cars.

Another way to achieve minimum camber change is to reduce the amount of dive and roll. The setups in Formula One cars are very stiff and the chassis does not move vertically very much. This reduces the amount of camber change. The reverse angle in an F1 car (meaning that the inside chassis mount is higher than the ball joint) of the upper control arms may be beneficial and cause a small amount of beneficial camber change. We will get more into that in the Level Two RCT course.

The third way we can minimize camber change is through the use of what we call bump stops. In modern race car design, some chassis are designed to travel a predetermined amount and then the shocks contact a bump material and basically stop motion.

These bumps can be made of hard rubber, polypropylene or actual steel springs. In each case, when the shock contacts the bump device, the vertical chassis motion is reduced to very small movements and the cambers change very little while the suspension is on the bumps.

Cars that use bumps include dirt late models, asphalt late models, stock car touring divisions, IMSA prototypes and Formula One. There are other benefits to using the bump setups, but reducing camber change is one of the most significant benefits realized by running on bumps.

Ultimate Camber – The Static camber settings (without dive and roll with the car is at ride height and not moving) are something teams experiment with to try to get the best Dynamic (after dive and roll) cambers that will provide the most Grip. Because all car designs and setups are different, so too are the camber changes for each car.

The ultimate Static cambers for one race car might be different than what is optimum for another in the same class using the same tires. The differences are caused by: 1) Variations in tire pressures, 2) Different front geometry designs that result in different camber change from similar chassis movements, and 3) Different stiffness in setups that will cause different dive and roll amounts.

We are going to cover suspension design, tire technology, and chassis setup dealing with stiffness in later Lessons in RCT Level One, but we want you to understand the basic characteristics of camber and camber change going forward.

Exam - In The Context Of This Lesson:

Tire/Wheel Camber Is The Degree of Tilting Of The Wheel From What?

1) Horizontal

2) A line at right angles to the racing surface

3) A Vertical Plumb Bob

4) A line through the ball joints

Negative Camber Is?

1) When It Moves The Wrong Direction

2) The Top Is Father Out Than The Bottom

3) When It Doesn't Change At All

4) The Top Is Closer To The Centerline of the Car

We Need Camber Because?

1) The Tire Distorts When Lateral Force Is Applied

2) The Suspension Points Move In Dive And Roll

3) Camber Helps Generate More Lateral Force

4) All Of The Above

Camber Change Occurs Because?

1) The Wheel Moves Up and Down

2) The Car Rolls

3) The Tire Distorts From Lateral Force

4) A Combination of 1 and 2.

What Changes To The Suspension Will Affect The Amount of Camber Change?

1) Changing the Spring Rates

2) Installing different length control arms

3) Changing the angles of the control arms

4) All of the above

How Much Camber Changes Do We Need?

1) Enough to get what the tire needs

2) An amount equal to the tire roll over

3) As much as possible

4) As little as possible

What Are Common Ways To Minimize Camber Change?

1) By running very stiff setups

2) Through special design of control arm angles

3) With the use of bump devices that limit suspension movement

4) All of the above

The Ultimate Cambers We Need Are Determined By?

1) The Amount of Lateral Load

2) The Degree of Camber Change In Our Suspension

3) The Stiffness of the Tire Sidewall

4) All Of The Above

Race Car Technology – Level One

Lesson Four – Caster: Definition and Effect

Caster and Camber are often talked about together, but they are much different in their design and use. And Caster can cause camber change, but that is not its normal use.

Caster is a design function in a vehicle to promote directional stability. It helps the vehicle to want to roll straight ahead and not deviate from a straight line.

Caster is used in many different types of vehicles, even grocery carts. In bicycles and motorcycles, it is the slant of the forks where the front wheel axle is well forward of the fork tube just below the handlebars. When weight is put on the bike, the wheel wants to point straight ahead.

If we turn the bike wheel, it tends to want to raise the bike. This goes against or opposite of the weight pushing down on the wheel from the weight of the bike plus your body weight. The weight and angle of the forks will resist any wheel movement from straight ahead. That is what caster does in a bike, and that is what caster does in a race cars front steering system.

What Caster Looks Like - Caster is set in a double A-arm front suspension by positioning the ball joints at an angle looking at them from a side view. There is Positive and Negative Caster.

In the side view, if the top ball joint is to the rear of the car in relation to the bottom ball joint, we call this Positive Caster. If the upper ball joint is forward in relation to the lower ball joint, we then have Negative Caster.

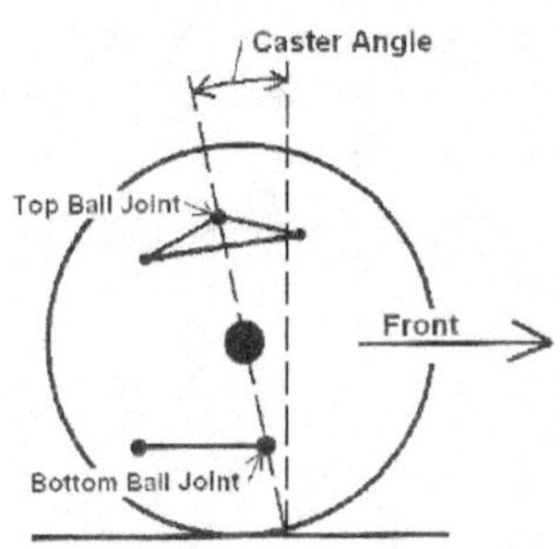

Caster is created when we design the suspension in a way that places the upper ball joint to the rear of the lower ball joint. This is called Positive Caster. If the upper ball joint were to the front of the lower ball joint, then we would call that Negative Caster.

Caster is measured in degrees of caster. This is the exact number of degrees that a line through the upper and lower ball joints centers of rotation makes with a vertical line as viewed from the side. It can be measured directly or with a tool called a camber gauge. We'll explain how to use this gauge later on.

Caster in a race car can make turning the steering wheel harder. The more caster, the harder it is to turn the steering wheel. With the use of power steering in a race car, this effect is reduced or eliminated.

How Much Caster? - We usually set the caster amounts, or degree of caster, the same for each front wheel in a road racing situation. The trend for road racing cars is to set high amounts of caster and it is thought that this enhances other effects of caster. More on that later.

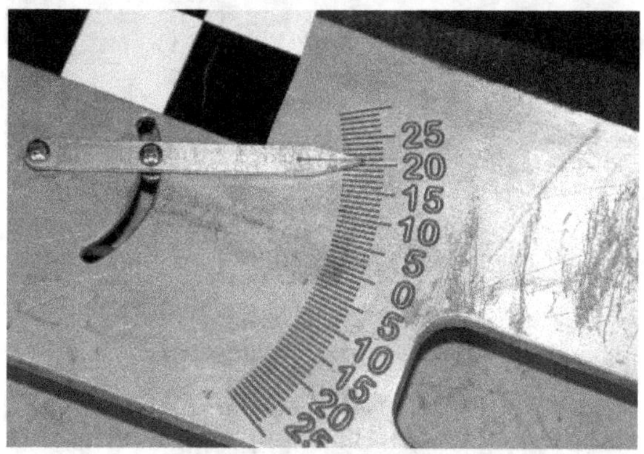

To measure caster, we first need to use a turn plate or other means to establish the number of degrees we are turning the wheel. The accepted SAE, or Society of Automotive Engineers, method is to turn the wheel twenty degrees one way from straight ahead and note the camber reading. Then we turn the wheel to twenty degrees in the opposite direction from straight ahead and remeasure the camber. The difference in camber amounts is the caster.

For circle track racing where the cars always turn one way, the caster settings for the inside wheel is made less than what we set for the outside wheel. The inside wheel caster may even be Negative, or with the upper ball joint forward of the lower ball joint. This is commonly called Caster Split or even Caster Stagger.

We stagger the caster settings in circle track racing to cause the car to want to turn into the corner. For circle track racing in the U.S. where the car turn left, the caster is always less on the left side than the right side, if you were to release the steering wheel when rolling forward, the car would naturally turn left on its own. The more caster split we design into the suspension, the more it will want to turn left.

Caster Changes - Caster Change is caused by wheel movement when there is a difference in upper and lower control arm angles when viewed from the side. If the upper control arm is angled from the side view and the lower control arm is either level or angled the opposite way, the caster will change when the spindle moves vertically.

If you use a camber gauge, when you turn the wheel out twenty degrees, you set zero in the gauge as shown. This is the right front wheel on a stock car and it is turned out.

We then turn the wheel in twenty degrees from straight ahead and re-read the camber gauge and this tells us the amount of caster.

When there is a difference in side view angles of the control arms associated with one side of a AA-arm suspension, the ball joints will move forward or rearward relative to each other and that changes the side view angle of the line running through the ball joint centers which defines caster.

When this line changes its angle, then the caster is also changing. We need to be aware of these changes when designing our race cars. If the right caster is increasing with vertical movement and the left caster is decreasing, or even staying the same, we are gaining caster split and the increasing the effect of the caster split.

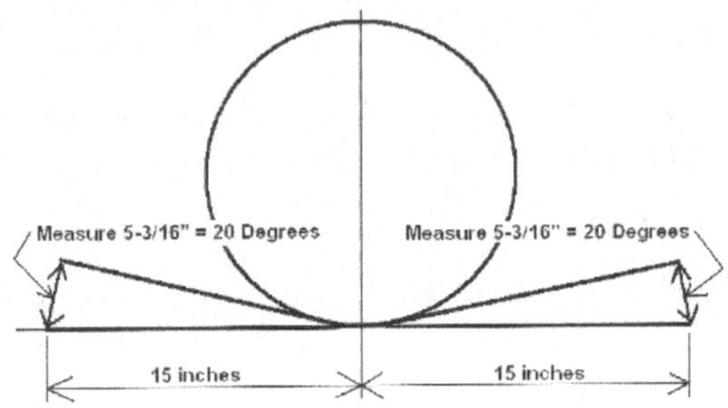

When you measure caster using a gauge, you need to know how far 20 degrees of rotation is. If you don't have turn plates to read the degrees of rotation, then this sketch shows you how to mark the 20 degree line on the floor so that your turning the tire the correct number of degrees.

Camber Change Due To Caster - As we turn the steering wheel, caster causes a change in camber for each wheel. The amount of camber change depends on the amount of steering and the amount of caster.

If we turned the wheel both ways 90 degrees from straight ahead, the camber would change the exact number of degrees as the caster amount. This is how caster gauges measure caster. They are designed to record in a similar way to what we described above. With the gauge, we turn the wheel out from straight ahead by exactly twenty degrees, set zero in the gauge, then turn the wheel to be pointed in from straight ahead and then read the gauge. It will read the degree of caster.

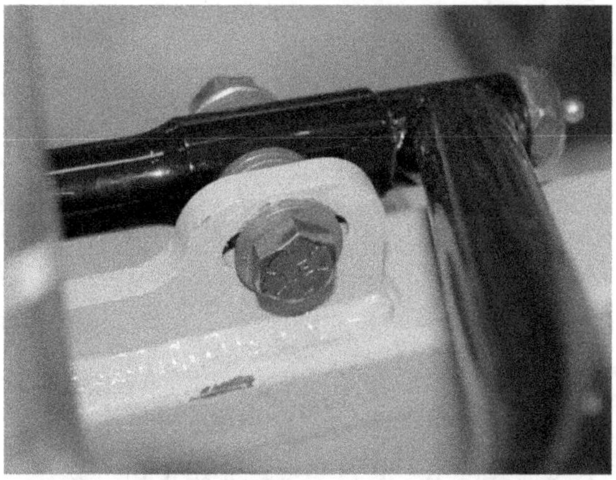

This upper control arm mount has a slot in the mounting plate that allows changes to the caster. By sliding the upper control arm shaft fore or aft, you can reduce or gain caster.

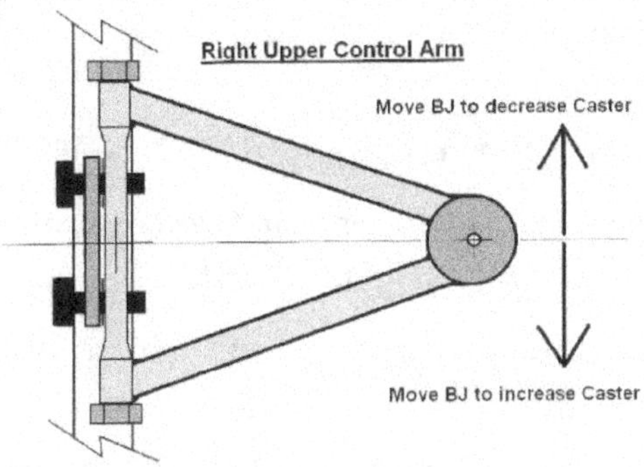

We make changes to caster in a stock car by moving the upper ball joint forward or rearward. Moving it to the front reduces the caster and moving it towards the back of the car increases the caster. Teams can install upper control arms with offset ball joint mounts to create the desired amount of caster.

There are other effects associated with caster, but we will cover those in our Lesson on Spindles later on. But to re-cap for you, Caster is a self-aligning effect that helps keep the wheels pointed straight ahead. Caster is caused by mounting the ball joints so that the upper ball joint is positioned to the rear of the bottom ball joint. This is referred to as Positive Caster.

If we have different stagger amounts on each side of the car, the car will want to steer in the direction of least caster. Running less caster in the left front of a left turning circle track car will cause the car to want to steer left on its own, without driver input.

And, caster causes camber change when the wheel is turned, along with other settings in the front suspension. The combined net amount of camber change is dependent on the other factors in the design of the suspension.

Exam - In The Context Of This Lesson:

Caster Is The Degree of Tilting Of The Ball Joints From What?

1) Horizontal In A Front View

2) Horizontal In A Side View

3) Equal to Camber

4) Vertical In A Side View

We Need Caster Because?

1) It makes the car turn easier

2) It adds directional stability

3) It helps the tires gain Grip

4) 1 and 2

Caster Is Measured By Doing This?

1) Measuring the height of the ball joints

2) Knowing the control arm angles

3) Measuring the degrees of sideview angle formed by the ball joints from vertical

4) Using a caster gauge

5) 3 and 4

Caster Split Is Caused By?

1) The change in caster from wheel movement

2) The movement of the spindles

3) The difference in the left and right caster amounts

4) Turning the steering wheel

Caster Change Is Caused By?

1) The lengths of the control arms

2) Turning the steering wheel

3) Wheel movement with differences in sideview control arm angles

4) Chassis dive and roll

Lesson Five – Spindle Design

The Spindle is the connection between the chassis and the wheel acting through the control arms. The spindle is a solid component that attaches to the upper and lower control arms through two ball joints, upper and lower, that are free to rotate in three dimensions to some extent.

The wheel is attached to the spindle by being mounted on the spindle pin. On this pin are mounted two sets of bearings that are connected to bearing races within the wheel hub. This completes the connection between the chassis and tire contact patch.

The spindle can be designed in many different ways to affect camber change, caster and caster trail, and primarily scrub radius.

This is a common, stock type of spindle (right of the shock and left of the orange colored brake rotor)that attaches to an upper and lower ball joint. The other connection is the tie rod end that is bolted to the spindle steering arm shown at the very front of the spindle.

The modern fabricated spindle used on circle track and road racing cars is made from welding flat plate steel into the shape needed for the application. These pieces are lighter in weight than the older cast iron spindles.

Spindles can be made out of any material that is practical, like this one made from carbon fiber using aluminum bolt on parts to hold the wheel spindle pin, ball joints and steering links.

The variations in spindle design are made to accommodate different race car designs. There are three basic design aspects for a spindle.

• Ball Joint Spacing – the distance between the ball joints for each spindle design is set by the spacing between the ball joint mounting holes is a stock car spindle, or the spacing between the heim joint mounts in a formula type of race car.

• Spindle Inclination – When the spindle is viewed from the front, the ball joints or centers of heim joints are staggered from a vertical line. In every case, the upper ball joint is placed towards the center of the chassis more so than the lower ball joint. The line through the ball joints when viewed from the front is the Spindle Inclination or what is sometimes called king pin angle, a reference back to the days of straight axle cars. This angle can vary depending on the design goals for each race car.

• Spindle Pin Height – The height of the spindle pin in relation to the positions of the ball joints varies depending on the design goals for each race car.

The spindle on this prototype is made from high strength aluminum. The upper control arm is mounted to the spindle by way of a mono-ball joint, in place of the larger, stock type ball joint. We can also see the steering arm forward of the upper mono-ball that attaches to the steering tie rod.

Scrub Radius Defined - Scrub Radius occurs when a spindle is designed whereby the point where a line through the centers of rotation of the ball joints intersects with the driving surface is some distance from the center of the contact patch. When a force is applied to the tire such as braking, if there is a scrub radius, then that force will want to turn the spindle.

The design goal of ending up with minimum Scrub Radius comes from early designs of production automobiles. A wheel assembly (spindle, wheel offset, wheel diameter, etc.) that is designed with a large amount of scrub radius will produce a moment arm of sorts. If the vehicle were to hit a pot hole, a force would be created which tries to rotate the wheel backwards.

If the wheel assembly had some amount of scrub radius, then the wheel would in fact be pushed back and this would feed a force through the steering system and the driver would feel that. This was not a desirable condition and could cause the car to steer violently in the direction the wheels were forced to move to.

By designing a wheel assembly that had zero scrub radius, this undesirable condition could be eliminated. Now, in modern day racing, scrub radius is not considered an important design criteria. Modern race tracks don't have pot holes.

All of this sounds complicated, but it's not. Just remember that scrub radius is not a performance enhancing component of a modern race car. With the advent of smooth race tracks, even on dirt, and power steering systems, even if we do hit bumps, the feedback to the driver is minimal is any at all.

Ball Joint Location and Spacing - So, we need to concentrate on those parts of the spindle that do relate to performance. The car is supported, in part, by the spindle. Most race cars have the spring attached to the lower control arm and that extends out to the spindle. So, end of the lower control arm attaches to a ball joint that is mounted to the spindle.

Since the spindle pin is basically the center of the wheel, its position in relation to the lower ball joint vertically, sets the chassis ride height.

We can install spindles with different pin offsets (distance from the lower ball joint) to change the race car ride height. If we change the spindle pin location, and keep our chassis ride height the same, we have then changed the lower control arm angle.

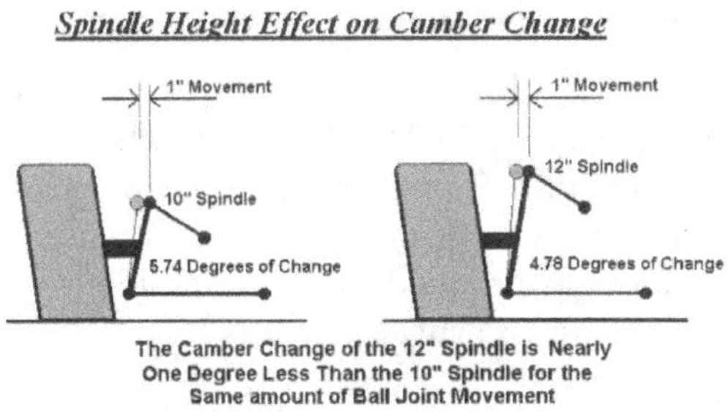

The height of the spindle has an effect on the camber change, a very important design function we will learn more about later on. This illustration shows now many degrees of camber change results from different spindle heights. The taller spindle changes less.

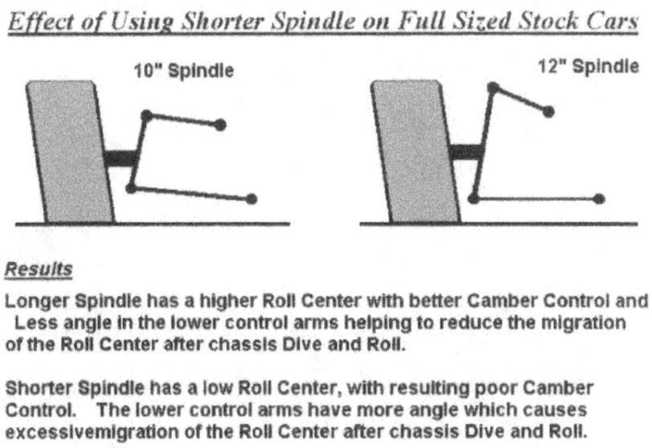

Longer spindles can offer a better design choice for several reasons. First, like we illustrated before, the camber change is better for the same movement of the ball joint. But the taller spindle also creates more angle for the upper control arm and that also has a positive effect on camber change reducing it further.

In a front view, as we discussed with scrub radius, the king pin angle is a design variable. We necessarily need different king pin angle designs for the right and left spindles in a circle track car. This helps reduce the camber change that results from steering the wheels.

Just as caster causes camber change, so does king pin inclination. These two combine their effects to end up with a net camber change per degree of wheel steering movement. All of this is taken into account in the overall suspension design and you don't need to completely understand all of that at this point in time. Just know that a spindle can be designed with different king pin angles.

Steering Arm - Since the spindle is attached to the wheel, and the wheel has to turn, a steering arm must be attached to and be a part of the spindle. This arm extends either forward or rearward and is attached to the steering system through a tie rod. This rod "ties" the spindle to the steering box or rack.

The steering arm for each spindle can be placed at a predetermined height, at different lengths and with the end different distances from the centerline of the chassis. The ultimate position and length of the steering arm coincides with the overall design of the steering system.

This is all you need to know about steering arms at this point in time. We will get into more detail about steering systems in future Lessons as a part of RCT Level Two and RCT Level Three.

A race team can create a taller spindle just by installing a ball joint with a taller, or longer, shaft. The height of the spindle for design purposes is the measurement between the centers of rotation of the ball joints. This longer ball joint shaft raises the center of rotation creating a taller spindle.

Mixing Different Spindle Designs - What we need to stress here is something that is important to understand from a basic perspective. We do not want to mix and match spindle designs without completely understanding how we might be changing our steering system, our front geometry layout and other considerations and design functions.

In the past, with some forms of racing, chassis manufacturers and teams have substituted spindles of different dimensions in order to save weight. In doing this, they inadvertently cause serious negative changes to the front ends of the cars and the performance suffered. It was some time before they all figured this out and redesigned the cars to work with the new spindles.

The spindle can be any design, including this very simple one made out of steel tubing and using simple heim joints for ball joints. This spindle is made for a scaled down stock car that weights much less than a full sized stock car and has much less loading on it.

Formula race cars have specific designs for their spindles where they cannot be interchanged. It is mostly with stock car racing where this occurs and a race team must be careful not to make undesirable changes with components without understanding how those parts will affect the suspension functions.

Spindle Recap - So, to recap, a spindle is a solid part that attaches to the control arms through the use of ball joints, or heim joints. These terms are interchangeable.

The spindle is the connection between the chassis and the wheel. It is attached to the wheel through a spindle pin that carries bearings. The bearings connect the spindle to the wheel hub and run on bearing races attached to the wheel hub.

The spindle helps support the chassis. Since the springs are mounted to the lower control arm, and the lower control arm extends to the spindle, it rests on the ball joint that is attached to the spindle. The load of the chassis is extended from the spring mount through the ball joint and down through the wheel to the tire contact patch.

The upper and lower ball joints that are attached to the spindle are the three dimensional connection of the spindle to the control arms. The front view angle of the ball joints from a vertical line is the spindle inclination or king pin inclination, interchangeable terms.

The other function of the spindle is to provide a means to steer the car by the use of a steering arm as a part of the spindle. The steering arm lengths, position laterally and height all need to coincide with the overall design of the steering system.

Exam - In The Context Of This Lesson:

The Spindle Connects Which Components?

1) The chassis to the wheel
2) The steering box or rack to the wheel
3) The spring to the wheel
4) The upper and lower ball joints
5) All of the above

Differences In The Following Affect Control Arm Angles

1) King Pin angles
2) Steering arm lengths
3) Spindle pin height
4) Wheel camber changes

Scrub Radius Affects The Following

1) Camber change
2) Caster change
3) Wheel Loading
4) Feedback through the steering system

Spindle Design Must Coincide With Which Other Components?

1) Control Arm Angles
2) Ride Height
3) Scrub Radius
4) Steering System Function
5) All of the above

Lesson Six – Control Arms

Control Arms are the connection between the chassis and the wheel. These are major components of the double A-arm suspension systems. These control arms can be found on either the front or rear suspension of the car. Most designs of control arms feature a long lower arm and a shorter upper arm. This is called the Short-Long arm suspension system.

Control arms make up a major part of the structure in a double A-arm type of suspension. Here we see the two arms, upper and lower, without the spindle attached. These are often called long arm/short arm suspensions because with most designs, the upper arm is shorter than the lower arm.

Control arms can be of a one-piece design, or be made up of different parts bolted together. The inner mounts will be bushings or heim joints and attached at two points to the chassis. The outer mounts are ball joints or heim joints (interchangeable terms) and those are mounted to a spindle.

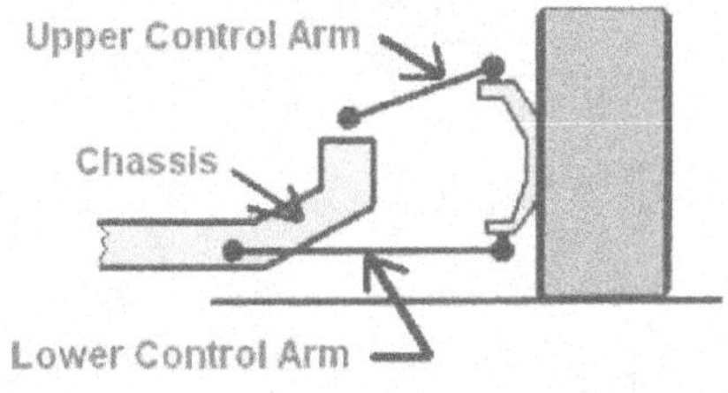

The control arms for a race car in a double A-arm suspension consist of an upper and lower arms that connect the spindle to the chassis. The outer mounts are called ball joints, or mono-ball, and the inner mounts are bushings, or heim joints. The angle of the arms is a critical component of race car suspension design.

Control arm angles are talked about in two separate views. From a front view, the upper control arm angle is usually with the chassis mount lower than the ball joint in stock cars and many prototype cars. The exception is formula cars such as Formula One. The angle of the upper control arm can be from five degrees to upwards of thirty degrees depending on the type of car and overall design goals.

The lower control arm, from a front view, can be either with the ball joint higher or lower than the chassis mount depending on the overall design goals. In either case, the angle is usually less than five degrees.

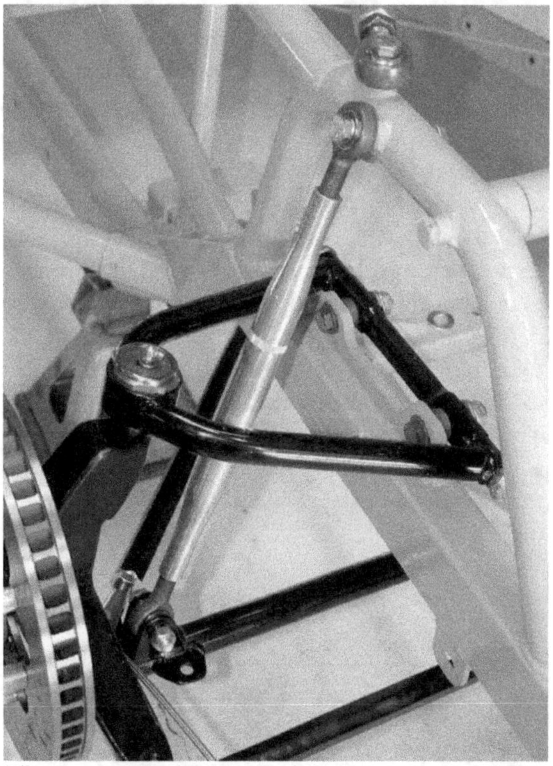

This is an example of a typical stock car, or tube frame road racing car, upper and lower control arms. The upper is one piece and adjustable for angle and camber by moving the inner mount through spacing or height of the control arm shaft. The silver link shown is a temporary link that replaces the coil-over shock and spring combination. Note that the springs that support the car are almost always attached to the lower control arm on almost every race car.

Control arm angles from a side view create anti or pro-dive, which is a function of braking force that we will get into later on. Just know that this

effect exists and the side view control arm angles create the effect.
The front view control arms positions and angles determine the suspension geometry components such as roll center and roll center movement, cambers and camber change, and anti and pro-dive.

This is an example of a double A-arm suspension on a prototype race car. The upper and lower arms consist of links attached to the spindle and chassis with mono-ball joints. The car is supported by springs mounted inboard on the transaxle housing and connected to the lower control arm/spindle with a push rod link.

In most cases with modern race cars, the spring that supports the chassis is mounted to the lower control arm. Where this spring is mounted determines the wheel rate and is the wheel rate is calculated using the motion ratio.
The lower control arm must be strong enough to support the forces generated by the spring through the amount of wheel travel that occurs on the race track.

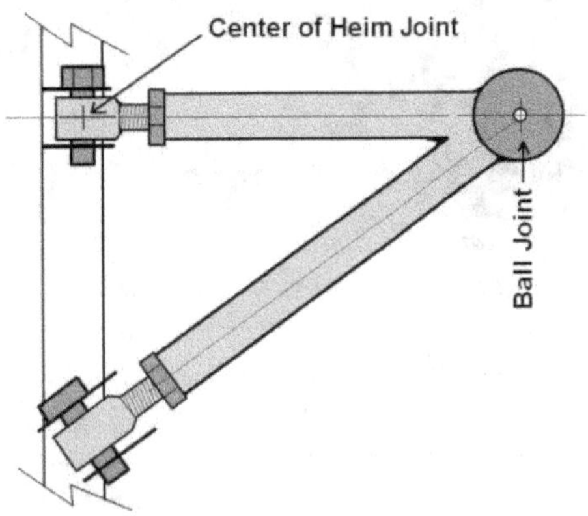

The upper control arm can consist of links mounted to the chassis with heim joints.

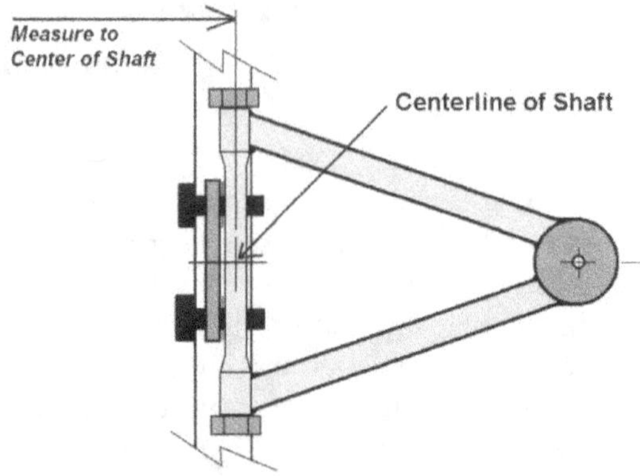

The most common upper control arm design for stock cars and many road racing cars is a one-piece design with a shaft attached to the chassis. The shaft can be moved in towards the centerline of the chassis to gain camber or out to remove camber. The shaft can be moved fore or aft to adjust the amount of caster.

Adjusting Upper Control Arm Locations

The upper control arms are usually adjustable for inner mounting heights and widths. If the control arm is solid, then the inner mount will be a shaft which can be spaced on the chassis mount in or out to make changes to the wheel camber.
The height of the upper control arm chassis mounts can be changed on any design and these changes are used to change the arm angles. The change in upper control arm angles produce changes in roll center location and migration. The upper control arm front view angles also affect camber change characteristics.
In most systems, the upper control arm can be moved fore and aft to create caster and caster split. If the upper control arm is a one-piece design, many of those are designed with slotted shafts where the team can loosen the attachment bolts and slide the arm fore or aft to make changes to the caster settings.

We can see the spacer plates positioned between the control arm shaft and the chassis mounting plate. The plate has slots and slugs with varying offset holes to move the shaft up or down to adjust the arm angle. Note too that the shaft is slotted to allow movement fore and aft to adjust the caster.

In the upper control arm mounts, we also usually see slots or slots with slugs so that the inner mounts can be moved up or down to change the front view angle of the control arm. The upper arm angle can also be changed by installing a ball joint with a longer or shorter stud. Some cars use mono-ball mounts where the ball can be spaced at different heights on the spindle mounting stud. From a side view, the mounting bolts for the upper control arm can be different heights. This creates anti or pro-dive. For anti-dive, the front mount is placed higher than the rear mount. Some upper control arms are adjustable for length using the same chassis mounting points and same ball joint heights. This adjustment is meant to speed up camber changes. Remember that when doing the camber change using adjustable length control arms that you are also changing the arm angle and with that the geometry of the front suspension.

The Lower Control Arms

The Lower control arms are usually less adjustable for front view angles than the upper arms. In side view, there can be changes to the angles to support anti and pro-dive designs. The length of the lower control arm compared to the distance the spring is mounted from the ball joint determines the motion ratio for that control arm.

The lower control arm on a stock car for circle track or road racing mostly supports the car by being attached to a spring or torsion bar. Here we see a stock spring lower control arm where the spring is resting in a round bucket that is a part of the lower control arm. The outer end of the arm rests on the lower ball joint.

Motion ratio is: the difference in distance that the wheel moves comparted to the movement of the spring. On most systems, the wheel always moves more than the spring. On some formula and prototype systems, the spring could move more than the wheel, but this is usually not the case.

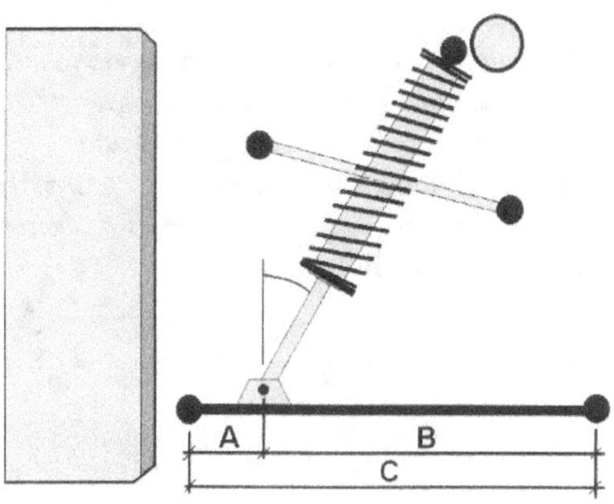

With the spring attached to the lower control arm, we can see where there is a motion ratio as the wheel moves. The spring always moves less than the wheel. The spring motion divided by the wheel motion is called the motion ratio. If the spring is mounted at an angle to the motion of the lower control arm, the spring motion is even less than the motion ratio.

As to anti and pro-dive, if the rear mount of the lower control arm is mounted higher than the front mount, we have anti-dive. If the rear mount is lower than the front mount, we have pro-dive. We will be discussing the anti and pro-dive functions in more detail in the next Lessons.

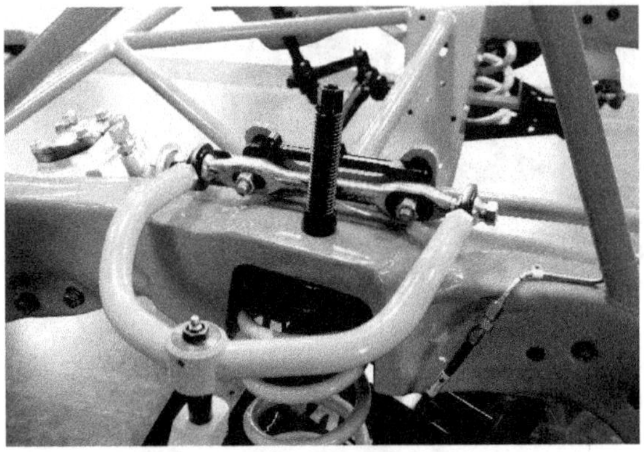

The angle of the control arm, upper or lower, is measured on a line passing through the centers of rotation of the ball joint and the inner mounts. This control arm is bent to avoid contact with the chassis as the wheel drops down, but the actual angle of the arm does not change.

Race cars with this type of upper chassis mount cannot make fine adjustments to the arm angle. The length of the arm, the cambers and caster settings are adjustable through the adjustable links that make up the arm. To change the arm angle, ball joints with longer or shorter shafts will need to be installed.

Exam - In The Context Of This Lesson:

Control Arms Connect Which Components?
1) The tie rd to the spindle
2) Chassis to the spindle
3) Wheel to he spindle
4) The braks to the spindle

Side View Angles Of The Control Arms Create?
1) Roll center locations
2) Wheel cambers
3) Anti or Pro-dive
4) Caster Settings

Changes To The Front View Angles Of The Control Arms Affect?
1) Roll center height
2) Roll center width
3) Camber change
4) All of the above

Which Suspension Component Helps Support The Car?
1) Upper control arm
2) Roll center height
3) Roll center width
4) The lower control arm

Using The Motion Ratio, We Can Determine?
1) The lower control arm angle
2) The ride height
3) Wheel rate
4) The upper control arm length

Race Car Technology – Level One

Lesson Seven – Anti and Pro Dive

Anti and Pro dive are design functions that either prevent the front suspension from diving when the driver is braking, or assists in diving when the driver is braking. To completely understand the Anti's and how they work, we need to understand what is happening to the car when we brake.

It is important to note that the Anti's will only function when there is braking force. The result of the braking force assists, and actually causes, the Anti's function.

There are various theories about how Anti's work in various teachings. In this school, we will deal with only one and that will deal with the actual forces in a practical way. This involves the force of deceleration and the torque created when we apply the brakes.

Sketch-01

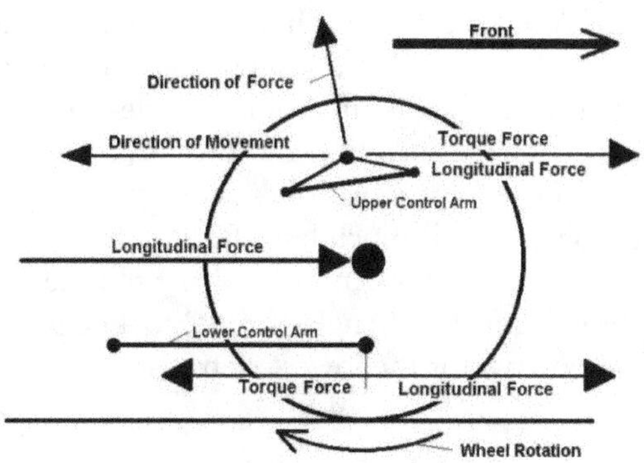

In this side-view sketch of the right front suspension, we see where the upper control arm is mounted at an angle. The rear arm bushing is lower than the front bushing. If we move the spindle up, or the chassis goes down, the ball joint will move in a direction that is to the rear, or back towards the rear of the car. Upon braking, the force of the brakes grabbing the rotor is trying to force the rotor and spindle that it is attached to in the direction of the rotation of the wheel.

Since these two motions are opposite, there is Anti-dive created. In addition, braking forces add to the torque forces and those two combine to create the Anti-dive effect. The chassis cannot, or is resisted from moving down by the forces acting to lift the chassis in opposition to diving.

On Braking – When we apply the brakes in a race car, there are several things going on. One, weight is transferred from the rear of the car to the front. This increase in weight on the front suspension causes the springs to be made to support that added weight, and so they compress.

When the spring compress, the front of the car goes lower to the track. Many teams don't desire this diving motion and so Anti-dive was created. It was found that by creating a side view angle with the upper and lower control arms, this tendency to dive was reduced or eliminated.

What Anti-dive does is lock up the suspension and force it not to move, or not to move as far. It does this by creating opposing forces to the motion of chassis dive. These forces are a combination of braking longitudinal forces and braking torque forces.

Looking at Sketch-01, we see where as the chassis moves down, or the spindle moves up, with the side-view angle in the upper and lower control arms, the ball joints will move in the opposite direction as the wheel rotation and the braking force and the chassis opposite to that.

Because there is a braking force trying to move the ball joints in the direction of the wheel rotation, and an opposite diving force trying move the ball joints in the opposite direction of the wheel rotation, these two forces oppose each other. This defines the torque aspect of Anti-dive.

Then too the longitudinal braking forces trying to drive the chassis forward puts a force on the ball joints as well as the chassis mounting points. When the control arms are angled, these forces act in a direction opposite of chassis dive, or upwards.

The opposing forces are the creation of Anti-dive. There is another effect that is the opposite of Anti-dive, and that is Pro-dive. If in our side-view angles of the upper and lower control arms, we mount the control arm angles in the opposite direction, we then can assist with chassis dive upon braking, if that is our desire.

Then the two forces are acting in the same direction and the chassis will dive quicker than if it had no Pro-dive angles in the control arms.

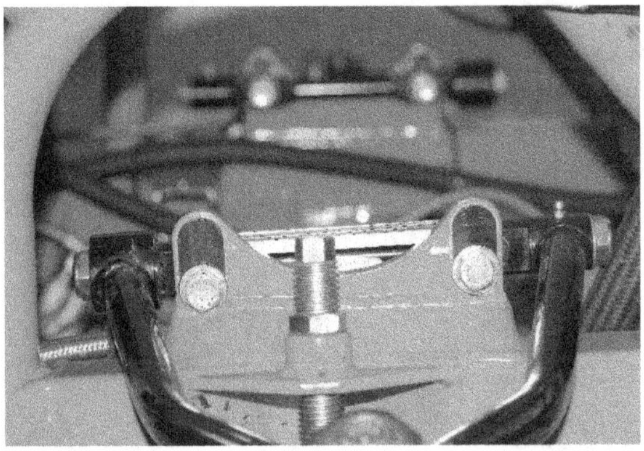

Here we see a right front control arm mount where the control arm shaft is not level and has some amount of Anti-dive. The front bushing is higher than the rear busing. If there were a corresponding angle in the lower control arm, then there would be Anti-dive in this system. To create more Anti-dive, we would raise the front mount by inserting a slug in the front slot that has less offset down, or offset up from the middle setting creating more sideview angle.

The only time we see this Pro-dive design used is on a circle track car, and on the left front suspension only. It is thought that it is desirable to cause the left front suspension to go quickly to a lower position by some setup designers.

It is not the intent of this school to comment, at this point, on the merits of certain setup routines and practices, so we won't. This school is created to teach the student about all of the chassis setup tools that are available and to understand the function of all of the components of a race car.

How Much Anti-Dive? – The amount of braking force that Anti-dive tries to control is directly in relation to the amount of braking that is taking place. Pressing lightly on the brakes cause less weight transfer to the front and therefore less dive. So, for light braking cars, there is less need for Anti-dive.

For heavy braking cars, there is obviously more need for Anti-dive and a greater amount of Anti-dive settings. Here is why we even need Anti-dive.

When a double A-arm suspension dives, it creates camber change in most cases. Camber change towards the negative will reduce the tire footprint and reduce grip. Less grip means the car does not turn in as well and it will take longer to slow the car.

For an average circle track car, or a road race car using a chassis similar to the circle track late model, a side view angle of the upper control arm of around 2 ½ degrees is considered moderate. Up to 5 degrees of angle in the uppers is considered acceptable.

This is a mounting plate for the right upper control arm. The rear offset slug has a ¼" offset down from center. The front slug has a 1/8" offset down from the center. This would create an 1/8" of difference in height, or about a 1.2 degrees of anti-dive angle. A half inch difference creates almost 5 degrees of anti-dive angle, about the maximum we would want to use normally.

The angle in the lower control arm is a definite part of Anti-dive and acts in the same way as the upper control arm. Its motion moves the lower ball joint much less than a similar angle would move the upper ball joint, and because the longitudinal and torque forces act in the opposite directions, there is much less effect with the lower control arm in Anti-dive than with the upper control arms. Angles from 2 to 4 degrees are acceptable for the lower control arm.

Exam - In The Context Of This Lesson:

Anti and Pro-dive Only Present When,?
1) The car is accelerating
2) The car is braking
3) The wheels are moving up or down
4) The car is turning

With Anti-dive The Upper Ball Joint Moves?
1) To the front when the chassis dives
2) The In the same direction as the lower ball joint
3) In the same direction as the wheel rotation
4) To the rear when the chassis dives

The Amount Of Anti-dive Needed Is Dependent On?
1) How much weight is transferred
2) The length of the track
3) How hard the driver is braking
4) The length of the upper control arm

With Pro-dive The Upper Ball Joint Moves?
1) To the front when the chassis dives
2) The In the same direction as the lower ball joint
3) In the same direction as the wheel rotation
4) To the rear when the chassis dives
5) 1 and 3

The Primary Benefit Of Anti-dive Is?
1) Keep the front end up
2) Cause less weight transfer
3) Reduce or prevent camber change
4) Help the brakes work harder

Race Car Technology – Level One

Lesson Eight – Steering Systems

As described in Lesson Two, steering systems are on the front of a race car for a reason. We use the steering to generate an angle of attack to the tangent of the curve we are driving around. We will describe the different types of steering systems used on race cars and explain the function of the various parts of the steering system.

As a part of this Lesson, we will also explain how different classes of race cars use and setup their steering systems. It is not necessarily the same for all classes. We'll tell you why.

Types Of Steering Systems - There are basically three different types of steering systems used in racing today. They are:

Rack and Pinion Steering – In this system, a rack is moved by a gear when the steering wheel is turned. The rack ends are connected to the inside ends of the right and left tie rods that are connected to the spindle to turn the wheels.

The rack and pinion steering system is the most common steering in most race cars and today in most production cars. It is simple in design and uses just the center rack and housing with a pinion gear that moves the rack back and forth to steer the car. There are various design parameters for installing the correct rack that will be covered in future Lessons. A power assist unit (at the front of the gold rack) can be added to the rack to reduce the force needed to turn the steering wheel.

Drag Link Steering – This is an older system used on many American production automobiles and adapted to stock car racing. It consists of a steering box that moves a link. On the link are mounted the two tie rod ends, one left and one right. The link can be designed to be in front of the center of the spindle or to the rear. The front is called a Front Steer car and the one to the rear, a Rear Steer car. The Front Steer is the more popular of the two in today's racing.

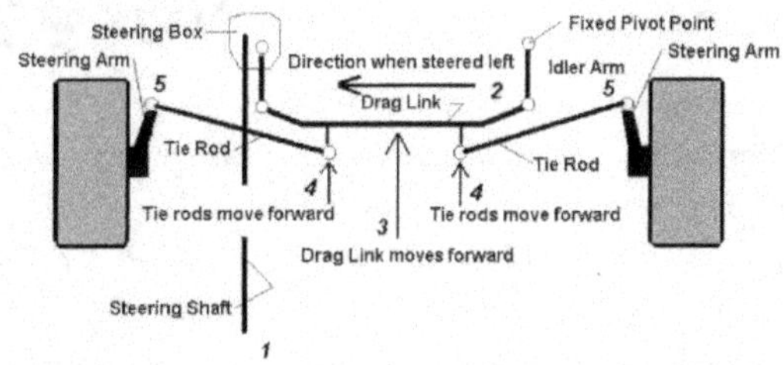

The Drag Link steering system is common on older production cars and some race cars. The layout is just as described, a steering box that when turned by the steering wheel, moves a drag link side to side. Attached to the drag link are the inner ends of the tie rods. As the link moves, so do the tie rods and wheels to steer the car.

Straight Axle Steering – On race cars, such as sprint cars, the front suspension is a straight axle design. To steer the car, a steering box is mounted just in front of the steering wheel and above the drivers legs. A shaft extends out of the right or left side of the car and a rod is mounted to an arm off the shaft. A tube link is then mounted to that arm and forward to a horizontal arm off the spindle on that side. The steering box moves the arm and tube link which moves the spindle arm turning the spindle. The other spindle is connected to the steered spindle by a cross link attached to steering arms that are a part of the spindles. This is a simple but effective steering system that has its flaws.

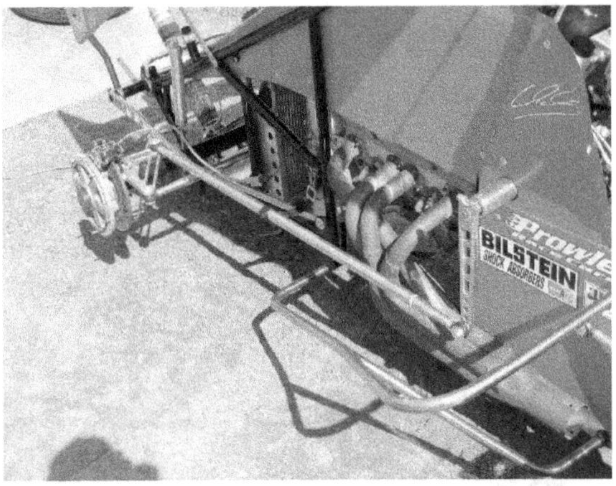

The steering on a straight axle is built differently than either the rack and pinion or the drag link systems. This unit uses a gear box located just forward of the steering wheel. This box turns a shaft that protrudes out of either side of the body. It has an arm at its end that moves a steering rod going forward to the spindle and a steering arm on the spindle that is parallel to the straight axle, as shown here. The steering arm also has a ninety degree arm connected to a tie rod that steers the other wheel.

Toe Defined - Toe is when one wheel is pointed in a different direction than the other at static ride height and with the wheel pointed straight ahead. We can have toe on both the front and rear wheels. Toe is used to stabilize the directional attitude of a vehicle and some production vehicles are designed with toe-in and some with toe-out in the front wheels.

To further explain what toe is, here is an example. If you could measure the width between the two tire centers of the tire tread at the front and at the rear of those tires, with the wheel pointed straight ahead, for toe-out, the front measurement would be more than the rear. Toe-in would be the opposite. In racing, we have various tools to measure toe.

Toe-out is most common for race cars for the front wheels. We change the toe in our race cars by lengthening or shortening the tie rods, which are designs to do this easily. Most tie rods have opposing direction threads on opposite ends of the rod end that attaches to the steering rack or drag link. When you turn the tie rod after loosening the jam nuts, the heim joints, or tie rod ends, move in or out together.

On many formula and prototype cars, the front wheels are set with toe-out and the rear wheels are set with toe-in. The toe can change as we negotiate turns and this change is caused by the suspension movement and the motion of steering the wheels. We'll talk more about those changes and why they happen later on.

When we measure the toe in our race cars, we can use toe plates that rest against the tires sidewalls, or we can use strings run parallel to the centerline of the car. There are also toe bars that offer a way to measure to the sidewall by first placing the bar in front, measuring an offset, and then to the rear to see how much more offset there is that is the toe amount.

We like to think of toe as the difference in the front and rear widths at the center of the tread. But if we measure at the sidewall, at the edge of toe plates, or even at the rim, which some teams do with the string method, we are not reading the toe amount at the tire tread, we are seeing something less.

This is because the width of where we are measuring is less than the diameter of the tire. If our measurement is less, then we are reading a number that is less than the real toe thought of as the difference in width of the centers of the tires at the contact patch.

To correct our numbers, we could divide the tire diameter by the width of the points we used to measure toe and then multiply that number times the toe number we get. For example, if we are measuring points 20 inches apart and our tire diameter were 27 inches, then dividing 27 by 20 gives us 1.35. If we measure our toe at 1/8", which is 0.125", then when we multiply 1.35 times 0.125, we get 0.169, or closer to 3/16" of toe.

And remember this too about toe, when we make changes to our wheel cambers, the movement of the upper control arm in or out does move the end of the steering arm, which changes the toe measurement. You must recheck and reset the toe after making changes to the cambers of the front wheels.

Steering Quickness - When we steer our cars, we need to have a steering ratio that is quick enough so that we don't need to take our hands off the steering wheel. And it should not be too quick so as to move the steering excessively with small movements of the steering wheel.

In stock classes where the teams are required to use stock steering boxes or racks, racing suppliers offer steering quickeners, which are a small geared box that speeds up the turning of the steering shaft.

Ratios of from 1.5/1.0 to 2.0/1.0 are offered. That means for a 2.0/1.0 ratio, if you turn the steering wheel 1.0 revolutions with the quickener installed, it would be like turning the steering wheel two times without the quickener. Now the driver doesn't need to cross over with the hands to turn far enough to steer the car.

There are mechanical effects that can cause the wheels to steer in directions other than what we intend to have happen. Those are:

Ackermann – When one wheel steers more than the other. When the wheels gain toe, it is called Ackermann. When the wheel lose toe, it is called Reverse Ackermann. We will cover the causes of Ackermann and how to fix it in Lesson 9.

Bump Steer – When the wheel steers due to vertical wheel movement, we have bump steer. Sometimes a very small amount of bump steer is desired, but normally we want near zero bump steer. We will cover what causes bump steer and how to measure it in Lesson 8.

How Much To Steer? - There is an optimum amount we need to steer. It is just beyond how much it would take to steer the car slowly through the turns. If our driver drove around through the turns at a slow speed, the amount of steering required to roll around the driving line would be nearly the same as when driving that same line at high speeds, but a little more.

If the driver finds that they need to steer a greater amount, then the car has a handling balance issue. This means that the front tires have less Grip than they need to keep up with the Grip of the rear tires. Changes to the setup must be made to bring the steering back to normal steering angles.

In dirt racing, many of the cars have a great amount of rear steer where the rear tires steer out away from the turn. This rear steer is so pronounced that the front wheels must be steered to the right to compensate and to keep the front tires tracking along the correct line. This is unique and very different than anything we see with racing on asphalt.

Steering As A Grip Generator - Steering creates Grip. When we steer the car, we are creating an angle of attack that produces a force we need to counter the lateral forces trying to push the car off the turns. The steering system is therefore a variable Grip generator. The more we steer, the more Grip we produce, up to a point.

As the angle of attack increases with increased steering angle, the Grip increases until we have gone too far. Then at that point, the tire can no longer hold to the track surface and it slides. The level of Grip the tire is producing drops off dramatically at this point and the front of the car will lose most of its Grip and slide.

This is why a car that is unbalanced and where the setup is tight (car is understeering), can still be made to turn and make it through the turns. The excess steering needed to do this forces the front tires to work harder than they would if the car were more neutral in handling balance.

Exam - In The Context Of This Lesson:

The Most Common Types of Common Steering Systems Is?

1) Tie Rod System

2) Caster System

3) Rack and Pinion

4) Rear Steer System

How Much Do We Need To Steer?

1) One full turn of the steering wheel

2) Till the rear stops sliding

3) Enough to track just inside the tangent of the arc we are driving

4) 1.5 times what we would on the street

Steering Quickeners Do What In Our Steering System?

1) Prevent excess toe out

2) Make the lap faster

3) Increase turn speeds

4) Make the car steer more with less steering input

Bump Steer Helps The Car Do What?

1) Use less camber

2) Become less stable

3) Increase turn speeds

4) Ride out bumps better

Ackermann Is?

1) When the inside wheel turns less than the outside wheel

2) When the car loses toe from steering input

3) When the car gains toe from steering input

4) A way to improve the steering system

Steering Generates Grip By What Method?

1) The creation of Ackermann

2) By reducing bump steer

3) Through increasing the steering quickness

4) By giving the tires an angle of attack

Race Car Technology – Level One

Lesson Nine – Bump Steer, Causes and Cures

When a wheel moves vertically, we say it bumps. This probably goes back to the day when our roads had a lot of pot holes and bumps. In that situation, if the car ran over a bump, automotive designers did not want the car to steer as the wheel moved over the bump. Thus came the term, "bump steer".

It is also true of our race cars that we don't want or need bump steer for the most part. In some more involved suspension designs, some amount of bump steer may be desirable, but here we will stick to the basics and say that bump steer is a bad thing.

Attaining zero bump steer is nearly impossible because of the complexities in our steering system, but we can design for near-zero bump steer that is so small it won't be noticed by the driver.

There are a few very complicated and sophisticated machines that measure bumps steer on race cars under chassis loading situations. We can come to understand all we need to know about bump steer by learning just a few important facts about bump steer.

What Creates Bump Steer? - The angles of the upper and lower control arms, meaning a line extending through the center of rotation of the ball joints and inner mounts of each arm, intersect at a point we call the Instant Center. In order to have near zero bump steer, the tie rods on each side need to be pointing towards the instant center for its side. This is one of two criteria for near zero bump steer.

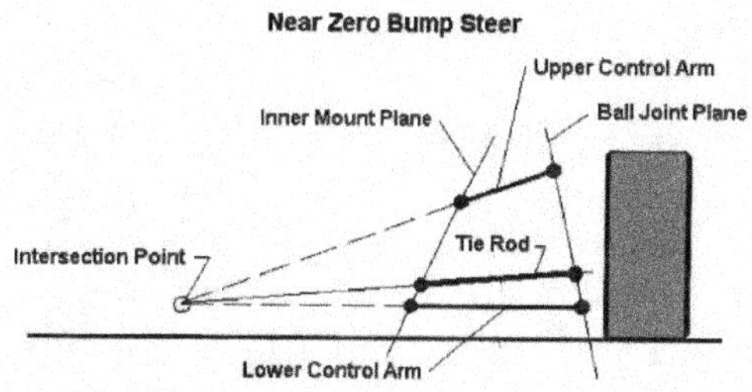

The basic criterial for having near zero bump steer is shown in this sketch. The tie rod is pointing at the instant center formed by the extension of the upper and lower control arms. And, the tie rod is the correct length, being the distance of its lines intersection with the two planes, one through the upper and lower chassis mounts, and the other plane through the ball joints.

The other thing we need is for the tie rod to be a specific length. That length must be equal to the distance formed by a line extending through the centers of rotation of the tie rod ends, and that lines intersection with two planes, one formed by the upper and lower ball joints and being parallel to the wheel, and two a plane that passes through the inner upper and lower chassis mounts. This can get a little complicated because it may be hard to visualize these two planes.

And, the tie rod does not necessarily need to physically fall within those two planes, it just needs to be that length between the points of intersection with each plane.

If we have the tie rod pointing to the instant center intersection of the upper and lower control arms and it is the right length, then we will have near zero bumps steer.

What Creates Bump Steer – When the tie rod is not aligned with the instant center and/or the length is wrong for the system, we have bump steer. As the wheel moves vertically, the wheel will either steer left or right. We will refer to the direction from a driver's perspective and for a front steer car only, in this discussion. By front steer, we mean that the tie rods are in front of the ball joints.

If the tie rod was pointed so the tie rod line passes below the instant center, then the wheel will bump in (towards the centerline of the car) as the wheel travels up and bump out when the wheel travels down. If the tie rod line passes above the instant center, then we will have bump out as the wheel travels up and bump in when the wheel travels down.

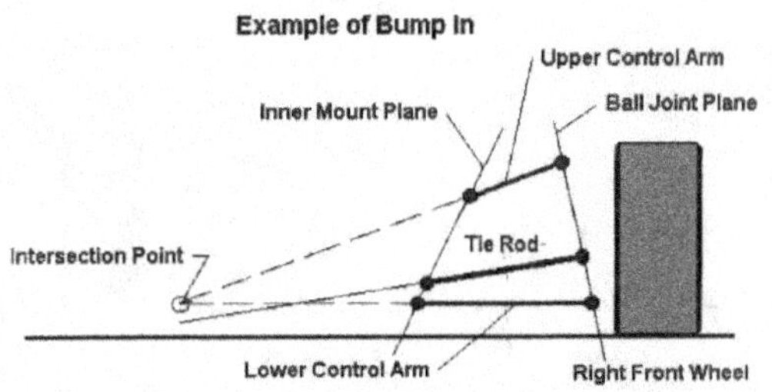

If the tie rod were pointed below the instant center, then when the wheel bumped up, the wheel would steer in, or left in this case. Other suspension design features can interfere with the bump steer layout and cause bump steer. This is caused by suspension movement that changes the alignment of the tie rod.

If the tie rod were too short, we would have bump steer in when the wheel travels in both directions from the static ride height position. If it were too long, then the wheel would bump out as the wheel traveled in both directions from ride height.

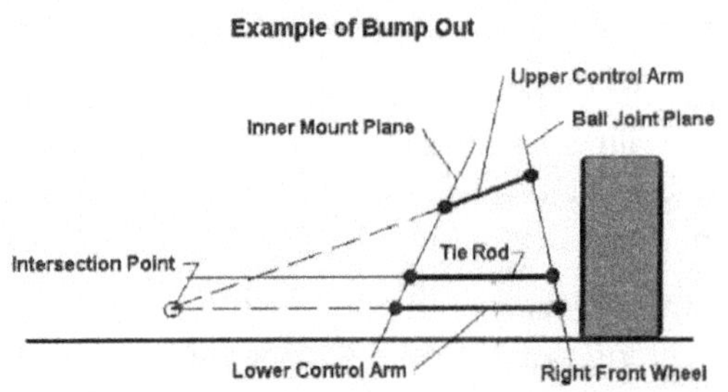

If the tie rod were pointed above the instant center, then when the wheel bumped up, the wheel would steer out, or to the right in this case. Changes to correct bump steer include movement of the ends of the tie rod either up or down, until it is aligned with the instant center.

We can easily check the race car for bumps steer by using a bump steer gauge. This measures differential movement of the sides of the plate bolted to the wheel hub as the spindle is jacked up from a preset position. This is a two dial gauge. There is also a one dial gauge where a roller replaces the one dial and the operator only reads the one dial. On the left front wheel, if the front dial was showing movement away and the rear dial showing movement to the operator, then this would be bump in.

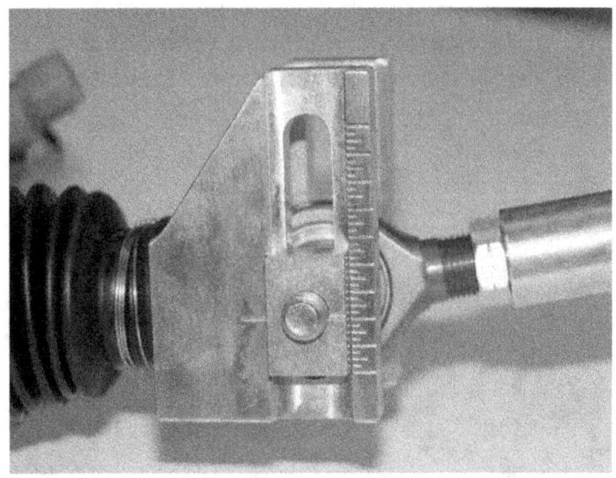

To correct bump steer on this rack and pinion steering system, we need only move the inner tie rod end up or down in this easy to read fixture. Drag link systems use spacers to move the drag link up or down, or offset slugs where the tie rod mounts to the drag link.

If we start making changes to the control arm angles, or other changes that would affect components needed for near zero bump steer, then after making those changes, we would need to re-adjust our bump steer settings.

For example, if we install extended lower ball joints to take angle out of the lower control arms, we change the angle of the lower arm and move the instant center height. After making that change, the tie rod may not now intersect with the instant center and we will have introduced bump steer. When making changes to the arm angles, we need to re-align the tie rod so that it stays pointed towards the instant center.

Another example is when we change our Anti-dive settings. In both dirt and asphalt racing, anti and pro-dive is used in various degrees. Introducing these effects cause changes to our bump steer. This is because with anti-dive for example, when the wheel travels up, the upper ball joint moves towards the rear of the car and this rotates the spindle from a right side view, counterclockwise. This rotation moves the outer tie rod end upwards and changes the angle of the tie rod. Now it no longer points towards the instant center.

Where we had near zero bump steer before with no anti-dive, we now have bump steer when the right front wheel travels up. With pro-dive, we see a similar affect, the tie rod end moves down with vertical travel and again the tie rod is miss-aligned with the instant center.

On this car, a spacer (red colored aluminum piece) has been placed between the end of the tie rod and the steering arm. In this case, the tie rod was pointing below the instant center by quite a bit. This is a large spacer.

Steering Affects Bump Steer – When we steer our front wheels, we change the angles of our tie rods due to caster, camber and degree of spindle on both sides. The tie rod ends travel in an arc that is not parallel to the ground. This changes the outer tie rod end height and therefore the B/S. It is for this reason that we recommend doing your B/S with the wheels both straight ahead and then again with the wheels turned equal to mid-turn steering at the track you will run.

For checking tie rod length on a rack and pinion system, you can usually look to see if the inner end of the tie rod is lined up with the inner mount for the lower control arm. The final check is the bump steer gauge.

Exam - In The Context Of This Lesson:

The Most Common Types of Steering Systems Is?

1) Tie Rod System

2) Caster System

3) Rack and Pinion

4) Rear Steer System

Which Is One Of The Two Elements Of Bumps Steer?

1) Tie rod angle

2) Tie rod length

3) Control arm angles

4) Control arm lengths

5) 1 and 2

Changes To Which Will Affect Bump Steer?

1) Tie rod angles

2) Tie rod lengths

3) Anti-dive

4) Caster settings

5) All of the above

Lesson Ten – Ackermann Steering

In every steering system, there is a design effect called Ackermann. It is the difference in turning rate for the inside and outside wheels. It is named for the man who came up with the effect and intended it to be used to track the front wheels of very early automobiles through gravel driveways without causing ruts. In today's race cars, there is very little need for the Ackermann steering effect.

Ackermann can be very destructive to a setup. When Ackermann effect is present in the steering system, the two wheels and tires will not track correctly and want to go in different directions. In this lesson we will explain what Ackermann is, how it is created and how to eliminate it.

Every steering system has within its geometry the means for a layout that produces Ackermann. For true Ackermann, the inside wheel (meaning to the inside of the turn) will always turn more per degree of steering input, than the outside wheel.

When this happens, the toe of the front wheels is increased. For production vehicles that may run toe-in, there is a loss of toe-in and the front wheels may well go into a toe-out condition. Reverse-Ackermann is the opposite condition, where the outside wheel turns more than the inside wheel. This produces reduced toe.

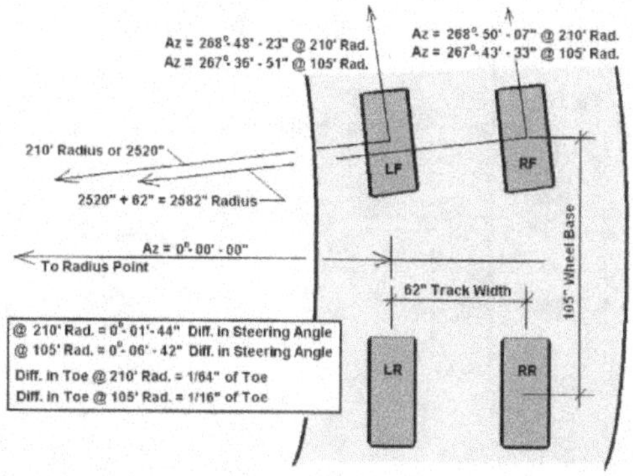

This calculation shows how little gain in toe we need in order for the wheels to track in their own path around the arcs that are slightly different from the inside wheels to the outside wheels. One degree of Ackermann for a typical race car going around an average radius turn is one half inch of toe. In the Lesson on Alignment, we will explain how that is a lot of toe out.

For most race cars that run toe-out, this would be increase. For average diameter tires run on race cars, one degree of Ackermann, which is a one degree of difference in steering direction of the front wheels, equals a half inch (0.500") added of toe-out. This is excessive.

For normal radius turns, we need fractions of a degree of Ackermann in our race car steering systems. Calculations tell us that for a 200 foot radius turn, we need about 1/64", or 0.015" of added toe. For a turn with a shorter radius of 100 feet, the added toe would be about 1/16", or 0.0625". If we set our car with 1/8" (0.125") of toe for a 1/3 or ½ mile track normally, if we go to a very short ¼ mile track, we only need to add 1/16" of static toe to have the needed 1/16" of Ackermann toe for that short radius turn. Then we don't need to redesign out steering system to produce the 1/16" of Ackermann.

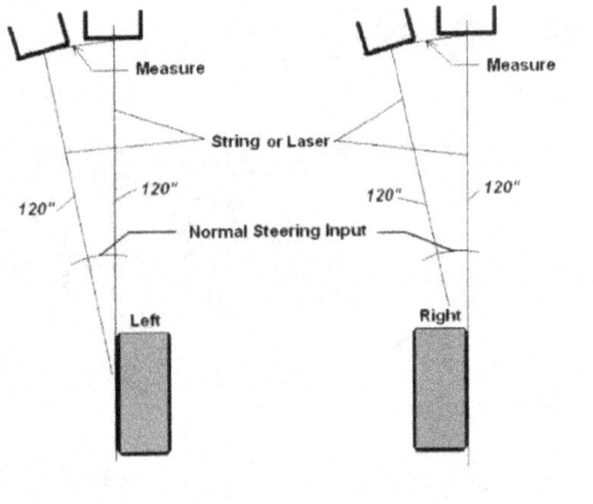

This sketch shows now we test for Ackermann. We set the wheels straight ahead and mark where they point at a distance from the center of the hub, in this case, ten feet or 120 inches. Then we turn the steering wheel usually the same amount as when we race the car. Two more marks are made where the wheels point after turning the wheel. If the measurement between those two sets of marks is different, then we have Ackermann if we have gained toe, or Reverse Ackermann if we lost toe.

How Steering Systems Produce Ackermann - Every steering system can be designed to create Ackermann, whether we want that effect or not. So, we have to be aware that when we assemble our car, or when someone else has assembled it before us, it might just have the tendency to produce unwanted Ackermann.

If we look at our steering system from a top view, we begin to see how Ackermann is produced. For a front steer, rack and pinion system, as the wheels are turned, the outer ends of the tie rods move backwards in relation to the rack. If the steering arms were mounted straight ahead on the spindle, and most of them are, and the tie rods were mounted so that the outer ends were ahead or forward of the inner tie rod ends, then as the wheel were steered, the outer tie rod ends would move back, the tie rods would straighten out, and that motion would push the outer ends apart. This produces Ackermann and increases the toe-out.

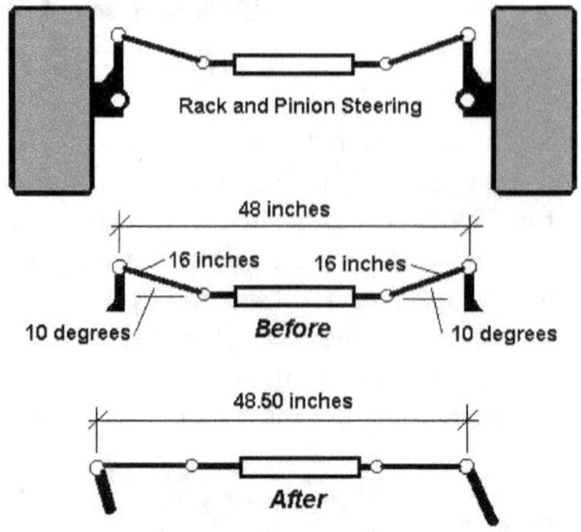

For a rack and pinion system, we can regulate and adjust the system for Ackermann by moving the rack forward or back. This does what we need for setting the Ackermann for turning both left and right as in road racing and when running on a dirt circle track. In this example, both steering arms are the same length.

The opposite would happen if the tie rods were mounted inline. Then as the outer tie rods ends moved back and the tie rods gained top view angle from originally being inline, the tie rod ends at the spindle would become closer together. This would cause Reverse Ackermann and we would lose toe.

The drag link steering systems work in a similar way. The big difference is that the steering arms are designed to be mounted at a top view angle on the spindle so that the outer tie rod end is closer to the centerline of the car than the lower ball joint. This angle in and of itself would produce Reverse Ackermann. When this design is used with a system that produces Ackermann, the two oppose each other and we get, as a result, near zero Ackermann.

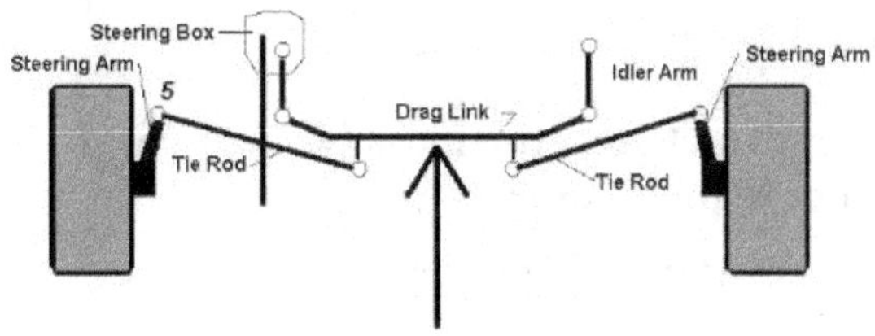

For a drag link system, the adjustment is similar to when we adjusted the rack and pinion. Instead of moving the rack forward and aft, we can regulate and adjust the system for Ackermann by moving the drag link forward or back. This also does what we need for setting the Ackermann for turning both left and right as in road racing and when running on a dirt circle track. As in the rack example, both steering arms are the same length.

As a little history lesson, most of the stock cars produced in the eastern U.S. in the 1990's were drag link steering designs which used top view angled steering arms. When the stock cars of the mid-west, that used mainly rack and pinion steering systems, went to lighter spindles, some manufacturers in the east switched spindles and mounted the lighter rack spindles, with top view straight ahead steering arms, on their cars to save weight. The two steering systems were not compatible and the drag link east cars with the mid-west rack spindles were producing a lot of Ackermann and not turning too well.

If we only turn one way, we can adjust the length of one steering arm to tune the Ackermann amount. On most circle track cars, the left spindle has a slot for moving the tie rod end fore or aft to quicken or slow the turning of that wheel. For road racing cars, the steering arms on both spindles would need to be slotted and the adjustments made the same for both sides in order for the Ackermann to be the same for turning left or right.

Be sure to look over your system to make sure it does not produce excess Ackermann. There is a simple string test you can do to check your Ackermann. If you have one, you can also use a laser system to do this check, but by all means, check it before you go racing. We'll tell you more about how to do this check in the Advanced RCT course.

Exam - In The Context Of This Lesson:

Ackermann Is Defined As?

1) The loss of front wheel toe

2) When the wheels bump steer out

3) A gain in toe when the wheels are steered

4) A loss of toe when the wheels are steered

Which Steering System Can Produce Ackermann?

1) Straight axle steering

2) Rack and Pinion steering

3) Drag link steering

4) All of the above

How Much Ackermann Is Needed For A 200 Foot Radius Turn?

1) One degree of Ackermann

2) A half degree of Ackermann

3) 1/64 inch of added toe

4) 1/16 inch of added toe

Rack and Pinion Steering System Require?

1) Top view straight ahead steering arms

2) Top view angled steering arms

3) A drag link attached to the tie rods

4) A shorter left steering arm

For Road Racing Cars, We Adjust Ackermann By?

1) Shortening the left steering arm only

2) Shortening the right steering arm only

3) Moving the rack or drag link forward or backwards

4) 1 and 2

Lesson Eleven – Front End Wheel Alignment

One of the most critical elements of race car setup is alignment of the wheels. The direction they are pointed at any point on the race track helps determine how well the car will handle and what the driver feels. We start with the alignment of the wheels when the car is at static (not moving) ride height and with the front wheels steered straight ahead.

We covered the concept of wheel toe in Lesson Nine because we needed to explore the Ackermann effect that can change the toe settings in our race cars. So, that being done, we'll get on to what the basics of alignment are.

Basic Alignment involves setting the suspension so that the tire contact patches line up when at static ride height and at mid-turn attitude. Alignment is different for circle track cars and road racing cars. For circle track alignment, we align the right side tire contact patches when the track width is different at each end of the car. For road racing, we align the contact patches to be equidistant from the centerline of the chassis. We have equipment available that can accurately measure the wheel alignment using lasers.

The most basic areas of study for alignment are:

- ***Overall Wheel Alignment*** – The alignment of each wheel in relation to the centerline of the car, or to a datum line we establish for our race car.
- ***Tire Pair Alignment*** – The alignment of the right side tires and the alignment of the left side tires to each other.
- ***Rear Alignment*** – The alignment of the rear in both static and dynamic conditions
- ***Tire Pair Alignment Dynamically*** – The way the side tire pairs align after the car has dived and rolled

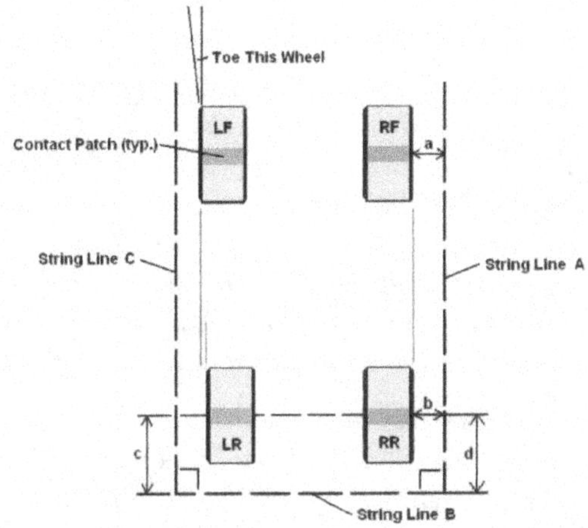

If we were to look down on the layout for alignment, it would look something like this. We concentrate on the tire contact patches because we often have camber in our wheels and aligning the tires at the hub height will not tell us if the tire contract patches are in alignment for cars with a-symmetrical cambers. This shows how a race team can use either strings or lasers to form a box around the outside of the car in order to measure for alignment.

Basic Understanding Of Alignment Components - We need to know how our wheels are aligned both when driving straight ahead, and when we are driving through the turns. The motion of the chassis can change the alignment we see at static ride height and if it does, we need to be able to measure that and adjust for the change.

When we discuss alignment, we are talking about the alignment of the tire contact patches. We often measure for alignment using the outside of the tire or rims at hub height. If the wheels have camber, then the contract patch is somewhere other than where the center of the tread is at hub height.

So, when we measure for alignment at hub height, we need to compensate our measurements so that we are aligning the contact patches, or the center of the tire tread at the ground.

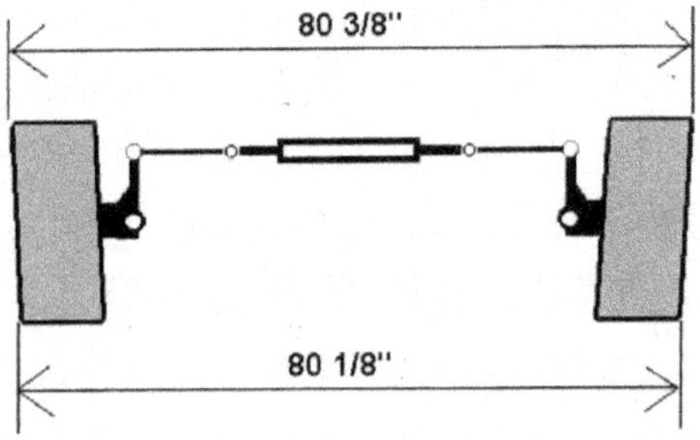

Toe is, as covered in previous Lessons, the aligning of the front, or rear, tires so that they are either closer together at the front than the back of the tire (Toe In), or the other way around (Toe Out). For most racing applications, the front wheels are Toed Out and the rear wheels are Toed In. Alignment also involves toe settings, but we set zero toe when we are aligning the four tires. Then after the four have been positioned correctly, we can then re-set the toe.

The front tire alignment is associated mainly with toe and Ackermann. We have studied both of these. But when we are checking alignment in relation to the rear tires, we need to eliminate the toe temporarily. The toe can be reset after we are finished checking and adjusting the alignment.

What we want as a goal for alignment is to make sure first of all that the rear tires are tracking in the same path as the front tires. This is the same process that is used on production cars at the alignment shop.

The alignment goals are a little different between a circle track car that only turns one direction and a road racing car. If the track widths (width of the sets of tires at each end of the car measured from center of contact patch to center of contact patch) on the front and rear of the car are different, then here are the requirements.

The use of lasers has become commonplace for checking and setting alignment and toe. The use of lasers can speed up the process of alignment and also make your work much more accurate than when using strings, which is an older method of alignment.

For circle track racing, we mostly line up the outside (right side on a left turning car) contact patches at static ride height. The rear track width is usually different and less than the front, so the inside tire contract patch will track to the right the front tire contact patch for a left turning car.

For road racing cars, we split the difference and align both ends so that the tire contact patches are equidistant from the centerline for each pair of tires on each end of the car. If the cambers are symmetrical at the front and rear, then we can align the car measuring at the hub height.

We can see where a laser line can be easily read on a ruler and be projected far beyond the ends of the car.

Alignment Can Change From Chassis Movement - The front track width can change due to chassis dive and roll. If the chassis moves enough to change the track width more than an 1/8th of an inch, then you might consider compensating for the movement when you are aligning the tire contact patches.

The best way to do this is to travel the chassis as much as it does going through the turns and then align the wheels. Many companies now offer pull-down rigs that will pull the chassis to replicate the shock travels that exist in the turns. This does two things, one it places the tire contact patches where they truly are when the chassis travels. We can then know that we are truly aligning the contact patches to be correct, or where we want them to be, at mid-turn.

When checking for alignment, we can install solid links in place of the shocks or coil-overs so that the chassis doesn't move and so that we can raise the car in order to work under it. If we are using a pulldown rig to check alignment at dynamic ride height, we install the springs and shocks that will be used when racing.

The other important thing a pull-down rig does is load the suspension just as it is loaded on the race track. Then any deflection of the chassis components that might affect alignment are properly deflected and we can then know our alignment is true.

The front alignment can change when we steer the wheels and when the wheels move vertically. The rear alignment can change, and often does, when the car dives and rolls. For cars with a straight axle rear suspension, the links that position the rear fore and aft as well as laterally, can move the rear tires side to side as the chassis moves.

When the straight axle rear moves side to side, it can also change its direction of alignment. This is called rear steer, because the rear tires are actually steering, just like the front tires. This action can be very detrimental, or helpful, to the setup. It all depends on what our goals are.

Exam - In The Context Of This Lesson:

Alignment Is?

1) Toe-out from bump
2) The direction the wheels are pointed
3) Amount wheels steer
4) Toe-in from bump

Alignment Can Change Due To?

1) Bump Steer
2) Setup changes
3) Chassis movement
4) Ackermann steering
5) All of the above

The Primary Goal For Alignment Is?

1) To balance the setup
2) To make the car neutral in handling
3) To make the tires track inline
4) To keep the tires from wearing

For Circle Track Racing, We Alignment The?

1) Front tires
2) Left side tires
3) Right side tires
4) Rear tires

For Road Racing, We Alignment The?

1) Front tires
2) Left side tires
3) Right side tires
4) Rear tires
5) Front or Rear tires equal distance from centerline

Lesson Twelve – Sway / Anti-Roll Bars

In race car setup, we control chassis roll by using what are called sway bars or anti-roll bars (the terms are interchangeable), the later better describing what they do. Not all race cars use sway bars. Dirt cars mostly don't use sway bars. The actual roll angle a chassis will roll to is variable and determined by:

- ***Spring Stiffness*** – Simply said, the stiffer the spring package is, the less roll we will see.
- ***Magnitude of G-force*** – The greater the lateral force through the turns, the greater the roll angle, in most cases.
- ***Track Width and Height of Center of Gravity*** – Roll angle is calculate knowing the Center of Gravity height, the track width, the lateral force and the weights of the race car. A wider track width reduces roll angle and a higher CG increases roll angle.
- ***Track Banking Angle*** – The greater the banking angle, for the same G-force, the less roll angle
- ***Anti-Roll Mechanism Stiffness*** – The stiffer the anti-roll mechanism, the less roll angle

The anti-roll, anti-sway, or roll bar systems in a race car are designed to restrict and reduce chassis roll as the car goes through the turns. In this way, the ride springs can be softer for ride considerations while the sway mechanism helps control the chassis roll. Excess roll can affect our cambers and other settings and is not considered beneficial.

These systems can be constructed in many different ways, like this example of a prototype rear suspension that uses links (white bars) that connect the spring rocker arms to the sway bar and blades at the rear behind the transaxle.

A sway bar is intended to limit the roll of a chassis so that the springs that are installed don't need to be overly stiff in order to resist roll. Excess roll can be very detrimental to a race car because it causes more suspension movement which can affect alignment, camber settings and weight distribution.

The actual sway bar mechanism can be designed in many ways. In all designs, the bar or material used in the mechanism, must bend or twist as the chassis rolls, not in dive. The vertical movement of the chassis is resisted by the springs. The lateral, or roll, movement of the chassis is resisted by both the springs and the sway bar mechanism.

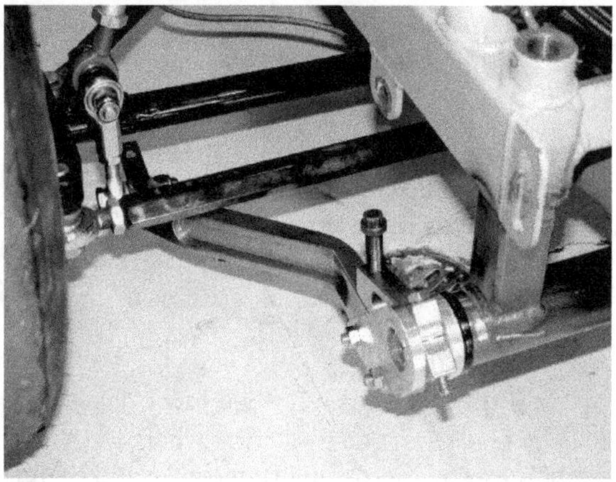

In a modern stock car, the anti-roll system consists of a roll bar linked to the lower control arms by way of an arm that is attached to the control arm. This bar has an adjustable end on the part of the arm that attaches to the bar. This is called a three piece sway bar, the actual straight bar and two arms.

What Determines The Sway Bar Stiffness? - Sway bar stiffness can be determined. Here are the factors that determine the stiffness of a system.

- *Sway bar diameter* – The outer diameter of the bar, and if hollow, the wall thickness of the bar both help determine the bar stiffness. The larger the diameter and the thicker the wall, the stiffer the bar will be.
- *Sway bar length* – The length of the portion of the bar that will twist helps determine the bar stiffness. The longer the bar, the softer or less stiff the bar will be.

• *Sway system material hardness* – The material the twisting and/or bending parts are made of help determine stiffness. Of course, the harder the material, the stiffer the system will be.

• *Length of Arms and/or Blades* – The length of the arms and/or blades relate to the stiffness of the system. The longer the arm or blade, the less stiff the system will be. This is because of the leverage affect.

• *Angle of Blade to Motion* – In most prototype systems the blades can be turned in relation to the direction of motion of the links to the suspension parts. If flat to the link, the system will be the softest in roll resistance. If turned more edge wise, or parallel, to the link, they will bend less and offer more resistance to roll.

• *Motion Ratio of System* – The motion ratio of all these systems relate to the overall stiffness. On stock cars, if the sway bar arms are mounted closer to the lower control arm chassis mounts, there will be less movement of the arm and less roll resistance.

In the prototype, the farther away from a one-to-one motion ratio (the link moving less than the wheel), the less roll resistance. In more of a one-to-one ratio where the link moves nearly as much as the wheel, the roll resistance will be greater.

Stock Car Sway Bars - The bar in a stock car may be made as a one-piece type where a straight portion extends across the frame rails, and then is bent to angle back to the be mounted under the lower control arms. Another design utilizes a straight bar with serrations on the ends. Separate arms are then clamped onto the ends of the sway bar and then extend back to be mounted under the lower control arms.

Sway bar systems can be tested for rate. This setup uses a spring rating device to measure pounds of force verses inches of movement of the end of the sway bar arm. It was discovered that the aluminum arm was bending causing loss of rate for the system. A similar steel arm did not bend and provide a higher system rate.

In stock car racing, the sway bar is mounted in front of the spindles and through the use of sway bar arms, it is connected to the lower control arms. As the chassis rolls, the right lower control arm moves up in relation to the sway bar arm and since the end of the sway bar arm is mounted to it, move up too.

In addition to the movement on the right side in a chassis roll, the left side lower control arm wants to move down, but since it is also attached to the other end of the sway bar through its own arm, it cannot easily do that. In this way, the sway bar limits the roll.

The sway bar must twist in order for the chassis to roll. If we use a large enough sway bar, there will be very little roll. It is more effective to place the lower control arm mounts for the sway bar closer to the ball joint where there is more vertical movement.

Eliminating chassis roll completely is not something that we design for. In the recent past, stock car teams have installed very large diameter sway bars that were very stiff in roll resistance. This design did not work very well with the other components of the suspension and setup. Now most teams have returned to using reasonable sized sway bars on stock cars.

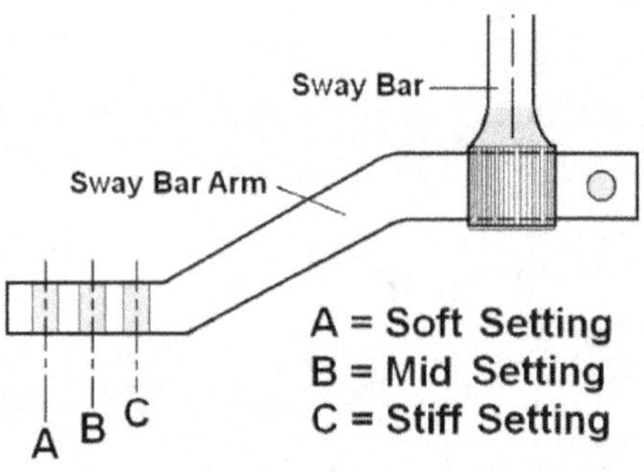

Of the components of the sway bar system that helps determine the rate is the length of the sway bar arm. We can see that the "A" setting provides the longest arm and is the softest setting. As the arm gets shorter by moving the mount towards the "C" hole, the system gets stiffer. The length of the sway bar and the diameter and wall thickness all have an influence on the stiffness of the systems.

Formula Car Anti-roll Mechanisms - For formula and prototype race cars such as the IMSA prototype classes, and Formula One, sway bars are not necessarily bars at all, or may be a combination of a bar and blades. These are linked to the lower control arm movement through a connection to the rocker assembly that transfers the load of the car to the springs.

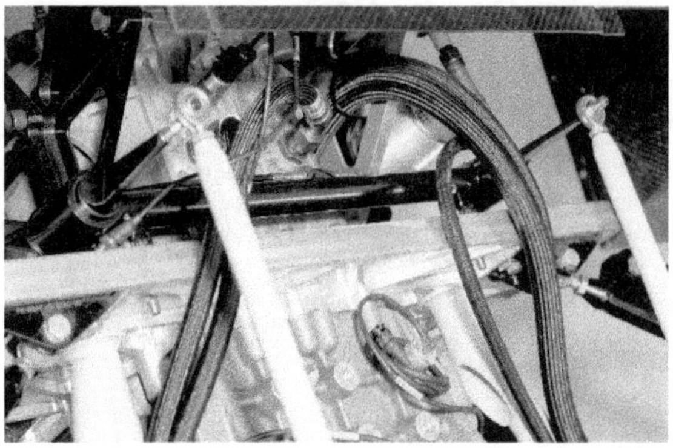

On this prototype anti-roll system, the links coming from the suspension rockers that move in opposite directions in chassis roll are connected to blades. These blades act much like the arms on stock car sway bar systems. In addition to acting as an arm, the blades will bend and help determine the stiffness of the system. The blades can also be turned. When turned in-line with the links, they will bend very little and offer the stiffest setting. Blades with different thicknesses can be installed to create different roll resistance.

When the chassis rolls, the links on each side move in opposite directions and twist or bend the sway "bar" components. Teams can install different diameter and wall thickness bars and/or different thickness blades to adjust the stiffness of the bar system.

Front Or Rear, Or Both - Circle track stock cars usually don't have sway bars in the rear of the chassis, just in the front. But most road racing and formula cars do have front and rear sway bar systems. That is because the formula cars have rear AA-arm suspensions, a lot similar to the front suspension, that must have roll control.

The stock cars that have rear sway bars are usually racing on road courses where it is hard to control rear roll like they can on oval, circle tracks. The roll stiffness of the sway bars must be coordinated with the spring stiffness.

The adjustment for pre-load, or twisting the bar at ride height, on a late model circle track car is a simple process. This screw jack puts the bar in a bind which loads the bar. Pre-load adds roll resistance before the chassis actually starts to roll.

Exam - In The Context Of This Lesson:

The Difference Between a Sway Bar and Anti-Roll Bar is?

1) One resist sway and the other does not
2) The Sway bar limits wheel movement
3) There is no difference
4) The Anti-roll bar does not roll

A Stock Car Sway Bar Arm Mounts To?

1) The upper control arm
2) The spring mount
3) The chassis frame
4) The lower control arm

Chassis Roll Is Determined Using Which Of The Following?

1) Spring stiffness
2) Amount of lateral force
3) Track width
4) Center of gravity height
5) All of the above

Which Race Cars Usually Don't Use Anti-Roll Bars?

1) Asphalt stock cars
2) Prototype IMSA cars
3) Dirt cars
4) Formula One cars

Which Of These Changes Creates A Stiffer Anti-roll System?

1) Increasing the sway bar diameter
2) Shortening the arm length
3) Increasing the sway bar wall thickness
4) Shortening the sway bar length
5) All of the above

Race Car Technology – Level One

Lesson Thirteen – Double A-arm Geometry

The AA-arm geometry systems in a race car have many components. Here we will describe the parts and how they function. In the Level II Advanced RCT we will get more technical in explaining how each can affect the others. So, let's take a look as the various parts.

• *Chassis Mounts* – These mounts connect the upper and lower control arms to the chassis.

• *Spindle* – The spindle is the connection between the control arms and the wheel. Its function is to allow the wheel to turn and move vertically.

• *Control Arms* – The control arms connect the chassis to the spindle and are free to move with the spindle within the range of motion the car is designed to operate within.

• *Anti-Pro Dive* – Anti and Pro Dive are effects that resist or promote the vertical motion of the chassis when braking.

• *Roll Centers* – The roll center can be two different points. One point is the center of rotation of the chassis and it is referred to as the kinetic roll center. It is formed by the motion of the chassis and has very little to do with race car design.

The other point is the bottom of the moment arm and its location is determined by the control arm angles. This is called the dynamic roll center, or moment center since it represents the bottom of the moment arm for a double A-arm suspension system. The Kinetic roll center is often nowhere near the Dynamic roll center.

• *Jacking Force Theory* – Jacking Force is a theory that states that forces acting through the tire contact patches are the predominant influences in chassis dynamics. This theory has been tested by the author and the results are non-conclusive as to its usability for race car design and setup.

• *Roll Angle Analysis* – Roll angle analysis is a method of comparing the roll stiffness of the front and rear suspension systems on any race car. It is this relationship that controls the load distribution after load transfer occurs during cornering. It is a critical process in determining the actual tire loading values for the four tires.

This is a typical layout of a double A-arm front suspension on a stock car used for circle track racing. We can see the upper and lower control arms are attached to the spindle and the chassis.

Chassis Mounts - The control arms are mounted to the chassis and the spindle. The springs are mounted to the lower control arms. The location of the spring mounts vary with the design. The coil-over (a spring mounted over a shock body to form one piece) chassis mount is above the upper control arm mount and mostly centered between the control arm links for a stock car.

The spring can be mounted as either a coil-over mount, as a large diameter spring housed in a bucket built into the lower control arm, or by way of a push rod that is attached to the lower control arm and extends up to the upper chassis and is connected to a rocker arm that is attached to the spring.

Both the stock big springs and the coil-over springs supporting the front of the chassis are mounted to the lower control arm and therefore the lower control arm must be strong enough to handle the loads and forces it will endure.

The lower control arm chassis mounts can be height adjustable in some race cars. This allows changes to the arm angle from a front view as well as a change to the mounting angle from a side view to assist with anti or pro dive. The upper control arm chassis mount can be adjustable for width, height and side view angle, again for creating anti or pro dive.

The other primary mount at the front of the car is the sway bar mount for a circle track car. These can be loose bushings housed in clamps bolted to the frame for a one-piece type of bar, or a sway bar tube with bearings or bushings at the ends for a three piece bar with the straight bar attached to an arm at each end.

Spindle - The spindle is the connection between the control arms and the wheel. Its design can be matched to the chassis mounting points spacing to accommodate the desired control arm angles. We spoke about spindle design in Lesson Five, but here we will talk about it in relation to geometry.

The primary parts of the spindle are, the ball joints (what the upper and lower control arms are attached to), the spindle snout pin (what the wheels are mounted on), and the steering arms (what the tie rods are attached to).

The spindle pin location from the lower ball joint center of rotation will determine the lower control arm angle aside from the small adjustments some chassis have for the lower chassis mounts. This is called the spindle pin offset. A pin that is placed higher than the normal or customary location is said to be a dropped spindle because it drops the ball joints.

The height of the spindle between the ball joint mounting holes helps determine the upper control arm angles once the spindle pin height has been decided upon. These angles are important in determining the Moment Center location and the camber change characteristics.

Control Arms - In most AA-arm suspension systems, the upper control arm is shorter than the lower control arm. These two arms provide vertical movement for the chassis and resist fore and aft movement of the wheels. The inner mounts are bushings or bearings and the outer connections to the spindle are referred to as ball joints, although the actual mounts can vary from balls to heim joints.

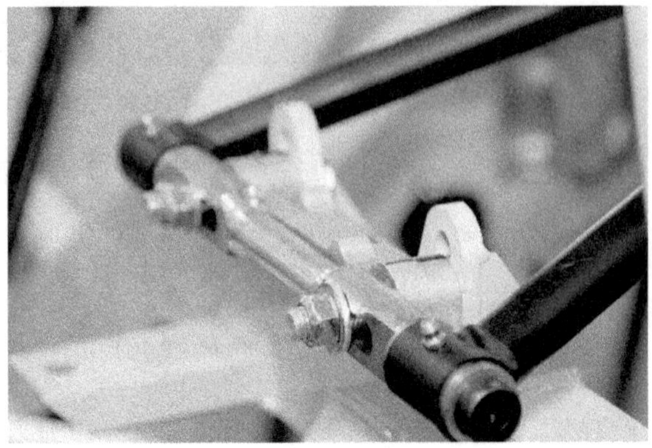

Many race car builders are incorporating slotted mounting holes in the chassis so the racer can tune their arm angles and therefore the MC locations. This Lefthander upper chassis mount has a series of overlapping holes cut in it so that the bolt can be located at different heights to change the upper control arm angles.

The control arms not only connect the chassis to the spindle, the front view angles they have relative to horizontal determine the moment center height and width. They can be different lengths and offset with the ball joint set rearward from symmetrical to accommodate setting caster.

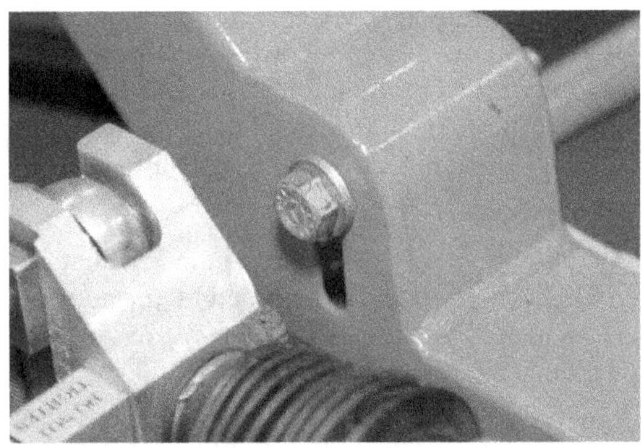

This lower control arm chassis mount is slotted so that the lower control arm angles can be adjusted. Note that this one is all of the way up. This can help compensate for the high amount of travel associated with the Bump Stop/Spring setups.

Either the upper or lower control arms can be one piece or parts bolted together. If the forward part of the arm is more perpendicular, or at a right angle, to the centerline of the chassis, it could be called a strut instead of an arm.

The upper control arm on this prototype road race car is very similar to the ones we see on stock cars. The "ball joint" is actually a mono-ball joint. The travel of the spindle on this car is much less than we usually see on stock cars.

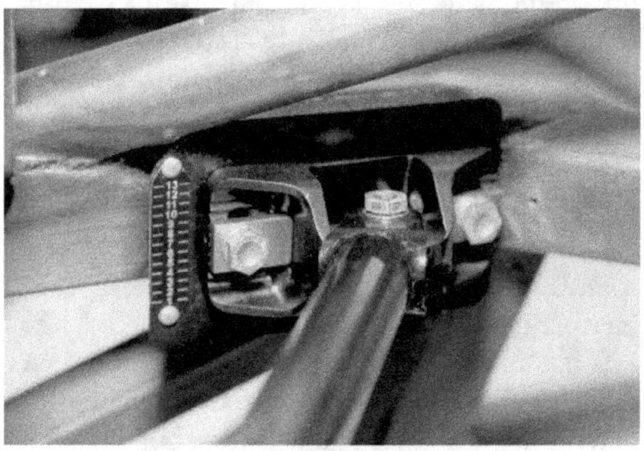

The chassis mount on this prototype race car for the upper mounts in the rear are adjustable for height which will change the control arm angle and the roll center location. The divisions are clearly marked.

Anti and Pro Dive - Anti-Dive is when the forces from braking that are trying to rotate the spindle are used to resist downward motion of the chassis. This effect only exists during braking. In the sketch, we show how when the wheel moves up, as the chassis moves down, with the side view angles in the upper and lower control arms, the ball joints want to move in the opposite direction the force is trying to move them.

It is this contradiction in forces and direction of movement that creates anti-dive. The instant center theory of anti-dive is not relevant to what is actually happening. If it were, then the instant center in most cases is so far to the rear of the front suspension that the lifting force would have little effect.

Pro-dive is when we use these same forces created by braking to assist the suspension in diving. We won't comment on the usefulness of this application, only that it exists. Teams that race on asphalt at circle tracks have been known to use Pro-dive on the left front suspension and anti-dive on the right front suspension.

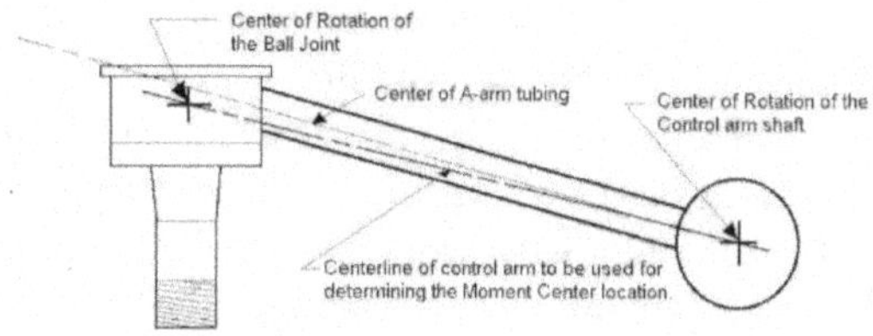

When we refer to the arm angle, we are referring to the angle of a line crossing through the centers of rotation of the ball joint and the chassis mount. The tubing used to construct the control arm may not be inline with a line crossing through the centers of rotation. Therefore, we cannot trust the angle of the tubing used to make the control arm in any roll center calculations.

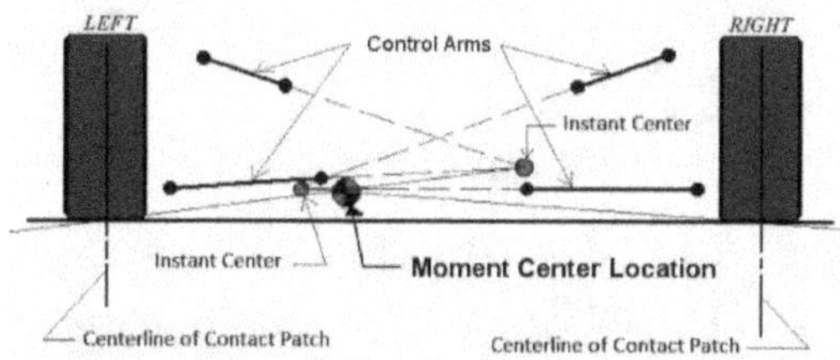

The front Moment Center location is the result using the intersections of lines extended from the centers of rotation of the control arms to form the points we call Instant Centers. The moment center is at the intersection of lines drawn from the left and right IC's to the centers of the contact patches of the corresponding tires.

Roll Center - To be perfectly clear, there are two roll centers engineers talk about for a AA-arm suspension. One is the kinetic roll center that is the point about which a suspension rolls. If you could look at a cross section of the suspension, it would be the point that doesn't move as the chassis dives and rolls going through the turns. That is not the roll center we need to know about.

The other roll center is calculated using the control arm angles, instant centers those angles produce, and intersections with the contact patches of the tire. This roll center is the bottom of the moment arm for that suspension.

Since it is part and parcel with the overturning moment that makes up the dynamic properties of this suspension, we like the term, Moment Center, because it better defines this point and differentiates it from the kinetic roll center. So, from now on in this course, we will refer to this point as the Moment Center, or MC.

The MC is a calculated point in a Double A-arm suspension. It is a point whose location helps to define and control the dynamics of the suspension it is derived from. It is probably the most important part, setup wise, of this type of suspension. Knowing where it is located in both the static (when the car is at rest at ride height) and dynamic (at steady state mid-turn with body dive and roll) locations is a part of the design of a race car just as choosing springs and weight distribution is.

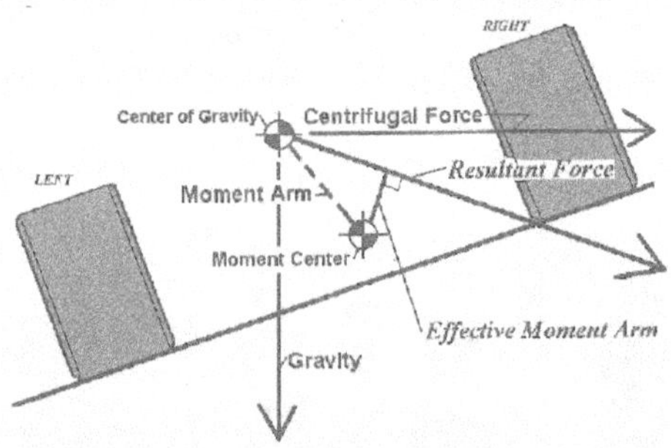

The roll center, or what is better described as a moment center, is the bottom of the moment arm. The top of the moment arm is the center of gravity. The difference in height of the two is the length of the moment arm. This sketch shows how the forces of Gravity and Centrifugal force work in a race car on a banked track.

The MC height represents the bottom of the moment arm in a AA-arm suspension. The top of the moment arm is the Center of Gravity of the sprung mass. More on that later. You have access to a lot of information about roll centers and there are a lot of opinions. I call them opinions because literally none of them have been proven, they just sound right to some people. There is no more controversial part of chassis dynamics than when talking about roll centers.

There has been a lot of study and testing connected with what we just said. In our testing, we have come up with some things we know to be true, and have dis-proven other theories about how the AA-arm suspension works dynamically.

Here is an explanation of how the front MC location is derived and what affects its location. The sketches are showing both the static and dynamic locations. Moment Center geometry software will help you determine the dynamic location for your MC.

On each side of the car we have upper and lower control arms, in a double A-arm suspension system, and those have pivot points at each end being a ball joint at the spindle and bushings or heim joints at the chassis mounts. If we draw a line through the upper control arm pivot points and again through the lower control arm pivots, these two lines will intersect at a point called an Instant Center.

For most stock cars, these IC intersections lie to the chassis side of the spindle. Each set of control arms on each side of the car has their own IC. If we also draw a line from each IC to the corresponding center of contact patch of the tire on the same side as the control arms we used to create the IC, then the intersection of the two lines from the left and right IC's defines the Moment Center location.

MC Movement – As the chassis dives and rolls through the turns, the control arm angles change, and therefore so does the MC location. We have found, as have many hundreds of race teams and race car builders, that the lateral location of the MC has an important affect on how well the AA-arm suspension performs. Current testing has not determined why exactly this is true, but we think it is due to the camber change associated with the angles that place the MC in a certain location laterally.

Location design for the MC is dependent on which type of race car you are working with. In a road racing car, we have found that if you can keep the MC close to the centerline of the chassis through dive and roll, the car performs much better. Theses cars turn both left and right, so it must work for both turning directions.

If we are working with the front end AA-arm suspension, in a road racing car, we would design the MC to be at centerline statically, and stay close to the centerline dynamically.

For cars that only turn one way, we can design the MC location to migrate farther off the centerline of the chassis and it has been found to work very well, on the race track, this way. So, historically, designers have come up with configurations of control arm angles that will place the MC more to the inside of the turn. When they have done that, the car turns much better than if the MC were more to the outside of the turn.

One many late model dirt cars and fabricated road racing cars, the chassis mount for the upper control arms are uprights of square tubing. To set the height of the chassis mount, holes are drilled in the tubing. To change the upper control arm angle, we only need to drill new holes in the proper location.

Control Arm Angles vs. MC Location - Control arm angles are measured in degrees from horizontal. Therefore, level would be zero degrees and straight up would be 90 degrees. The lower control arm angles largely control the amount of lateral movement of the front MC and the upper control arm angles mostly control the lateral location of the two MC's, static and dynamic.

Excessive lower control arm angles (chassis mounts being lower than the ball joints) cause the front MC to move a greater distance as the car goes through the turns. Increasing or decreasing one or both of the upper control arm angles moves the MC side to side.

Low or reverse (the chassis mount being higher than the ball joint) angles in the upper control arms cause excessive camber change in the front wheels, especially in the right front wheel where it affects the handling the most. In some formula cars, this is common. Because those cars are sprung very stiffly, there is little chassis movement, and therefore less camber change.

In Level II Advanced RCT we will explain how to measure for the MC location and how to redesign a current race car to make changes to the location of the MC.

Jacking Force Theory - The concept of Jacking Force has been around for many years and was referenced in a book titled Race Car Vehicle Dynamics. In that book, the jacking force was in reference to a swing arm rear suspension and not necessarily focused on a AA-arm suspension.

Nonetheless, certain individuals ran with the concept and alluded to it being the whole influence on AA-arm dynamics. The truth is, it isn't. Not only is it not a measurable influence on suspension dynamics, it really doesn't apply at all.

What we are presenting here is not a theory to counter other theories. We have done our own testing, in conjunction with, and in the presence of, some of the primary proponents of the jacking force theory. What we saw, together, was that there is little if any influence on vehicle dynamics in a AA-arm suspension due to jacking forces.

So, we are left to concentrate on what we do know to be true from model testing and on track evaluations done over many years. We just wanted to address this area of dynamics that is getting a lot of press at the present time.

Exam - In The Context Of This Lesson:

A Coil-over Spring Is Which Of The Following?

1) One that is over the top of the chassis

2) A part of the sway bar package

3) A spring and shock combination

4) Useful for preventing chassis roll

A Chassis Mount Is?

1) The way the engine is connected to the chassis

2) Where the inner ends of the control arms are mounted

3) Anything attached to the chassis

4) 1 and 2

Which Is A Primary Part Of A Spindle?

1) The ball joints

2) Steering arm

3) Snout Pin

4) All of the above

Why Are Control Arms Adjustable For Angle?

1) To assist with anti and pro-dive design

2) To allow changes to the Moment Center location

3) To provide better camber change

4) All of the above

Anti and Pro-dive Only Happen When?

1) The race car is accelerating

2) When the driver is on the brakes

3) When we turn the steering wheel

4) When the roll center is in the correct location

The Moment Center Is?

1) The point around which the chassis rolls

2) Where the control arm angles meet

3) The bottom of the moment arm

4) Used to find shock travels

Race Car Technology – Level One

Lesson Fourteen – Types Of Rear Suspensions

There are two primary rear ends, and accompanying rear suspensions, used in racing. One is the straight axle type that is popular in stock car racing for both circle track and road racing, and the other is the independent rear end used along with a AA-arm suspension that we have already talked a little about.

Each of these have different types of rear end have their unique way that they are connected to the chassis and their own unique methods of spring and shock mounting designs. Let's talk a little about each one starting with the straight axle, so we can become more familiar with the terminology and the various designs.

This is an independent rear suspension. It is independent only because the rear wheels move independent of the gear box. The actual gear case is attached to the chassis and becomes a structural part of the car.

The straight axle rear end is just that, a straight piece consisting of a set of gears, a differential (the part that transmits power from the engine to the wheels), axles mounted to each wheel, axle tubes that house the axles, brake mounts, and control arm mounts. These types of rear end are not a part of the chassis, but rather move independently of the chassis.

There are two different kinds of straight axle rear ends available for stock car racing, the production based stock rear end and the quick change designed primarily for race cars. The most popular stock version rear end is based on the Ford 9 inch design which is considered much stronger. All of the top classes of stock cars touring the U.S. use this 9 inch Ford design.

This rear end unit must be attached to the race car so that it cannot move unrestricted laterally or forward and backwards. This is done through the use of links, or in some cases leafs. The angles these links form can cause movement in all directions of the straight axle when the chassis moves. We'll talk about how that happens later on.

The straight axle has a center section that houses the gears and the differential. This set of gears takes the power from the motor, through the transmission and driveshaft, to the pinion at the front of the differential housing. The pinion is a gear that turns on a ring gear that is attached to the axles through the differential.

The differential is a mechanical device of different designs. It compensates for the different speeds the wheels rotate when we are going around a curve, or through the race track turns. Various designs are named Open differential, Posi-traction, Detroit Locker, Tru-Trac, etc. All of these will help both tires to have Grip and still turn at different speeds from one another.

When a differential is locked with both axles connected together, a spool is used, or the open differential is welded so that the gears are locked. Th spool is a one piece fitting where both axles are contained so that both wheels will turn at the same rotational speed all the time.

The spool would not be of much use on a road course where the rear tires are the same circumference. On oval circle tracks where cars turn the same direction, the spool can be used successfully in some cases because the teams use different size tires for what is called stagger.

In a left turning car, the outside rear tire would be larger in circumference than the left rear tire. The stagger must be matched to the radius and banking angle of the track.

This shows the insides of a quick change type of rear differential. The primary gears are shown, and to the back (left in the photo) are contained a set of easily changed gears that determine the final gear ratio. The primary gear ratio might be something like a 4.11:1 ratio. Then if the quick change gears are say 1.00 to 0.900, then the final ratio would be found by multiplying those two together, for a 4.56 final ratio (1.0 / 0.90 = 1.111 x 4.11 = 4.56). Each gear set has instructions for finding the final ratio using a chart.

The gearing in a rear end can be changed so that for a particular sized track, the engine can operate within a range of RPM (revolutions per minute) that produce the best power. The quick-change rear end allows removal and replacement of the gears without much effort. That is why they are called quick-change. This OEM style rear end must be built to a specific gear ratio and cannot be changed easily. The race team would stock multiple rear ends with ratios they think they might need.

The Straight Axle Rear Suspensions

The various straight axle rear ends use different designs of rear suspensions to locate them within the chassis. The motion of the chassis when entering and driving through the turns will move the links that attach the rear end to the chassis.

This causes changes in the direction the rear wheels are pointed. This is called Rear Steer. Rear steer is a major consideration in the design of a race car, any race car. The Angle of Attack is critical for the rear tires, just as it is for the front tires.

In the rear we have no way for the driver to change the angle of attack like with the front wheels. The only way we can regulate and dictate the amount of rear angle of attack is through design of the angles of the links from the rear end to the chassis.

The most common types of straight axle rear suspensions are:

- ***Three Link*** – This suspension has three links, with two at each side, usually under the axle tube and mounted to the tube and the chassis forward of the tube to restrict fore and aft movement. A third link is attached to brackets mounted above the rear center section forward to the chassis. A panhard bar runs from the rear axle tube laterally to the chassis to restrict lateral movement.

The side view angles the two outer links end up at will largely determine the direction the rear end points when going through the turns. Since the rear tires already have an angle of attack naturally, we just need to fine tune this angle, not radically change it. We will get into this in much more detail in RCT Level Two. The three link suspension is most common on asphalt late model cars.

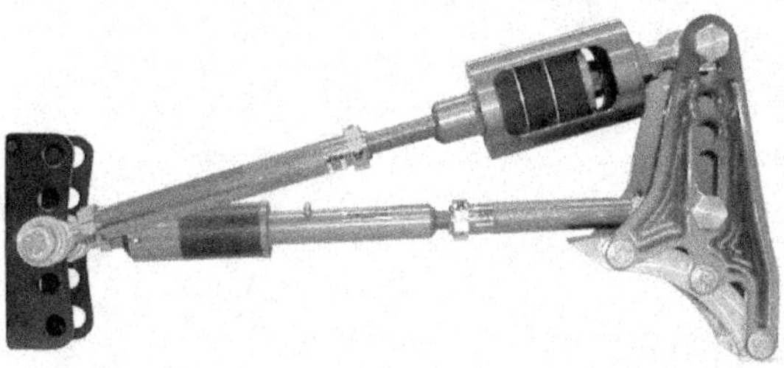

The part of a three link suspension that resists the rotation of the rear end under acceleration and deceleration is the third link which can either be a single link mounted above the rear end, a lift arm mounted to the rear end solidly, or a pull bar that is much like the single link, but which extends using a spring or rubber disc.

• ***Four Link*** – A four link is similar to the three link, except it has two links on each side, one on top of the axle tube and one underneath the axle tube, instead of the just the one on each side for the three link. The four link also has a panhard bar, or what dirt racers call a J-bar, (these two terms are interchangeable) because it is bent into a J shape to go over the drive shaft.

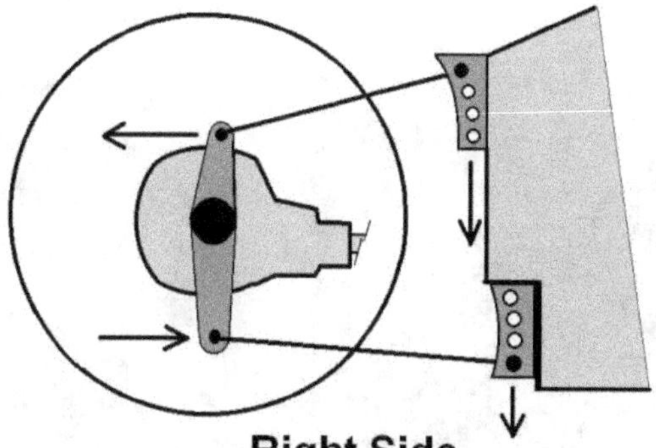

This type of rear suspension is common on dirt cars. Just like the three link, the link angles of the four link are critical in order to get the rear end angle we desire. That ultimate angle is very different for dirt cars than for asphalt cars.

• ***Z-Link*** – A Z-link is a four-link rear suspension where the bottom link runs forward to the chassis and the top link runs back to a mount behind the rear end axle tube. This type of suspension is common on dirt cars, both modifieds and late models. This type also has a third link or lift arm or pull bar to limit the rotation of the rear end. It also has a panhard bar to limit lateral movement.

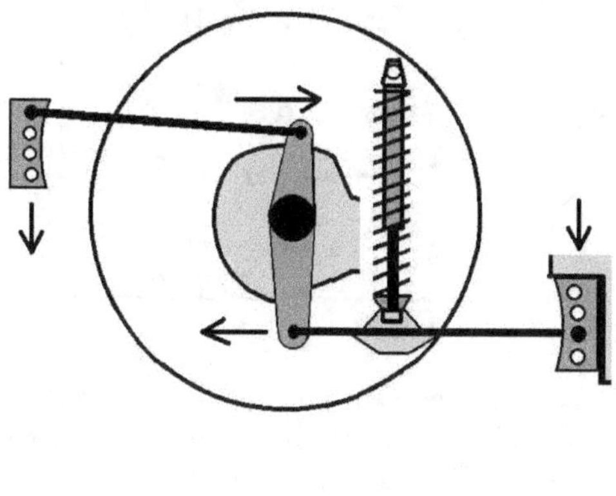

A unique feature of this type is that the spring, or coil-over, is mounted onto the forward link. When this is done, there is a motion ratio created and the chassis for setup purposes, feels less spring rate than the rate of the installed spring.

• ***Truck Arm Link*** – A truck arm rear suspension is based on a mid-1960's Chevy pickup truck suspension. It consists of two arms mounted solid to the rear axle tubes, one on each side, and running forward to bushings that are the chassis mounts. These resist fore and aft movement. This system uses a panhard bar to locate the rear end laterally.

The rear springs are mounted onto the truck arms just forward of the rear axle tubes. Mounting the springs in this way creates a motion ratio much like the Z-link we have already discussed.

• ***Leaf Spring*** – An older, but still used, type of rear suspension is the leaf spring. This system has two springs, one on each side, that attach solidly to the rear axle tubes, and mount to the chassis using two eyes at each end of each leaf.

The leafs can be single leaf or multi-leaf, meaning a layer of different length leafs are stacked to produce a desired spring rate.

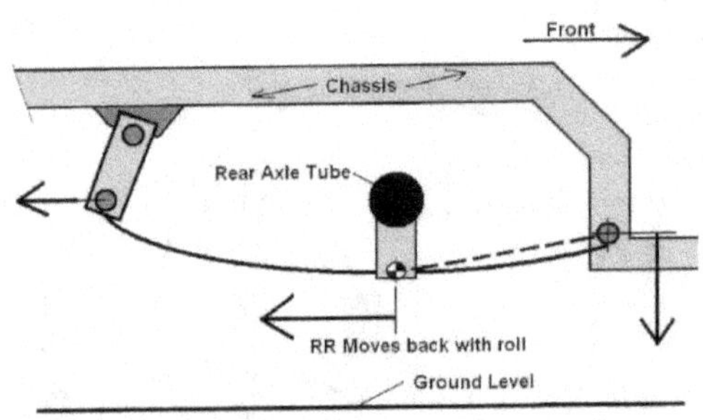

The leaf system provides the springs the car needs in the rear, as well as the lateral locating system for the car in most cases. In some cases, a panhard bar may be used as a lateral locating device.

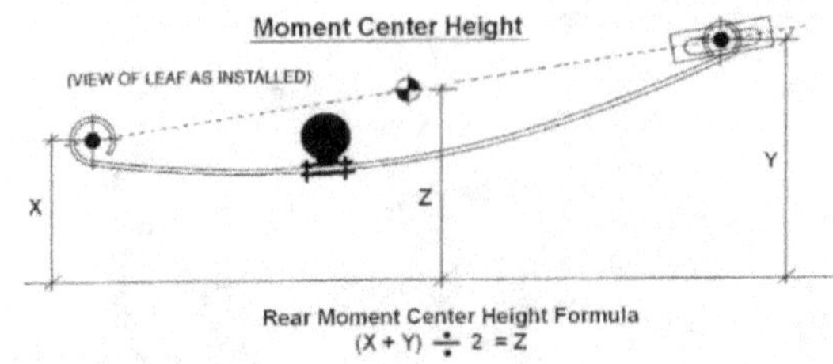

The leaf syste also provides a means to resist rotational movement of the rear end under acceleration, much like a third link, pull bar or lift arm that are used on a three or four link rear suspension. With the leaf system, there is no need for a third link type of attachment.

• ***Metric Four Link*** – This suspension type has been used on many 1980's and 1990's production automobiles and is mostly found in stock classes of racing. It consists of four links, two that are mounted on top of the rear end center section, and two at the bottom of the axle tubes that are mounted out farther towards the wheels. Both the upper and lower links run forward to the chassis.

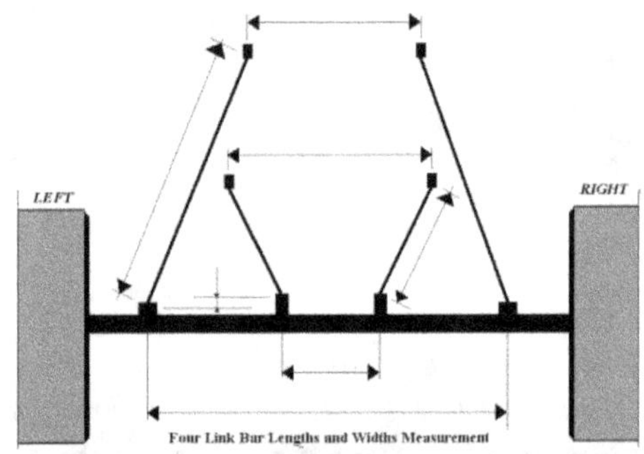

The chassis mounts for the upper links are mounted wider apart than the rear end mounts. And the chassis mounts for the lower links are mounted closer together than the mounts attached to the axle tubes. This matrix of top view angles creates a resistance to lateral movement and this suspension type does not have or need a panhard bar.

Panhard Bar Function

In the straight axle suspensions, the links determine the fore and aft location of the rear wheels. The panhard bar (J-bar used by some classes, but meaning the same) determines the lateral location of the rear end. Because the chassis moves in dive and roll, the angle of the panhard bar is constantly changing.

When the bar angle changes, is moves the rear end slightly one way or the other. We can design movement in to our system to cause the rear wheels to steer, just as we can steer the rear wheels with link angles.

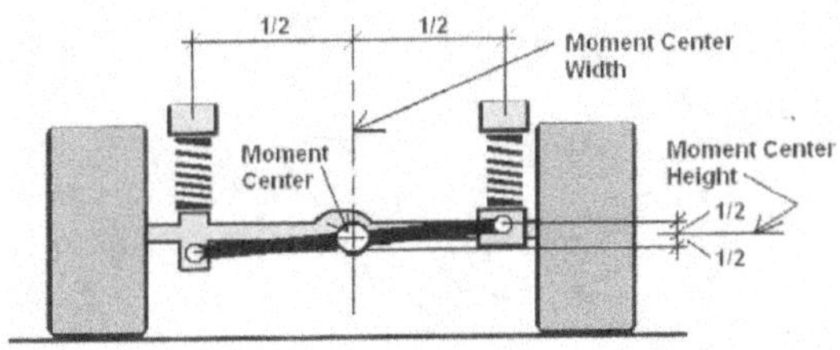

The panhard bar can be mounted to the chassis on either the right side, which is very common for circle track asphalt race cars that use panhard bars. Or, it can be mounted on the left side, which is very common for circle track dirt cars.

For road racing, it can be mounted to either side, but in turning one way, the reaction of the rear end lateral movement will be different when the car steers the other way. So, the panhard bar is not thought to be ideal for road racing applications. This is what is:

Watts Link

The Watts Link is a link that is mounted to both sides of the chassis, one link on top of a center pivot pin, and the other on the bottom of that pin. A plate device is mounted onto the pin and allowed to rotate. When the chassis moves vertically, the plate is free to rotate to compensate for the change in length from the rear pin to each side of the chassis. In this way, the rear end stays centered within the chassis and there is no lateral movement of the rear end that might cause the rear end to steer.

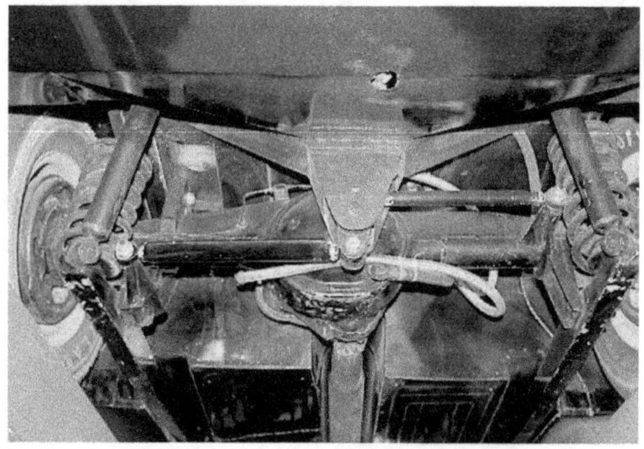

This Watts link was designed and mounted in an early Cup car built by the legendary Smokey Yunick. This car was outlawed at Daytona in 1968 for being way ahead of its time.

Independent Rear Suspension

In most formula and prototype race cars, the rear suspension and rear ends are of the AA-arm type, much like a front AA-arm suspension. The transmission is included within the structure of the rear end and handles the duties of the differential as well. It is commonly called a trans-axle, being a transmission and axle housing. A trans-axle can be ridged in the way the two axles move vertically or independent.

For the independent trans-axle, the most common type, the axles are not connected or housed in a common structure, and can move independently of one another. The difference in rotational speed of the two rear wheels is handled through a differential device that is a part of the transaxle structure.

Just as in a front AA-arm suspension, the rear AA-arm suspension has camber change, a moment center and anti and pro-dive properties. What the rear AA-arm suspension does not need are steering arms or caster, since the rear wheels don't turn.

While the rear wheels don't steer in a rear AA-arm suspension, we do set various toe amounts and there could be bump steer we need to address and minimize in these systems.

Just as much engineering goes into the arrangement of the rear suspension link angles and other geometry parts as for the front suspension. In RCT Level Two, we will explain in more detail how each system works and how to arrange the link angles for maximum performance.

Exam - In The Context Of This Lesson:

The Two Most Common Types of Rear End Used In Racing Are?

1) Lift arm rear end

2) Pull bar rear end

3) Straight axle rear end

4) Independent rear end

5) 3and 4

The Differential Transfers Power How?

1) From the rear tires

2) Through the third link

3) Through the pull bar

4) From the engine to the wheels

A Straight Axle Rear End Is Used In What Type Of Race Cars?

1) Formula One

2) Dirt late models

3) Asphalt stock cars

4) Stock car touring series cars

5) 2, 3 and 4

An Independent Rear End Is Used In What Type Of Race Cars?

1) Formula One

2) Dirt late models

3) Asphalt stock cars

4) IMSA Prototype cars

5) 1 and 4

Lesson Fifteen –Rear Roll/Moment Center

The rear suspension has a roll center, just as the front does. And like the front, as we have already discussed, we like to call this point a Moment Center because it represents the bottom of the rear moment arm.

The rear MC is determined differently depending on the type of rear suspension. We will look at the three, four, Metric and Z-link systems, as well as the leaf system designed for the straight axle. And we'll examine the rear independent AA-arm system for its MC design.

The Panhard Bar Moment Center

With a suspension that uses a panhard bar as a lateral locating device, the rear MC height is the average height of the centers of the two bolts and/or heim joints that mount the panhard bar to the rear end and the chassis. This height might change when the chassis is moving.

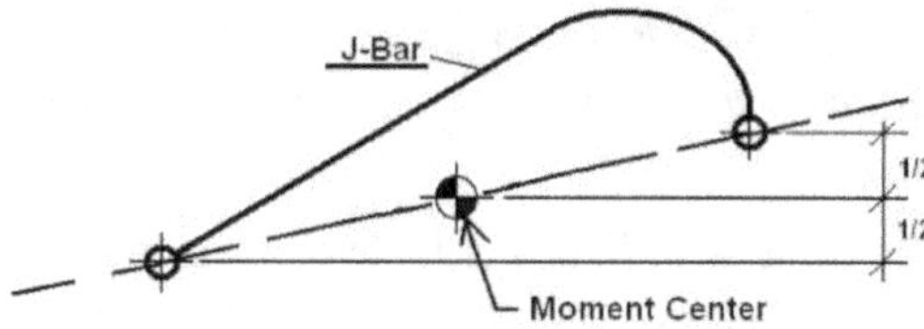

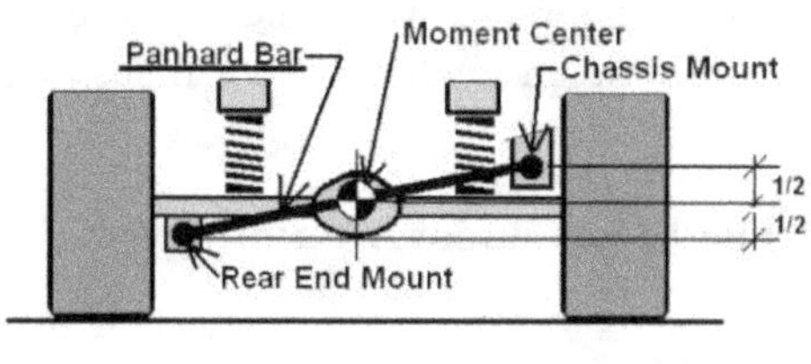

With the Panhard bar or J-bar type of lateral locating device, one end is mounted to the rear axle of the rear end and the other to the chassis. The average height of the two ends is the height of the rear Moment Center, or the height of the front moment center for a car with a straight axle front suspension.

If the panhard bar were mounted to the right side of the chassis, and the car were turning left, then the chassis mount would be lower in the turns due to the motion of the chassis on that side. Since the left side of the chassis moves very little compared to the right, or outside side, the overall MC height will be lower in the turns.

On this rear axle tube mount, there is a serrated plate with a matching through bolt that makes the height of the rear end mount adjustable for height. In this way the team can fine tune the rear moment center height. There is usually a similar adjustable mount on the chassis side.

Remember that the rear MC for a panhard bar system is the average height of the two ends. If one end is lower in the turns, then the average height of the MC will then also be lower.

If the panhard bar were mounted to the left side of the chassis, in a left hand turn, then both sides would remain at nearly the same height and the height of the MC would not change.

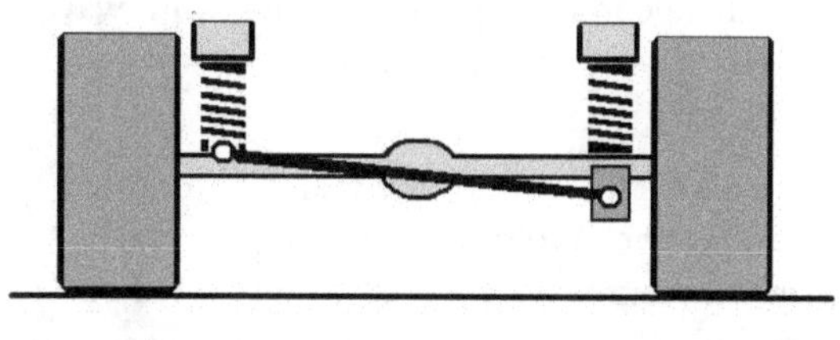

The current trend in circle track racing where the cars turn left is to mount the right side of the panhard bar lower than the left side axle mount. This puts the bar more inline with the combined forces of Gravity and lateral force through the turns. It provides a more straight pull on the bar which reduces possible adverse jacking forces.

For road racing applications, in a left turn, with a left chassis mount, the MC height would stay the same, but turning the other way, it would change and be lower. This causes different handling balance from left turns to right turns, which is not ideal.

The Leaf Spring Moment Center

With the leaf spring system, the rear roll center is calculated in the following way. We average the height of the two mounting eye bolts at the front and rear of each spring. Then we average the two right and left average heights of the eye bolts and that is the height of the rear MC for leaf springs.

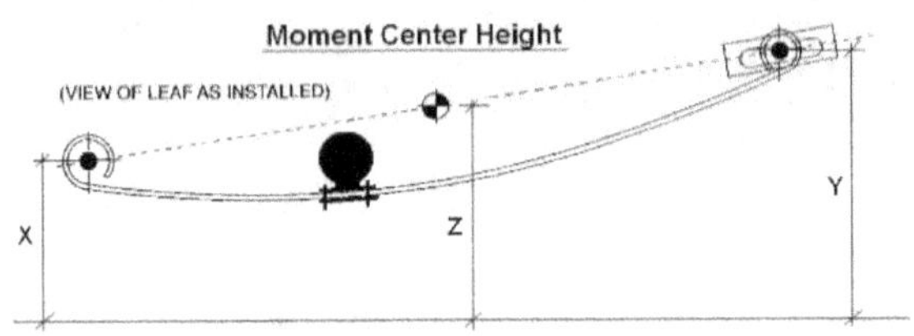

This is the formula for calculating the moment center height of a straight axle, leaf spring suspension. The rear moment center height is the average of the heights of the two springs individual moment center height, in case they are not the same. As the chassis rolls and the outside chassis moves lower, that side will produce a lower point in the moment center package which will lower the overall moment center height, much like a panhard bar does when its chassis mount moves lower on the right side.

With the leaf system, we also have a change in MC height when we are going through the turns much like a panhard bar, except that we have the same change in height when turning both ways. The car stays consistent between left and right hand turns.

To change the height of the rear MC for a leaf system, we need to change the height of the spring in relation to the ground, and we do that by spacing the spring differently at the mount to the rear axle tube.

If we want the rear MC to be higher, but the chassis to remain the same height, then we need to make two changes. First we raise the leaf spring in relation to the rear end axle tubes. Then we raise the spring mounts on the chassis by the same amount. We have then raised the MC height without raising the height of the chassis.

The Watts Link Rear Moment Center

The Watts link rear suspensions Moment Center is unique in that it stays very consistent in height. This is because the link is mounted to the rear end and stays relatively at the same height throughout the lap. We change the moment center height by changing the height of the swivel link, and ideally the chassis mounts of the links from the centerpiece to the chassis so that they remain level.

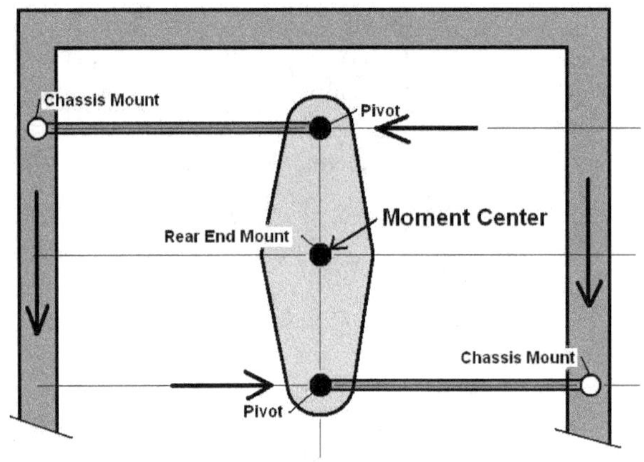

The Watts link has a moment center located at the center of the pivot bracket. This point can, in some designs, be moved up or down to tune the height of the rear moment center.

Most designs of Watts link have incorporated these adjustments into the overall design of the rear suspension. Otherwise, there is no way to change the moment arm in the rear to tune the balance between the front and rear suspensions.

The Independent AA-arm Rear Moment Center

The MC for an independent rear suspension using a AA-arm geometry is the same as when we are measuring a front end AA-arm suspension. We use the upper and lower arm angles to produce an instant center for each side. Then the lines from the instant centers to the contact patch on each side are crossed. The point where they cross is the rear MC location.

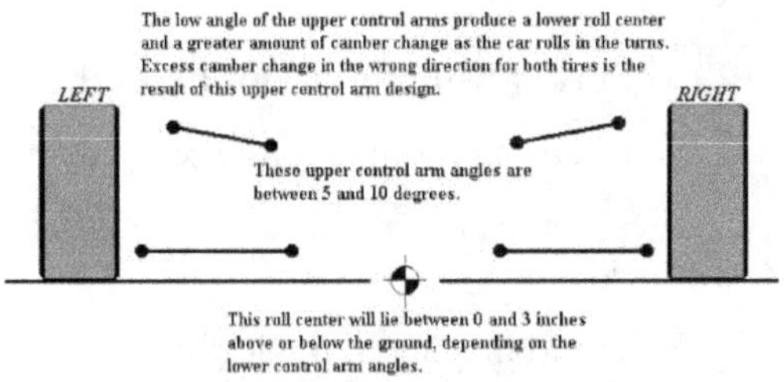

In a Double A-arm rear suspension, there are control arms much like the AA-arm front suspension. The moment center is adjustable in most of these systems by changing the angles of the control arms at the chassis mount.

For design purposes, we want the rear MC to be centered at static ride height and to not move very far from the centerline when the car is going through the turns. We can do this by arranging the arm angles so that after plotting or computing the MC location after dive and roll, the MC stays near the center.

Exam - In The Context Of This Lesson:

The Rear Moment Center Is?

1) The point around which the rear of the chassis rolls

2) The top of the rear moment arm

3) The bottom of the rear moment arm

4) The center of the rear chassis at static ride height

The Panhard Bar Moment Center Is?

1) Likely to move in the turns

2) Very stable in height when the chassis moves

3) The average height of the end mounting bolts

4) 1 and 3

The Leaf Spring Moment Center Height Is?

1) Where the spring mounts to the rear axle tube

2) Averages of the center of the eye mounts for the two springs

3) The front end heights of the leaf mounts

4) The center of the axle tubes

The Independent AA-arm Rear Moment Center Is?

1) The average of the instant center heights

2) Always located at the center of the chassis

3) Calculated by the same method as the front AA-arm MC

4) Should always be to the outside of the turns

Lesson Sixteen –Alignment and Toe

Alignment for the rear suspension is looked at a little differently than the way we looked at the front. This has a lot to do with what we talked about in Lesson Two, Why A Race Car Turns. If you remember, our rear wheels are already pointed to the left of the tangent line we are following around an arc, or turn of a race track.

Rear Steer - First off, the rear wheels do not turn, or at least the driver cannot steer the rear wheels. The links that attach the rear end to the chassis can cause the rear wheels to steer when the chassis moves. This movement is called Rear Steer.

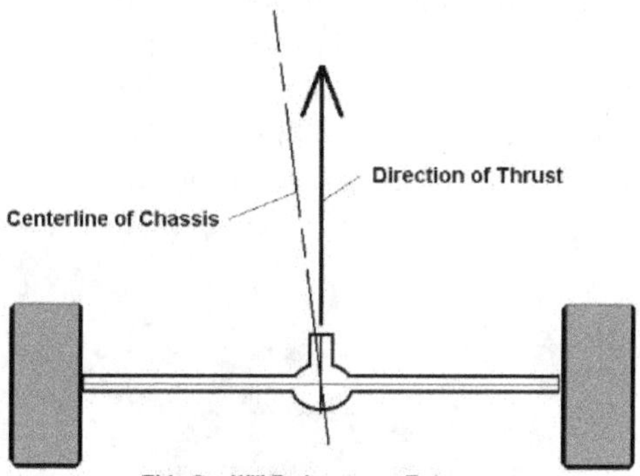

Rear steer can be detrimental to performance. We can use some amount of rear steer to our advantage, but excess steer is never an advantage. Steer to the right will make a circle track car loose, or in the case of a road racing car, loose in the left hand turns and tight in the right hand turns. This miss-alignment can come from static settings or from movement of the suspension parts causing the steering.

The steering direction can be either to the outside, or left, of the tangent line, or to the right of it. When the steering is to the right, we call this Rear Steer to the right and the opposite when the wheels are steered to the left of the tangent line.

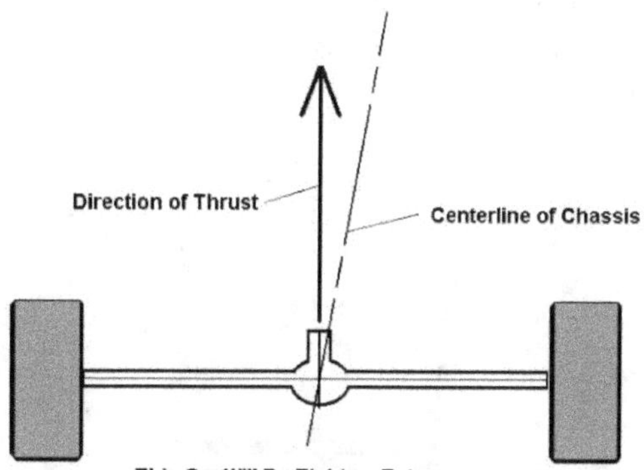

For most applications, a small amount of rear steer to the inside of the turn can help gain grip. If we overdue the steer in this direction, we can make the car tight. The example is related to left turning circle track racing.

For the effect of RS, we need to think about which way the car is turning. In a left turn, such as in circle track racing in the U.S., RS to the right would make the car loose, or oversteer depending on the amount of the change in the steering angle.

For left hand turns, RS to the left would tighten the car, or make it understeer is the amount of steering angle change were more than the amount the rear wheels needed to resist the lateral force, or centrifugal force, which tries to push the rear of the car to the outside of the turn.

For turning only one way, we can easily design RS into our race car that will maximize and optimize our rear wheel/tire angle of attack to work in conjunction with the front wheels/tires. This helps to balance the setup not only for handling balance, but for dynamic balance too, our primary goal for race car setup.

In RCT Level Two, we will get into a lot of detail about how different systems produce RS and how we can design and tune the different suspensions. We will help the student understand the limits to RS and how the designs are different for different classes of race cars.

For now, just know that there is a fine line between too much and too little RS. We want to develop a good amount of RS to produce the highest tire angle of attack without going too far and losing all of our rear grip.

Rear Alignment – The rear alignment for race cars is fairly simple, but there are variations that teams choose to follow. Simply put, for circle track racing where we are only turning one way, the outside tire contact patches should follow the same line.

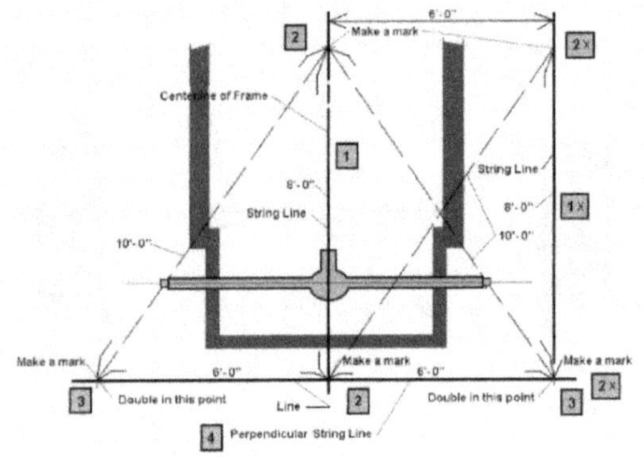

To setup for alignment of the rear end, we need to establish alignment of the chassis and find the centerline or some datum line to use as a reference. Here, in the example we use a perimeter car that is symmetrical. Once the centerline has been established, we setup a line that is ninety degrees off of the centerline to check the rear end for square.

If the front and rear track width (distance between the centers of each wheel contact patch on one end of the car) are different, then we line up the outside tire contact patches and the inside contact patches will be miss-aligned.

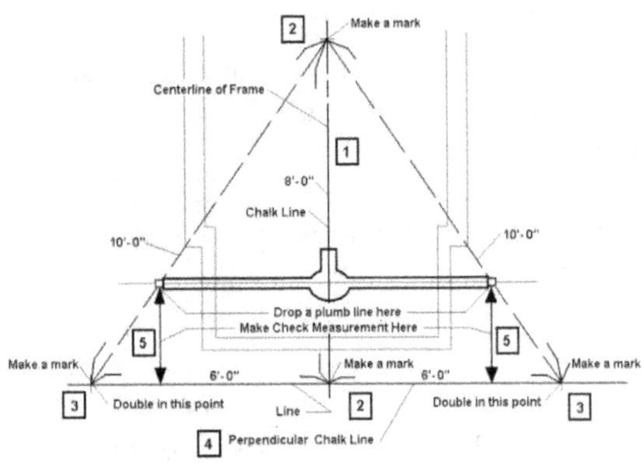

With the rear line established off of the centerline, we measure to the outer ends of the rear end to see if it is square. Adjustments to the lengths of the suspension links will bring the rear end into square.

We do that because the outside tires will carry the most loading in almost every case. The more heavily loaded tires will perform better when they are inline and running on the same arc.

For road racing cars, we center the front and rear tire contact patches on the chassis centerline. If the front and rear track widths are different, then the outside and inside tires will be miss-aligned equally. We also make sure that all four wheels are pointed parallel to each other. All tires would be parallel to the centerline of the chassis.

Rear Toe – Rear toe for a straight axle rear suspensions, running on circle tracks, is zero in most cases. We want to toe these wheels straight ahead so the tire angle of attack is consistent for both tires.

For road racing, we often see where teams will toe the rear tires in (the front of the tire is narrower than the rear of the tire). This is opposite of the way we mostly toe the front tires, being toe-out. This is done for stability mostly. On high speed long straights, the rear of a road racing car can dance around somewhat.

Rear toe can be an issue as well. If the rear wheels are not parallel, then we can have toe-in or toe-out in the rear wheels for a straight axle suspension. This rear end has a brace installed that can help it keep in alignment and correct any miss-alignment that may occur.

This design goals for toe-in in the rear for road racing cars is easy to do with a AA-arm rear suspensions. There is a tie rod link that positions the rear spindles so that the rear wheels are pointed where we want them to be pointed. We can easily make changes to the tie rod lengths to make changes to the rear wheel toe settings. Normal toe amounts for the rear are from zero toe to 0.050" of toe per wheel. We usually don't see much more that that used.

To create toe for a straight axle suspension, we need to bend the axle tubes. We can do this by heating them on one side, an art form at least, or by installing toe links that will pull on the front of the axle tubes to produce toe-in. These links will keep the toe consistent as the car is raced and can be easily adjusted.

Exam - In The Context Of This Lesson:

What Mostly Miss-aligns The Rear Of The Race Car?

1) The driver using his foot on the throttle

2) Incorrect lengths of the rear suspension links

3) Rotation of the rear end

4) Squatting of the rear suspension

Rear Alignment To The Left Would Do What?

1) Cause the car to be slower

2) Make the car faster

3) Point the tires left in relation to the front tires

4) Make the car looser

Rear Alignment To The Right Would Do What?

1) Cause the car to be slower

2) Make the car faster

3) Point the tires right in relation to the front tires

4) Make the car looser

Rear Alignment Can Do What?

1) Help the rear to gain grip

2) Cause the rear to lose grip

3) Help the rear track more inline with the front tires

4) All of the above

For Road Racing, We Align The Rear Tires How?

1) Line up the outside tire contact patches

2) Line up the inside tire contract patches

3) Miss-align both rear tires the same

4) Make both rear tires line up with the front tires

Rear Toe Is Usually?

1) With the tires pointed out at the front

2) With the tires pointed in at the front

3) Straight ahead

4) 2 and 3

Race Car Technology – Level One

Lesson Seventeen – Rear Steer, Roll Steer and Bump Steer

Alignment for the rear suspension is looked at a little differently than the way we looked at the front. This has a lot to do with what we talked about in Lesson Two, Why A Race Car Turns. If you remember, our rear wheels are already pointed to the left of the tangent line we are following around an arc, or turn, of a race track. So, we don't really need to steer the rear wheels unless we think we can gain performance by doing so.

The most extreme case of rear steer can be seen in a dirt late model or dirt modified. There are advantages that can be realized on dirt where these same tactics would never work on asphalt race tracks. Here we will discuss the conditions of rear steer and roll steer and how they occur.

Rear Steer - First off, the rear wheels do not turn, or at least the driver cannot steer the rear wheels. The links that attach the rear end to the chassis can cause the rear wheels to steer when the chassis moves. This movement is called Rear Steer.

We talk about Rear Steer mostly in relation to circle track cars. For road racing cars, there may be some slight amount of rear steer incorporated into the designs of the rear suspension, but nothing compared to the amount used for some circle track cars, mostly dirt cars.

This is a dirt car rear suspension left rear link system. There are numerous holes in the chassis to allow the team to change the angles of the links. Depending on the angles, when the chassis moves in bump or roll, the rear end on that side of the car will move forward, backward, or not at all. These types of suspension add a high degree of complexity to the setup of dirt cars not seen in any other type of race car.

The steering direction can be either to the outside, or left, of the tangent line, or to the right of it. When the steering is to the right, we call this Rear Steer to the right and the opposite when the wheels are steered to the left of the tangent line.

For the effect of RS, we need to think about which way the car is turning. In a left turn, such as in circle track racing in the U.S., RS to the right would make the car loose, or oversteer depending on the amount of the change in the steering angle.

This right rear chassis mount for the three link rear suspension on an asphalt late model car is adjustable for height. This adjustment changes the angle of the link. When the chassis rolls, the angle of the link can either pull the right rear wheel forward, backwards or have it remain where it was after the car has fully rolled. During the roll process, this link will, in most cases, push the wheel back before it pulls it back towards the front.

For left hand turns, RS to the left would tighten the car, or make it understeer is the amount of steering angle change were more than the amount the rear wheels needed to resist the lateral force, or centrifugal force, which tries to push the rear of the car to the outside of the turn.

For turning only one way, we can easily design RS into our race car that will maximize and optimize our rear wheel/tire angle of attack to work in conjunction with the front wheels/tires. This helps to balance the setup not only for handling balance, but for dynamic balance too, our primary goal for race car setup.

In RCT Level Two, we will get into a lot of detail about how different systems produce RS and how we can design and tune the different suspensions. We will help the student understand the limits to RS and how the designs are different for different classes of race cars.

For now, just know that there is a fine line between too much and too little RS. We want to develop a good amount of RS to produce the highest tire angle of attack without going too far and losing all of our rear grip.

Roll Steer – Rear roll steer is much the same as bump steer. In a rear suspension, the links that hold a straight axle in place, or the control arms in a AA-arm, might steer the rear wheels when the car rolls going through a turn. We can eliminate this roll steer, or we can use it to our advantage to gain grip or overcome other handling problems.

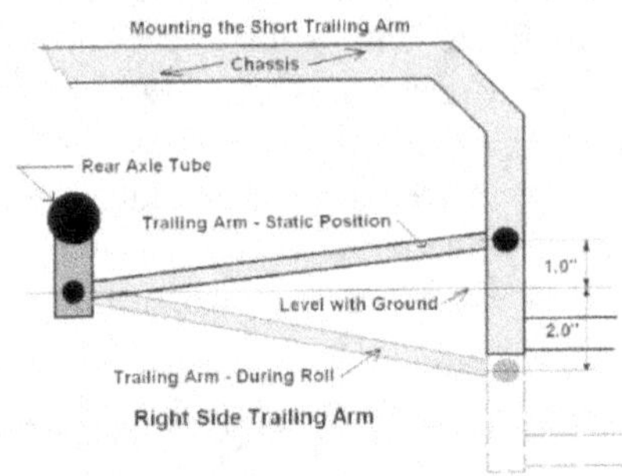

A common way to set the right rear link, or trailing arm as it is called, chassis mount is to set it up from level by one third of the total distance it will travel due to chassis roll. This will provide a slight amount of rear steer to the left for more angle of attack for the rear tires and more grip.

Depending on how the links are setup, we can cause rear steer to develop in any direction, right or left of centerline. From what we have already learned in Lesson 16, steer to the right would loosen the car and steer to the left (in a circle track car turning only left) would tighten the car.

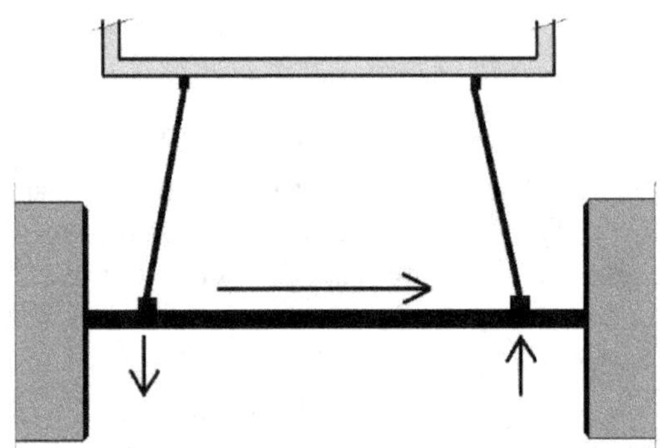

On a race car with a panhard bar, the bar does not remain level during chassis roll. Because it is attached to the chassis, and the chassis moves vertically, it will pull the rear end one way or the other. On some designs of rear suspension, the links are not parallel from a top view. So, when the rear end will move side to side as the panhard bar moves through its travel, creating rear steer.

Bump Steer – Bump steer in the rear suspension is much the same as in the front AA-arm suspension. When the chassis moves vertically, the wheels steer if there is any inherent bump steer. We try to eliminate bump steer in the front suspension, but in the rear, we might want bump steer in some cases.

If we desire a straight ahead alignment when going through the turns then the roll steer characteristics must be such that no steer occurs. But if we want rear steer coming off the turns, then we can introduce bump steer to steer the rear end when the rear squats under acceleration. This is accomplished with the inside links on a circle track car. We'll get more into detail on how that is done in RCT Level Two.

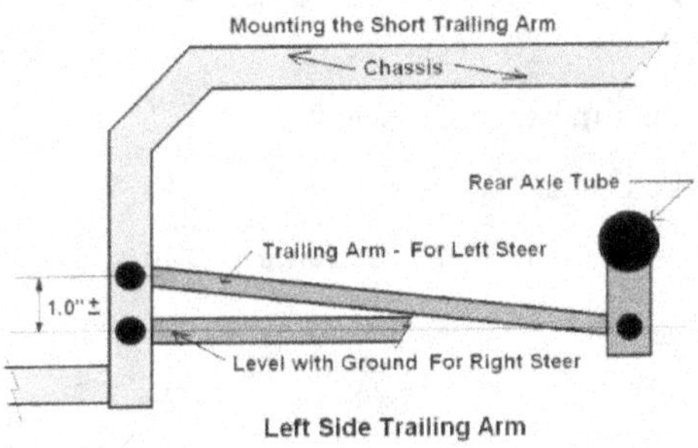

We get bump steer in the rear suspension from the car squatting upon acceleration off the corners. The rear of the car will squat and travel down some amount depending on other design features, and when it does, we can produce rear steer if the links are arranged in a certain way. If the left rear link in a three link suspension is set with the chassis mount higher than the axle mount, when the car squats, it will push the left rear wheel back. This creates rear steer to the left of centerline.

Rear Alignment – The rear alignment for race cars is fairly simple, but there are variations that teams choose to follow. Simply put, for circle track racing where we are only turning one way, the outside tire contact patches should follow the same line.

There are many ways to introduce rear steer in a race car. One way that some teams use is to include a link in the right rear that is allowed to compress under acceleration. This compression shortens the link and pulls the right rear wheel forward to introduce rear steer to the left of centerline to tighten the car.

If the front and rear track width (distance between the centers of each wheel contact patch on one end of the car) are different, then we line up the outside tire contact patches and the inside contact patches will be miss-aligned.

We do that because the outside tires will carry the most loading in almost every case. The more heavily loaded tires will perform better when they are inline and running on the same arc.

For road racing cars, we center the front and rear tire contact patches on the chassis centerline. If the front and rear track widths are different, then the outside and inside tires will be miss-aligned equally. We also make sure that all four wheels are pointed parallel to each other. All tires would be parallel to the centerline of the chassis.

Rear Toe – Rear toe for a straight axle rear suspensions, running on circle tracks, is zero in most cases. We want to toe these wheels straight ahead so the tire angle of attack is consistent for both tires.

For road racing, we often see where teams will toe the rear tires in (the front of the tire is narrower than the rear of the tire). This is opposite of the way we mostly toe the front tires, being toe-out. This is done for stability mostly. On high speed long straights, the rear of a road racing car can dance around somewhat.

This design goals for toe-in in the rear for road racing cars is easy to do with a AA-arm rear suspensions. There is a tie rod link that positions the rear spindles so that the rear wheels are pointed where we want them to be pointed. We can easily make changes to the tie rod lengths to make changes to the rear wheel toe settings. Normal toe amounts for the rear are from zero toe to 0.050" of toe per wheel. We usually don't see much more that that used.

To create toe for a straight axle suspension, we need to bend the axle tubes. We can do this by heating them on one side, an art form at least, or by installing toe links that will pull on the front of the axle tubes to produce toe-in. These links will keep the toe consistent as the car is raced and can be easily adjusted.

Exam - In The Context Of This Lesson:

What Mostly Steer The Rear Of The Race Car?

1) The driver using his foot on the throttle

2) Incorrect Angles of the rear suspension links

3) Rotation of the rear end

4) Squatting of the rear suspension

Rear Alignment To The Left Would Do What?

1) Cause the car to be slower

2) Make the car faster

3) Point the tires left in relation to the arc tangent line

4) Make the car looser

Rear Alignment To The Right Would Do What?

1) Cause the car to be slower

2) Make the car faster

3) Point the tires right in relation to the arc tangent line

4) Make the car looser

Rear Steer Can Do What?

1) Help the rear to gain grip

2) Cause the rear to lose grip

3) Help match the front grip levels for more balance

4) All of the above

For Road Racing, We Align The Rear Tires How?

1) Line up the outside tire contact patches

2) Line up the inside tire contract patches

3) Miss-align both rear tires the same

4) Make both rear tires line up with the front tires

Rear Toe Is Usually Set This Way?

1) With the tires pointed out at the front

2) With the tires pointed in at the front

3) Straight ahead

4) 2 and 3

Race Car Technology – Level One

Lesson Eighteen – Anti-Squat

Anti-Squat is just what the term implies. It is an effect caused by the force that is accelerating the race car, and that force uses the links that position the rear end to provide Anti-Squat (A/S). There has been a lot written about anti-squat and we are going to try to explain how it really works. This seems complicated, but in reality, it is a simple concept. Refer to the drawings often to try to picture what is going on here.

When the engine is powered up to full throttle (or something in-between), the torque goes through the transmission, drive shaft and rear end gears to turn the axles and rear tires to propel the car forward and to make it speed up.

In every rear end that is based on a straight axle design and where the transmission, or transaxle is mounted to the chassis or engine, this effect is a part of the geometry. So, we are talking about circle track cars as well as road racing cars.

Anti-squat is a mechanical effect in the rear suspension of a race car that initiates a force that counters the tendency to squat, or compress the rear springs, upon acceleration. As the car accelerates, weight is transferred to the rear suspension and that causes the rear spring to compress. Anti-squat reduces that motion.

Straight Axle Rear Ends - In those cars with straight axle rear ends, the drive shaft connects to a pinion shaft and pinion gear that turns a ring gear. The ring gear is connected to the axles and those are connected to the wheels. It is the action of the pinion gear turning the ring gear that creates part of the force needed for AS. The rest of the force is provided by the thrust, or pushing forward as the car accelerates.

To propel the car forward, while the pinion gear is turning the ring gear, it is trying to climb the ring gear. In doing so, it is also putting a Torque force on the rear end trying to rotate it so that the pinion shaft would move up, if it could. There is a lot of force involved with this action.

Note that we said this only happens when we are applying power from the engine through the drive shaft and rear end. As the pinion tries to climb the ring gear, it wants to move up causing a powerful force that is also trying to lift up on the pinion and rotate the rear end in a clockwise direction when viewed from the left side. Try to imagine this force for a moment.

At the same time, the Thrust force is pushing on the suspension links. It is the combination of those two forces, engine torque to the rear end in a straight axle car, and forward thrust in all types of rear ends. that cause A/S to occur.

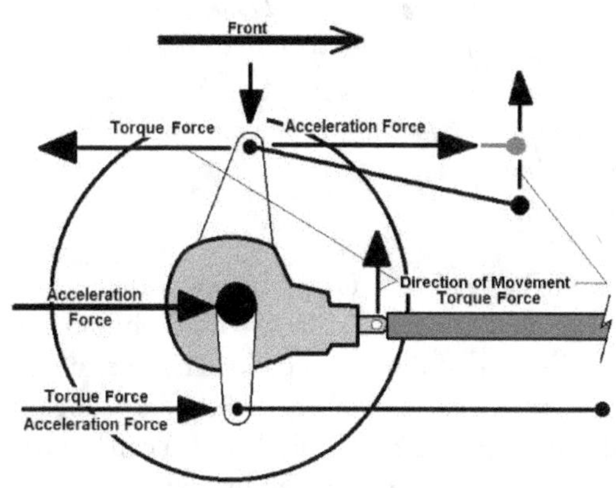

This shows how anti-squat works in three link rear solid axle suspension. As the car accelerates, the torque of the engine is transferred to the rear end and tries to rotate the rear end. As this rotational force is applied, the bottom links are pushed into compression and the top third link is put in tension, or pulled against. At the same time, the Acceleration forces are pushing against both the upper link(s) and the lower links. At the top, the forces are opposed and at the bottom they combine.

If the links we attach to the rear end to keep it from rotating are arranged in a certain way, this force will try to lift the rear of the car. The force trying to lift the car has to overcome the weight of the rear portion of the sprung weight as well as one other load.

The other load is this. When we accelerate the car, we create load transfer from the front to the rear. This load is easily calculated, although we won't be doing the calculations here. We just want you to understand the process and what is involved.

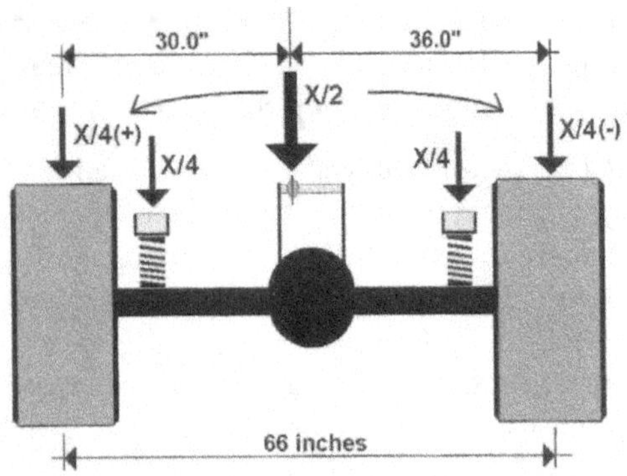

As weight is transferred to the rear upon acceleration, the anti-squat effect lifts on the front of the three links in a three-link suspension and pushes down on the rear end at the point where the three links are mounted to the rear end bracket. If we move the third link mount left, the, its load is applied more to the left and closer to the left rear tire. The same applies to the lower links (not shown) in that they cause a lifting force to help overcome chassis squat upon acceleration and put an equal loading onto the rear axle housing.

So, we have load transfer to the rear when we accelerate and that added load would obviously cause the rear of the car to be lower since the springs will have to compress when that added load is put upon them. With A/S, we can prevent the springs from compressing by taking the added load onto the links through A/S geometry.

Third Link - The link that helps cause AS can be a "third link", a lift arm, a lift bar, or other types of link that prevents the rear end from rotating. In the case of the third link, it is mounted with the front end lower than the rear end. The front is mounted to the chassis and the rear connection is mounted to the rear end housing.

As the rear end tries to rotate from acceleration, it pulls on the third link. Some of these links that use springs to allow them to expand are actually called pull bars. As the force is put on the third link, it wants to become straight to the direction of the force. In doing so, it creates a lifting force at the chassis end that makes it want to move up.

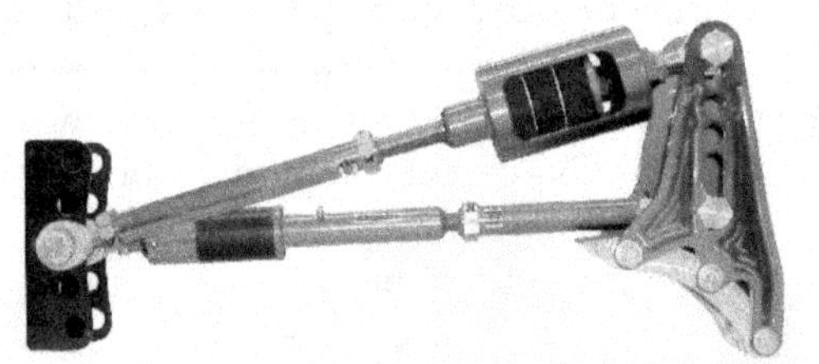

Various designs of third links are available and do different things. This one applies the anti-squat forces, reduces the torque shock to the rear end, and reduces adverse wheel hop when the car is braking. Note the angle of the pull-bar on the top. This angle is what creates the anti-squat force. The bottom link is used for compression resistance when decelerating.

The front end of the third link mounted is lower and wants to move up. This is a part of the lifting action that reduces rear squat that is the A/S component in a three link rear suspension. The rear of the third link wants to go down in order to become straight to the pull and this puts loading on the rear end that is centered where the link is mounted.

So, A/S not only helps prevent squat, it also puts some of the transferred load onto the rear end and not onto the springs where it would go if there were no A/S. We can use this A/S effect to re-distribute loading on the rear tires while the car is accelerating. By moving the third link sideways, and/or changing the angles of all of the links, we can change the load distribution on the rear tires. We will get more into that in RCT Level Two.

Lift Arm/Bar - With the Lift arm type of third link, we have a different application of the force referred to the rear end. We still have the torque through the driveshaft and pinion, but when the rear end tries to rotate, it is resisted by an arm mounted solid to the rear end housing that tries to lift the sprung weight while putting additional loading onto the rear end at the same time.

This type of link is more adjustable for the amount of force that represents A/S by being adjustable for length. A longer arm or bar creates less A/S force. This is pure leverage at work. The two different types of link will provide A/S in similar ways, but it is the adjustability that differentiates them.

The lift arm (black tubing beside the drive shaft) is another design of anti-squat used mostly in dirt cars. As the car accelerates, the arm, mounted solidly to the rear end, tries to lift the chassis at the point where the spring and shock are mounted. The opposite direction of force is applied to the rear end again at the point where the lift arm is mounted.

The Lower Links – While all of that is happening with the upper links in a three link system, the lower links, if the angles are pointing up going forward, will themselves provide A/S forces. The Torque of the engine transferred to the rear end will provide a force onto all of the three links. The Thrust forces also act on the three links, but for the lower links in a three link suspension, the Torque and Thrust forces combine and this makes the A/S forces and effect more significant for the lower mounts.

In the case of a lift arm car, all of the Thrust is taken by the four links in this suspension system. Since the lift arm controls all of the torque forces, then all of the Thrust forces are taken by the four links. It is the arrangement of the link, or bar, angles that determines the A/S effect for those cars.

It is possible, and quite common, for four-link cars to create more than 100% A/S and lift the rear of the car higher than it would be at ride height. Many dirt late model setups lift the left rear corner higher than the right rear corner due to the angles of the left side bars. This puts a lot of loading, sometimes most of the rear loading, onto the left rear tire.

Exam - In The Context Of This Lesson:

Anti-squat Is Caused By?

1) Braking into the corners

2) The torque of the motor when accelerating

3) Forward thrust due to acceleration

4) Stiff rear springs

5) 2 and 3

Anti-squat Effect Is Common In?

1) Prototype race cars with transaxles

2) The front suspension of a race car

3) Straight axle rear end suspensions

4) Cars with high centers of gravity

5) 1 and 3

Anti-squat Produces What Effect?

1) It tries to lift the rear of the car

2) It causes an effect that adds load to the rear tires

3) It helps the car to turn the corner

4) It creates a force that resists the effects of load transfer

5) 1 and 4

The Length Of A Lift Arm Affects What?

1) The amount of load transfer under acceleration

2) Acceleration off the turns

3) The amount of anti-squat force

4) Torque transferred to the rear end

Lesson Nineteen – Driveline Alignment

Over View - A Driveline is a connection between the engine and the rear end in a race car with a straight axle rear suspension. Because the rear end moves in relation to the chassis, and because the engine is mounted solid to the chassis, there must be a connection that can adjust to the motion between the motor and the rear end.

Drive Line Terminology – The component names for the driveline parts are as follows:

Slip Yoke – this is the yoke at the front of the driveshaft that goes into the transmission and can slip to allow adjustment for chassis movement.

Weld Yoke – is the yoke at the each end of the drive shaft that accepts the U-joint. These attach to the both the pinion yoke at the rear end and the Slip yoke at the front. They are bonded to the shaft tubing.

Universal Joint Kit or U-Joint (UJ) – is the actual part that forms the rotational connection between the driveshaft and both the transmission and rear end.

Tubing – is the metal or composite tubing between the yokes.

Pinion Yoke – is the yoke that is attached to the pinion shaft at the rear end.

How A Drive Line Should Work - A drive shaft has two Universal joints, called that because they can move universally, in any direction up to a certain degree of angle. The front UJ is connected to a yoke that slides over the transmission output shaft. The rear UJ connects the drive shaft to the pinion shaft at the rear end.

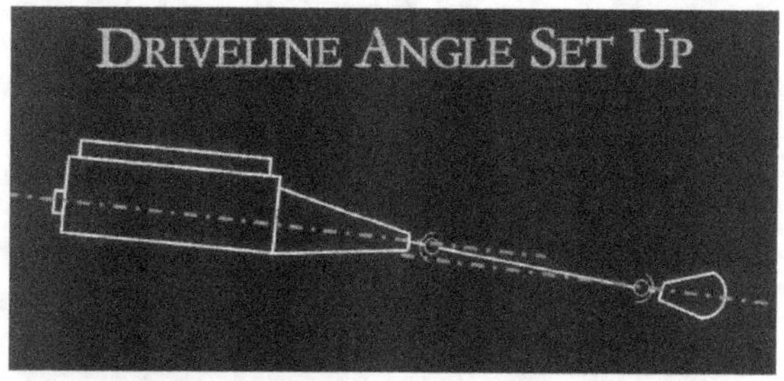

Driveline alignment is a very simple concept. It has been firmly established long ago what the correct alignment should be in order to reduce driveline vibrations and parts fatigue. It is, the engine/transmission must be inline and parallel to the pinion shaft in the rear end. And, the angles should be minimal between the engine/transmission/pinion shaft and the drive shaft.

The angles the transmission output shaft and the pinion shaft make with the driveshaft is critical. One of the most important design parts of a high performance race car is drive line alignment. And it is very simple in concept.

Ideal Drive line alignment features are: 1) Having the transmission output shaft parallel to the pinion shaft; 2) Making sure the angle created between both the transmission output shaft and the pinion shaft to the driveshaft is minimal. It is that simple, but again, very important.

Drive Line Angles Gone Wrong – When the UJ operates with any amount of drive line miss-alignment, it creates a problem. The bearings speed up and slow down twice per revolution of the driveshaft. This causes an oscillation in the power train. The more miss-alignment angle that we have, the higher the peaks of oscillation we see and therefore the greater chance of vibration.

Drive line miss-alignment angles are a cause of vibration and power loss. If your race car must have driveline angles from a design standpoint, the angle of the drive shaft to both the transmission output shaft and the pinion shaft should be equal and also opposite. The angle should also be kept to a minimum when at all possible.

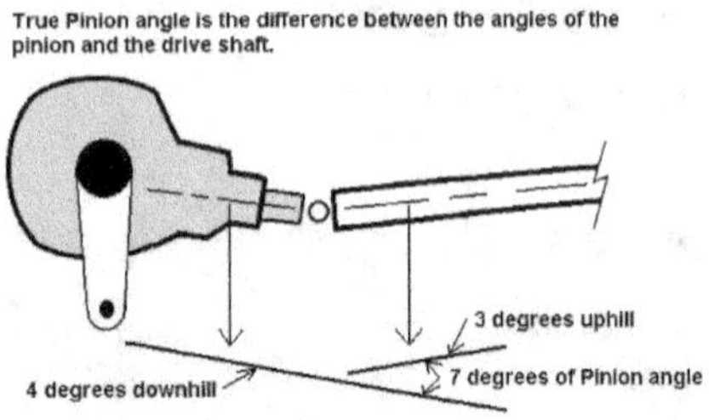

The pinion angle historically has been considered the angular difference between the sideview angle of the pinion shaft and the drive shaft. This angle, while needing to be minimal, has nothing in and of itself to do with correct driveline alignment. And, it does nothing to promote forward bite as sometimes spoken of. This angle must be the same and in the opposite direction at the transmission to drive shaft angle. Then the transmission output shaft and the pinion shaft will be inline and parallel.

Drive shaft angles are not only measured from a side view, but also from a top view. Some offset late model cars can have as much as a 1 ½ inch displacement of the rear of the drive shaft from the front. That equals almost 2 degrees of drive shaft angle at both the transmission and the pinion. So, we can align the drive shaft from a side view to zero angle and still have 2 degrees of drive shaft angle present in the system, which is plenty.

The latest recommendation might come as somewhat of a shock to most racers, but strictly for racing applications, near Zero Drive Shaft Angles are encouraged. Based on recent studies and testing, a race car with zero drive shaft angles at the transmission and pinion will not harm the u-joint bearings. The natural vibrations that are produced by the race car will cause the u-joint bearings to rotate enough to stay lubricated and not flat-spot.

In the real world, the drive shaft will never really maintain a zero angle configuration through the diving and rolling of the chassis as we lap the track. Starting at zero means that we will stay very close to zero at all times. For extreme applications, such as when racing at over 180 MPH at Daytona or a similar type of track, the teams will align the drive train for zero angles with the car placed at the on-track race attitude.

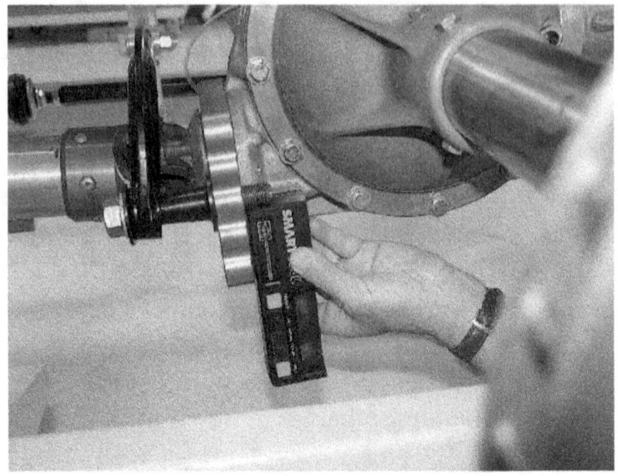

To achieve correct driveline alignment, we physically measure the pinion angle and the transmission angle. The first goal is to make those two parallel. Then we want to reduce the angle they make with the drive shaft to less than two degrees.

The mockup phase of car assembly is a good time to begin checking Driveshaft alignment. You have positioned the engine to meet all of its location constraints for best weight distribution and the rear axle location to meet the wheelbase requirements.

It is one of the things we do when going over our race car to create what we will call the total package that will be referred to a lot in the coming Lessons. There are certain "musts" we put on our list of things to do, and this is one of those.

NOTE: For overall consideration in drive line angle design, number one is to keep the angles equal and opposite at each end of the drive shaft (the transmission output shaft and pinion shaft are parallel) and to keep the driveline angles as small as possible.

Exam - In The Context Of This Lesson:

The Slip Yoke Does What?

1) Helps the tires to slip

2) Connects the pinion to the drive shaft.

3) Allows the drive shaft to turn without binding up

4) Attaches the drive shaft to the transmission

The Driveshaft Angles We Want Can Be Created From?

1) As a result of the motion of the rear end

2) From a top view angle due to miss-alignment of the engine and pinion shaft

3) From a sideview angle miss-alignment of the engine and pinion shaft

4) All of the above

Ideal Driveshaft Alignment Involves?

1) Having the driveshaft line up with the transmission output shaft

2) Installing the rear end with down angle in the pinion shaft

3) Making the transmission output shaft and the pinion shaft parallel

4) Mounting the engine/transmission and rear end so that there is less driveshaft angle

5) 3 and 4

Race Car Technology – Level One

Lesson Twenty – Types of Springs, How To Rate

Glossary:

Spring – A device that supports a load in either compression or extension. Springs can be a wound wire of different wire size and diameter of coils, a flat leaf design supported at each end and that bends to support weight, or a bar that is made to twist as load is applied – called a torsion bar.

Spring Rate – The actual rate of resistance to weight applied to it. We usually refer to a spring rate in pounds or kilos per inch. A 100 lb/in rate would deflect one inch if it were made to support a force of 100 pounds.

Spring Split – A difference in spring rates for the two springs on the same axle, or suspension system. Spring split is usually used in circle track racing where the cars only turn one way.

Motion Ratio – A difference in movement between one part of a suspension and another. It is usually used in reference to the wheels vertical movement in relation to the movement of the spring. When a spring is mounted on a lower control arm, it will move less than the wheel.

Force – The total pounds or kilos of work the spring does to support a measured load. A spring rated at 100 lb/in would develop a force of 400 pounds if compressed four inches. (100 pounds per inch rate compressed 4.0 inches; 100 x 4.0 = 400 pounds of force)

Weight – The actual measure of the loading on the tires measured in pounds or kilos of the parts of the car caused by gravity.

Sprung Mass – That part of a race car that is supported by the springs. The components that attach the sprung mass to the wheel structure are counted as part of the sprung mass using half their individual weight.

Over View – A car is supported by devices called springs. These come in many different varieties, but they all do basically the same thing. The provide a rising rate of force necessary to overcome first of all, gravity, and also the forces of aerodynamics, weight transfer and mechanical downforce due to cornering.

In modern day circle track racing, springs are being used like never before. Race teams now more than ever desire a low attitude and very little chassis movement. This provides very good aero qualities, a low center of gravity, and very little camber change, which the car likes. The result is a race car that is easier to setup and that runs faster.

In this Lesson, we will identify the different types of springs and tell you how they work to support the car and all of the loads related to performance driving. In developing a racing setup, we need to know the spring rates and we'll tell you how to find those rates accurately.

There is a sort of revolution going on in the industry today. Springs are being used in ways never before imagined. Combinations of springs, softer to stiffer springs, and stacked springs are being used in all forms of racing. So, this is not an ordinary lesson, it is the basis of much of what goes into race car design and setup.

The circle track teams are now learning what road racers have known for some time, that a stiffer setup that yields less chassis movement is a more stable package. What road race teams might now learn from circle track teams is that they could be running much lower to the track for better aero and a lower center of gravity. The circle track racers have perfected the low attitude setups.

Types of Springs:

Coil Springs – A coil spring is a bar that has been wound around a mandrel. The spring will have a designed wire diameter, an inside diameter of the mandrel, a specified height and a number of coils. The rate is determined by; 1) the hardness of the steel used to make the wire, 2) the wire diameter, 3) the diameter of the coils (used to find the length of the bar) and, 4) the number of coils (again related to finding the length of the bar).

The coil spring is by far the most common spring used in racing today. Other types less used are torsion bar systems, leaf springs, bevel springs, air springs, and bump stops and bump springs. Any way you look at it, springs support the car and dictate most of the dynamic properties of the race car.

Most coil springs are fairly linear in rate, meaning that as they support more load, the number of pounds or kilos it will support per inch or centimeter of travel stays consistent. For example, if I put 200 pounds of weight on a coil spring rated at 200 lbs per inch, then the spring will compress one inch (200 / 200 = 1.0")

We must know the rate of the springs we are using. This simple spring rater tells the team how many pounds (or kilos) a spring will support per inch (or centimeter) of spring travel. What we are getting into in the industry is also knowing the force a spring provides at a known distance of travel.

The if I add another 200 pounds to the weight supported by the spring, it should compress another inch for a total of 2.0 inches. If it were to compress more or less than an inch, it would be progressive (gaining rate), or digressive (losing rate).

It is common in some types of racing to use progressive springs to intentionally gain spring rate while the suspension travels. We almost never see digressive coil springs.

Torsion Bar Springs – Torsion bar springs are used on a variety of race cars, including every class from sprint cars to formula cars. These are simply coil springs that have never been wound. They work by being mounted solid at one end, having an arm attached to the other end, and a load being applied to the end of the arm causing the bar to twist. The resistance to twisting creates the spring rate.

Other types of springs include the torsion bar spring shown here on a sprint car. The bar is housed inside a tube (at the back end of the aluminum arm) and the spring rate is measured where the other end of the arm is attached to the axle tube.

This torsion bar spring, also on a sprint car, runs back from the front of the car to the straight axle tube. It rest on the tube instead of being bolted to it. The rate is the force at the point where the arm meets and rests on the axle tube.

The rate of a torsion bar is calculated using the following, being very similar to the coil spring; 1) the hardness of the steel used to make the torsion bar, 2) the diameter of the portion of the bar that twists, and 3) the length of the part of the bar that twists.

Gale Force Suspension offers a rating tool that can measure the sway bar rate and force while on the car. With so many different materials used to make sway bars, and the many different installation options we see, this rater takes the guessing out of sway bar rating.

To changes spring rates in a torsion bar system, you would simply install a bar with a different diameter, or change the length of the arm which changes the motion ratio causing the bar to twist more or less for the same amount of chassis movement.

Leaf Springs – Leaf springs are a flat bar, or layers of flat bars, that bend when load is applied to them. They have eyes at each end to mount them to the chassis and one end is mounted so as not to move and the other is mounted to a shackle and moves to compensate for becoming longer or shorter as the spring moves under load.

The leaf type of spring is common to older, vintage style, race cars. This one has a jack screw used to adjust the loading on the four tires. All leaf springs need a slider, or shackle, at one end to allow for length changes as the spring is loaded and un-loaded.

Leaf springs are a product of earlier production cars and trucks and are still used in some trucks and some race cars. The mono-leaf has become very popular for racing. This spring has one leaf, or flat bar, and can be designed to have different thicknesses, thicker closer to the axle and thinner out towards the eyes. Leaf springs are more common on dirt cars.

Half leaf designs may also be used. It is similar to a full leaf spring cut in half. In this configuration, the thicker part of the leaf is mounted solid to the chassis and the eye end is what supports the wheel assembly, usually the rear straight axle or independent suspension.

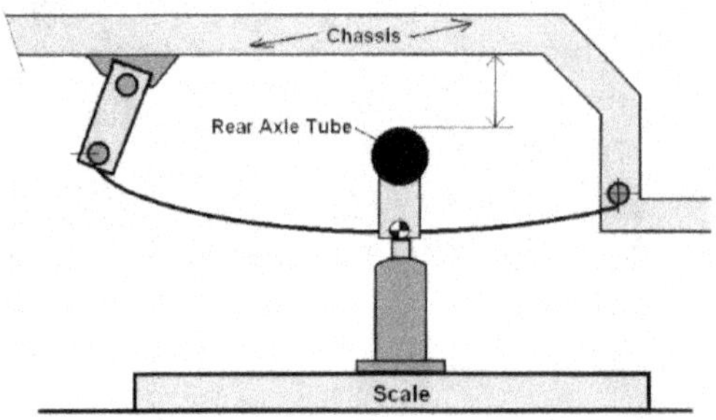

Race teams can rate the leaf spring while it is on the car. Rating leaf springs used to be nearly impossible, but now with this method, anyone with a scale can find the rate. Just measure the distance from the top of the axle tube to the frame. Jack the car us until the distance is less than it was by an inch or so. Then divide the number of pounds gained on the scale by the distance it moved closer to the frame and you will have a rate per inch.

The leaf spring is rated the same way a coil spring is. The rate is the number of pounds or kilos of load divided by the deflection in inches or centimeters to find the spring rate. These too are mostly linear in rate, although when using multiple leafs, the rate becomes progressive.

Bellows Springs – These springs are rarely used, but are available. The work by causing a deflection in a disc of specified material that resists being deflected. You usually stack a series of bellows springs together and the more discs in the stack, the softer the rate of the bellows springs.

The spring material could be anything from steel to carbon fiber. The uses include replacement of coil springs and use as a bump spring, which will be covered shortly. Just know that this type of spring exists and is not widely used.

Bump Stops and Bump Springs – The springs we have discussed above are referred to as ride springs. The car rides on these springs. There are supplemental springs that are used to add to the rate of the ride springs. These are bump stops and bump springs.

Bump stops and Bump Springs have been popular in the racing market place for some time. They are used in almost every form of racing from circle track to Formula One. They limit the travel of the suspension and provide a higher spring rate once a certain amount of travel has been reached. This shows a bump spring, a type of bump that provides an actual linear spring rate whereas most bump materials are linear in rate gaining a lot of rate in a very short distance of travel.

Both of these are usually mounted on the shock shaft of a coil-over spring (a spring mounted on a shock body). In some cases, the bumps are mounted separately from the coil-over. In any case, the bump device will add to the rate of the ride spring when it is "in-play".

For most applications, the term in-play means that the suspension has traveled far enough to engage the bump device. Then how much the suspension travels beyond that point determines the amount of force, or load, the bump supports.

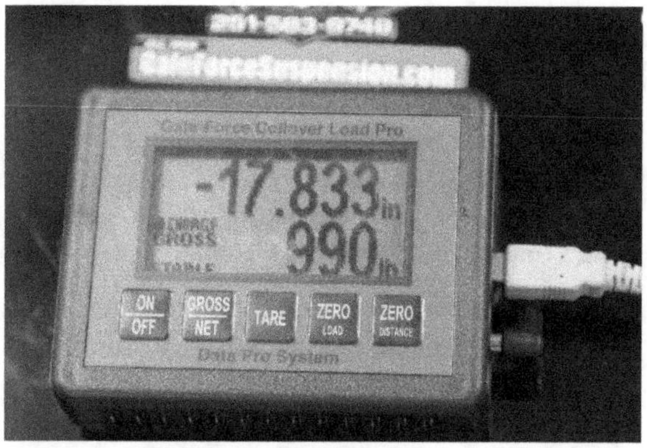

The newest and most useful information race teams are using in today's racing is Forced Based analysis. This means that teams are finding the force on the spring at a known distance of travel measured on the race track. This way they can know how much force is on the tire in order to re-arrange the corner heights to develop a better force distribution. We cover that in detail in Levels Two and Three of the RCT Course.

The bump stop is different than the bump spring in one important aspect. The stop is linear in rate, gaining rate at a very fast pace. The bump spring is mostly linear and gains rate at a consistent and predictable pace.

Bump devices are widely used in circle track racing, and prototype and formula racing. They provide a stable platform that can limit chassis travel, fix cambers and provide a much stiffer spring rate for roll control. There is a wide variety of rates and sizes available as well. The bump device is a major step forward in race car engineering for setups.

Air Springs – A fairly new and not yet popular spring is the air spring. It is the resistance to loading that comes from a cylinder that compresses air as it moves. The rate of the air spring is very non-linear and c can be easily raised or lowered by adjusting the air pressure in the cylinder.

The current uses of the air spring is for coil spring replacement and in use as a bump spring. Because is it so progressive, it may be hard to determine what rate to use in calculating roll angle and roll resistance. And temperature changes in the air can cause fluctuations in the spring rate.

Rating A Spring – We need to know the rate of a spring we are installing in our race car. It is the relationship of the spring rates that determines our setup balance. Here is how to rate a spring for a race car.

First, we must rate the spring in the range of motion it will be working within. A spring will very often provide a different rate as the suspension moves. This is not necessarily because the spring itself is progressive or digressive, it is usually because the spring angles, motion ratios, or bumps come into play and cause changes.

Carefully rating a spring is essential to having the correct setup. Most top race teams will rate their springs on a consistent basis and never install a spring that has not been rated. And always rate your springs within the range of travel they will experience while on the race car going around the race track.

When we rate a spring, we use a spring rating tool. There are many varieties on the market, but here is how they work. A spring is installed in the tool, and then compressed. The load is shown and recorded as the spring is compressed. We rate the spring by noting the gain in rate verses the amount of travel.

The first step in rating a spring is knowing the installed height of the spring and the total amount of travel the spring will encounter on the race track. This provides a defined range of motion and the area we need to know the rate within.

If a bump device is used in conjunction with the ride spring, then we can also rate the combination of ride spring and bump to find the new spring rate. No matter what the overall loading on the spring or spring/bump combo is, there is still a spring rate per inch or centimeter that we need to know for setup purposes. And, there is another way to look at this for further analysis.

Force Measurement – In modern racing, it is becoming popular and useful to measure the spring loading within the range of motion of the spring. This differs from spring rate in that it shows the force on the spring. This is used to compare loading to the companion spring at that end of the car.

The Load Pro Coil-over force machine is the latest advanced tool for the racing industry. It records and prints out for small increments of travel the force on the spring. Knowing that and knowing the actual travel of the spring on the race track will give you valuable information about the force on the four tires in relation to each other.

The force on the spring has a direct relationship to loading on the tires. A way to balance a setup is to find a balance in the loading of the springs. This then translates to proper loading of the tires and can indicate problems with the setup where the tires are not properly loaded. More on that in future Lessons and Courses we will offer.

All of these different types of springs can, and are used in racing. The types can be mixed and matched depending on the overall goals for the type of race car and where it is intended to be raced.

Characteristics Of Springs In Setups – When we talked about setups, we talked about roll angles and how the front and rear must work together. The springs play a major role in this dynamic. Here are the ways a spring can and does influence the roll angle: 1) Spring rates combined per axle, 2) Spring split (i.e. spring rate on one side different than the rate on the other side), 3) Spring base (how wide apart the springs are mounted on a straight axle), Motion ratio (where the spring is mounted on a AA-arm lower arm).

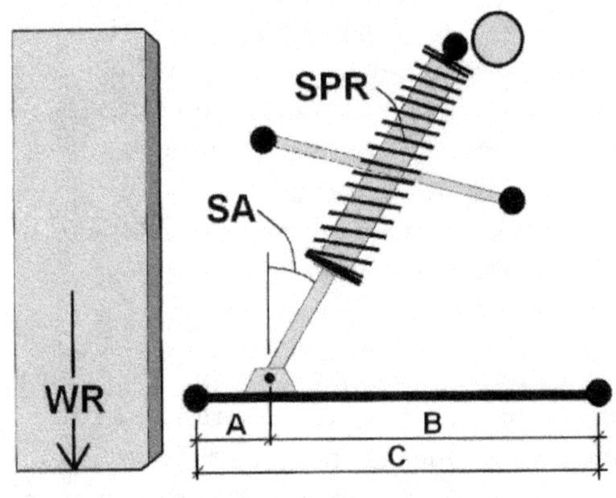

Each installed spring has a motion ratio of its motion to the wheel motion. The force it takes for the spring to support the wheel is much greater than the force of the wheel upon the ground. We can measure the force on the tire using scales. But on the race track we must use force based equipment to know what the tire contact patch loads are through the turns.

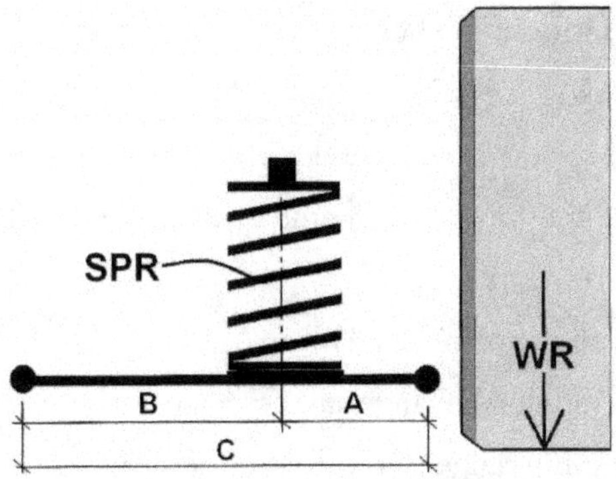

The "big spring" cars using more stock type of lower control arms and larger stock based springs must generate more spring force in order to develop the same wheel rate force. This is because the motion ratio is so much different and less than the coil-over sprung cars.

We are interested in finding the influence of the springs in creating the roll angle. This can be compared to roll stiffness, but there is more to the story than just that. As the chassis moves in dive and roll through the turns, the spring mounting geometry might change and that will affect the spring rate of the suspension. This gets somewhat complicated and we'll go over that part of springs in RCT Level Two.

Exam - In The Context Of This Lesson:

Springs Provide A Force To Overcome Which?

1) Gravity
2) Mechanical downforce
3) Aerodynamic downforce
4) Weight transfer
5) All of the above

A Spring That Is Linear Means?

1) It is straight and not bowed
2) Is mounted straight up
3) Is consistent in rate as it compresses
4) It is inline with the direction of travel

Coil And Torsion Bars Are Rated Based On?

1) Length
2) Wire or bar size
3) Hardness of the material
4) All of the above

A Bump Device Is Used To?

1) Smooth out the bumps in the race track
2) Overcome gravity
3) Add spring rate to the ride spring
4) Help the shocks overcome load changes

We Rate A Spring By Doing What?

1) Measure load changes in increments of travel
2) Using the range of motion of the spring
3) Measuring the travel of the spring on the track
4) All of the above

Springs Affect Setup Through?

1) How far they are mounted apart
2) Motion ratio
3) Changes in the spring rates
4) The difference in rates on a straight axle
5) All of the above

Spring Split Is?

1) When a spring is cut to make a different rate
2) A difference in spring angle for two springs
3) A difference in spring rate on the same axle
4) Different motion ratios for a AA-arm suspension

Race Car Technology – Level One

Lesson Twenty - One – Motion Ratio and Wheel Rate

Glossary:

Spring Rate – The actual rate of resistance to weight applied to it. We usually refer to a spring rate in pounds or kilos per inch. A 100 lb/in rate would deflect one inch if it were made to support a force of 100 pounds.

Motion Ratio – A difference in movement between one part of a suspension and another. It is usually used in reference to the wheels vertical movement in relation to the movement of the spring or shock. When a spring is mounted on a lower control arm, it will move less than the wheel.

Force – The total pounds or kilos of work the spring does to support a measured load. A spring rated at 100 lb/in would develop a force of 400 pounds if compressed four inches. (100 pounds per inch rate compressed 4.0 inches; 100 x 4.0 = 400 pounds of force)

Load – The actual measure of the weight on the tires measured in pounds or kilos of the parts of the car caused by gravity and other effects.

Over View – It is useless to know the rate of the springs used in our race car if we don't have other pertinent and important information. Remember our ultimate goal in race car setup is to provide the best load distribution on the four tire contact patches. In order to do this, we need to know how the loads on the springs end up on the tires.

We can calculate the wheel rate for a double A-arm suspension using simple formulas. We want to stress here that a solid axle suspension does not have a wheel rate. While it does in theory, that does not enter into consideration for calculating dynamic roll stiffness or roll angle analysis. We will cover this in much more detail in the RCT Level Two course.

So, for the purpose of this exercise, we are talking about wheel rate for a double A-arm suspension, or a swing arm rear link suspension, as you will see, or any other suspension with a motion ratio between motion of the spring and the wheel.

This photo shows how motion ratio works. In this rather complicated design, the pushrod (lower right quadrant) connects the spindle to a rocker arm (mid-photo). The rocker arm is then attached to the chassis forming a pivot point. Above this pivot point is the end of the coil-over shock mount. So, the wheel moves up, the push rod moves in, the rocker swings around and the spring compresses. The motion ratio is the ratio of the movement of the wheel to the amount the spring is compressed.

From the last Lesson, we can easily determine the loading on the springs if we know the range of motion the springs operate within. We just compress them to those ranges and then record the loads. But it doesn't stop there, that loading is transferred to the wheel and tire. How it gets there is what we will learn now.

Motion Ratio – In a AA-arm suspension, the spring is almost never mounted to move on a one-to-one ratio with the wheel movement. In almost every case, the spring is mounted so that it moves less than the wheel and this is called a motion ratio because we can calculate the ratio.

If, on a AA-arm suspension, the spring were mounted one-quarter of the way from the ball joint to the chassis mounts, then we would have a motion ratio of 0.75:1, the spring moves 0.75" and the wheel moves 1.0". That is 75 percent movement of the spring to the movement of the ball joint, or wheel. Pretty simple, right? Not so fast.

In this example we see a stock type of spring mounted in a bucket designed into the lower control arm. The wheel moves more than the spring in vertical motion. It is the division of the spring movement by the amount the wheel moves that is the motion ratio.

Motion ratios are not limited to double A-arm suspensions. Some race cars have rear suspension parts and links that create a motion ratio that affects the spring rate the car utilizes, or "feels". In this example, we see a three or Z-link rear suspension link going forward from the straight axle tube. The coil-over spring is mounted onto the link some distance back from the point where it attaches to the axle tube. When the chassis moves vertically, the spring moves less than the chassis. The spring may also be mounted at an angle to the link which creates another motion ratio.

Now we have the motion ratio of the movement of the ball joint to the spring mount. But, in many cases with coil-over springs especially, the spring is not mounted at ninety degrees to the control arm. Now as the spring compresses, its motion is not a one-to-one motion, it is yet another motion ratio.

If the spring were mounted at 20 degrees, then as the bottom of the spring moved up, the spring would compress 94.0 percent of the amount the bottom moved. So, our combined motion ratio in this AA-arm suspension system is 94 percent times 75 percent which looks like this. We take 0.94 times 0.75 = 0.705. This means that our spring compresses 70.5 percent of the distance the wheel moves.

In a straight axle suspension system, there is a motion ratio of the springs to the wheels, but that motion ratio is not useful in dynamic calculations. This is because the dynamics of the straight axle tell us that spring base determines roll resistance. This very important to know now because we will discuss this relationship in future Lessons.

Wheel Rate – We use the motion ratio in a AA-arm suspension to determine the wheel rate, or spring rate at the wheel. This is something we definitely need to know in order to plan out our setups. The way to find the wheel rate knowing the spring rate is simple. It is the combined motion ratio (mount motion ratio plus spring angle ratio) squared (motion ratio number times itself).

If we know the overall motion ratio, which combines the control arm motion ratio plus the spring angle motion ratio, we just square that number. In our example above, we had a combined motion ratio of 0.705. If we square that, we get 0.497, or 49.7 percent. The wheel rate ends up being about half the installed spring rate.

Taking a spring rate of say, 300 lb/in (pounds per inch), times 0.497, and we get a wheel rate of 149.1 lb/in. If we know the wheel travel, we can multiply the wheel travel times the wheel rate to find the tire loading. In this example, let's say the wheel travel were 4.0 inches.

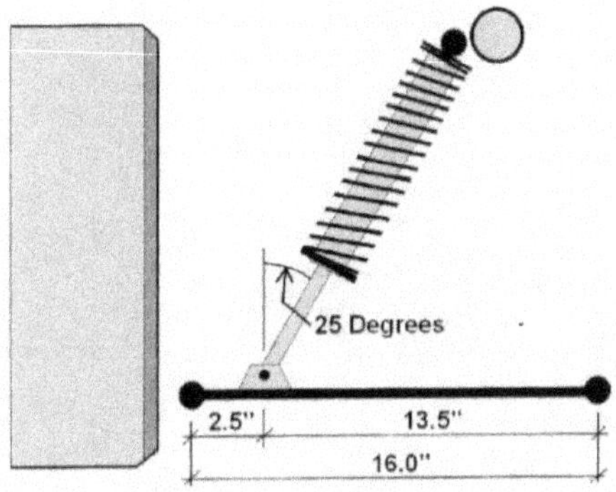

This is a typical coil-over mount we see on dirt or asphalt race cars. The bottom of the coil-over is mounted to the lower control arm some distance from the ball joint and the top of the coil-over is mounted to the chassis. There is a motion ratio created by the mounting points and another motion ratio created by the shock angle.

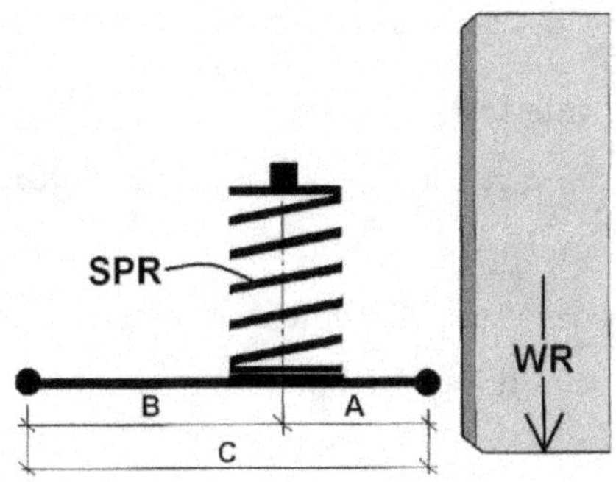

In the big spring type of suspension, the spring is mounted mostly at right angles to the lower control arm. So, in this case, we only have the actual mounting motion ratio to calculate to find the wheel rate.

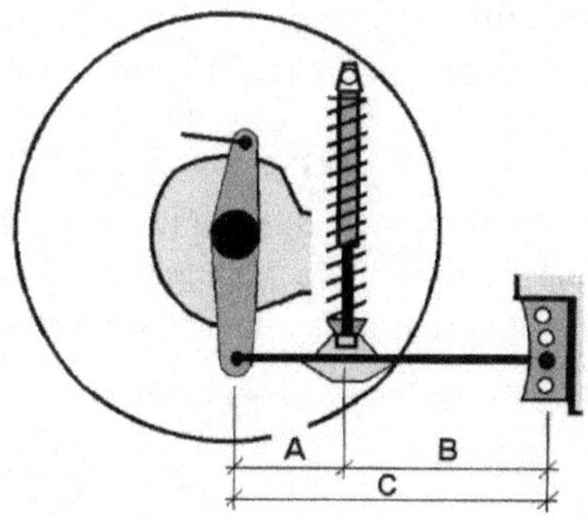

This is a more detailed sketch of a rear link where the spring is mounted on the link. We can see where as the chassis moves vertically, the spring will move a smaller amount. To visualize the "wheel rate" so to speak, we can imagine this to be a double A-arm setup with the axle mount acting much like the wheel. Then it is just a matter of doing the math.

If we multiply the wheel rate of 149.1 times the wheel travel of 4.0, we get an additional wheel loading of 596.4 pounds over that which existed at static ride height. Remember, we already had a wheel load holding the car up that was resisting gravity before we traveled the 4.0 inches.

This number, 596.4 lb, represents the additional load on the tire at that wheel travel. If we add the static loading, which can be directly measured with a scale, of say 550 pounds, this extra loading is added to that for a total loading of 1146.4 pounds at mid-turn.

How To Easily Measure Motion Ratio – Measuring the lengths of the control arms and the distance from the ball joint to the spring mount takes time and can be inaccurate. And, measuring the shock angle is difficult as well and can also be inaccurate. There is a much more simple way to find your motion ratio.

To find the motion ratio the simple way, start by supporting the chassis at nearly the normal ride height. Remove the spring or coil-over. Then measure the length of the spring mount. For a coil-over, just measure the centers of where the ends are bolted. For a stock type of big spring car, measure from the bottom of the bucket in the control arm up to a point directly above the bottom point.

Now place a block of wood or metal that is close to the amount the chassis will travel on the race track under the tire. Note how thick the spacer under the tire is. Then remeasure the spring length. Divide the amount the spring traveled by the spacer thickness and you have the motion ratio. To find the wheel rate, square the motion ratio and multiply that times the spring rate.

Motion Ratio and Wheel Rate are important concepts to understand. We don't need to go into detail about how they are ultimately used in race car dynamics, we'll get into that in RCT Level Two. Now that you have been introduced to this, some of what we will be getting into later on will make more sense.

Exam - In The Context Of This Lesson:

In A AA-arm Suspension, The Spring Moves?

1) Separately from the wheel
2) A different distance than the wheel
3) More than the wheel
4) Half the distance of the wheel

Motion Ratio Is Defined As?

1) How far the spring moves
2) How far the wheel moves
3) Relationship of spring movement to wheel movement
4) How far the chassis rolls

Wheel Rate Is?

1) The distance the wheel moves in the turns
2) How much load is on the tire
3) The shock rate times the motion ratio
4) Motion ratio squared times the spring rate

Mid-turn Tire Loading Is?

1) The gravity loading plus spring rate
2) Gravity loading plus wheel travel loading
3) A product of motion ratio
4) More than the spring loading

Lesson Twenty-Two – Spring Base and Choosing Springs

Glossary:

Spring Rate – The actual rate of resistance to weight applied to it. We usually refer to a spring rate in pounds or kilos per inch. A 100 lb/in rate would deflect one inch if it were made to support a force of 100 pounds.

Spring Split – A difference in spring rates for the two springs on the same axle, or suspension system. Spring split is usually used in circle track racing where the cars only turn one way.

Motion Ratio – A difference in movement between one part of a suspension and another. It is usually used in reference to the wheels vertical movement in relation to the movement of the spring. When a spring is mounted on a lower control arm, it will move less than the wheel.

Force – The total pounds or kilos of work the spring does to support a measured load. A spring rated at 100 lb/in would develop a force of 400 pounds if compressed four inches. (100 pounds per inch rate compressed 4.0 inches; 100 x 4.0 = 400 pounds of force)

Load – The actual measure of the weight on the tires measured in pounds or kilos of the parts of the car caused by gravity and other effects.

Over View – As covered in the previous Lesson, the spring base for a AA-arm suspension is the width of the tires. This is because the spring rate is transferred out to the wheel contact patches. It is as if the car were riding directly on the springs.

For the straight axle suspension system, it is a little different. The chassis "feels" the spring base in those systems as the width of the springs, or more appropriately, the width at the top of the springs. So, here is how that looks.

Our choice of spring rates will determine, in part, our roll stiffness and roll angle for a suspension system. It is the calculation of roll angles that is at the heart of setup for any race car. In the past, the straight axle suspension system was looked at differently and the spring rates were translated out to the wheels. This is not the correct way to look at the vehicle dynamics of this suspension system, and I say that having proof developed by years of testing.

For a double A-arm suspension, the spring base is the width between the centers of the tires. For a solid axle suspension, it is different. The chassis "feels" the spring base at the point where the chassis rests on the springs. For a coil-over design, the top mounting point is the top of the spring, so to speak. It is the distance between the top of the two springs that forms the spring base for the solid axle suspension.

Spring Base – The spring base is the width of the springs where the chassis mounts to them in a straight axle suspension system. For a AA-arm suspension the spring base is a little different. Because we translate the spring rate out to, basically the ball joint or end of the lower control arm, that ball joint becomes our spring as such. So, the spring base is the width of the distance between the centers of the two lower control arm ball joints, one on each side.

This concept of spring base is important to understand because the roll stiffness and ultimate roll angle is a product of: 1) the spring base width, 2) the moment arm length, 3) the spring rates, 4) the magnitude of the lateral force, 5) spring split. It is the last of these that we need to explain a little more about.

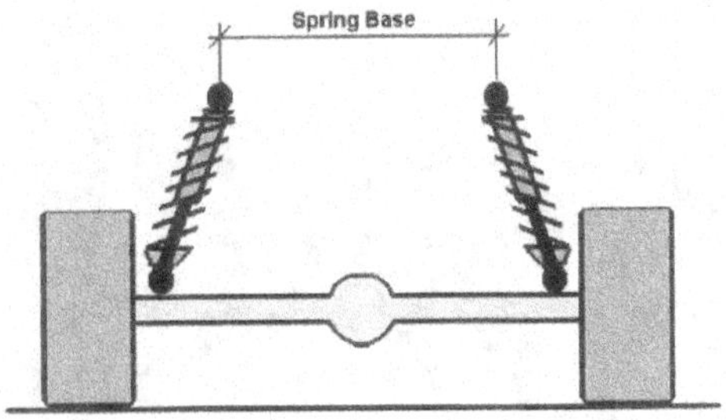

This illustrates the spring base for a solid axle suspension. The base is located where the chassis attaches to the springs, or coil-over assembly. This sketch closely resembles a dirt late model configuration with a lot of spring angle. The car does not know where the tires are and the width of the tires has no effect on the dynamics of the suspension. It's as if the chassis were sitting on springs attached to the ground. And the wider the spring base, the greater the roll resistance, or stiffness of the chassis.

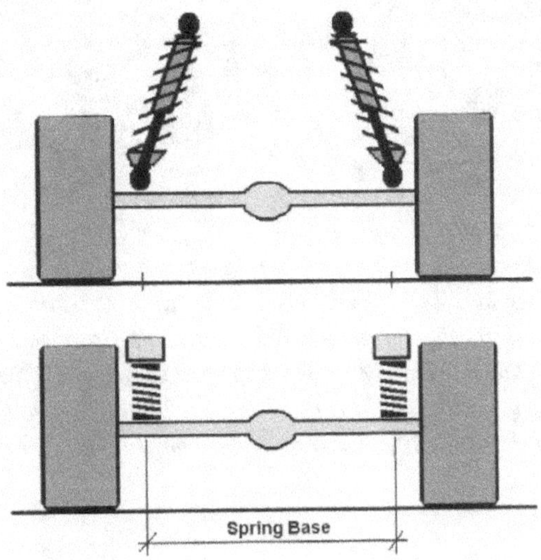

This illustrates the two basic designs for a straight axle suspension. Some designs place the springs on top of the axle tubes and use stock type of springs, some mount to truck arms in a similar way (top series Cup cars use this design) and some are the coil-over design we already talked about. Note that the spring base for the stock spring mounted on the axle tube is wider than the coil-over design used on dirt cars which have a lot of spring angle to them. Asphalt late model cars incorporate coil-over springs that are much more vertical and therefore have a wider spring base than what is shown.

Spring Split - The spring split mentioned in 5) above means that for a circle track stock car, we can and often do, install springs of different rates on each side of the car. The RF spring might be stiffer in rate than the LF spring. And the RR spring might be stiffer than the LR spring. This is almost never the case in road racing where the dynamics must be the same for left and right hand turns.

The benefit to using spring split is this. When we are turning and especially when the turn is banked, the forces on the car will do two things. They will try to roll the car over, and they will try to compress the car. The combination of the two forces, gravity and lateral force, combine for a net force that has a magnitude and direction that will roll, as well as compress, the chassis. The compression is called mechanical downforce.

It is this compression that plays into spring split. When we compress a suspension that has different spring rates on each side, we create a roll angle. The softer spring compresses more than the stiffer spring. So, in a left turn on a high banked track, if the LR spring were less rate than the RR spring, the compression would cause a roll angle to the left, or what we can call a negative roll angle.

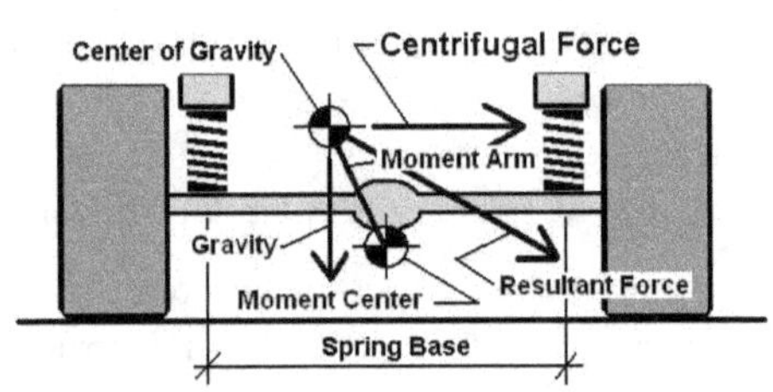

The dynamics of the straight axle suspension are pictured here. We have a lateral force called the Centrifugal force that is a result of traveling on a arc at high speeds. We also have Gravity which is always present. When we combine these two forces, we get a Resultant Force. Note the angle of this force. One a banked track, it will be pointed well between the two tires and provides what we call mechanical downforce pulling down on the chassis. With spring split we first talked about in Lesson Twenty, we can create more or less roll angle with this type of suspension easily.

If the lateral force were creating a positive roll angle to the right in a left turn, then we would add the two angles, and if let's say the spring split angle were -1.0 degrees and the lateral force roll angle were +2.0, the result would be a net roll angle of +1.0 degrees. This really does happen in a race car using spring split.

The spring split concept was developed to counter the roll forces for a straight axle rear suspension. It became hard to limit the rear roll angle as much as we could limit the front roll angle in a stock car. In the AA-arm suspension, we can install larger sway bars, stiffer springs, and/or bumps to reduce the roll angle to very low numbers.

In order for the rear suspension to be matched to the front for roll angle, we cannot, or should not install very stiff springs. Raising the height of the rear moment center can reduce roll angle, but there are limits to how high we can and should go.

The answer to the problem was spring split. Race teams found that by installing springs whereby the inside spring were softer than the outside spring, the roll angle was significantly reduced. By varying the amount of split, the team could effectively tune the roll angle balance of the chassis and create the perfect setup.

Choosing Circle Track Spring Rates – So far we have been talking mostly about stock cars and straight axle suspensions, but choosing spring rates is a topic of discussion meant for any race car, no matter the design. So, we have split the discussion into circle track and road racing sections.

We choose our spring rates for circle track racing with the dynamic balance and roll angle harmony as the primary considerations. And, for some aero sensitive cars, the rake of the chassis and body has a lot of importance. But the primary goal is dynamic balance.

For CT racing, we have more tools in the tool box than for other types of racing. We have, as discussed above, spring split. We also have a variable on roll center design and camber layout we can use. Whereas the road racing cars need to be more symmetric in their design left to right, the CT car does not.

The circle track cars can run different cambers left to right, different roll center location than being centered in the chassis, and they can run different spring rates on each side. So, it then becomes almost too easy to balance a CT car for roll angles. You just have to know what to install for spring rates, etc., the tools are all there. We only need satisfy one direction of turning.

The other thing we can change to affect the roll angles is the sway bar. And in CT racing, we can either have just a front sway bar, or a front and a rear. For rules that limit rear spring split, as has happened in the past, the team can install a rear sway bar to help limit rear roll.

Road Racing Spring Selection – For road racing cars, including stock type cars, prototypes and formula cars, the springs must be symmetric in rates. These cars must turn in both directions and so any spring split could work one way, but not the other. Its effect would be reversed when turning the other way.

Another major consideration for road racing cars is pitch attitude. Many of these cars are very aero sensitive. The angle of the chassis fore to aft, is critical and any change in pitch angle can have significant effect on the amount of downforce and downforce balance front to rear. This can cause a huge change in front to rear grip and handling balance.

To produce the ideal pitch angle and to help maintain that pitch, the teams will choose spring rates and bumps that will put the chassis at the correct pitch. This is sometimes done at the expense of dynamic balance, or as we call it, matching of the roll angles. This is where the sway bar, or anti-roll bar comes into play.

The sway bar can make up for deficiencies in the spring stiffness to help match the race cars roll angles. Suppose that the rear suspension was too soft and was rolling too much and more than the front. If the front angle seemed reasonable, then the team could increase the rear sway bar stiffness to more evenly match the front and rear roll stiffness and ultimately the roll angles.

In modern prototype cars and formula cars, there are methods that are used to adjust the stiffness of the bar. If the bar arms are made of bendable blades, then these blades can be rotated. The range is from ninety degrees to the attachment link (flat and producing the least resistance to bending) to zero degrees, or inline to where there is no bending for full stiff positioning.

The bar itself can be switched out to a smaller or larger bar, whichever is desired. In our example, we would install a larger diameter bar to try to work within the adjustment range of the blades if full stiff were not enough.

The selection of spring rates has to be done in conjunction with other settings and parts on the car. The spring rates are just one part of the dynamic mix that controls the roll angles at the front and rear of the car.

Exam - In The Context Of This Lesson:

A Straight Axle Spring Base Is?

1) The bottom of the springs

2) The width of the tire contact patches

3) The distance between the top of the springs

4) The spring rates times the track width

The Roll Angle Is A Product Of Which Of The Following?

1) the spring rates and spring base

2) the moment arm length,

3) the magnitude of the lateral force,

4) spring split

5) All of the above

Spring Split With Inside Spring Softer Causes?

1) Less roll angle

2) More roll angle

3) The car to be tight

4) The car to be loose

Circle Track Race Cars Can Do What That Road Racing Cars Can't?

1) Handle high banked turns

2) Run different rate springs side to side

3) Run softer rear springs

4) Run a stiffer overall spring rate

Road Racing Cars Can Best Adjust Roll Angles By?

1) Using different spring rates at each end

2) Making the spring stiffness more or less

3) Making changes to the sway bar stiffness

4) Moving the moment center side to side

Racecar Technology – Level One

Lesson Twenty-Three – What Shocks Do and Don't Do

Glossary:

Load – The actual measure of the weight on the tires measured in pounds or kilos of the parts of the car caused by gravity and other effects.

Compression – The motion of the shock when the shaft is pushed in and the shock becomes shorter.

Rebound – The motion of the shock when the shaft is pulled out and the shock becomes longer.

In Transition – A car that is either slowing or accelerating. Transition portions of the race track are the entry during braking and the exit during acceleration.

Over View – A shock is a control and timing device. It works in conjunction with the springs and other components to control the motion of the springs and to regulate the distribution of loading on the tire contract patches. There are things a shock can do, and things it cannot do. Here is a list of those:

What Shocks Do Not Do – 1) Shocks do not and cannot support the car, i.e. they are not springs, 2) Shocks do not increase or decrease the amount of load transfer, 3) Shocks do not influence chassis dynamics if they are not moving, 4) Shocks are not a cure-all for basic handling problems.

What Shocks Actually Do – 1) Shocks control the speed of the vertical motion of the part of the suspension they are bolted to by controlling the speed of the rebound and compression of the spring, 2) Shocks with varying designs of resistance allow more or less rapid movement of a suspension corner than opposing corners, 3) Shocks regulate the amount of time it takes for a corner of the car, while in transition, to assume a new ride height, but load transfer happens immediately, 4) Shocks can be used to redistribute the loads at the four corners of the car as the car is in transition on corner entry and exit, 5) and, shocks can be used to overcome the ride springs on one or more corners of the car when running bump setups.

This road race car is at mid-turn and has stopped slowing and not yet started to accelerate off the turn. At this point on the race track, the shocks are doing no work. Shock do not have a dynamic function if they are not moving. This is the first important principle you must learn and know about racing shocks.

This Pro Late model race car is entering a turn and is under braking. The shocks are managing the transfer of weight from the rear to the front as a by-product of rapid slowing. The rear shocks are managing the rebounding of the rear springs and the front shocks are managing the compression of the front springs. This is where shocks can make a difference in the dynamics of the chassis.

Let's go over these one at a time. It is thought by some that a shock can help tune the setups of the car at mid-turn, steady state conditions where the car is neither slowing or accelerating. That is not true. Shocks may help or hurt how you got to mid-turn, but once there, shock no longer play a part.

The reason a shock cannot do that is because of a simple fact about shocks. The do not do any work unless they are moving. A spring works all the time it is loaded. Even if it is not moving, a spring is working providing force that holds up the car. Not so with a shock. A shock does no work when it is not moving.

When a weight is put onto a spring and it has to compress, it resists that motion. When it releases as load comes off of it, it wants to release that energy quickly and move very quickly. A spring resist compression motion and promotes rebound motion. The shock, more than anything, resist the rebound motion of the spring. In a much smaller way, it can control the compression of the spring.

When the race car is at steady state condition, meaning it is not accelerating nor slowing, the shocks have no influence on the distribution of loads on the four tires. It may as well be sitting on the trailer. One other thing about this photo is the angle of the shock. It will move at a slower speed than the spring as the chassis moves because of the extreme angle. Therefore, its rate must be higher than if it were mounted straight up in order to do the same work.

The stiffer the shock is, the slower the motion of the spring is, in both compression and rebound. If we install shocks of different rates of compression and rebound at the four corners of the car in relation to the spring rates, we can cause the loads on the four wheels to change while the car is in transition.

At two points on the race track, the weights on the four tires are steady, on the straights and in the turns at mid-turn, steady state (not slowing, not accelerating). The weight has transferred and has settled onto the springs and tires.

We setup our cars so that the grip generated load distribution is such that the grip at the front and rear are the same. But when the car is in transition, we might need a different load distribution. Since the shocks work while in motion, and when braking and accelerating the suspension is in motion, the shocks can help us out.

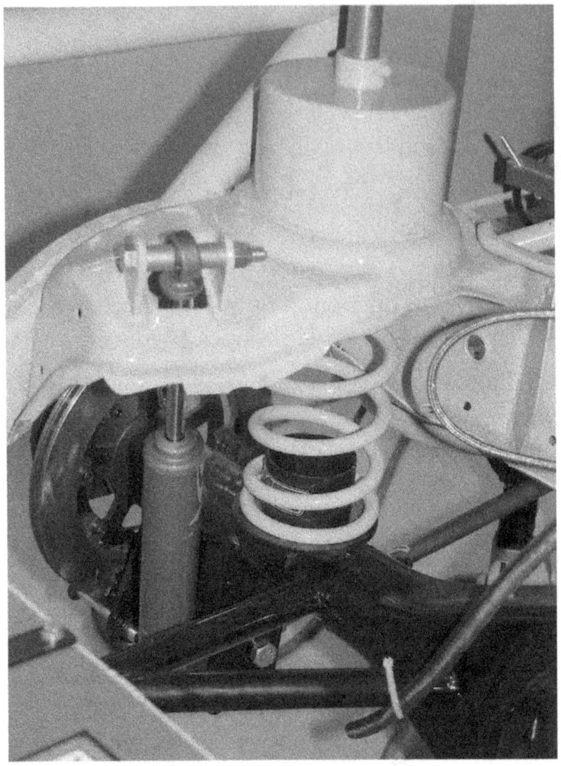

This shock is mounted where it will do the most good along with the motion of the spring. It is mounted vertically and in nearly the same distance from the centerline of the car. Its motion ratio is nearly one to one.

By altering the amount of control the shocks have in either compression or rebound, we can control the load distribution during those times when the suspension is in motion.

When we add load or take load off a spring, it wants to change to a new height. This is what happens on the corners of the car as weight transfers during braking, cornering, and acceleration. A shock controls that motion and dictates how long it takes for the spring to assume the new height.

This is the essence of what a shock does. It is up to us to design our shock package to time the movement of the springs and suspension so that the load distribution is what is needed for maximum performance. We will cover how race teams do that in RCT Level Two.

Exam - In The Context Of This Lesson:

What A Shock Does Not Do?

1) Support the race car
2) Add load to the tires
3) Promote load transfer
4) Change the handling at mid-turn
5) All of the above

What A Shock Does Do?

1) Adds load to the rear tires
2) Regulates amount of load transfer
3) Resists motion of the springs and suspension
4) Tunes mid-turn handling

Shocks Redistribute Loading On The Tires By?

1) Promoting faster or slower movement of each spring
2) Allowing quicker load transfer
3) Providing more compression than rebound resistance
4) Causing changes in the handling

The Suspension Is In Steady State When?

1) On turn entry
2) On turn exit
3) At mid-turn
4) Half way down the straight away
5) 3 and 4

A Spring Moves To A New Height When?

1) The shock is stiffer
2) Weight is added
3) The shock is softer
4) When weight is removed
5) 2 and 4

Race Car Technology – Level One

Lesson Twenty Four – How A Shock Works

Glossary:

Load – The actual measure of the weight on the tires measured in pounds or kilos of the parts of the car caused by gravity and other effects.

Shock – A device that controls spring movement. This is also called a dampener, but the terms are interchangeable.

Twin Tube Shock – A shock design that uses two tubes, one to hold a gas bag to separate the shock oil from the gas, and one to hold the shock piston, oil and shaft.

Gas Pressure Shock – A shock that uses pressurization of the gas to separate the oil from the gas.

Shock Body - The tube that holds the fluids and other parts including the shaft, separator piston and valve attached to the end of the shaft. The end of the shock body is attached to the chassis or control arm, depending on which way the shock is mounted.

Separator Piston – A free floating piston that is used to separate the shock oil from the pressurized gas.

Shock Piston – A part that is attached to the end of the piston shaft and to which is attached valve discs that regulate the flow of shock oil as the shock moves in compression and rebound.

Shock Valve – A series of discs of different thicknesses and diameters that are mounted on the shock piston that control the amount of fluid that passes through it when the shock moves in compression or rebound.

Bleed – A small hole, or needle valve, that allows the free passage of oil from one side of the shock piston to the other.

Over View – A shock works to control movement of the springs. It does this by making a fluid move through openings of various sizes and it takes longer and more effort for the fluid to move through a smaller opening than a larger one. That is the short explanation. Here is a longer one that will help you completely understand how they really work and how we can change how efficiently they work.

How Shocks Work - Shocks resist motion by using a piston that must move through a fluid (thin oil) as the suspension moves so that the fluid must pass through holes, valves and slots. Varying resistance is created when the oil is forced through different sized openings. The resistance is usually different for each of the two motions, compression (inward motion) and rebound (outward motion).

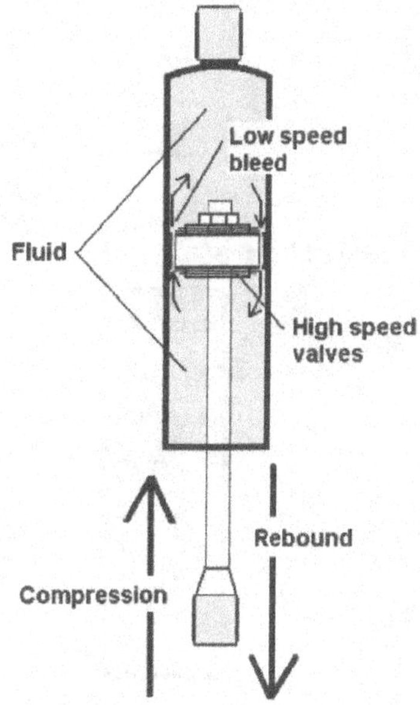

THE INSIDE OF A RACING SHOCK

This is basically how a shock works. It takes energy and force to move a shock because when it moves either in or out, fluid must pass through holes and ports. The smaller the holes and valve ports, the harder it is for the shock to move. So, the larger holes and valve ports will allow more rapid movement of the shock. And smaller holes and valve ports will slow the movement, or require more force to move it the same speed. In this way we can control different spring rates.

All racing shocks are of two basic designs – twin tube and mono-tube and can be either gas pressurized or "low" pressure. The twin tube has literally two tubes, the inside tube is where the work is done and the outside tube is a reservoir that holds extra fluids and a flexible gas filled bag.

The mono-tube design is almost always pressurized to some PSI (pounds per square inch pressure). Most every dedicated race car now uses the mono-tube, gas pressure shock.

Shaft Displacement – An important feature related to the design of a racing shock is called shaft displacement. When the shock shaft is pushed into the shock body and into the fluid, it takes up space. Suppose we pull the shock shaft out as far as it would go, fill the shock body with oil only, and then reseal the shock body. If we tried to push the shaft into the shock body and into the volume of oil it would not go because the shaft would be trying to compress and displace some of the oil and oil will not compress.

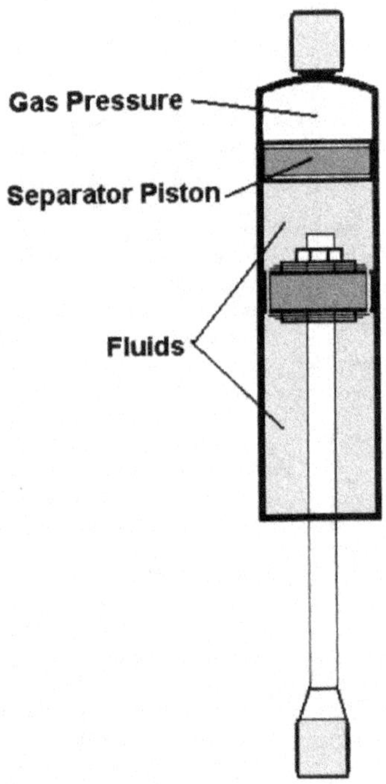

In reality, every shock must have a chamber that contains a gas. Gas is compressible whereas fluid is not. When the shaft is pushed into the shock body, it takes up space and displaces shock fluid. There must be a way to allow the volume of the shock shaft to exist within the shock body. That way is by way of gas compression. As the shaft moves in and takes up space, the gas compresses to provide the room for the shock shaft. A piston separates the gas from the shock fluid.

We need a space inside the shock filled with a substance that will compress. Gases such as air and nitrogen will compress. Every shock needs to have a certain volume of gas, usually nitrogen, along with the fluid, in order to allow for the shaft to move into the shock body and take up space. Nitrogen is a dry gas that suits our purpose and is widely used in racing shocks.

Keeping the Oil and Gas Separate – The Nitrogen gas we put into our shocks to compress and allow for the volume of the shock shaft must be separated from the fluids in all shocks. The gas can be contained inside a plastic bag or separated by another piston.

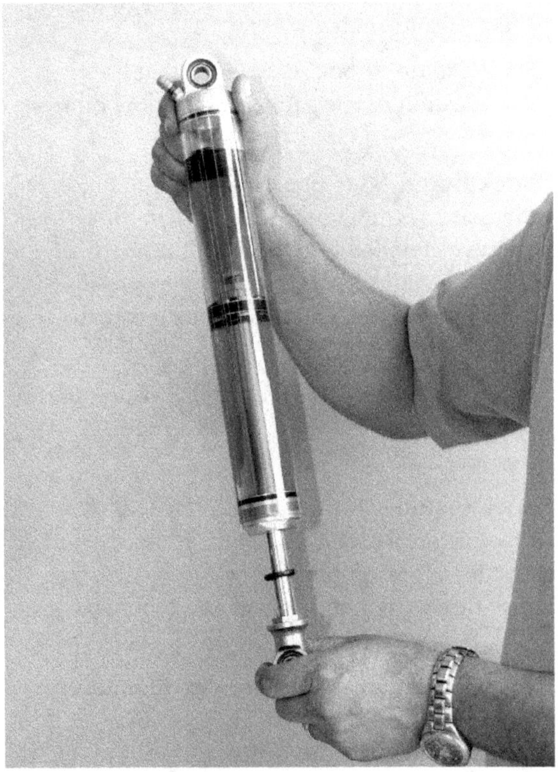

This is a demonstration shock with a clear body showing the internal parts of a gas pressure shock. Here we can clearly see the red shock fluid, the shock piston that controls the flow of fluid in compression and rebound, and the separator piston between the red shock fluid and the clear compressible gas at the top.

Here we see a closeup of the shock piston and the valves lying under the nut. There are different designs of pistons that react to the movement of the shock differently. The primary two are the Digressive and Linear piston designs. Along with those different pistons, we have many different thicknesses of valve discs that produce many different levels of resistance force to control different rates of springs.

Because the bags used in twin-tube designs are not pressurized, these shocks are referred to as non-pressurized shocks. In truth, as the shock shaft is pushed into the shock body, some amount of pressurization must take place due to the displacement of the shaft and the compressing of the gas inside the body.

In mono-tube shocks, separation is accomplished by installing a second separator floating piston, which provides a seal that separates the fluid from the gases. A Schrader valve that is installed in the shock body at the gas chamber end of the shock allows us to pressurize the gas inside the shock. This pressure ensures that the gas will be forced to separate from the fluids at all times. That is because the seal on the piston will usually allow the gas to seep past the piston from the fluid side to the gas chamber, but seals the heavier fluid from escaping into the gas chamber.

Compression and Rebound - The compression control side of what the shock does resists: 1) the bump movement of a corner of the car when we hit bumps (rises in the track), 2) the bump movement of the front shocks due to the transfer of load to the front end during braking/deceleration, 3) the bump movement due to the transfer of load to the rear upon acceleration, 4) and the speed at which the right side springs compress when the lateral forces are applied as we deviate from a straight line and turn into a corner.

The rebound control side of the shock resists the following: 1) rear chassis rebound, or vertical movement, when decelerating, 2) front chassis rebound movement upon acceleration, and 3) left side

rebound vertical movement caused by load transfer from the inside to the outside of the turn as we negotiate the turns.

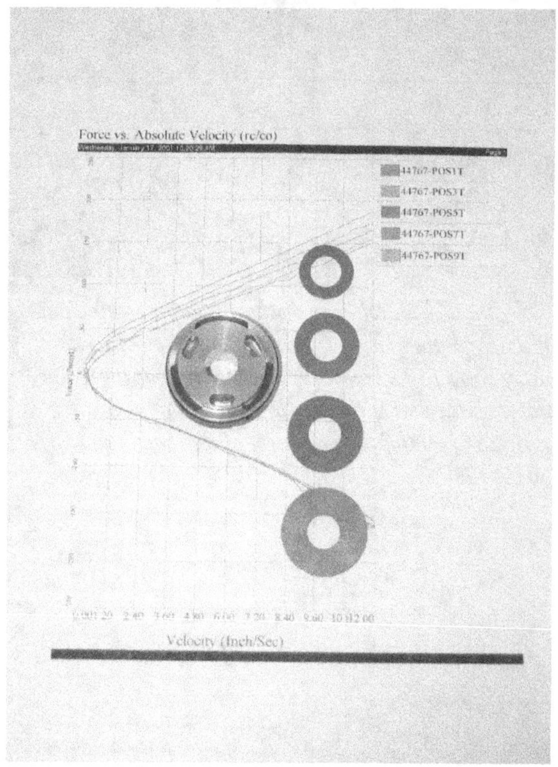

The shock valve and the valve discs are shown here. Fluid flows through the slots in the piston and must force the valves open in order to pass through. We can stack different numbers of discs and use different thicknesses of discs to create the resistance force we need.

The amount of resistance that the shock provides with each movement, compression and rebound increases with the speed at which the shock is forced to move. Low speeds create low resistance and high speed movement creates higher resistance.

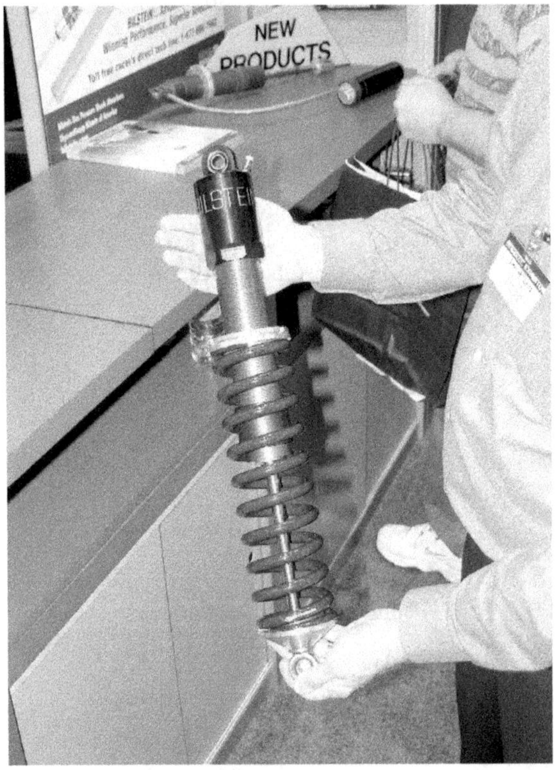

This is a modern coil-over shock and spring combination. We can see the black gas chamber at the top of the assembly. The shock body is threaded so that the spring nut atop the blue spring can be adjusted to preload the spring and control the ride height of the chassis and force on the tire.

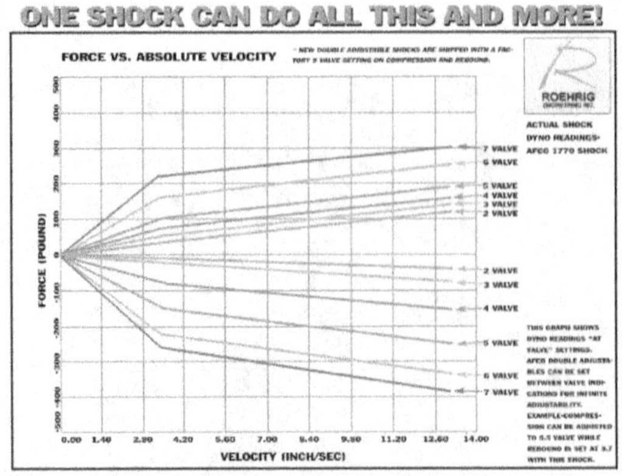

Older non-adjustable shocks were built to specific resistances and numbered as to the stiffness. In this graph we can see the approximate force values for each number of shock. A one (1) shock was the least stiff shock. The left side shows the force in pounds and the bottom scale shows the speed of the shock movement in inches per second. As speed increases, the force also increases. This example has a digressive shock piston installed. It digresses in force after about 3.0 inches per second.

Low Speed Control – Low speed shock movement is defined as shaft speeds that are between 1 to 5 inches of movement per second. Many racers use 3 inches per second to evaluate the shocks resistance to normal entry and exit movements. The lower speeds are mostly associated with suspension movement caused by chassis roll and chassis dive at turn entry where the loss of speed is moderate. The low speed control dictates much of the handling side of the shock design.

Each shock has a piston mounted on the end of the shaft and one or more small "bleed" holes in the piston to allow fluid inside of the shock to flow freely from one side of the piston to the other. The size of the bleed hole regulates how quickly the fluid will flow back and forth and that is primarily how the different levels of resistance are created for low speed control.

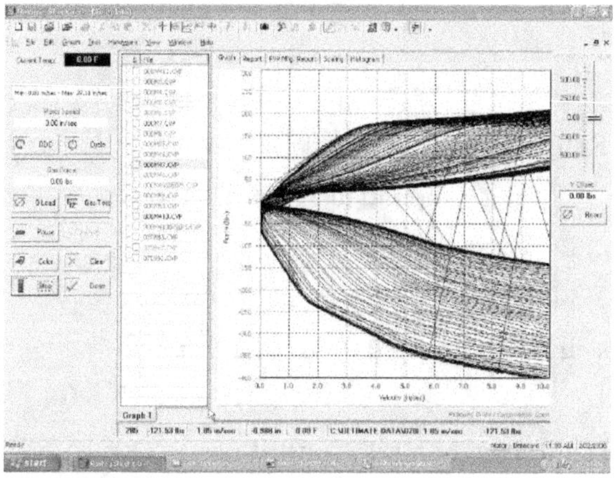

In a modern day adjustable shock, we can create almost infinite numbers of force settings. Race teams use these shocks to fine tune the transitions of corner entry and corner exit. Remember that the shock has control only when it is moving.

High Speed Control – As we experience the greater velocities of shaft movement we go into what is called high speed control with shaft velocities of from 5 to 10 or more inches of movement per second. Types of suspension movement that cause the higher shaft speeds in our shocks are: 1) bumps or holes in the racing surface (creating very high shaft speeds), 2) the driver stabbing the brakes on entry and hard on the throttle on exit, or 3) a sudden change in banking angle such as transitioning from banking onto the apron of the race track.

The piston mounted to the end of the shaft also contains a valving mechanism that allows the fluid to flow through slots that are designed into the piston. These valves consist of disks that open as the pressure increases due to more rapid shaft movement in either compression or rebound. These disks are used to control the dampening rate associated with higher shaft speeds.

Linear and Digressive Designs – There are two basic piston designs used in most race cars. One is the Linear piston that has a high flow rate at low shaft speeds and hence little resistance, and increases resistance as the shaft speed increases. The rate of the shock continues to increase as long as the speed increases.

The other design is called a Digressive piston and has a lower flow rate at low shaft speeds which provides more resistance and control, and then the resistance rate of gain decreases with an increase in shaft speed to a pre-designed level. Then the resistance tapers off and stays relatively the same as the shaft speed continues to increase.

This "pop-off" characteristic works well to keep the shock from building excessively high amounts of resistance usually associated with sharp increases in shaft speeds.

A rut or hole in the track could cause the shock to move at rates well above 10 inches per second up to 20 to 30 inches per second or more. If that happened with a linear valved shock, it would feel like the suspension were solid with no suspension spring and the bump would jar the car considerably causing damage or loss of control.

Special Design Notation - If we study the automotive design books related to shock design for production automobiles, we would learn that the criteria calls for un-equal resistance in each direction, compression and rebound, in combination with the action of the springs.

Because springs naturally resist compression and promote rebound, we should not use a true 50/50 rated shock where the resistance is the same for both rebound and compression.

As the spring rate increases, the compression rate can be less and the rebound rate must increase. With bump devices with associated high spring rates, the rebound resistance must also be high in order to control the high spring rate.

Exam - In The Context Of This Lesson:

The Basic Primary Function Of A Shock Is?

1) Provide spring rate
2) Control motion of the suspension
3) Resist compression of the spring
4) Resist rebound of the spring
5) 2,3,4

A Shock Controls Motion By?

1) Adding spring rate to the suspension
2) Moving weight around on the four tires
3) Regulating fluid flow through the shock piston
4) Being pressurized

Compression and Rebound Are?

1) Equal in a perfect shock
2) Terms used to explain the motions of a shock
3) Made the same as the spring rate
4) The two types of pistons used in shocks

Compression Rates Should Be?

1) Equal to rebound in a perfect shock
2) Less than the spring rate
3) More than the spring rate
4) Intended to push down on the tire

Rebound Rates Should Be?

1) Equal to rebound in a perfect shock
2) Less than the spring rate
3) More than the spring rate
4) Intended to push down on the tire

High Speed Shock Control Is?

1) To control higher speeds of the car
2) The measure of control at faster shaft speeds
3) Less than that of low speed control
4) More than that of low speed control
5) 2 and 4

A Linear Shock Valve Does?

1) Keeps the shock shaft inline
2) Causes the force to be less at higher shaft speeds
3) Increases force with increases in shaft speed
4) Lines up the spring and shock on a coil-over

A Digressive Shock Valve Does?

1) Provides more control at higher shaft speeds
2) Causes the force to be more at higher shaft speeds
3) Diminishes the force with increases in shaft speed
4) Keeps the suspension from bottoming out

Lesson Twenty Five – Different Types of Shocks

Glossary:

Twin Tube Shock – A shock design that uses two tubes, one to hold a gas bag to separate the shock oil from the gas, and one to hold the shock piston, oil and shaft.

Gas Pressure Shock – A shock that uses pressurization of the gas to separate the oil from the gas.

Shock Body - The tube that holds the fluids and other parts including the shaft, separator piston and valve attached to the end of the shaft. The end of the shock body is attached to the chassis or control arm, depending on which way the shock is mounted.

Separator Piston – A free floating piston that is used to separate the shock oil from the pressurized gas.

Shock Piston – A part that is attached to the end of the piston shaft and to which is attached valve discs that regulate the flow of shock oil as the shock moves in compression and rebound.

Shock Valve – A series of discs of different thicknesses and diameters that are mounted on the shock piston that control the amount of fluid that passes through it when the shock moves in compression or rebound.

Bleed – A small hole, or needle valve, that allows the free passage of oil from one side of the shock piston to the other.

Shock Speed – The speed of movement of the shock shaft measured in inches per second. Generally speaking, as the shaft speed increases, so does the shocks resistance to movement.

Over View – All shocks are not of the same design. All shocks have a primary function to control the speed of suspension movement by regulating the flow of fluid that must flow through a valve system as the shock moves in compression and rebound. Simple shocks utilize simple methods to accomplish the required task.

All shocks must contain a volume of compressible gas to allow for the space taken up by the shock shaft as it is pushed into the shock body when the shock is in compression. As the shaft is pushed into the body, the gas compresses whereas the shock fluid cannot compress.

The most common shock used in today's racing is the gas pressure, fully adjustable shock. This one has an external gas canister that has its own valves to control compression settings. This type of shock can be different lengths and configured with different spring packages, including bump stops and bump springs. The functions remain the same though.

It has been shown and proven that high speed and repeated motion of the shock will cause mixing of the gas within the shock with the fluid if it is not made to be separate. As stated above, we need a volume of gas inside the shock body to allow the volume of the shock shaft to displace space within the shock.

If we don't separate the gas from the fluid, then they will mix and this affects the efficiency of the fluid passing through the shock piston and valves. The shock becomes less reliable as to the amount of force it provides. When the fluid is mixed with the gas, the combination flows more easily and therefore the force needed to move the shock the same speed becomes less and we lose efficiency.

This old style mono-tube shock seen here on a stock application is a rare sight in todays racing. "Stock" type shocks are now made for passenger cars that are of the gas pressure design, just like race shocks, and do a much better job of controlling the springs.

Shock Consistency – One of the main properties we require of our racing shock is that it stays consistent in how it provides resistance to the movement of the suspension. When the gas mixes with the fluid, the density of the mixture becomes less and then the resistance to the force also becomes less and not consistent.

Pressurizing the gas forces it out of the fluid and keeps the two substances separated. The gas is also physically separated by a floating piston in the design of a gas pressure shock, or by use of a gas bag in a twin tube shock.

The twin tube shock, although not formally pressurized, will have gas pressure as the shock shaft is pushed into the shock body and takes up space. When this happens, the gas in the bag must compress and therefore we have some amount of gas pressure in a twin tube shock.

Stock replacement shocks that are gas pressure are made by several shock companies like Bilstein and QA1. These do a very good job of controlling movement and staying consistent.

Differences Between Gas Pressure Shocks and Twin-tube – There is a distinct difference in the design and construction between the gas pressure shock and the twin-tube design. The gas pressure shock has only one tube to hold the fluid, the shock shaft and piston, the pressurized gas and the separator piston.

The twin tube shock has an inner tube that holds the fluid, the shock shaft and the piston, while another tube that surrounds the inner tube holds more fluid plus a gas bag. The fluid flows freely between the two tubes and the gas bag compresses as the shock shaft is pushed into the inner tube.

Primary Shocks Used For Racing – Now that we know the two primary types of shock design, the most commonly used shock design for racing is the gas pressure shock. So, we will concentrate on how that shock is designed for different applications. Much of what we discuss about the gas pressure shock will relate to the twin tube shock as well.

In the category of gas pressure shocks (GPS), there are different applications based on the different types of racing. This is directly related to what the shock is asked to do in each application. We can easily understand that the shocks will move much slower when installed in a race car that runs on a smooth, asphalt racing surface as opposed to a rough and rutted dirt track.

On asphalt, we can expect slower shock shaft movement in the range of 3 to 10 inches per second. These are the speeds where we design our asphalt shock rates to operate within. On dirt tracks, we can often see a range of shock shaft movement of from 10 to 25 inches per second or more.

The reason why the dirt shocks move so quickly is because the wheel moves that quickly when it hits a bump or hole in the track that may be inches deep. So, the shock built for dirt applications must necessarily be built differently.

Types Of Shock Piston - Within the description of different types of shocks we have different designs of shock pistons. The two basic designs are Digressive and Linear. Let's discuss the Linear piston first.

A linear design of shock piston starts out at a very slow speeds with very low resistance to movement. As the shock shaft speed increases, the rate of resistance increases in a linear way. In other words, if the speed doubled, then the rate also doubled. The rate doesn't have to double, but as the speed increases, so does the rate.

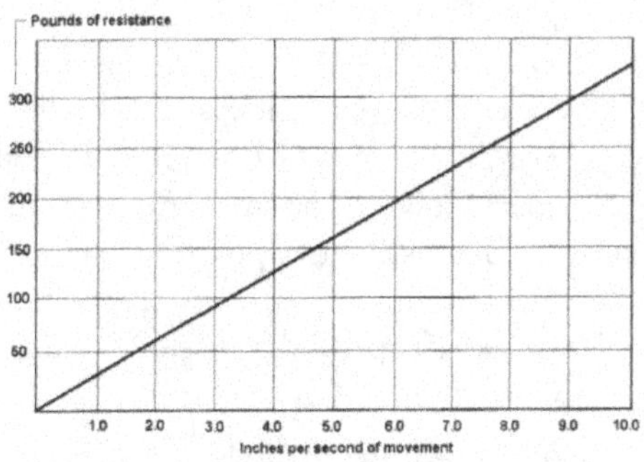

Example of a Linear Shock

One of the two main types of shocks and shock valving is the linear design. We can see the nearly straight line the force verses speed charted makes. As the speed of the shock shaft increases, the rate also increases proportionately. These shocks can achieve a very high rate of resistance if made to move very quickly. It can virtually feel like the shock bleed shut off and then the suspension becomes very stiff. It is best to only use this type of shock on a smooth race track.

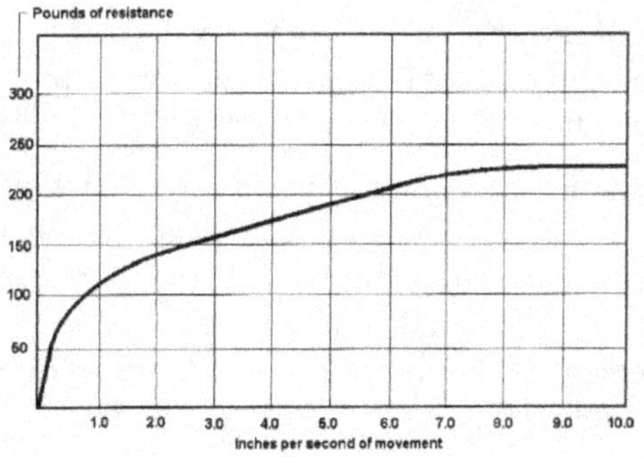

Example of a Digressive shock

The other type of shock and piston design is the digressive shock. For this design, as the shaft speed increases, the rate of the shock in pounds of resistance does not build proportionately. It will tend to taper off as the speed reaches a certain predetermined point. This allows the shock and wheel to ride over bumps that produce very high shaft speeds without producing high resistance. This design is much better for dirt applications where the tire rides over bumps and holes.

A digressive design of shock piston, when compared to a linear piston, has a higher rate of resistance to movement at slower shock speeds of say 1 to 3 inches per second. Somewhere around 5 inches per second (ips) the gain in rate begins to taper off and beyond 6 to 8 inches per second the gain in rate is minimal. Of course, we can design any rate curve we want, but generally speaking, this is the way a digressive shock piston works.

This is important to understand, because when we think about designing shocks for a dirt application, we have to think about those very high shock shaft speeds. Our shocks cannot keep building rate up to 25 ips or more shaft speeds or the shock would lock up. The fluids cannot move that fast from one side of the shock piston to the other.

If we use the digressive design of shock piston, then the rate would stabilize at some point and not gain and there would be a more free flow of fluid to allow the wheel to move over the bump. Otherwise, the suspension would lock up and the car would leave the racing surface and take off losing all grip.

Other Design Parts – The other designs available to make a shock what we want it to be for our application are many. The discs we stack onto our piston can be any number, and size and any thickness. The more disc and the bigger the size and thickness, the more they will resist flexing and the less fluid that is allowed to pass through the piston. The rate goes up.

Besides the valves on the piston that allow fluid to pass from one side of the piston to the other, we have what we call Bleed. Whereas the valves will be closed until they are forced open by movement of the shock shaft, the bleed will always allow fluid to flow freely from one side to the other.

At slower shaft speeds, there may not be enough force to open the valves whereas the bleed will always allow the flow of fluid, even at very slow speeds. So, the bleed controls the slow shaft speeds and the valves control the faster shaft speeds.

On many adjustable shocks, the bleed is a needle valve located up near the shock piston. A rod is placed inside the shock shaft and extends to the heim joint where the shock connects to the control arm, or links thereto. We can see in the slot above the eye of the heim the bottom of the rod. This end has indentions where the team will insert a pin to rotate the rod and regulate the opening of the needle valve to adjust the low speed bleed.

The designs of bleed are varied. The bleed can be a simple hole drilled through the piston and the size of the hole determines the slow speed shock rate of resistance. Since the hole is permanent, it cannot be adjusted easily. You can drill a bigger hole, but you cannot make an existing hole smaller.

So, shock manufacturers decided to introduce a new way to bleed fluid. They make a hollow shaft that allows fluid to flow through it from one side of the piston to the other and the flow can be changed by adjusting the height of a needle valve within the shaft.

The use of shocks on race cars takes many shapes and forms. Even on this prototype rear suspension, the coil-over shocks with external canisters do the same job other shocks do in other applications. The only difference being the amount of travel and speed of movement between different applications.

A simple screw adjustment moves the needle valve closer or farther away from a hole to make the space around it larger or smaller. For asphalt applications, obviously this type of adjustment works very well because we are working within a range of shock shaft speed that is slower and within the working range of the bleed.

For dirt applications, the bleed becomes less useful and we would need to work more so with the valve design to adjust for the higher speeds we see on rough dirt tracks.

Shock Graphs – The way we know how much work a shock is doing is by running it on a shock dynamometer. This device or machine measures the force of resistance of the shock at different speeds of shaft movement. It will record the forces in pounds or kilos and then produce a graph that can be printed out.

We would then look at the graph and note the force numbers for different shaft speeds. We want to match our shocks to the spring or wheel rates we have installed in the race car. Remember that shocks are primarily used to resist the spring wanting to rebound and the springs help the shock to resist compression. In almost every application, the shock will be deigned with more rebound rate than compression rate for the same shaft speed.

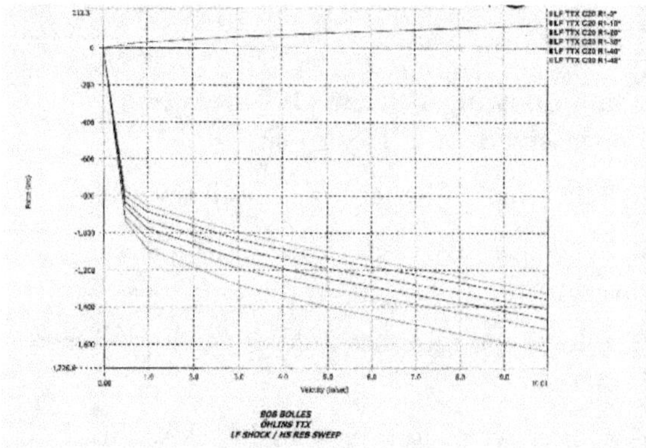

The use of a shock dynamometer and the printouts they provide are essential tools racers use to fine tune the setups of their cars in today's racing. In order to know exactly how much control a shock has verses the spring rate the shock will have to control, we need to look at the rate verses the speed of the shock in relation to what that corner of the car does on the race track.

So, to review, we have two primary types of shocks, the twin tube and the gas pressure. Within those two types, we have different internal designs of pistons, valves and bleeds. The type of piston, linear or digressive, and the size and number of valve discs and bleeds are chosen to control movement for the springs we have installed in our race car and the track conditions we will encounter.

And the shock dyno will tell us exactly what rates our shocks have for each shaft speed by producing a dyno graph we can read. Asphalt race teams will be more interested in the rates in the lower shaft speed range and dirt race teams will be looking at the rates in the higher shaft speed ranges.

Exam - In The Context Of This Lesson:

The Two Primary Types of Shock Are?

1) Twin tube

2) Gas pressure

3) Dirt applications

4) Asphalt applications

5) 1 and 2

What Is The Difference Between A Gas Pressure Shock And A Twin Tube?

1) The gps has a higher rate of resistance

2) The twin tube is better for dirt applications

3) A Gps has a single tube

4) The Twin tube shock has a gas bag

5) 3 and 4

The Following Is True Of Linear and Digressive Piston Designs?

1) The Linear is better for smooth tracks

2) The digressive piston allows faster shaft movements

3) One is good for dirt and the other good for asphalt

4) Digressive pistons have higher slow speed control

5) All of the above

Which Best Controls Slow Speed Shaft Movement?

1) The bleed design

2) The number, size and thickness of the piston valve discs

3) A linear piston

4) A digressive piston

Shock Graphs From A Shock Dyno Tell Us?

1) How much low speed control we have

2) How much high speed control we have

3) Resistance along a range of shaft speeds

4) If our shocks are good for dirt or asphalt

Lesson Twenty-Six – How To Control A Spring

Glossary:

Shock Piston – A part that is attached to the end of the piston shaft and to which is attached valve discs that regulate the flow of shock oil as the shock moves in compression and rebound.

Shock Valve – A series of discs of different thicknesses and diameters that are mounted on the shock piston that control the amount of fluid that passes through it when the shock moves in compression or rebound.

Bleed – A small hole, or needle valve, that allows the free passage of oil from one side of the shock piston to the other.

Shock Speed – The speed of movement of the shock shaft measured in inches per second. Generally speaking, as the shaft speed increases, so does the shocks resistance to movement.

Over View – The very reason we need shocks on a vehicle is to control the springs. Springs resist compression, or pushing them together. Once they are compressed, they want to expand to their free height, or the length they are when they are not loaded and compressed.

In a race car, or any car for that matter, the springs resist compression of the suspension and promote rebound of the suspension. If we didn't have shocks, the suspension would oscillate up and down freely. So, it has been known for a long time that we need to control the springs against this oscillation tendency.

For race cars, the primary function of shocks is to control the spring rate we have installed in the car. But there are other functions beyond that simple task that are useful. More on that later.

With racing shocks, we are trying to control the installed spring. In this design, the shock is mounted at a different motion ratio than the spring. Where it is mounted, it needs to control the spring as if that spring were converted to a coil-over spring, or much less than its actual rate. If we were to install a coil-over spring to replace the stock mounted spring, to maintain the wheel rate, we would need to use a much lower rate of spring. So, the shock mounted where a coil-over spring would be mounted needs to control an equivalent spring rate.

What We Are Trying To Control - When we refer to controlling the spring rate, we are mostly talking about the rebound of the spring. Since the spring resists compression and promotes rebound, the rebound motion is what the shocks primary job is to control. We don't need much compression control since the spring is helping to resist that motion and in fact doing most of the work.

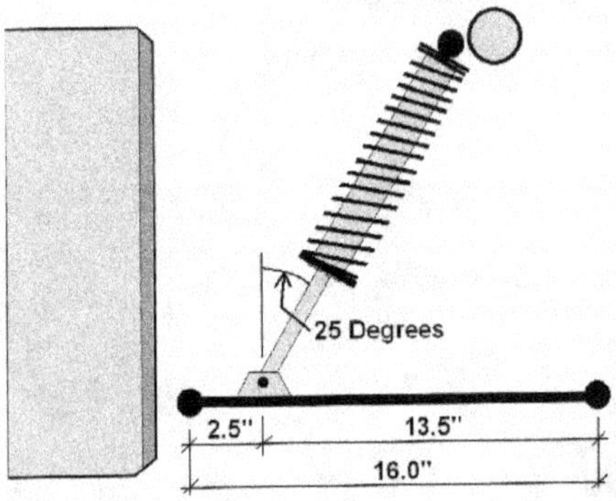

The shock that is a part of the coil-over system needs to control the spring on a one-to-one basis since it is mounted parallel to the spring and moves at the same speed and distance as the spring. This is much easier, unless we incorporate a bump, or dual spring system into the equation.

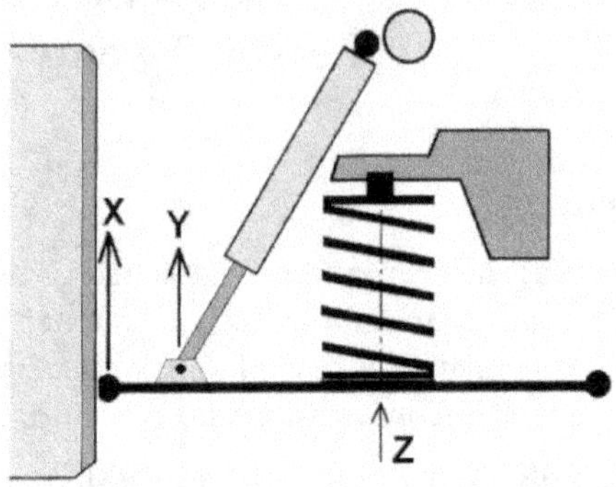

As stated in the first photo description, the shock in a stock system moves at a quicker speed than the installed spring and therefore can be setup to a lesser rate in order to control the stiffer spring moving slower. As the wheel moves at X pace, the Y end of the shock will move less and the Z center of the spring will move even less. The angle of the shock to the motion of Y reduces the shock speed even more.

When we talk about controlling the spring, we are referring to controlling the specific spring rate. Naturally the higher the rate of the spring, such when going from a 100 pound per inch rate to a 200 pound per inch rate, we would need to double our shock rate of resistance to rebound.

There is no set or average shock rates, only what will control a specific spring rate. So, if we install very stiff springs such as with road racing cars or formula type race cars, the shocks must have sufficient rebound rate to control the very stiff spring. In other words, the shock rebound rate must also be very high.

Compression Control - Conversely, if the spring rate is high, then that spring will resist compression at a high rate too. Our shock compression rate would be very small because the spring is doing all of the work resisting the compression motion.

Another example of high spring rate in a race car is the current trend of stock cars running on bumps. Bumps are devices that provide another spring that is higher than the ride spring, and in many cases, much higher. The bump can be either a bump stop using very stiff material or bump springs which are coil springs rated very high in pounds or kilos per inch.

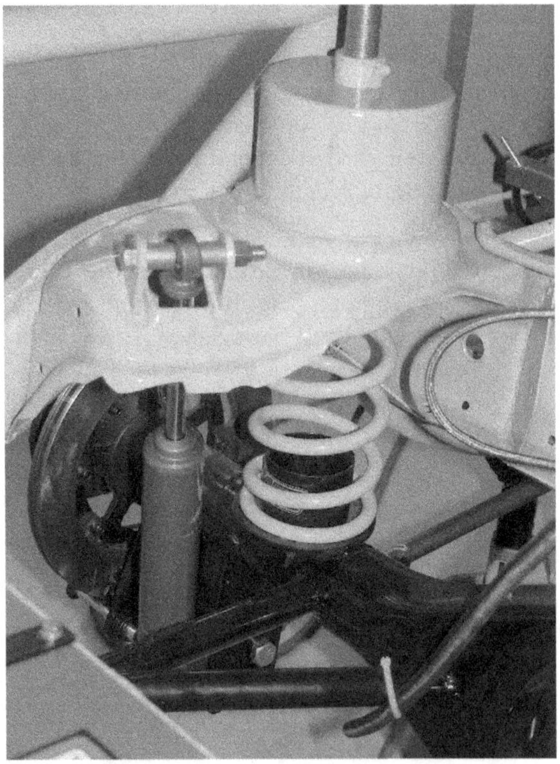

In this design, the shock is moving as if it were a coil-over shock, inline with the spring and at the same speed and distance in chassis movement. So, the team would look at the proper speed of movement for that type of track and then choose a shock rate of rebound resistance that matched the spring rate.

The shocks these teams use in conjunction with the bump setups have to control the high spring rate of the bump and so they are built to have a lot of rebound resistance. These shocks are wrongly referred to as tie-down shocks because they overcome the much softer ride spring and keep the chassis down on the bump device. In reality, even though they do hold the chassis down, their real function is to control that very stiff bump device and its tendency to rebound more aggressively.

So, we control a spring by using a shock that is rated in rebound resistance equal to the rate of the spring. In compression, the spring is doing most, if not all, of the work and therefore we need much less compression control in our race shocks.

This is a stacked spring. It works like this. The two springs when stacked and working free are a lower rate than either of the springs individually. The top spring has a higher rate than the bottom spring. When the shock has compressed a pre-determined amount as the car goes through the turns, the divider ring hits a stop and only the top spring is then in use. Since it is a higher rate than the combined rate of the two springs together, the overall spring rate on that corner of the car goes up. And the shock must control the higher rate.

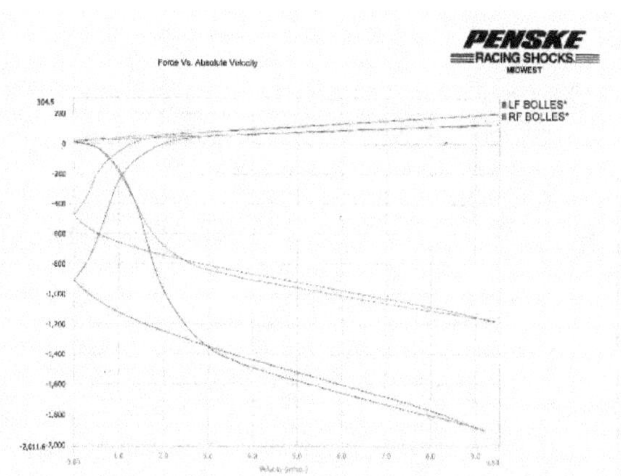

For bump setups, or stacked springs, as the spring rate spikes, so does the shock rate in rebound. The lines below the zero (0) represent the rebound settings and rate. This shock has a rate of resistance for the Right Front (RF) shock of nearly 500 pounds before the shock even moves. This is called "nose". This keeps the shock inside the spring rate range of the bump. Note that the left front shock has a nose rate of 900 pounds to control a heavier bump device. Note also that the compression rate, the lines above the zero on the left scale are next to nothing. The heavy bumps resist compression, so the shock doesn't have to do any work.

The Gale Force Load Pro measures the force of the spring as it travels through a range equal to what it does on the track. It records and shows the rate per each 1/16th of an inch throughout the travel range. In this way, the user can see exactly when the rate begins to rise and when it spikes as the bump nears maximum compression.

Exam - In The Context Of This Lesson:

The Primary Reason We Need Shocks Is?

1) Slow the motion of the chassis
2) Provide loading at one or more corners
3) To control the installed spring rate
4) To provide compression resistance

What Needs More Control?

1) The front end on braking
2) The rear on acceleration
3) Rebound of the springs
4) Compression of the springs

Running High Spring Rates Or Bump Setups Require?

1) Less rebound shock control
2) More compression shock control
3) Less compression shock control
4) More rebound shock control
5) 3 and 4

Race Car Technology – Level One

Lesson Twenty-Seven – Motion Ratio Affects Shock Rate

Glossary:

Lower Control Arm – A part of a Double A-arm suspension. In most cases, the springs are mounted to the lower control arms.

Coil-Over Shock – A shock where the spring is mounted over the shock body. The spring moves the same distance and at the same speed as the shock when the wheel moves.

Wheel Rate – The installed spring rate translated out to the wheel. The wheel becomes a spring rate as if it were an actual spring and it is always less than the actual installed spring rate.

Motion Ratio – A ratio that defines the difference in movement between two parts in a suspension. Common motion ratios of interest in a race car are: wheel to spring, and spring to shock ratios.

Over View – We now know how a shock controls the spring rate in the race car. We know the shock resists rebound and the spring mostly resists compression. But what if the spring and the shock are moving at different speeds? How do we compensate for the difference in speed of movement? That is what you are about to learn.

Although this spring mount has many motion ratios through the rockers, etc. the spring and shock are still in a one to one ratio. The movement of the shock/spring divided by the movement of the wheel equals the wheel rate motion ratio. On this type of car the shock could move less or more depending on the design, but the shock to spring ratio is still one to one.

In almost every race car, the wheel moves at a different speed than the spring and shock. If we want to control the speed of movement of the wheel, then we need to know what to do with our shock rates to accomplish that.

In this example, we have a shock that moves at nearly the wheel speed and a spring mounted well inboard of the wheel that moves much slower. So, the shock will be more efficient than if it were mounted to where it moved at the same speed as the spring. If it took a certain rate to control this spring on a one to one ratio of movement, then with the shock moving faster, and having a greater rate due to that speed, we can use a shock rated at a lower value to do the same work of controlling the spring.

Motion Ratio Explained – A motion ratio is when one part moves more or less than another connected part. In the case of springs and shocks, the wheel always moves a greater amount than the spring or shock. And the shock may move a different amount than the spring. If the spring were mounted on the shock, then the spring to shock ratio would be one to one. But the spring and shock are almost never mounted to where there is a one to one ratio to the movement of the wheel with one exception.

On some prototype and formula cars, there are rockers and pushrods that transfer the spring rate to the wheel. The ratio of the rocker could conceivably cause the spring and/or shock to move farther and faster than the wheel.

So, we can assume that for every inch the wheel moves, the shock and spring are moving either more or less. For stock cars, the closer the spring or shock is mounted to the chassis mount and away from the wheel (ball joint), the less it moves in relation to the wheel.

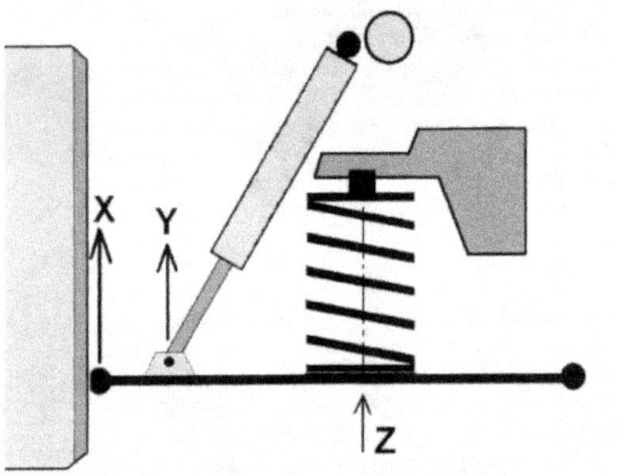

Here we see the actual condition described in the last photo. The wheel moves at X speed, the shock moves at Y speed, a little slower than the wheel, and the spring moves at Z speed, much slower than the wheel or shock. If we could find the motion ratio between the movement of the shock and that of the spring, we could multiply that number times the shock rating to find the rate needed to control the spring with this motion ratio. We will do exercises where we determine the exact shock rate to use in RCT Level Two.

For a shock that is mounted half way between the ball joint and the inner control arm mounts, the motion ratio for wheel to shock would be 2:1, or for every two inches the wheel moved, the shock would move one inch. Inversely, the shock motion ratio to the wheel would be 1:2, or one inch of shock movement to two inches of wheel movement.

This is an important consideration when we are designing our shocks to control wheel movement. And, if the shock is mounted at a different motion ratio than the spring, that again is a consideration when designing the shock to control the spring rate.

Which Direction To Go – For race cars running coil-over shocks where the spring is mounted on the shock body and moves at the same speed as the shock, we basically need to control the spring with the same shock rate as the spring rate for a certain predetermined shock speed. There can be variations, but the statement is basically true.

This is a very good example of how a spring moves at a different speed than the shock. On this car we call a swing arm, or Z-link, the spring is mounted onto the lower control arm. The device holding the spring is a slider to locate the spring and it is not a shock. The shock is mounted onto the rear axle tube on more of a one to one ratio of movement to the wheel. In this case as in the last example, the shock moves quicker than the spring and therefore needs less rate to control the spring than if it were a coil-over design.

If the shock were mounted away from the spring, then we need to study and know the motion ratio of the spring to shock movement. This design is common on stock cars where a large spring supports the car and is mounted to the lower control arm and the shock is mounted away from the spring and usually closer to the ball joint.

In this case, the shock would move a greater distance and at a higher speed than the spring. Since shocks gain rate with higher speeds, we would necessarily need to reduce the rate of the shock from a what we would require for a one to one ratio to something less.

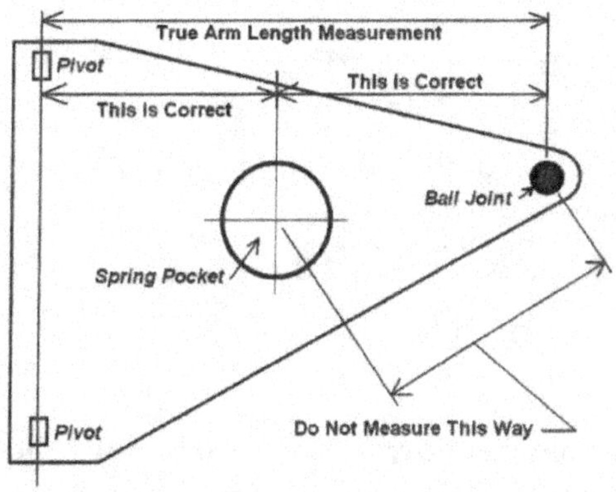

When measuring for motion ratio calculations, be sure to measure correctly. We want the length of the control arm to be the rotational radius of the arm. So, we measure the distance from the pivot line to the center of rotation of the ball joint at a ninety degree angle, as shown.

In some prototype cars, it is now customary to mount the shocks in a different location than the springs. Again, we must know if the shock is moving faster or slower than the spring in order to rate it to properly control the spring rate.

If the spring were moving faster, then the shock would need to be rated higher than the spring rate. If the spring were moving slower, the opposite would be true, we would reduce the rate of the shock.

We now know that motion ratios affect how the shock controls the spring and wheel movement. In designing a race car, we must know the motion ratios and how those affect the control of the spring rate.

Exam - In The Context Of This Lesson:

A Motion Ratio Is?

1) How far a wheel moves in the turns

2) The amount the wheel moves versus the spring

3) The ratio of shock to spring movement

4) 2 and 3

A Wheel To Spring Ratio Could Be?

1) One to one

2) 1:2

3) 2:1

4) Less than the shock movement

5) All of the above

If The Spring Is Moving Faster Than The Shock?

1) The rate of the shock must be less than the spring

2) The rate of the shock must be more than the spring

3) The motion ratio is 1:2

4) The motion ratio is 2:1

For Some Prototype Cars The Spring/Shock Moves Faster Than The Wheel Because?

1) The motion ratio of the spring to wheel is less than one to one

2) The rocker design motion ratio causes quicker spring/shock movement

3) The spring/shock is mounted away from the lower control arm

4) The push rod is mounted at an angle to the lower control arm

Race Car Technology – Level One

Lesson Twenty-Eight – Race Tire Basics

Glossary:

Tire Camber – The angle a tire makes from vertical relative to the racing surface. Positive camber is when the top of the tire is tilted to the outside from the chassis. Negative camber is when the top of the tire is tilted in towards the chassis.

Tire Temperatures – The readings of the core temperatures across the face of a race tire after it has been run a sufficient number of laps to reach operating temperatures.

Tire Contact Patch – The part of the tire that contacts the track surface. The area and loading of the contact patch is the most important consideration for race teams and everything else we do in race car setup and design is intended to maximize the tire contact patch size and loading.

Over View – This entire Course is about race car technology, defining the parts that make up a race car and how those parts fit together. The ultimate goal for race car engineers and crew chiefs is to find the most Grip so that the car will go faster through the turns.

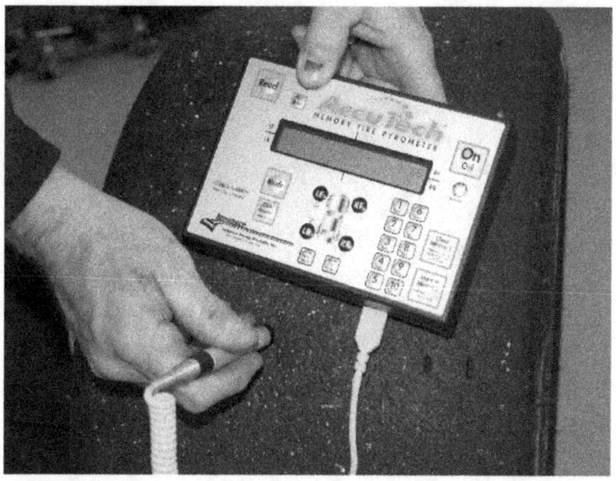

We evaluate how our tires are working on the race track by taking tire temperatures, as shown here, taking tire pressures and by evaluating the tire wear. We use this data to determine what settings to use for pressures, camber and setup that will produce the greatest performance.

In working with the race tires, we learn how to develop more grip for the whole car. Speed gained through lateral grip in the turns will be add to the average speed over the entire lap. That statement is probably the most important thing to understand and grasp of the entire course.

Tire Characteristics – Tires have more grip as they become more heavily loaded, or have more weight on them, and when the contract patch is made larger. So, we try to design our race car so that we can maximize the amount of loading on the four tires and provide the largest contact patches for those tires.

One important thing to understand about tires is this: Tires gain traction or grip when adding load in a non-linear way. If we double the amount of load on the tire, it gains less than double the grip. So, for a pair of tires on the same axle, or same end of the race car, the most grip we can obtain for that end of the car is when both tires are equally loaded.

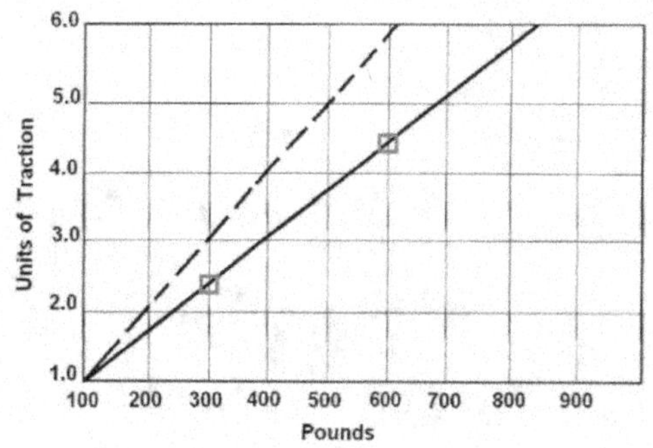

When we add load to a tire, the grip, or traction, the tire has does not increase at the same rate as the increase in load. The grip increases less. So, if we double the loading on the tire, the grip increases by less than double. This concept is one of the most important to understand and is the essence of understanding race car setup and balance.

For most race car applications, we cannot equally load the tires, so we try to get them to be the closest to equally loaded. With load transfer happening in the turns, the inside tires end up with less loading than the outside tires in almost every case. The more load we can retain on the inside tires, the more equally loaded the two tires will be and the more grip we will have.

As for contact patch, our tire pressures mostly dictate the optimum contact patch area that will distribute the loading in the most efficient way. Then we will be able to utilize the loading we have produced by designing the best setup.

Tire Temperatures – Tire temperatures tell us a lot about how the tire is working and how much grip we can expect from our tire contact patches. We usually measure the temperatures of the tires as quickly as possible after the car has completed enough laps to properly heat the tires.

The proper way to take tire temperatures is to read the temps in three places across the face of the tire with a tire temperature probe. This instrument has a needle like end that is inserted into the rubber to get the inside temperature. It should be inserted at a 45 degree angle into the rubber and at the edges about one inch inside the very edge of the inside and outside of the tire and at the middle of the tire.

The proper way to insert a temperature probe is to push it in at a 45 degree angle. Take the outside temperatures in from the outside edge of the tire approximately 1 to 1 ½ inches. Allow enough time for the reading to settle before recording and moving on. Then take a middle of the tire reading, and then an inside of the tire reading again about 1 to 1 ½ inches in from the edge.

Give the instrument time to react to the temperature of the tire. Once the reading stabilizes, record the amount and move on to the next reading. Don't be in too much of a hurry to do this because a lot of decisions will be made based on the temperature readings. They must be accurate.

Tire Pressures – For every type and use of a racing tire, there is an optimum pressure that will provide the best possible contact patch area and therefore the maximum amount of grip. Most race teams will use Nitrogen gas to inflate their race tires because it contains very little moisture. Moisture when heated expands and causes a greater increase in the tire pressures.

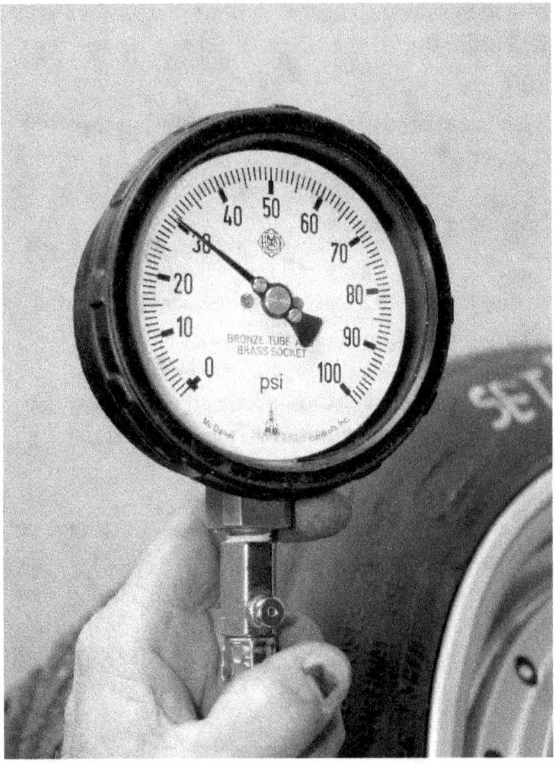

For taking tire pressures, always try to use a large dial gauge. Teams will sometimes set tire pressures in half pound increments and a larger dial makes this easier to do. Tire pressures should be taken a soon as the car is off the track. We can read the tire temperatures to tell if our tire pressures are correct.

When race tires are run at high speeds through turns, they heat up. The rubber heats the gas inside the tire and it expands. When it expands, the shape of the tire changes. Since our contact patch shape is so important, we want to make sure that when the pressures have increased that the final pressure will create the best contact patch shape. Excess gains in the tire pressures will cause a reduction in the contact patch size.

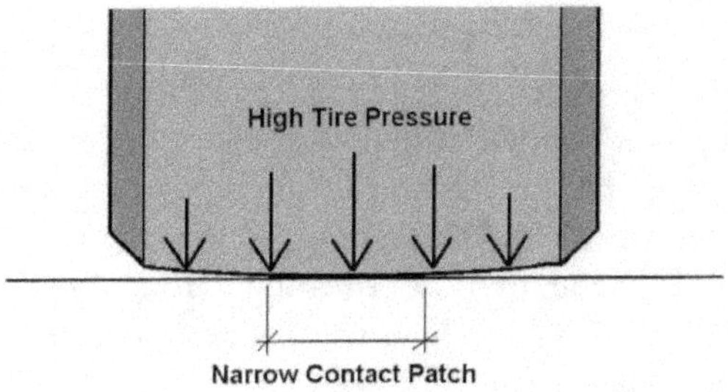

If the tire temperatures are too high at the middle of the tire, the tire pressures are too high meaning that the middle is working harder than the outsides of the tire. This causes a reduction in the size of the contact patch resulting in loss of grip. The middle temperature should be the average of the two outside temperatures.

We try to predict the pressure changes for each tire so that we can start out with lower pressures than optimum with the goal of ending up with the correct pressures and greatest area of contact patch. This takes time and experimentation because each car and track is different in the way the tires heat up and expand.

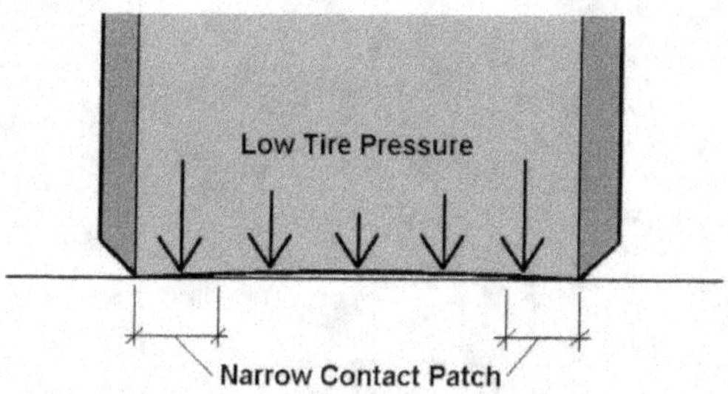

If the middle temperature is too low, the tire pressures are then too low. This also causes a loss of contact patch area and loss of grip. We need to add pressure until the middle temperature is an average of the outside tire temperatures. The problem is not that all of the width of the tire is not on the ground, it is the fact that whole area of the contact patch is not exerting the maximum pressure on the racing surface like it would be if the pressures were correct.

Running and Evaluating the Tires – When we run on a set of tires at race speeds, we need to look at three things to determine how well the tires are working. We look at the tire Temperatures, tire Pressures, and tire Wear.

Chart 1

Session #3

Left Front

Loc 12	Loc 11	Loc 10
146	154	150
Avereage Temp 150.00		
Pr. Cold	Pr. Hot	Pr. Gain
20	23	3

Right Front

Loc 1	Loc 2	Loc 3
211	204	207
Average Temp 207.33		
Pr. Cold	Pr. Hot	Pr. Gain
24	29	5

Front Average = 178.67

Left Rear

Loc 9	Loc 8	Loc 7
162	168	158
Average Temp 162.67		
Pr. Cold	Pr. Hot	Pr. Gain
20	25	5

Right Rear

Loc 4	Loc 5	Loc 6
216	214	210
Average Temp 213.33		
Pr. Cold	Pr. Hot	Pr. Gain
24	32	8

Rear Average = 188.00

In this chart of tire temperatures, we can see several problems. At the left front, the Loc 11 tire temperature is higher than the two outside temperatures for this tire. We would need to lower the tire pressure. At the right front, the Loc 2 temperature is cooler than the average of the two outside tire temperatures and this tire would need more tire pressure. As far as setup, the left front tire average temperature of 150.0 deg. Is cooler than the left rear average temperature of 162.67. This indicates a setup that is unbalanced and needs correction. In RCT Level Two we will tell you how to do that.

The last two are fairly easy to read for both asphalt and dirt teams, but the first may be difficult for dirt racers to get due to the way heat bleeds off the tire after a run. None the less, if possible, get all three to help understand all that the tire is doing.

Chart 2

Session #4

Left Front

Loc 12	Loc 11	Loc 10
148	152	157
Avereage Temp 152.33		
Pr. Cold	Pr. Hot	Pr. Gain
18	22	4

Right Front

Loc 1	Loc 2	Loc 3
212	208	204
Average Temp 208.00		
Pr. Cold	Pr. Hot	Pr. Gain
26	32	6

Front Average = 180.17

Left Rear

Loc 9	Loc 8	Loc 7
164	162	159
Average Temp 161.67		
Pr. Cold	Pr. Hot	Pr. Gain
18	23	5

Right Rear

Loc 4	Loc 5	Loc 6
214	212	209
Average Temp 211.67		
Pr. Cold	Pr. Hot	Pr. Gain
26	34	8

Rear Average = 186.67

In evaluating this tire temperature and pressure chart for a circle track race car turning left, we can see more problems. At the left front, the inside tire temperature at Loc 10 is hotter than the outside tire temperature. This is reverse of what we want to see for this type of car. The right front looks fairly good with the middle temperature about average of the outside temperatures and the inside temperature is higher than the outside temperature. Also, we see where the right rear tire pressure is higher than the right front meaning the car is a slight bit loose causing the right rear tire to be hotter.

The temperatures tell us how well the tire footprint is working. It would seem that the heat ideally should be close to even across the face of the tire. This all depends on the type of race car and where it is racing.

Chart Four

Left Front

Loc 12	Loc 11	Loc 10
163	163	162
Avereage Temp. 162.66		
Pr. Cold	Pr. Hot	Pr. Gain
17	21	4

Right Front

Loc 1	Loc 2	Loc 3
204	204	203
Average Temp. 203.66		
Pr. Cold	Pr. Hot	Pr. Gain
25	32	7

Front Average = 183.16

Session #6

Left Rear

Loc 9	Loc 8	Loc 7
164	163	159
Average Temp. 162.00		
Pr. Cold	Pr. Hot	Pr. Gain
17	21	4

Right Rear

Loc 4	Loc 5	Loc 6
204	202	201
Average Temp. 203.00		
Pr. Cold	Pr. Hot	Pr. Gain
25	32	7

Rear Average = 182.50

Reading this chart, we see what appears to be a perfect setup and tire cambers. The averages for each side are the same, the pressures for each side are the same, and the tire temperatures across the face of each tire are nearly the same. This used to be what we looked for, but more recent experimentation has led us to a different view of front tire temperatures that we will discuss here.

In modern stock cars racing on circle tracks, most teams prefer to see more heat on the inside (towards the inside of the race track) of the tire, but the progression of the heat should be steadily increasing from the outside to the inside in any event.

The higher heat on the inside part of the tire shows that it is more heavily loaded. When that part of the tire is more heavily loaded, it expands or flattens out. This creates a shape that produces a larger contact patch area than if the tire temperatures were even across the tire.

This scenario only works with high profile tires where the sidewalls are tall and fairly soft as opposed to low profile tires with stiff sidewalls. In road racing, where we see the later type of tires used, this practice would not apply.

How To Make Changes – The team should not make major changes to their tire pressures or front cambers based on the tire temps. until they have had a chance to run the car at full speed for more than five laps. We need to get maximum temperatures in the tires in order to see the true picture of what the tires are doing.

If the temps are lower or higher in the middle of the tire, it is an indication of lower or higher than adequate tire pressures. A low temperature at the middle means the pressure is too low and the middle of the tire is not pressing hard enough against the race track surface to create the ideal heat.

If the middle temperature is too high, that means the tire has too much pressure and the middle is working harder than the outside edges of the tire. We try to maintain an even spread of temperatures across the face of the tire, not equal temperatures necessarily.

The tire wear also tells us a lot about how a tire is working. If we measure the depth of the tire grooves or wear slots across the face of the tire, we can see if the cambers and pressures are correct. This is especially useful for situations where taking tire temperatures is not practical, such as when racing on dirt.

For some types of racing it is very difficult to take tire temperatures. So, we can take readings of the tire tread depth to get an indication of how the tire is working. More wear on one side of the tire verses the other tells us we need to make changes to the cambers. If the middle is wearing less than the outsides, then we need to increase the tire pressure to make the middle work harder and wear more as a result.

The depths can be read similar to how we read tire temperatures. If the middle of the tire is worn more so than the outsides, the pressures are probably too low. If the outside of the tire is wearing more than the inside, the cambers are probably too little and the top of the tire needs to be leaned in towards the inside of the turns.

Tire Cambers Vs Contact Patch Size – The final layout of the tire temperatures and the resulting tire contact patch shape and size are dependent on the tire camber and tire pressures. These two conditions work together to produce the contact patch loading for each tire.

Race teams are constantly experimenting by adjusting the camber angles and the tire pressures to find the best grip that will cause the greatest speed through the turns. The pressures cause the contact patch to be consistent in shape across the face of the tire. The camber angle optimizes the size of the contact patch.

Lesson Three tells us about cambers and camber change. This area of technology for race cars is one of the most important to understand in todays racing. More grip has been found by working with cambers and the geometry that maximizes cambers than with any other area of race car technology.

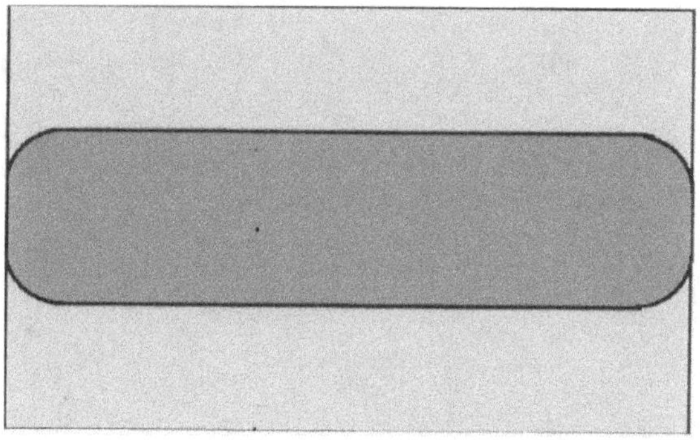

When we record even tire temperatures across the face of the tire, this is what the tire contact patch looks like. For road racing applications, this may be as good as it gets for low profile tires with harder sidewalls. For circle track racing and for any application where the tires used are taller and more compliant, there might be a better design.

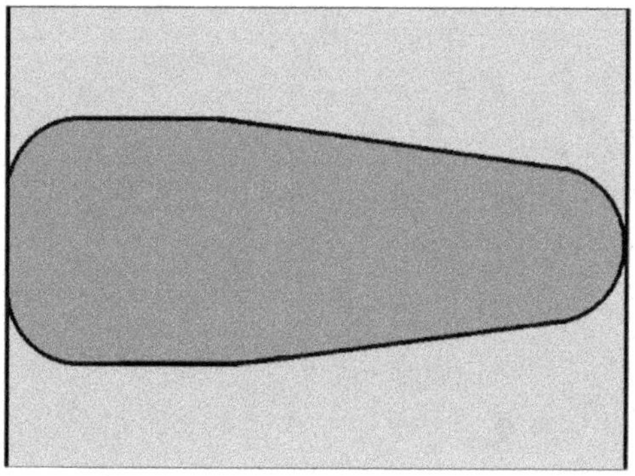

In the recent past for circle track race cars and formula cars with high profile tires and fairly compliant sidewalls, we liked to see the inside tire temperatures higher than the outside. This created a larger tire contact patch and helped the car to turn better. As teams experimented more with cambers, this process went a little farther.

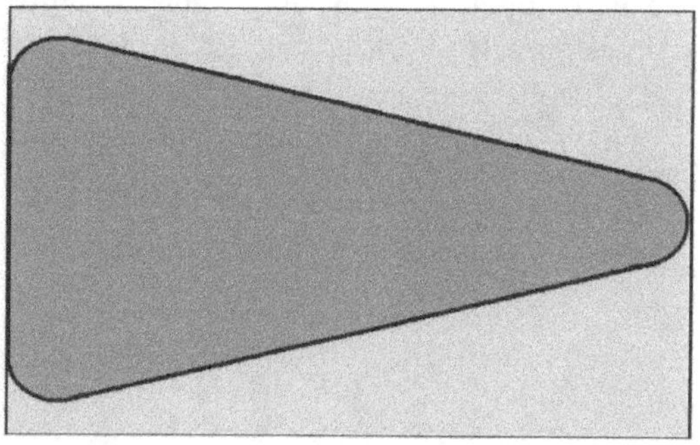

For tires with taller sidewalls that are more compliant, we can run much more camber than was thought possible in years past. This creates more loading and heat on the inside of the tire (towards the radius of the turn) which produces the contact patch shape shown here for a left turning car. This contact patch area is larger than what we would see if the tire temperatures were even across the face of the tire. It has a larger area and therefore creates more grip to help the car turn.

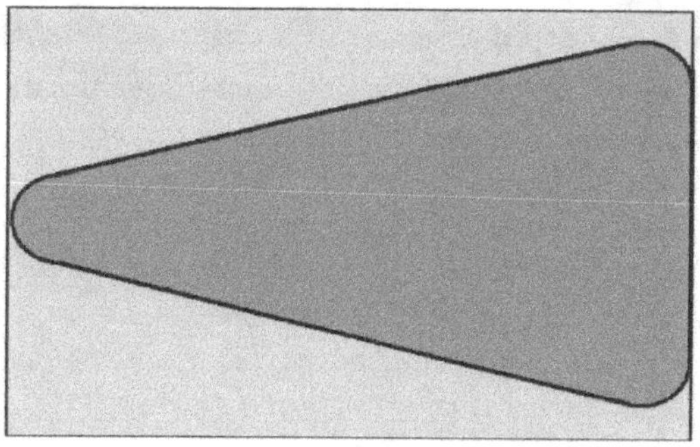

The rear tires on a straight axle suspension cannot be easily adjusted for camber angles. So, if there is no adjustment, how can we maximize the contact patch shape. If we allow a lower pressure to be run in a tire on a straight axle turning left, then we might end up with this contact patch shape. The temperatures must still be an even progression across the tire face with the middle being an average of the outside temperatures.

Storing Tires – Tires will lose their grip properties quickly with age. The chemicals that produce a softer tire compound evaporate starting right after they are manufactured. It is best to cover new tires with a wrapping and store them in a dark, cooler place to delay the aging process.

Used tires can always be run in practice or testing providing we understand that they will provide less performance than newer tires. If we know the difference, then we can accurately determine if we are competitive without needing to compare lap times directly.

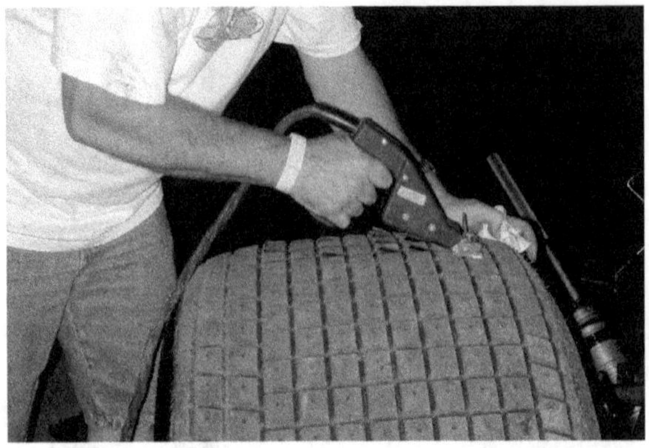

There is a lot more technology associated with racing tires and depending on the use and type of racing, more to teach about. In RCT Level Two and Three we will discuss in detail how each class of racing works with their tires.

Tires used for racing are softer than those used for road car applications. The compounds and chemicals that make this possible will evaporate over time. So, from the time a tire is produced until it is used, it degrades and becomes harder. A harder tire has less grip than one that is softer, to a point. A tire can be too soft and there is a point where maximum grip is reached and the tire will still hold up. These tires are wrapped and covered to prevent degradation and loss of grip properties.

Exam - In The Context Of This Lesson:

When We Double The Load On The Tire, We?

1) Double the grip
2) Create more than double the grip
3) Create less than double the grip
4) Can run stiffer springs

Tire Pressures Help The Tire How?

1) Develop the best tire contact patch
2) Keep the temperature of the tire down
3) Creates more tire temperature
4) Maintain equal loading across the face of the tire

Ideal Camber Settings Help The Tire How?

1) Bring equal temperatures across the face of the tire
2) Keep the temperatures lower
3) Create a larger contact patch
4) Keep the middle of the tire cooler

Tire Pressures Change Because?

1) The valve stem leaks
2) The track surface changes
3) The cambers are not correctly set
4) The tires get hot

Low Profile and Hard Sidewall Tires Need?

1) To have a hotter inside temperature
2) Lower tire pressures
3) Even tire temperatures across the face of the tire
4) Higher tire pressures

Tires With A Tall And Softer Sidewall Need?

1) To have a hotter inside temperature
2) Lower tire pressures
3) Even tire temperatures across the face of the tire
4) Higher tire pressures

The Best Time To Make Changes To Tire Settings is?

1) After the first run of the day
2) Right before the race
3) After getting the tires up to optimum temperatures
4) Once the air temperature stabilizes

The First Thing To Make Changes To Is?

1) Tire cambers
2) Cross weight
3) Toe settings
4) Tire pressures

Race Car Technology – Level One

Lesson Twenty Nine – Tire Stagger

Glossary:

Circumference – The distance, or measurement, around the outside of a tire to determine how far it would roll in one complete revolution.

Rear Differntial – A mechanical device that compensates for the difference in radius of the outside and inside tires when going through a turn and allows the two tires to turn at the correct speed.

Tire Stagger – A difference in circumference for two tires mounted on the same axle, or end, of the car. Stagger is a design function to allow for different radii of the inside and outside tires when going through a turn.

Spool – A mechanical device that locks the two axles in a rear end so that they turn at the same speed.

Over View – We learned early on in this course that the ultimate performance of a race car is dependent on how well the tires Grip the race track. We will now provide you with information on how we work with the tire stagger to gain the most Grip.

This information is primarily intended for race cars that only turn in one direction such as circle track race cars that run in the U.S. and other parts of the world. For road racing cars, we will most often select tires that are equal in circumference on each side at the same end of the car. So, stagger is not an issue with those cars.

Selecting tire sizes and predicting changes in stagger is the first step in working with race tires. For race cars that only turn one way, we utilize stagger in the rear end,, or whichever end drives the car, so that each tire will be rotating the correct speed to match the radius of the arc they will follow around the turn.

There are various rear differentials that race teams use to lock and un-lock the rear axles when entering (un-lock) and exiting (locked) the turns. When the axles are locked, the tire circumferences must be correct or one or both tires will slip and that end of the race car will lose traction.

On circle tracks, when going through the turns, the outside tires turn at a different speed than the inside tires if they are the same diameter. If the tires are connected to a solid axle system, they would fight each other at different speeds and cause handling problems. Because we want the tires to turn at the same speed, we need to introduce stagger, or a difference in diameter of the tires, to allow the tires to roll around their own radius easily. The outside tire would be larger than the inside tire.

Matching Stagger to the Race Track - One of the most important things to know is that rear stagger should match the race track. Using more or less stagger than the race track requires will cause loss of grip.

It was at one time common to try to fix handling problems by adjusting the stagger. This is not recommended and is considered a crutch, or false way to solve a race car handling problem.

The tire sizes cold and with the same pressures are marked on the tires in chalk for these circle track racing tires. Stagger is achieved by selecting different sizes and using different pressures for the left and right side tires.

Each circle track racetrack has a turn radius (average radius of both turns) that will require a certain rear stagger amount. That stagger number represents the difference in circumference of the two tires on the same axle, the left being the smaller one (for left turns), that would result in both tires turning the same number of revolutions through the middle of turns.

Type of Rear Differential - Stagger influences a car that is using a spool rear differential (both axles locked together at all times) on entry to the turns as well as through the middle and off the corners. With a "Detroit Locker" type of rear differential, one of the rear wheels is unlocked when entering the turns. The Detroit Locker will then lock both rear wheels under acceleration and stagger mostly affects handling from mid-turn and off the corners.

Less Stagger - Too little rear stagger will try to turn the car towards the outside wall and cause a push on exit. That is the two tires are trying to follow a larger arc that is projected to somewhere off the race track. If we can overcome this tendency, then the rear tires will fight each other and there will be loss of grip.

More Stagger - Excess stagger causes the tires to want to follow a smaller radius arc and because of the centrifugal force cannot force the front to follow that arc. Deficient stagger will usually show up as a loose condition because the rear tires are turning at different speeds, and one or both rear tires will slip causing loss of traction.

Important Considerations - So, we can see that correct stagger is a must. To arrive at the best tire sizes, we need to consider: 1) what is the correct stagger amount for the track we are running, 2) how much might each tire grow and how soon, and 3) How can I quickly solve a stagger problem using the four tires on the car.

Correct Stagger – This is the amount of stagger that will produce equal revolutions of the rear wheels when driving around the turns. For rear wheel drive cars, stagger is not as serious a consideration at the front of the car because the tires are not connected by an axle like the rear tires are.

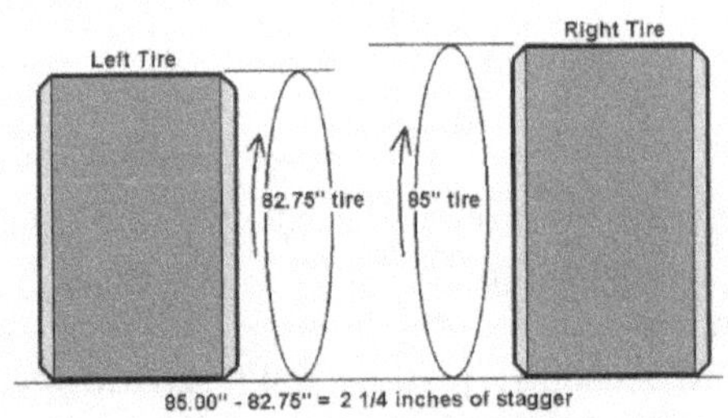

Stagger is the difference in circumference of two tires on the same end of the car. Take a measurement around each tire near the middle of the tread. Subtract the two measurements to find Stagger or "Roll Out" as it is sometimes called. Always inflate the tire to race pressures before measuring.

We can calculate what our stagger needs to be, or refer to charts that show the correct stagger for the radius of the turns, the track width of the two tires, and the average tire sizes.

Front Stagger - A definite consideration for front stagger is stability under braking. Because there is a lot of load transfer to the front under braking, the front tires and brakes work harder than the rears. If both wheels are turning the same number of revolutions, then the car will brake mostly in line with the direction the car is traveling.

If the Left Front wheel is turning faster (i.e. a smaller tire circumference and therefore more stagger) than required, the car may pull to the left on entry under hard braking. If the front stagger is too little, the car may not pivot as needed and develop a push on entry.

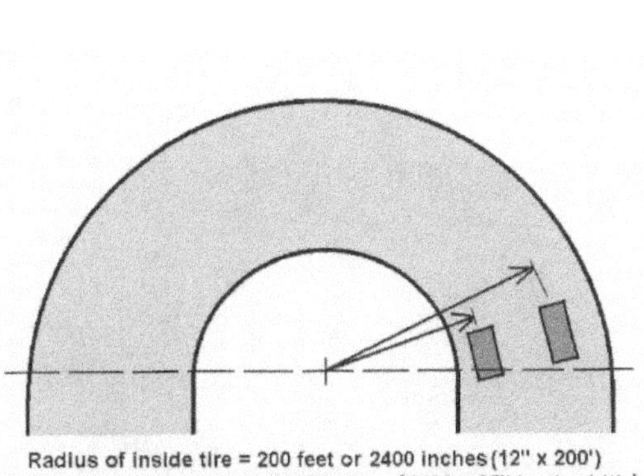

Radius of inside tire = 200 feet or 2400 inches (12" x 200')
Radius of outside tire = 2465 inches (2400 + 65" track width)
Outside tire travels 2465 x 2 x PI (3.1416) ÷ 2 = 7744 inches
Inside tire travels 2400 x 2 x PI ÷ 2 = 7540 inches
Outside tire = 85 inches in circumference
Inside Tire = 85 x (7540 ÷ 7744) = 82.75
Correct Stagger = 85 minus 82.75 or 2 1/4 inches.

This shows how to calculate the stagger the car needs in order for the inside and outside tires to rotate at the same speed and follow the arc of the turn. This is for a flat track and the track banking angle does affect how much stagger we will need. The banking reduces the stagger amount. To illustrate that, for a track banked 90 degrees, we would need zero stagger.

The general rule is to maintain a front stagger amount that is close to the correct stagger needed for equal revolutions of the wheels in the radius the wheels are running in. Remember that the radius on entry is much more than at mid-turn, so you will need less stagger than at the rear of the car.

Anticipating Tire Growth – Part of the "art" of tire stagger selection is being able to predict how much each tire will grow due to increases in pressure as the race progresses. As a rule, both of the right side tires (for a left turning car) will grow more than anticipated. But because tire growth is mostly influenced by pressure increase due to heat, it is possible that the left rear could grow more than the right rear tire if it is slipping and gaining excess heat.

Example: If we have a car with too little stagger in the rear, but one that turns well enough to overcome the tendency to push, then the left rear tire will slip and heat up. That is because, in most cases, the right tire is the more heavily loaded tire at mid-turn and is less likely to slip. One of the rear tires must slip, so it ends up being the left rear.

	Right Rear Tire Circumference								
Turn Radius	80"	82"	84"	86"	88"	90"	92"	94"	96"
100''	4.24	4.34	4.45	4.56	4.66	4.77	4.87	4.98	5.09
125'	3.39	3.48	3.56	3.65	3.73	3.81	3.90	3.98	4.07
150'	2.83	2.90	2.97	3.04	3.11	3.18	3.25	3.32	3.39
175'	2.42	2.48	2.54	2.60	2.66	2.72	2.79	2.85	2.91
200'	2.12	2.17	2.23	2.28	2.33	2.38	2.44	2.49	2.54
250'	1.70	1.74	1.78	1.82	1.87	1.91	1.95	1.99	2.03
375'	1.13	1.16	1.19	1.22	1.24	1.27	1.30	1.33	1.36
500'	0.85	0.87	0.89	0.91	0.93	0.95	0.97	1.00	1.02
1250' Daytona	0.34	0.35	0.36	0.36	0.37	0.38	0.39	0.40	0.41

This chart shows typical stagger amounts for a sample car with a 65 inch track width and running on a twelve degree banked circle track.

Handling Affects Stagger - Most of the time, tight or loose handling will determine which of the right side tires will grow the most. A severe push will heat up the right front tire and cause it to grow beyond normal. The rear tires may maintain correct stagger because neither of the rear tires is being abused.

A much more common problem is the tight / loose syndrome. That is when the car is somewhat tight, but with increased steering input, the car reverses handling balance at mid-turn and goes loose off the corner. The right rear tire suffers abuse as it slips and gains heat and pressure. Stagger will grow as a result.

To find stagger the easy way, follow this simple method that I learned from a friend many years ago, and you will never need to subtract tire sizes again. Measure the larger tire first. Then remember the number.

Wrap the tape measure around the smaller tire and note where the large tire stagger number falls near the end of the tape. Read the measurement from the end and you have the stagger. You read the stagger directly instead of having to subtract each time.

Working With Stagger – Stagger is a very important part of race car setup for those cars that need it. Understanding what stagger is and what influences it helps when it comes time to work with a real race car. But understand that every car is different and every track has its own characteristics, so finding the right stagger is mostly a product of time and experience.

Exam - In The Context Of This Lesson:

Tire Stagger Is?

1) When one tire is ahead of the other at one end of the car

2) The inside tire having a larger circumference than the outside tire

3) The inside tire having a smaller circumference than the outside tire

4) Not important for cars that turn the same direction

Tire Stagger Is Measured How?

1) By knowing the width of the tires

2) Measuring the track width, center to center of the tires

3) Knowing the radius of the track

4) Finding the difference in circumference of the two tires on one end of the car

Excess Stagger Results In?

1) Loss of Grip

2) A car that may be loose into the corner

3) A car that is loose off the corner

4) All of the above

Too Little Stagger Results In?

1) A push off the corners

2) A car that is loose through the middle of the turns

3) A hotter inside tire temperature

4) All of the above

Correct Stagger Is A Product Of Knowing?

1) The radius of the turns

2) The track width of the tires on the same end of the car

3) Average tire sizes

4) All of the above

Race Car Technology – Level One

Lesson Thirty – How Brakes Work

Glossary:

Brake – A system in a race car that slows and stops the car. Most every system in todays racing is a disc and caliper system.

Brake Fluid – A hydraulic fluid used in brake systems that is usually glycol-ether, mineral oil or silicone based. Brake fluid is not compressible.

Brake Disc – A disc of various diameters, thicknesses and designs that is mounted to the wheel hub and is grabbed by the brake caliper to slow the motion of the wheel and therefore slow the race car.

Brake Caliper - In a disc brake system, a clamping device that attaches to the spindle for a double A-arm suspension or the axle tube for a straight axle rear end. Fluid is forced by the brake master cylinders into the caliper to push pistons against the brake pads and the brake rotor to slow the car.

Brake Pads – In a disc brake system, a pad that is positioned between the caliper piston and the brake disc to apply friction to the disc to slow the car.

Brake Bias – The tuning of the amount of stopping force of the front and rear brakes. There are various ways to create this bias.

Cooling Vanes – Air slots cast within the rotor to direct air in and through the rotor to cool it. Air comes in at the center of the rotor and exits at the outer edge by the use of centrifugal force as the rotor rotates.

Over View – Speed in a race car is not only about accelerating off the turns and getting through the turns, but involves deceleration when going into a turn. The longer it takes to decelerate, the longer the lap times and the less performance the car has.

Brake systems for race cars must be designed for the type of racing they will operate within. A system designed for high heat and severe applications won't work very well in uses where light braking takes place over shorter periods of time. We'll explain all of that.

Just know that there are certain undisputed characteristics of race car braking systems that you must know. First let's describe a modern race car braking system.

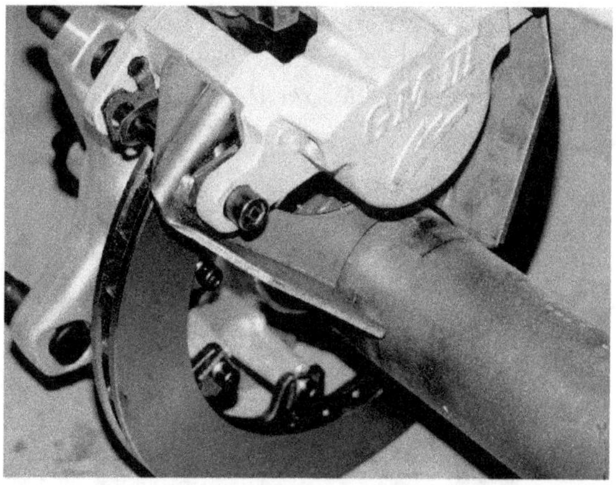

Most modern brake systems are disc systems. A disc is clamped by a caliper that holds brake pads. The pads have a friction property that grabs the disc slowing the car. The size of the caliper, the makeup of the pads, and the sizes of the master cylinders all help to create and determine the braking force.

Brake Components – Most modern brake systems for race cars are of the disc design. Most modern road cars use the very same systems with variations on the size and properties of the components. We first have a brake pedal that the driver pushes on when they want to slow the car. This pedal then pushes on a master cylinder mounted somewhere behind the brake pedal.

The master cylinder applies a force to non-compressible brake fluid through a brake line (tube that the fluid can flow through) into a brake caliper near the wheel.

The caliper has one or more pistons that are forced out against a set of brake pads. These pads are made of a material that has friction and rubs against a brake rotor, or disc that is attached to and rotates along with the wheel.

It is the clamping force on the disc that slows the car. The greater the force applied to the fluid, the greater the clamping force. The greater the friction the pads have, the more grip they will have on the disc and the greater the stopping force.

How Much Braking Force – We can either have too much or too little braking force for the weight of the race car and other factors such as tire grip, loading on the tires, track surface grip, etc.

For dirt tracks for example, if we use too much braking force, the tires will just slide across the track and not slow the car. This dirt surface provides very little grip for the tires. So, a lighter touch on the brake pedal and/or less grip in the brake system helps the tires to maintain grip when we brake.

For high grip surfaces like asphalt, if we have good tires, and aero downforce adding a lot of load to the tires, we can brake very hard and slow the car much faster. This is how it works with Formula One cars. They can decelerate from near 200MPH to 50MPH in just a few seconds.

Brake Master Cylinders – As the driver pushes on the brake pedal in the cockpit, a rod connected to a master cylinder is pushed and that pushes on a piston. This piston in-turn pushes on the brake fluid which is an oil based fluid that will not compress. This fluid passes through brake lines to the brake calipers.

Most race car braking systems use two brake master cylinders, one for the front brakes and one for the rear brakes. In this way, we can tune the braking force between the two ends of the car. More on that later.

Brake Calipers – As the brake fluid is pushed into the caliper, it applies a force to the pistons within the caliper. These pistons then push on the back of the brake pads to apply a force to the brake rotor to stop the car.

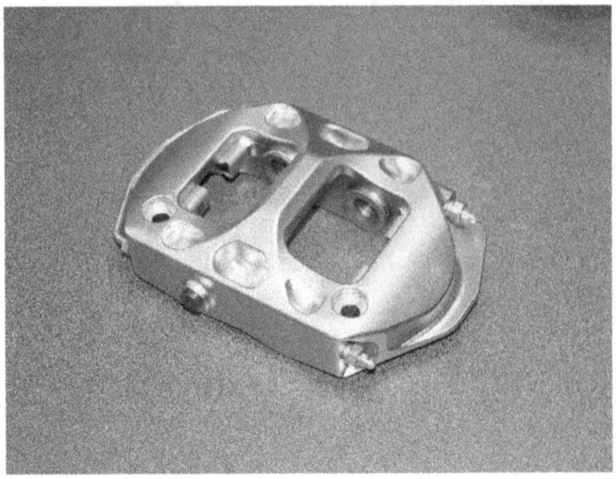

Modern race car calipers are one-piece designs that incorporate two pistons, or more, on each side to press two brake pads against both sides of the brake rotor.

Once the pedal is released by the driver, the fluid moves back into the brake lines and into the master cylinder waiting for the next application.

Brake Pads – Companies that make and sell brake parts have a wide range of brake pads available. The ones a particular race teams needs will depend on how they are to be used. Considerations include: 1) how much braking will take place for what period of time?, 2) What speeds are we braking from and to?, 3) How much grip does the track surface provide?

A brake pad that has a lot of grip might not survive high heat and repeated and prolonged use because we can think of them as softer. The material grips better because it sticks to the rotor better. Think of rubber and the grip it has verses steel.

If we put load on something that is coated with rubber and try to drag it across the floor, it will be very hard. If that same load were put onto something lined with steel, it would move much easier. In the same sense, the rubber would wear out quickly whereas the steel would last a long time.

In race car braking, the above example relates to the brake pads. We might want better braking force from a softer pad, but if it wears out before the end of the race, we end up with no brakes.

So, the choice of brake pads is a compromise between having more grip and wanting the pads to last. There is a happy medium for every type of race car.

Brake Rotors – The brake rotor is attached to the wheel and rotates with the wheel. To stop the car, the caliper with the brake pads must grab the rotor and apply friction that will stop it from spinning. That is how brakes work.

There are different thicknesses of rotors, different diameters for different uses, and different overall designs for the face (sides) of the rotor and the cooling vanes (slots within the rotor that route air to cool the rotor).

The brake rotor is attached to the wheel hub and rotates with the wheel. Note the cooling veins in the middle of the rotor. In this design, they are angled to the rear at the top. This left front disc is cooled by forcing air from the center of the rotor to the outer edge through the use of centrifugal force.

Rotors can be different diameters and the larger the diameter, the more efficient the braking system. That means, the caliper and pads can apply more stopping force to a larger diameter rotor. It is a product of leverage. The rotor size versus the tire diameter dictates the leverage force.

What is applying the force is the wheel and tire. The lever is the radius of the tire. It's length compared to the radius of the rotor creates a motion ratio. The larger the ratio, the more force the tire puts on the brake rotor. When the rotor is larger, the motion ratio is smaller and the tire has less leverage to counter the braking force and we stop faster.

Which Rotor To Use - We use thinner rotors for applications where we use less brakes. An example would be a car racing on a very fast track at a fairly constant speed like at Daytona. Even at Indianapolis or maybe Charlotte, where the cars must slow to go through the turns, the drivers use engine braking and not the actual brakes.

Thicker brake rotors are used for short track applications and road racing where there is constant and heavy braking that cause the rotors to experience a lot of heat and wear. It is in this application where we spend most of our time evaluating and designing brake systems.

Braking Force – The amount of braking force we will have in our braking system is dependent on the following factors: 1) amount of pressure applied to the brake pedal, 2) brake pedal leverage to the master cylinders, 3) the surface area of the master cylinder piston face, 4) the total surface area of the brake caliper pistons, 5) the brake pad surface area, 6) the friction properties of the brake pad compound, 7)the rotor diameter size, i.e. leverage ratio, 8) rotor surface friction, and 9) the heat generated by the system when in use.

The amount of braking force a driver can use is directly related to how much grip the tires have. If the driver over-brakes, then the tires will slide. The driver must apply as much braking force as is possible without causing the tires to lose traction.

The cutaway of a brake master cylinder shows the piston. The shaft to the right connects to the brake pedal and the brake line to the calipers comes out of the left end in front of the piston. The spring helps the piston return to the resting position after the braking is completed and the force is off the brake pedal. The brake fluid reservoir is on top over the piston and bore.

For race cars with high aero downforce, the braking force can be much higher in the initial phase of braking where the speeds and aero downforce is the greatest. As the car slows, the aero downforce becomes less and the loading on the tires also becomes less. Then, later in the turn entry when the tire loads are less, we must reduce the braking force to prevent the tires from losing their grip on the racing surface.

A race car transfers weight when it is decelerating, as when braking, and so the front tires will have more loading and grip than the rear tires in most situations. Therefore, we can use more braking force on the front tires than on the rear tires.

If we used the same amount of force for both the front and rear tires, one of two things would happen, either the front tires would not be braking to their maximum potential, or the rear tires would lose traction and slide.

If we design our braking system so that the front brakes will generate more braking force than the rear brakes, we can then maximize our braking system.

Brake Bias Explained – Because we cannot, in most cases, apply equal braking force to the front and rear tires, we must create brake bias, or a differential in braking force for the front verses the rear systems. The front brakes will have more force than the rear brakes in almost every case.

Brake bias can also be applied between opposing brakes on the same axle, or end of the car. On dirt cars, the designers recognize that the right front tire must always maintain grip with the track surface in order to turn the car, so they sometimes either reduce the braking force to the right front brake, or lock it out and eliminate any braking force to that corner of the car.

In a typical installation for most race cars, the three master cylinders are shown. The closest is the clutch master. The other two are the brake master cylinders, one for the front and one for the rear brakes. We can install different master cylinder bore and piston sizes for each master to create brake bias for the front and rear brakes.

This differential in braking force could also be applied to a road racing, or formula car where there are high G-forces in high speed corners that generate a lot of load transfer off of the inside tire. That tire cannot generate as much braking force as the outside tire without losing it grip on the racing surface.

If the designers could create a differential braking system that reduced the braking force to the inside wheel, then the driver could brake harder without causing brake lockup of the inside wheel.

How To Create Bias - There are several different ways to create brake bias. Since we want to apply as much braking force as possible, we can divide the force applied by the driver between the front and rear brakes.

Brake master cylinders come in different sizes as to the diameter of the pistons and the bore size associated with the piston size. The smaller the piston, the greater the force on the fluid for the same pedal pressure applied by the driver. This is a very important concept to understand.

Pressure in our braking system is measured in pounds (or any other measure of weight) per square inch. If we apply 50 pounds to a one square inch piston, we have applied 50 pounds per square inch of force to the calipers.

If we double the size of the master cylinder piston to two square inches, we will be providing only 25 pounds per square inch of force to the calipers. Again, the smaller the size of the piston in the master cylinder, the greater the force on the calipers.

So, we can use a larger diameter master cylinder for the rear brakes than we use for the front brakes to get the brake bias we need that will provide greater braking force to the front verses the rear brakes.

Another way to create brake bias is to use different sized brake calipers. The piston sizes and numbers in the calipers dictate how much brake line force will be applied to the brake pads. The force verses caliper piston size is the inverse of how the master cylinders work.

The larger the piston in the caliper, the greater the force on the brake pads will be. If our line pressure were 50 psi (pounds per square inch) and our caliper piston areas was 2.0 square inches, we would be applying 100 pounds of force to the brake pads.

If the caliper piston area were 3.0 square inches, we would be applying 3.0 times 50psi = 150 pounds of force to the brake pads. If we install larger calipers in the front than at the rear, we can create brake bias using the calipers.

When we install the brake calipers, we need to make sure that they are positioned correctly in height and that they are parallel to the disc rotor. Clamping the caliper to the rotor before welding the mounts ensures a perfect fit.

Brake Bias Balance Bar - Once we have established a good ratio of braking force for the front to rear brake bias, we can use what is called a brake bias balance bar that is attached to the brake pedal. This bar can be adjusted by the driver to apply a bias of force between the front and rear master cylinders.

This device is a fine-tuning tool only and should not be used to create the initial brake bias we need due to tire loading differentials caused by load transfer on braking.

Brake Fluid – One thing that is important to know about brake fluid is that it can absorb moisture from the air. Any water that finds its way into the fluid will expand and boil when the fluid heats up causing partial or complete brake failure.

Heat is a bi-product of braking. The primary cause of the heat is the brake rotor rubbing against the brake pad and heating to many hundreds or thousands of degrees. The rotor transfers that heat on to the fluid through the brake pads and calipers.

Race teams need to regularly change the brake fluid in the system and add fresh, non-contaminated fluid that does not contain water. Other regular maintenance includes inspection and changing out of the O-rings that seal the master cylinder pistons and the caliper pistons.

Exam - In The Context Of This Lesson:

The Amount of Braking Force We Can Apply Is Regulated By?

1) The grip of the tires
2) The frictional properties of the racing surface
3) The loading on the tires
4) The force applied to the pedal
5) All of the above

The Master Cylinder Provides More Force When?

1) It is larger than the rear master cylinder
2) It has a smaller piston
3) The brake fluid lines are larger
4) It has a larger piston

The Caliper Provides More Force When?

1) It is smaller than the rear master cylinder
2) It has smaller pistons
3) The brake fluid lines are larger
4) It has larger pistons

Brake Rotors Have More Stopping Power When?

1) They are smaller in relation to the tire diameter
2) The disc is thicker
3) The disc is thinner
4) They are larger in relation to the tire diameter

Why Do We Need Brake Bias?

1) Because of load transfer when accelerating
2) To keep the brake force equal on all four tires
3) To provide greater braking force on the rear tires
4) To compensate for unequal loading on the tires

Brake Bias Can Be Created By?

1) Using different sized brake calipers
2) Using different sized brake master cylinders
3) Using brake pads with different compounds
4) Using a bias balance bar
5) All of the above

Race Car Technology – Level One

Lesson Thirty-One – Summation RCT Level One

Congratulations, you have completed the Race Car Technology course, Level One. This final Lesson will outline all that we have accomplished in this course. In Race Car Technology Level One we have introduced you to all of the parts and pieces of the race car and talked about what each ones role is with the car.

We first outlined all of the goals of Level One. We proceeded to go through all of the systems on a race car and help you understand how each one works.

We first taught you about the Goals in setting up and racing a car and what we wanted to achieve in Level One.

We then told you what makes a race car turn. This is very important to understand because all we do is centered on first making the car turn well.

We described the various geometry part and pieces and how they might interconnect.

We discussed the steering systems and how things like Ackermann can affect the way the car turns.

We told you about Alignment and how this works with the other setup parameters to make the car perform at it peak.

We went through all of the Anti's in a race car and described how that works. We talked about springs, motion ratios, and the influence of springs.

Then we told you how shocks control the spring rates to control chassis motion and weight distribution. We described all of the parts of a modern racing shock and how they control movement of the shock, and in conjunction with that, the motion of the chassis.

You learned about racing tires, how pressures, camber, and temperatures affect the tires. Everything we do in race car engineering is intent on making the tires work as hard as possible under the existing conditions.

And finally, we learned about racing brake systems. The faster you can make the car slow, the faster you can go.

Level Two goes on to explain and teach how to work with all of these parts and pieces. Each of the suspension parts will need to be worked with and set so that their contribution helps make the car faster and more consistent.

Racecar Technology – Level Two

Lesson One – The Goals Of RCT Level Two

Now that we know all of the parts and pieces of a race car, knowledge we learned in RCT Level One, we can get on with actually setting up a race car. In RCT Level Two, we will, in the logical order of importance, be setting up a race car. The methods we will use and the order we will progress in is mostly common with all race cars from short track street stocks to Late Models, to Prototype road racing, and even with the Formula One race cars.

Race cars don't just come ready to race. You might be buying a brand-new race car, starting to work with a used race car you just acquired, or maybe you are building your own race car. In any event, you must go through the car and setup each system of that car properly and in a specific order.

No one gets away with not doing any of this. You cannot be successful in racing if you don't work with these setup routines. And even if the race car manufacturer or previous owner says it is good to go, you cannot trust anyone but yourselves. We must verify each and every setup parameter of the race car.

If you are racing now, or have watched others race and have asked how they do it to make their cars so successful, it isn't magic and no one every stumbles onto a winning setup by accident. No matter how easy it all looks, a lot of effort and smart thinking goes into every winning race car. Here is what you can expect from RCT Level Two.

What We Will Be Setting Up – In this Level Two Course, we will go from simple, but necessary functions like determining ride heights. We'll set the weight distribution, and establish proper caster and camber for your application. We'll go through all of the various alignment settings and carefully explain why we are concerned with each phase and how those will make the car perform.

We will establish what spring rates to use, which sway bar will match those spring rates. We'll match our shock rates to the spring and sway bar rates too. We will choose tires and tire sizes and learn how to select tires and prepare them for competition. Then we'll put this package to the test and fine tune it at the race track.

When you get done with this Level Two Course, you will be able to take a modern day race car and set it up. When I first started into racing, I could find no good information to tell me how to setup a race car. Even today, information comes at us in parts and pieces and it is very difficult to put it all together into a winning combination. Now you can. This school is designed to provide all of the necessary information to guide you through the setup of your race car.

Our ultimate goals are as explained in RCT Level One. For those of you who forgot, or those who chose to begin your schooling with RCT Level Two, we urge you to go back and take the Level One course. In our introduction to Level One, we explained what our overall goals are for designing a high performance, winning race car. It bears repeating, so we will again explain.

The race engineer's goals are very simple in concept, but more complicated to carry out. On any race course, be it with circle track or road racing, on a dirt or asphalt surface, there are key areas of performance where your car needs to be made better than any of your competition and those are the following all associated with increasing grip.

The goals for designing our race cars are basically the same be it a formula road racing car or a late model circle track car. Performance gains in the slower portions of the track cause all of the speeds to increase around the track including on the straights.

Maximizing Grip - When non-chassis elements like powerplants are more or less equal, Grip is what will help you win races and championships. Motors accelerate you, brakes slow you, but Grip makes you faster through the slower turn portions of the track and relatively small gains in lateral Grip will produce huge gains in speed and performance.

What are the Key elements of Grip? The tire contact patch is where Grip is produced. Causing the tire contact patch to be larger increases Grip and causing the most vertical loading on that tire contact patch are the two key ingredients for maximizing Grip.

Maximizing the amount of traction that is available from the four tires on a race car, any race car, will make you as fast as you can be, all other things being equal. Everything we present in RCT L1 will ultimately lead to optimization of the race cars Grip and the use of that Grip to go faster.

What are the basic parts that make up Grip?

- Loading On The Tire – The more load we can put on a tire, the more Grip that tire will have, period. But, this is gain in grip is not linear as we will explain later on. Load can come from the weight of the car, mechanical downforce from banking, and aerodynamic downforce, all of which will be explained in detail later on.

- Contact Patch Area – The greater the size of the contact patch, the more Grip. If we can find ways to make the tire contact patch larger, then that tire will produce more Grip.

- Tire Compound – The physical and chemical makeup of the tire can provide more Grip. The softer the material, the more Grip within limits. We have rules that govern the softness of the tires, but we need to stay very close to those limits.

- Load Distribution – A pair of tires on the same end of the car, or same axle, will produce more Grip when they are more equally loaded. The most Grip from a pair of opposing tires will come when they are equally loaded. There is a variation to this concept for dirt racing that will be addressed later on.

- Angle of Attack – What is called Angle of Attack, or Slip Angle, is when a tire is pointed slightly to the inside of the arc it is following through a turn. If a tire were to follow the exact tangent line around a curve or arc of the turn, it would not generate any side force to counter the centrifugal force.

In consideration of the other items that make up Grip, it is fair to say that none of those would be useful if it weren't for the creation of Angle of Attack. No matter what amount of Loading or size of the Contact Patch area is, or how soft the tire compound is, or how equal the load distribution between opposing tires, the car would not stay on the course without the tires developing an Angle of Attack.

So, there you have it. Those five things represent the parts that help make the Grip we seek to make us faster so we can win races. As we go through each part of the race car in the other Lessons, we will explain how to optimize of those parts to enhance our Grip and make us faster through the slow speed turns. And we will understand how that will in-turn make our whole lap faster at every point around the course.

Why Does More Grip Make Us Faster?

When a race car turns, a lateral force called Centrifugal Force tries to push the car to the outside of the turn. The tire contact patches resist this force. The speed we can drive through the turn is limited by the amount of Grip we have in our tires. The more Grip, the faster we can drive through the turns.

The one often overlooked benefit of achieving faster turn speeds is this: the faster you exit a slow speed turn, the higher the speed at which you will start accelerating down the straight part of the track in-between the turns. So, speed gained in the slow speed turns will be carried down the straights too. It's not just that we gain speed in the slow speed turns with more Grip, we gain everywhere around a race track.

To give you an example, on a typical or average length Formula One track, one tenth of a second is about 17 feet on the race track. Some teams are a full second slower per lap than the fastest teams. That is 170 feet per lap that the faster cars move ahead of the slower cars each lap. In a fifty lap race, that equals 1.6 miles, or 2.5 kilometers.

We often think of the formula one cars as being the ultimate race cars and perfect in every way. If that were so, then there wouldn't be a two second gap between the fastest and slowest cars. There must be a difference in designs that causes that huge gap in speeds. We like to think it is lack of mechanical grip that makes not only the slow turns faster, but the entire lap.

We already know that some teams get lapped in a F1 race, and the average length of those tracks is around 3 miles. So, those lapped cars are on average 2.0 seconds per lap slower than the winning team. And they all run with the same choice of tires. That is hard to fathom.

Is the gain realized by the winning F1 teams all Grip? No, as we have read, some engine packages are down on horsepower and that is a factor in the difference in lap times. But what about the teams who have the same engine package as the winners? Why are they so slow? It could be that they lack mechanical grip in the slower portions of the race track.

Maximizing mechanical grip is one of the most important areas of race car engineering, even over and above aerodynamic grip. The reason is this; To gain maximum aero grip, you must have high speeds and the most gains from mechanical grip happen in the slower turns where the speeds and the aero downforce is lower for all of the teams. So, the only thing we can point to as the reason for the gain in performance for the winning teams is Mechanical Grip.

The Concept Of Mechanical Balance - If your car has more overall Grip than other cars early in the race, that Grip might not stay superior and you might end up being a slower car through the turns later in the race. This is due to the Balance Factor. There are two definitions of balance. They are not the same and they cannot be confused. One is handling balance which is defined as a car that is neither pushing (understeering) or loose (oversteering). Pushing is when the front set of tires has less Grip than the rear set of tires. Loose is the opposite.

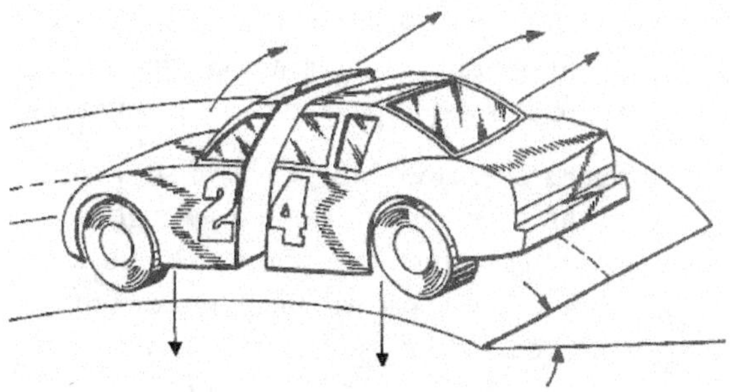

The Mechanical Balance concept is simple. We want both ends of the car to work in sync so that one does not try to cause the other to do what it does not want to do. When the two ends of the car are working in unison, we can achieve the goal of mechanical balance.

A team can attain handling balance fairly easily, but just the fact that the car is neutral in handling doesn't translate to speed or consistency. A car can have less overall Grip than other cars and still be neutral in handling. Although a neutral handling car is a goal, it is not the first and primary goal and not what we are discussing here.

The other balance is Mechanical Balance. Simply put, this is when the two suspension systems, front and rear, are working in sync and where the load transfer is predictable and maximized. This is what teams need to search for. The end result of all we do with chassis engineering must be towards the goal of Mechanical Balance.

Circle track race cars are complicated, but much easier to setup than a road racing car. This is because the setup in a circle track car can be asymmetrical, meaning the spring rates, cambers, etc. from side to side can be different. The circle track car is only turning in one direction and so the setup need only be correct for that direction of turning.

The Search and Desire for Mechanical Balance should be the foremost goal for every race engineer. A lot of good things come with achieving MB. The tires do not work as hard and will last longer. The car is much easier to drive. The car maintains higher turn speeds as the tires wear later in the race, or later in each stint for cars that are allowed to pit and take on new tires.

So, to reiterate what we have covered so far, our goals for this course are to gain knowledge, understand the terminology, and to learn how to setup a race car to maximize Grip in a way that utilizes a Mechanical Balanced state. In the Lessons to come, we will be presenting information you will need to understand and apply principles that will help achieve these two important goals.

And we are not leaving aerodynamic downforce out of the discussion, we just need to separate the mechanical Grip from Aerodynamic enhanced Grip. We need to do this because, many times in the search for the maximum aero properties, a team might deviate from the goal to achieve a Mechanical Balance. We think you can have the best of both worlds.

Exam - In The Context Of This Lesson:

The Primary Goal For Setup In A Race Car?

1) Make it easier to drive
2) Increase grip
3) Balance the two suspension systems
4) Create less chassis roll
5) 2 and 3

Having More Grip Makes Us Faster?

1) On entry to the corner
2) On exit off the corner
3) Through the middle of the corner
4) Down the straights
5) All of the above

Some Basic Elements Of Grip Are?

1) Contact patch optimization
2) Load Distribution
3) Tire angle of attack
4) All of the above

Mechanical Balance Is?

1) Drive line angles
2) Tires are balanced
3) Suspension systems are working together
4) Weights are correct

Racecar Technology – Level Two

Lesson Two – Weights and Placement Goals

Weight and Weight Placement Goals – Part of the design of a race car is the establishment of corner weights, or how much of the weight of the whole car each tire ends up supporting. We use weight scales to measure the weight, or loading, under each tire. The amount of Static Corner Weight each tire supports statically is critical to how load ends up on the four tires through the turns.

Race teams measure weight with and without the driver in the car. If they just want to maintain certain percentages after making changes, they don't really need the driver in the car, they just need to know what the percentages are with and without the driver.

Professional teams weigh their cars the same way we do. This Cup car at a test session at Daytona International Speedway is up on stands that are adjustable for height in order to make the scales level and on the same plane. The weights are critical to the handling of the car and are specific to the overall setup so that the load distribution on the four tires through the turns is ideal. These teams establish their weights and percentages with the driver in the car, and then record what they are without the driver. This way they don't have to chase down the driver in-between runs in order to get the baseline weights back after a spring or other change. The percentages taken without the driver are representative of those as if the driver were in the car.

We need to establish and record these categories, or divisions, of weight: 1) Total Vehicle Weight, 2) Front and Rear Percentages of the Total, 3) Side Percentages of the Total, and 4) Diagonal Weight, also known as cross weight or wedge/bite in dirt racing.

If we are doing design work and computing for setups or calculating weight transfer, the entire weight of the car, including the driver, must be used because that is the weight the car will compete at. A race car will lose some of the total weight, change its cross weight and front to rear percent when the driver steps out of the car.

Common wheel scales are used to find the weights, or loading, on the four tires. These individual weights are then calculated into percentages of: Front and Rear, Left and Right Side, and Cross Weight, or percent of the total vehicle weight that the RF and LR tires support. If we wanted to find Left Rear Weight, as is a common designation for dirt teams, we would subtract the RR weight from the LR weight.

Legal Weights – Most sanctioning bodies and all tracks have weight requirements. The cars must weigh a minimum weight with, or without, the driver, have a legal left or right side percent maximum or minimum, depending on which way you are turning, and sometimes a minimum rear percent.

So, the first task in developing a weight plan for your car is to place any moveable weights in the car to meet the overall minimum total weight, and the side and front to rear percentages of that total. You might find that you need to move weight around in the car to meet those rules. Then later on you can adjust your corner weights to adjust for cross weight percent.

Weight Distribution Related To Type Of Racing - In road racing, the side percentages are usually the same, 50-50% as is the diagonal percentages being 50-50%. This is because a road racing car must turn both ways, left and right, and therefore must have a more symmetrical weight distribution.

In circle track racing, where the cars turn only one way, left in the U.S. and sometimes right in other parts of the world, we can create different side percentages and weights that are different from 50% diagonal, or what is termed cross weight.

In circle track racing, the use of the term Cross Weight gives us an indication of the load distribution on the four tires. It is defined as the combined load that is resting on the Right Front and the Left Rear scales added together and then divided by the total vehicle weight. That math gives us a percentage number to relate to, i.e. "51.4 percent of cross".

Road racing cars must turn both ways, left and right. Therefore, the cross weight percent must be 50-50%, or the same for both diagonals. If it were different, then the load distribution for left turns would be different for right hand turns and the handling would be different.

Dirt Verses Asphalt Terminology – For asphalt cars, the percentage number for cross weight is used, and for dirt cars, the terminology might be a little different. Most dirt teams refer to the amount of "bite" or "left rear weight" in the car, which is the number of pounds of weight that the LR tire supports over the RR tire. By subtracting the amount of weight that the RR tire supports from the LR tire weight, we arrive at a number and call that the amount of bite in the car, i.e. "100 pounds of bite or left rear".

Nonetheless, that bite number has a cross weight number associated with it. A low bite number relates to a cross weight around 50%, and a high LR weight or bite number puts our cross weight up around 56-58% most of the time. Coincidentally, low and high cross weight numbers in asphalt racing are in the same basic range.

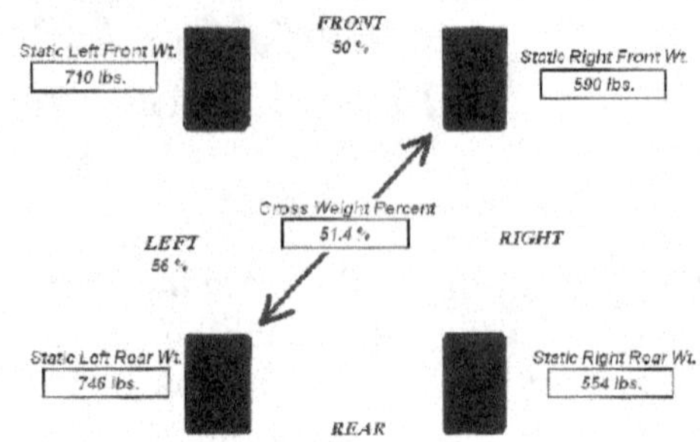

This is a typical layout of weights and percentages for an asphalt late model race car. The total weight of the car is 2600 pounds. The front to rear percent is 50% and the left side percent is 56%. This car running on a medium banked 1/3 to ½ mile track would require a cross weight of 51.4%. That is calculated as the weight distribution needed in order for the correct weight distribution to be placed on the four tires at mid-turn. If we can calculate the weight transfer during cornering, then back out what we need for static weight distribution in order to achieve those dynamic weights, then we would end up with this static cross weight for this car. The Left Rear weight for this car would be 746 minus 554 = 192 pounds of LR weight. Most dirt late models run a higher rear percent than 50%, so this does not represent a true LR weight for dirt cars, just an example of how to get that number.

How Cross Weight Changes - It is important to note that if we increase the load supported by one tire, say the LR, we also increase the load supported by the diagonal corner of the car, or the RF tire. As we make that change, the opposite takes place with the other diagonal in that those tires (the LF and RR) experience a decrease in the amount of load they support.

So, for example, when we increase the Bite in the car (left rear), we also increase the loading on the RF tire, and at the same time decrease the loading on the LF and RR tires. You cannot change one tires loading without changing the loading on the other three tires.

In The Past – Historically, these numbers meant little to the circle track racer related to different setups, different types of race tracks, or overall weight distribution. All we knew was that a certain range of cross weight or bite worked for certain conditions.

Today we have an entirely new understanding of cross weight and the use of bite as those relate to the setups of the cars. Beyond knowing generally that more cross weight makes the car tighter at mid-turn and less tends to loosen the car, we can now know exact numbers to be used with a balanced setup.

For simplicity, we will refer to the weight distribution of the cars, whether dirt or asphalt, as cross weight. Dirt racers as well need to understand the relationship of changing the LR weight and how that affects the other three corners of the car.

When we weigh the car at the track, the scales there may not be on the same plane or level. The readings for cross weight are often different than what we read at the shop on pads that have been corrected to be on the same plane. So, we can usually rely on the front to rear percent, the side percent and total, but not cross weight percent. So, just record what the track scales tell you for cross as a reference.

Remember that the load distribution in the car changes as we encounter lateral forces associated with cornering. Some of the load carried by the inside (the turn) tires will transfer to the outside (outside of the turn) and the overall load distribution will change. We need to set the static weight distribution in anticipation of this load transfer so that the four tires will have optimum loading through the turns.

Multiple Ranges - There are different ranges of cross weight percent that will make your car neutral in handling. The same car at the same race track may be neutral in handling in the range of 48 to 52 percent and also be neutral in a range between 58 to 60 percent of cross weight. Dirt racers often change the LR from less than 100 pounds of bite to upwards of 300 pounds of bite. That represents a range of from 48 percent cross to 60 percent cross weight, the same as with the asphalt cars.

The actual cross weight, or overall weight distribution, number that will work for your car is dependent on the front to rear weight distribution, the type of track and the setup. Generally, the flatter asphalt race tracks with less grip work better using the high range of cross weight percent with the associated high left rear weight, while the same car will work better in the lower cross weight range at the higher banked asphalt tracks.

Many teams who run on asphalt tracks will try to run the high cross weight range that they have successfully developed for flat tracks when they go to race at the higher banked tracks. They would do much better as far as being more consistent by switching to the low cross weight range on the high banked asphalt tracks.

Cross Is Related to Front to Rear Weight Distribution - The exact amount of cross weight that will make your car neutral is directly related to the front to rear weight distribution. The greater the rear percent, the more cross weight a car must have to stay neutral in handling. Again, this is directly related to the anticipation of weight transfer so that at mid-turn, the weights on the four tires are correct.

Cross Weight Related to Front Percent

Front Percent	46%	48%	50%	52%	54%
Cross	59.1% X	55.6% X	52.1% X	48.6% X	45.0% X
Rear Percent	54%	52%	50%	48%	46%

This chart is a general indication of how the ideal cross weight will change when we change the front to rear percent. These numbers are for a particular car and not necessarily right for every car. But they do provide an interesting view of the relationship. One very interesting thing to note is that with a race car with 50% cross weight, we would need to run a rear percent of 49.4%. If a road racing car ran anything but 49.5%, then the handling would be off if that car were dynamically balanced.

For dirt racing, with the associated higher than 50% rear weight distribution, the range of cross weight that will make the car neutral is opposite of that associated with asphalt cars. The high range is more normal to the high rear percent cars, while the low range also can be also yield neutral handling.

We adjust our weight distribution by adjusting the loading on the springs, or the spring height. This coil-over setup has a threaded body on the shock and a ring that can be adjusted for height to compress the spring to provide the correct loading. Changes in spring loading will also change the ride height of the corners of the car. Changes in wheel weights must be made at all four corners of the car at the same time. We'll explain how to do that.

Sway Bar Effect - We can cause the static Cross Weight to change by pre-loading the sway bar. Pre-load is when we twist the bar at static ride height through an adjuster mechanism. Pre-load on the sway bar puts load onto the RF and LR tires which increases the amount of static cross weight percentage. The larger the sway bar, the greater the effect of pre-load on the car.

If we want to pre-load the bar, we should know how much the cross percentage will change, then lower the static cross weight percent by that amount. Then when we pre-load the bar, we will have the exact static cross weight we desired and want, not something higher. The pre-load on the sway bar mostly helps provide bite off the corners and shouldn't be used to tighten the car through the mid-turn phase.

Weights Can Indicate Balance – When you run a setup that is balanced dynamically, that means the front and rear are working together. This also means that the load transfer is predictable and your car will need a predetermined cross weight percent in order to end up with the correct weight distribution at mid-turn. The cross weight percent we need to run in our cars to make them neutral in handling can be an indication of how balanced our cars are.

If the ideal cross weight for a particular application that is dynamically balanced is 51.4 percent, and we find that we need to reduce cross weight to 49.5 percent in order to be neutral in handling, then our setup must not be truly balanced.

As we change our setup from unbalanced to a more balanced setup, we will also need to change the cross weight percent to what works best for that dynamic balance. If your team insists on using a fixed cross weight percent that you "always ran", then your setup cannot be improved until you decide to change the cross weight along with positive chassis setup changes. Weight distribution and balance go hand in hand.

Weighing the Car In The Shop – When we setup our scales at the shop, or wherever we do it, we must level the scales. If the floor is relatively level, say in a commercial building, things will be easier. If you are in a garage intended to house a car, the floor will usually slope towards the garage door. This allows any spilled water to flow out the door and not accumulate.

In today's racing, it is necessary to have a set of scales and a good set of stands in order to maintain your wheel weights. Note that if you make changes in spring rates, camber settings, sway bar pre-load, etc. those changes will also change your race cars weight distribution.

It is less important for the scales to be level front to rear, or side to side. It is extremely important for the scales at each axle to be on the same plane, just like we talked about in the Ride Height Lesson. So, if we have put the scales on the same plane and the front to rear were off level by less than an inch, it's not the end of the world. Side to side we can be off level by up to a half inch and be alright.

Weigh the car with all of the fluids and driver weights it will have in it when it is on the track. Many teams will simulate driver weight by putting lead in the driver's seat, but when doing this, you should put about 30% of the lead in the foot well and the other 70% in the seat. And you can place lead on top of the fuel tank to simulate a full tank. Fuel weighs about 6 pounds per gallon.

Bring all of the tires up to race pressures because they will grow differently front to rear from cold pressures and affect the height of the wheels which will affect the loading on the scales. And make sure the front wheels are pointed straight ahead.

If you have high rebound shocks on any corner of the car, wait till they stabilize before you take readings. For adjustable shocks, reduce the rebound all of the way while you are scaling the car. Disconnect the sway bar while you are taking scale readings. You can load and/or preload the bar later on. Now you can read the scales.

How To Change Weight Distribution – Changing the loads can be done one of two ways. Either you want to move or add weight, or you want to redistribute the weights that are already in the car to different wheel loadings. Moving weight or adding weight will change the loading on the four tires and change the front to rear percent, side to side percent and the cross weight percent.

If you want to change load distribution, but not side or front to rear percent or the total weight, there is a way to do that and not change the ride heights appreciably. I say that because there will always be a little tweaking needed to fine tune your ride heights after you have changed load distribution.

If you need to increase the cross weight (RF and LR increased as a percentage of total weight of the car), you will force down the right front and left rear springs and ease up on the left front and right rear springs. The most common device for doing that is a screw jack or adjuster rings on a coil over shock.

So, if the springs are similar in rate from side to side like at the front, you can add turns to the adjuster at the RF in the same amount as you take turns away from the LF spring. This same process can be used on a road racing car where the rear springs are the same. At the rear on a circle track car, the rear spring rates are usually different side to side. If you divide the softer spring rate by the stiffer spring rate, you get a percentage number.

If the LR were a 200ppi spring rate and the RR were a 400ppi rate, the percentage would be 0.50. If you want to turn the LR spring down by four turns, you would turn the stiffer RR spring the opposite direction by the percentage times that, or up two turns. That is because the same number of turns on the stiffer spring will move that corner twice as much as that number of turns on the softer spring.

Try to record how much the cross weight changes per turn of the adjuster as you make changes to the front and rear spring loads. That way, when you want to add cross at the track, you'll know exactly how much you need to turn your adjusters for the percentage of change desired.

After you have established your desired weight distribution and if one end or the other is higher or lower than the established ride height, just turn the adjusters for each spring on that end the same amount to get the ride heights back. That way the cross weight won't change.

If you follow this rule, your ride heights will remain close to what they were when you started. Now you can hook up that sway bar. If you want to preload the bar, do it now and see how much cross percent it adds per turn. Then you can make adjustments to return to the original cross weight if you add turns at the track.

Bump Setups and Cross Weight - If you are running the bump setups, when the car goes down onto the bumps, there is a good chance that your sway bar will gain or lose loading and affect the load distribution on the four tires. This would change your handling if not corrected.

Set the bar to neutral at ride height. Push down on the car till it is on the bumps and then note how many turns it takes to go to neutral again. If this gain or loss is not accounted for, then when we race the car with the front or rear down on the bumps, our weight distribution will be something other than what we want and need.

Also with bumps, if the bump device is not spaced correctly from the shock, when the two front shocks contact the bumps, the cross weight may change significantly. It is a good idea to remove the front springs and let the bumps support the front of the car. Then check the cross weight on the scales. If it is different than when on the ride springs, add or take away packers, or spacers, till the cross weight is the same as when you set it on the ride springs.

Weighing At The Track – The officials track scales are almost never level. Get your weight percentages at the shop and trust those numbers. When you get to the track, re-weigh the car on the track scales and note the difference in cross weight percentage and maintain that difference.

The front to rear and side to side percentages should not change much at all because we stated at the top of the article that it would take more than an inch or so of difference in level side to side, or slope in the floor, to make a noticeable difference in readings. It is with the cross weight percent readings that the unlevel scales really show a difference from your shop scales.

Summary – Before getting into the setup of the car, we need to establish our race car weights being: 1) Total Weight to meet the rules, 2) Front to rear and side to side Weight percentages, 3) Weight distribution on the four tires, or cross/bite numbers.

We need to maintain those numbers when we change cambers, pre-load the sway bar, make spring changes and/or scale the car at the track. Consistency starts with maintaining the weight distribution in the race car.

Exam - In The Context Of This Lesson:

Weight Distribution Includes Which?

1) The weight on the four tires

2) The front to rear percent

3) The side to side percent

4) The diagonal weight percent

5) All of the above

Before I Set Cross Weight, I Need To?

1) Set the total weight of the car

2) Set the front to rear percent

3) Set the side to side percent

4) All of the above

We Set Cross Weight So That?

1) The cross equals the left side percent

2) The cross weight is the same as the rear percent

3) The cross weight is 50-50 for road racing

4) The cross weight percent causes ideal dynamic tire loading

5) 3 and 4

"Bite" is The Same As "Cross Weight"?

1) True

2) False

The Following Can Change The Weight Distribution?

1) Sway bar pre-load

2) Front end travel

3) Travel onto the bump stops or springs

4) Camber Changes

5) All of the above

Lesson Three – Geometry Settings

In this Lesson, we will talk about double A-arm geometry, and on circle track and some road racing cars with rear solid axle suspensions, this will be only on the front end. For formula road racing cars, it will be front and rear. Geometry can cover many different aspects of the construction of the chassis, but for the purpose of this Lesson, we'll only be involved with what is historically called the roll center, or what I always like to call the Moment Center.

The reason I call this point the moment center, as stated in RCT Level One, is because it represents the bottom of the moment arm for the AA-arm suspension, being the front in a circle track or road racing car, and possibly the front and rear in a formula car.

The moment center, or what is typically called the roll center, is a point formed by the angles of the upper and lower control arms. The location of the Moment Center moves as the chassis dives and rolls through the turns because those actions cause changes to the control arm angles. We take measurements from the ground and from the centerline to find the location of the MC.

Theories About AA-arm Geometry - After much research and testing over the recent past by myself and others, it has been determined that the primary use and function of moment center design is to optimize the camber change and contact patch size in a AA-arm suspension, as well as to influence the roll resistance.

Other theories exist that talk about Jacking Force and that has also been tested. The conclusion the testing reached is that the control arm angles that produce the best camber change and contact patch area for both the inside and outside (the turn) wheels are basically the same for moment center and jacking force theories.

In RCT Level One, we told how the moment center is arrived at and what makes up the parts and pieces of roll/moment center geometry. We won't re-speak that Lesson here and if you for some reason did not enroll in the RCT Level One course and don't know how the moment center is arrived at, then I urge you to go back and take that Course.

What we will be learning here in this Lesson is how to design a better moment center and what works for different types of race cars. I can tell you through many years of research, consulting and testing that what you will learn here is not widely known in the industry, even at the highest levels of competitive racing. But it is very effective in enhancing performance.

Two Moment Centers – There are two moment center locations we are concerned with. First is the Static MC location and that is where the MC is located when the car is at ride height and still. The other is the Dynamic MC location, being where it moves to as the chassis moves to a new attitude going through the turns. The dynamic position is the most important location to design for because it influences the camber change and the contact patch area.

Stock type of cars can benefit from moment center re-design too. They are especially prone to poor design due to being stock and intended for street use. You can cut off the upper arm mounts and create a plate type of mount that is easy to adjust for camber and moment center. This example stock clip front end has an aftermarket upper control arm mount welded to the stock frame. If allowed in your rules,

this makes mounting an aftermarket upper control arm and adjusting your Moment Center location much easier.

We can look at both the static and dynamic MC locations in readily available geometry software programs. To find the MC location, we need to know the height and width of each point that makes up the AA-arm geometry for the suspension we are working on. Those points include the two ball joints on each side as well as the top and bottom chassis mounts for the control arms.

Those points create the control arm angles that in-turn form the MC location in the system. The MC will move once the chassis starts to move in dive and roll because those control arm angles will change along with that movement. So, we can track where the MC moves to with the chassis motion.

Circle Track Car Design – For circle track cars, we want the MC to start out to the inside of the turns and for cars turning left, that would be left of centerline from a driver's view. We also do not want the MC to move very far in most cases, although some experimentation here could help some situations.

If we get this design right, the camber change in the outside wheel (right front for left turning cars) will be very little. Race tires do not like camber change, especially at the right front, or outside tire. If the camber of that tire stays relatively unchanged through the dive and roll motions of the chassis, then that tire will have the most consistent level of grip possible.

As to the inside wheel/tire, there will always be camber change and the best we can do is minimize the amount of change. Designing the MC to be left of centerline in a left turning car does produce control arm angles that result in minimal camber change for both the inside and outside wheels.

Road Racing Car Design – In road racing, the cars turn right and left. It is also true that there is a combination of control arm angles that will produce the least camber change resulting from the cars chassis movement in dive and roll going through the turns.

If we get this right, there will be minimal camber change for the outside tire, but still like the circle track cars, the inside tire will always experience camber change. The best we can hope for is minimal camber change in the outside tire to maximize that tires grip and then we have to live with whatever grip results from the inside tires resulting camber.

How To Design The MC Location - If you want to know exactly where your moment center is located in both the static (at ride height and not moving) and dynamic (at speed going through mid-turn) locations, you will need to purchase a front end geometry software program. It's just not feasible to draw it out on paper. This discussion will assume you are using, or plan on using a program. If you don't ever get to the point, at least here you will learn how that is done and that information is important to any racer.

In order to find your Moment Center in both the static and dynamic locations, you will need a computer program. Teams have tried in the past to draw the MC out on a large sheet of paper, but drawing the dynamic location is very difficult, if not impossible to do. If you don't have a program, you might find a friendly team who does and use that to find your MC location.

Because I have been doing this for over twenty years now, I know that there is no trial and error method that works to find the best geometry design. The system is just too complicated. If you don't have one of these programs, or cannot afford one, see if a fellow team will work with you and let you use theirs. The idea of that may be a stretch in some racing circles, but most racers want to beat you on your best day, or not.

Getting Started With Measurements – First you will need to get measurements. We measure the height and width of each chassis mounting point and the centers of rotation of the ball joints on each side. Once you have your numbers, you will enter them into the program. The two important sets of numbers are height and width of the (1) static MC location, or SMC, and (2) the dynamic location of the MC, or DMC.

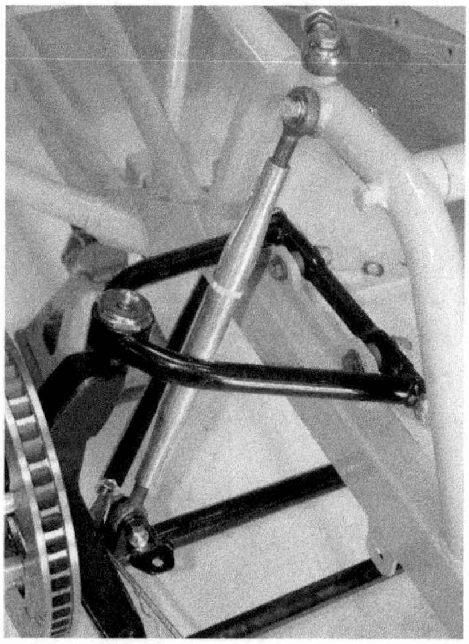

When measuring your car, make sure to fix the spindles in the same location as they would be at normal ride height with all of the weight in the car. This well made link that replaces the shock in a late model is adjustable for length and makes positioning the spindle quick and easy once the wheels are off and the car is raised up.

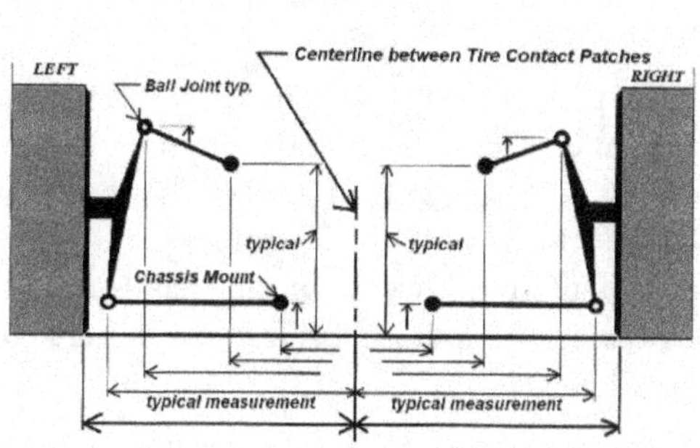

To find the MC location, we measure the height and width of each point used to establish the intersections. These are: 1) the ball joints on each side including the upper and lower ball joints, 2) the chassis mounting points for the upper and lower ball joints. The centerline we will use is the midpoint between the two tire contact patches.

The location of the DMC is most important because this is where our front end is doing all of the work. Over years of practice, experimentation and use, we have narrowed down the range of location of the DMC that most racing technology experts will agree works the best.

Coincidentally, what moment center advocates think is a good location for the DMC agrees with the proponents of the jacking force theory. Since we are all in agreement on the location, and not necessarily why that place is good, we can continue without controversy.

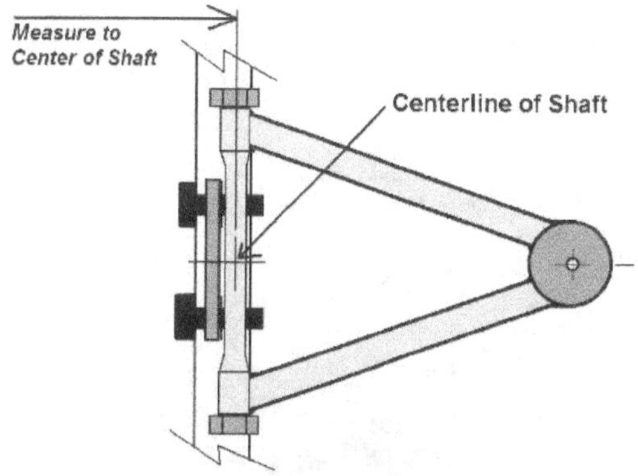

When measuring the upper chassis mount on a typical circle track stock car or some road racing cars based on those cars, we measure to the centers of the bushings, or centers of rotation. Some shafts are offset between the bushings, so be careful when you measure to not mistake the center of shaft at the middle for the center of rotation.

Not only is the location important, but the distance the MC moves from SMC to DMC is important. I always say the MC moves fairly quickly from SMC to DMC, but I still want the two locations fairly close together. This means that I want them less than eight inches apart. Yes, cars will work if they are farther apart, but may not work as well, especially on entry.

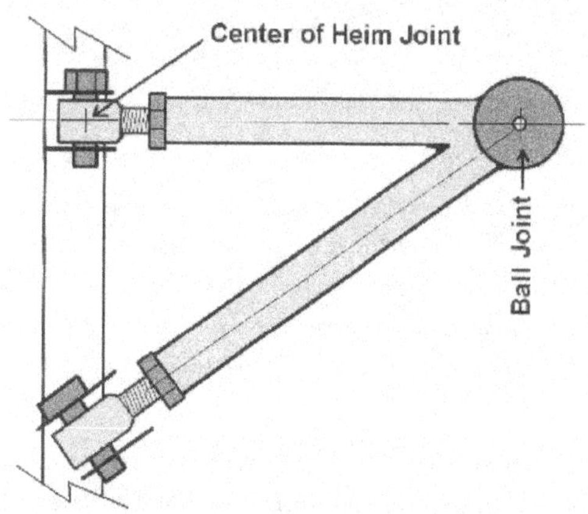

For upper control arms that are this shape, you can just measure to the chassis mount that is ninety degrees to the centerline from the ball joint. With the limited angular movement of the wheel for most applications, this is adequate for finding the MC location. For stock lower control arms where the line between the lower mounts is angled from a top view, the front mount is usually near ninety degrees off of centerline inline with the ball joint and you can also use just that mount for MC location.

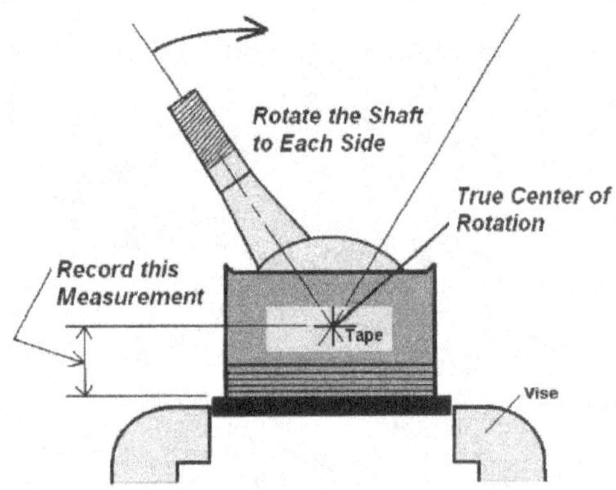

To find the center of rotation for a ball joint, you can go through this process. Mount the ball joint in a vise or similar holder, and rotate it all of the way to one side. Use a straight edge to project the center of the shaft onto a piece of tape on the side of the housing. Rotate the ball joint all of the way to the opposite side and again draw a line through the center of the shaft onto the housing. Where the two lines cross is the center of rotation and you can measure from the edge of the housing to that mark to use for all similar ball joints.

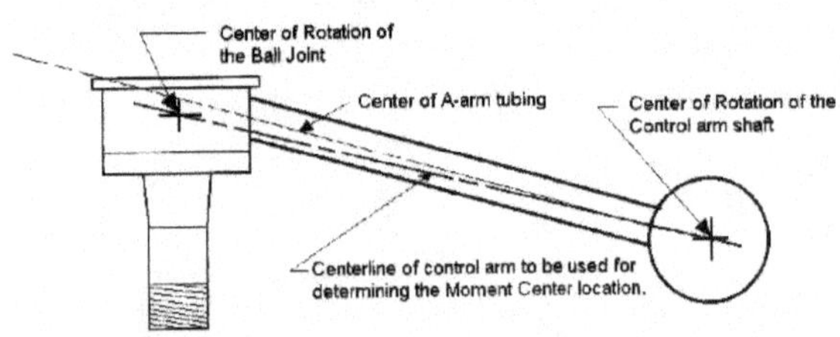

The upper, or lower for that matter, control arm angle is formed from a line projected through the centers of rotation of the ball joint and chassis mounting point. This line almost never follows or represents the center of the connecting shaft that joins the ball joint to the chassis mount.

Lower Control Arms– The lower control arms determine somewhat the lateral location of the MC, but mostly they determine the amount the MC moves from SMC to DMC. So, the angles are important. So we usually set these angles and forget them. Most of our work in MC design is done with the upper control arm angles.

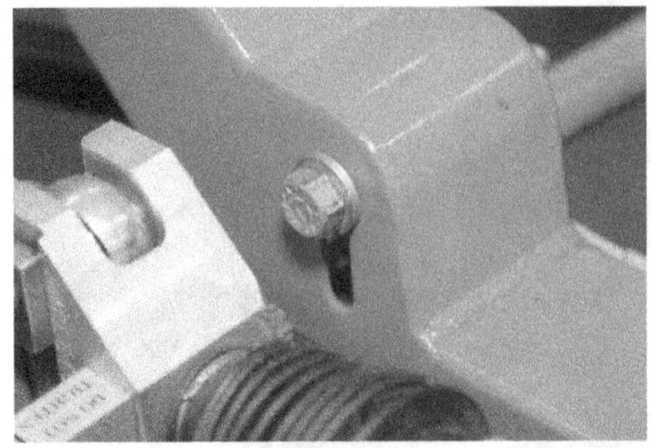

To make changes to the lower arm angles, most modern race car designs utilize slotted mounting holes. When you find the best angles for your use, tack weld the washer in place so that it doesn't move during use. You can also use different length ball joint shafts, like we used for the upper arm angles, to change and perfect your lower control arm angles.

From experimentation, for circle track cars, we have learned the following: A) Both lower control arms need to be positive, that is, angle with the chassis mount heights lower than the ball joint heights; B) you need to have more angle in the left lower control arm than the right one; C) The angles need to be between 1.0 to 2.0 degrees different, no more, no less.

What the proper design for MC does is limit the movement of the MC's to under eight inches from SMC to DMC in most cases. The reason we don't want the lower control arm angles to be negative (ball joint lower than the chassis mount) is mostly because that affects our camber change in a negative way

Upper Control Arms – The angle of the control arm is the angle from horizontal of a line formed by the centers of rotation of the chassis mounts and the ball joints, not the angle of the tubing connecting the two. The angles of the control arms dictate the locations of the SMC and the DMC. The upper control arms mostly determine where the MC's are laterally. We change these angles to move the lateral location of the two MC's. Here is how.

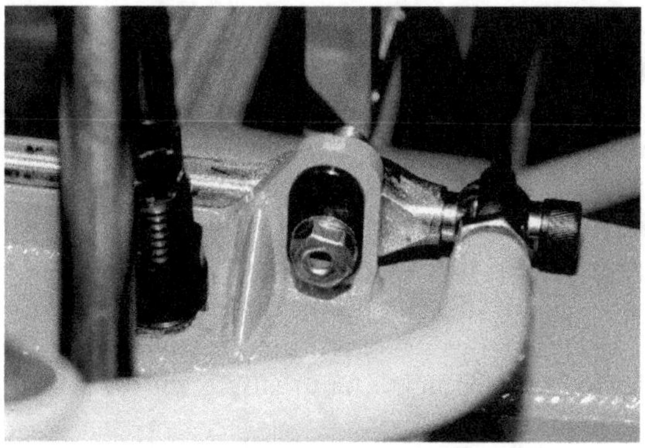

Most upper control arm mounts are adjustable for height. This one has slugs you can use in different offsets for the hole to move the chassis mounts up or down. You can also change the ball joint shafts to raise or lower the ball joint center of rotation. Either will c ause changes to the upper arm angles and move the Moment Center to where you need it to be.

The MC location is determined by the location of the instant centers, or IC's. We don't necessarily need to know where the IC's are located, we just need to know that when we change the angles of the uppers, we are changing the locations of the IC's. This again is all covered in RCT Level One.

Generally, putting more angle (degrees measured from horizontal) in either upper control arm will move the MC towards that side. So, if we increase the left upper control arm angle and decrease the right upper control arm angle, the MC will move to the left. That is because we have moved the IC's on both sides in that direction.

This design has only two mounting holes that don't provide enough fine adjustment to design your MC location. To change the arm angle, you can either change the height of the chassis mount to the other hole or change the height of the ball joint. You can install ball joints with different length shafts to raise or lower the ball joint center of rotation. Some modified cars use mono-ball ball joints that are easy to make height changes with. That is the essence of MC re-design associated with the upper control arm angles.

Modern aftermarket, or non-OEM ball joints can be ordered with different length shafts. This way you can move the center of rotation up or down to fine tune the arm angle and ultimately the MC location. Just the shaft can be purchased if you already have these type of ball joints and don't need the entire assembly.

Many dirt late models and some road racing race cars use square tubing uprights to mount the upper control arm to. On these designs, to change the arm angle, we only need to drill new holes into the uprights at the location that will produce the desired arm angle we need.

Road Racing MC Design - For road racing cars, the upper and lower control arm angles will be the same left to right. There are exceptions to this rule, but for the most part, they are the same. For optimum design, a lower MC works best and a MC that moves very little from centerline through dive and roll works best. That combination produces the best outside tire camber change, or what we call minimal camber change.

Road racing cars can benefit from MC redesign too. The angles of the upper and lower control arms will determine the camber change that takes place during cornering. With wide tires like those used on this class of prototype car, there is little room for error with the tire cambers.

The resulting upper control arm angles will be between 10 and 20 degrees in most cases for typical designs. The exception is in the Formula One cars where the arm angles are reversed with the ball joint lower than the chassis mounts. This is a special exception that we won't go into in this Lesson because those cars represent a fraction of the race cars in the world.

Camber Change Considerations – During early research into the front end geometry and the MC locations, we were working with conventional setups. The soft springs combined with bump technology we see today was still some ten years away. What we learned was that there was a combination of upper control arm angles that would yield near zero change in the outside tire camber through turn entry, middle and exit off the corners.

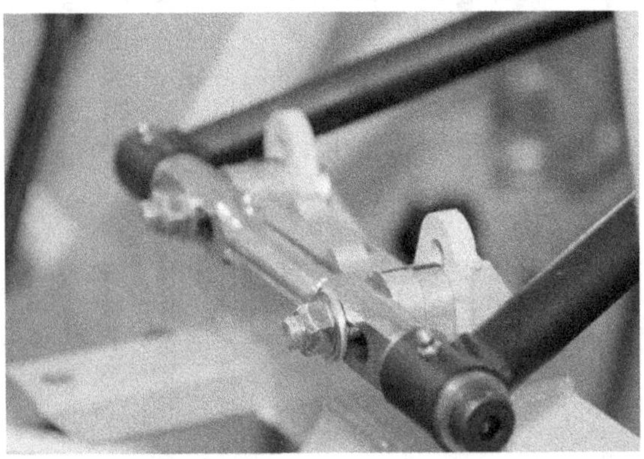

You can adjust your camber easily on most race car designs. This upper mount on an asphalt late model has spacers between the upper control arm shaft and the upper chassis mounting plate. When you optimize your MC design to reduce the camber change, don't forget to adjust the static camber. See the Camber Lesson to learn more about the correct way to do that.

With the modern circle track setups, the front ends travel more vertically now than before, so zero change is not now possible. The right front always gains negative camber. The left front wheel always lost much of its positive camber with the old school setups and still does with the newer soft spring setups. But now with a lot of front end travel, it loses even more camber and the right front gains a lot more negative camber. We cannot do much about that, it's the world we live in now days.

The positive thing about the bump setups is that the cambers, once the chassis has moved down onto the bumps, stay relatively the same with little change. Remember what we said about the older conventional setups, they did not like camber change. We now have systems that again produce little camber change.

Ideal Moment Center Locations – Now that we know how to re-design the MC and a little about arm lengths and angles, let's discuss where we need to end up with our DMC. We always talk about the DMC location because it is the only one that counts. Mid-turn dynamics represents 80-90 percent of our focus with chassis setup.

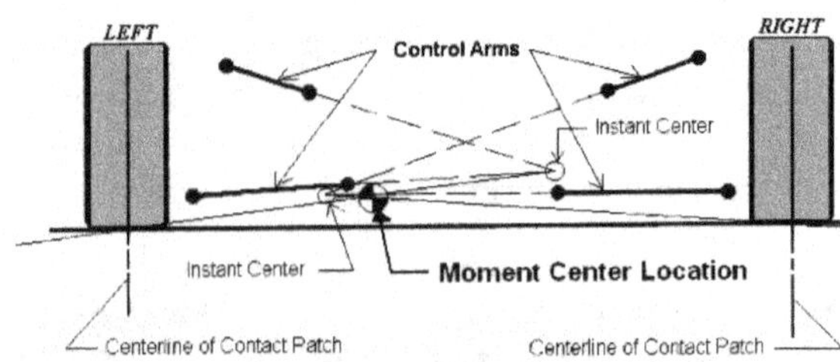

The MC is positioned at the intersection of the lines from the Instant Centers to the center of the contact patches. You can see in this illustration how that is done. The control arms change angles when the chassis moves in dive and roll. So, the lines change as does the MC location. For most applications for circle track racing, the MC is best located towards the inside of the turn. For road racing, we want the MC to stay very close to the centerline of the chassis through dive and roll.

The ideal MC location varies by class and type of racing. For street stock and most of the more conventional setups, the DMC needs to end up somewhere under the center of gravity. So, if you have a 60 inch track width and are running 54 percent left side weight, the DMC should be located left of centerline by about 2.4 inches.

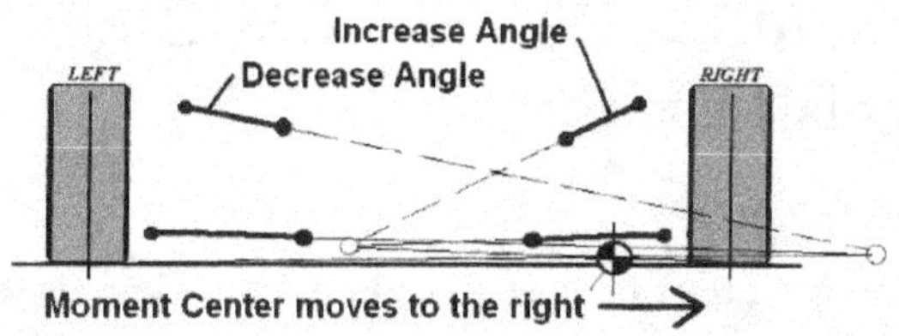

We can move the location of the MC by making changes to the control arm angles. Putting more angle in an upper arm, in this case the Right Upper arm, will move the MC location to the right. Decreasing upper control arm angle in an upper control arm, in this case the Left Upper arm, will move the MC towards the opposite side, or right side. This give you some idea of what to change to move the MC in the direction you need it to move.

You can experiment with the location by moving it farther left, but on those cars, with the high center of gravity, you can definitely overdue it. Keep the MC left of centerline, but not too far left and your car will turn very well, especially the dirt cars that are stock based.

For late models on dirt or asphalt, we see a much more aggressive approach to MC location. The DMC for soft conventional setups follows the above example in reference to the center of gravity. So, for a car with 66 inches of track width running 58 percent left side weight, the DMC should be located about 5.5 inches left of centerline to put it under the CG. This used to work very well and still does for those cars.

As to the super late models running on bump setups, the design can place the MC a little farther left for several reasons. One, with the high amount of dive and stiffness in the spring rates of the bumps, we can venture much farther left.

For most bump late models, the common location for the DMC is from 12 to 16 inches left of centerline. Most modern and successful teams, if they were honest about it, will tell you that this DMC location is where the car turns best and coincidentally that is where we find the camber change is the least.

For those types of cars, the fact that the car is down on the bumps when entering and going through the turns, there is very little vertical movement (chassis dive) and very little chassis roll, so the camber change is minimal regardless of the MC location. Nonetheless, we still design for the locations we stated above.

For road racing cars and formula cars, the best MC design has the MC starting at centerline in static location and then moving very little from centerline through chassis dive and roll. And, with minimal camber change in the outside tire. In early experimentation with both a AA-arm front and solid axle rear type of cars and later on with prototype cars designed with AA-arm front and rear suspensions, optimization of the control arm angles resulted in significant gains in grip through the turns.

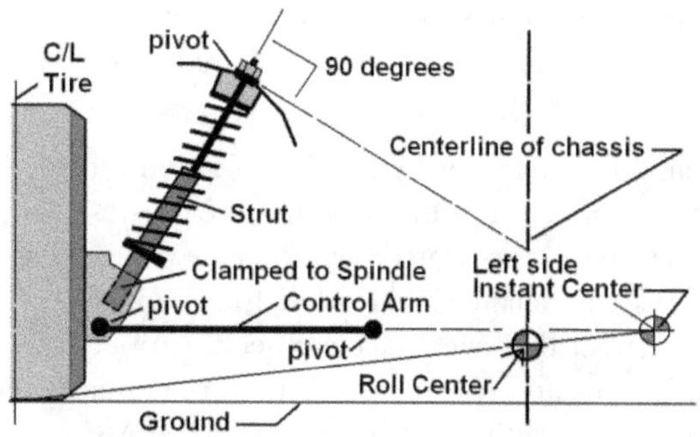

Stock type McPherson strut designs have Moment Centers too. The location is a little different than with a AA-arm suspension. The lower arm forms a line to the instant center, and a line at ninety degrees off the line of the shock, or strut, forms the other line to the instant center. These systems are almost always symmetrical, and so the static MC is always at centerline, or very close to it. If the cambers of each wheel are adjusted by moving the upper mount of the strut, then the line off the strut we use to find the instant center also moves causing the MC location to move. For Circle Track racing, we need more positive camber in the inside wheel and more negative camber in the outside wheel. This necessarily moves the instant centers and the MC location to the outside of the turn, which is not ideal by any means. This is why these cars are very hard to make work.

Summary – The geometry design is one of the very first and most important design features on any race car. Other geometry changes such as camber and caster change won't appreciably affect the MC location, so we can do this design early on in our overall chassis setup routine.

Changes made to move the MC location will affect the bump steer and other design parameters because those may be affected by the control arm angles. So be careful when you make changes to the MC that you don't inadvertently change other geometry settings.

I can tell you without a shadow of a doubt that MC design and optimization has the greatest effect on chassis related performance gains aside from setup balance and load distribution. Remember that every race car needs proper loading in the tires and maximum tire contact patch along with that loading. And your MC design helps achieve the best contact patch design.

Exam - In The Context Of This Lesson:

The Moment Center Is Called That Because?

1) It only takes a moment to find it

2) It is the bottom of the moment arm

3) It is different than the roll center

4) It is the center of the chassis roll

The Two Primary Moment Center Locations Are?

1) Right of Centerline

2) Static location

3) Dynamic location

4) Left of centerline

5) 2 and 3

For Circle Track Racing, The MC Should Always Be?

1) Towards the outside of the turns

2) Towards the inside of the turns

3) Higher than the ground

4) Lower than the ground

For Road Racing, The MC Should Always Be?

1) Towards the outside of the turns

2) Towards the inside of the turns

3) Above Ground

4) Close to centerline

What Are The Primary Measurements Needed To Locate The MC?

1) Distance from the centerline

2) Height off the ground

3) Distance from the front axle

4) 1 and 2

The Lower Control Arms Mostly Influence?

1) Where the MC is located laterally

2) The height of the MC

3) How far the MC moves static to dynamic

4) The camber change of the wheels

The Upper Control Arms Mostly Influence?

1) Where the MC is located laterally

2) The height of the MC

3) How far the MC moves static to dynamic

4) The camber change of the wheels

5) 1 and 4

We Can Change The MC Location By?

1) Raising or lowering the chassis mounts

2) Using taller or shorter ball joints

3) Changing the lower control arm angles

4) Changing the upper control arm angles

5) All of the above

Putting More Angle In The Left Upper Control Arm Moves The MC?

1) Towards the right

2) Lower to the ground

3) Towards the left

4) Higher off the ground

Putting Less Angle In The Left Upper Control Arm Moves The MC?

1) Towards the right

2) Lower to the ground

3) Towards the left

4) Higher off the ground

We Can Find The Control Arm Angles By Measuring The Angle Of The Tubing?

1) True

2) False

The Most Positive Thing About The Bump Setups Is?

1) We don't need shocks anymore

2) The car is lower

3) The cambers don't change very much

4) The car doesn't roll

Racecar Technology – Level Two

Lesson Four – Setting Caster

Caster settings are set relative to the type of car and the design of the race track as well as the drivers preference. When you determine the correct settings, you need to maintain those settings. We will discuss what affects caster has in our race cars as well as how to properly measure it.

It is important to know how to properly measure for the amount of caster in your race car. All teams need to learn the proper procedure for determining the correct caster that exists in the front end geometry of their race cars because it is many times a driver preference. Here is the definition of caster.

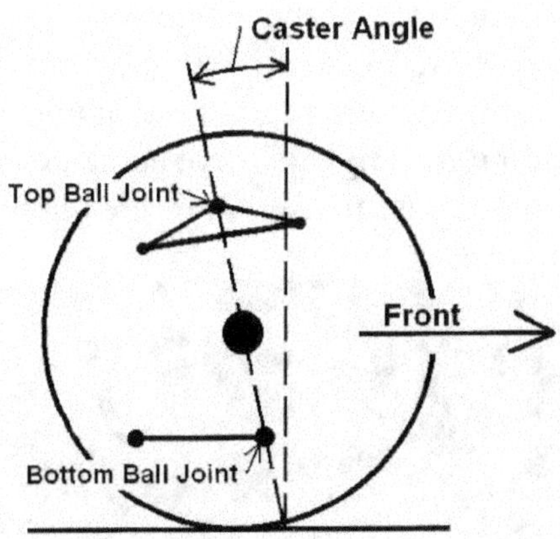

One of the most basic, but also most important, settings on the front of the car is Caster. This setting affects the drivers feel when steering the car through the turns. Caster occurs when the upper ball joint is mounted to the rear of the lower ball joint. This creates the caster angle. Positive aster in the front wheel assembly is created when the ball joints are offset, from a side view, so that the upper ball joint is farther to the rear of the car than the bottom ball joint. The degree of Caster is related to the angle in degrees that a line through the ball joints forms from a vertical line.

Caster Defined – Caster is a design condition that serves to cause a wheel to want to track straight ahead. A common example is a bicycle front wheel and fork assembly. The tube that the handlebars are attached to is mounted in a set of bearings above the fork and from a side view this tube is angled so that the bottom bearing is ahead of the top bearing. If we turn the front wheel to the side from straight ahead, it will want to return to straight ahead by the effect of caster. This is called positive caster. The same effect is present in the front wheel assemblies of our race car.

The car wants to return to straight ahead because when the wheel is turned, the positive caster effect raises both the bike and your race car. The weight of the car pushes down on the wheel and it naturally wants to return to straight ahead where the car is at its lowest.

What Caster Split Does - To ease the amount of effort it takes to turn the wheel in our circle track race cars, we introduce caster split into the design. Split means that we set different caster amounts into each front wheel assembly so that the car will want to turn only one way naturally and thereby reduce the amount of effort it takes for the driver to hold the steering wheel when negotiating the turns.

Proper split for circle track racing cars turning left means that the left front wheel will have less positive caster, or in some cases negative caster, as opposed to the right front wheel. Teams have been known to set negative caster in the LF wheel and positive caster in the RF wheel. If the driver took his/her hands off the steering wheel, the car would naturally turn left due to the caster split we have described.

For road racing cars, the caster amount will be equal on each side of the front suspension so the steering effort will be the same for right and left hand turns. The trend in road racing is to run greater amounts of caster because of the camber change that caster produces, gaining negative camber in the outside wheel and gaining positive camber on the inside wheel. These effects produce better contact patch shapes for greater front grip.

Because road racing cars have equal caster side to side, all discussion about caster split will relate to circle track cars that turn in only one direction.

How To Measure Caster - To measure caster in each wheel, we use a Caster gauge. This tool attaches to the wheel hub. To check the amount of caster, we need to follow these instructions:

1) Attach the Caster gauge to the Right Front wheel hub first.

2) Turn the steering wheel to the right so that the right front wheel has turned exactly twenty degrees from straight ahead.

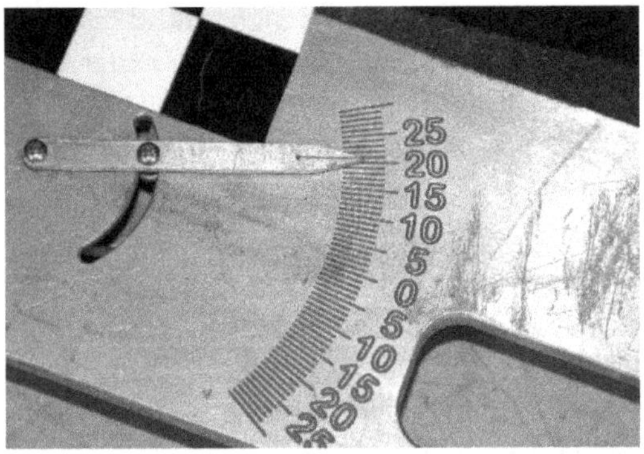

It is a good idea to use turn plates in order to know exactly when you have turned the wheel twenty degrees, the standard amount for caster measurements according to SAE standards, to which the gauges are designed to accurately measure caster.

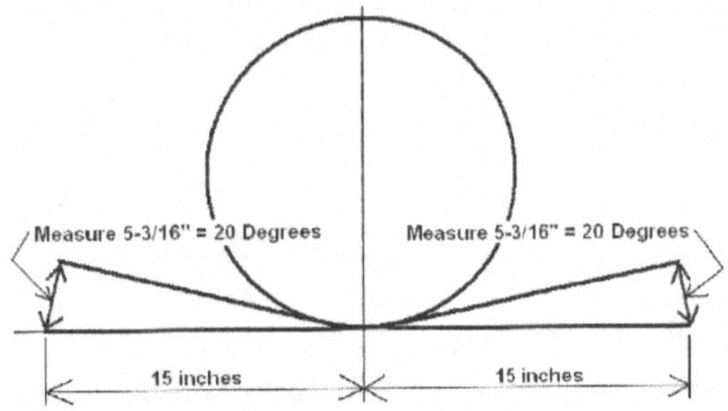

To measure on the floor to get exactly twenty degrees of steering, cut a 30 inch piece of 1X6 wood or similar straight, flat piece and lay it against the tire. Mark mid-way on the wood (15 inches) and line that up with the hub. Mark the outside corner of each end of the wood. Turn the steering wheel until the ends have moved 5-3/16 inches and you will have turned the wheel 20 degrees.

To measure caster with a bubble gauge, level the gauge and set the bubble at zero on the Caster side of the gauge with the wheel turned right exactly 20 degrees. Some teams use one edge of the bubble so that reading the angle is more accurate than trying to estimate the center of the bubble.

For a digital gauge, turn the wheels to the right so that they turn twenty degrees, level the gauge, and then push the "zero" button to set zero in the display.

3) For a bubble gauge, level the gauge and set the adjustable Caster bubble vial so that the bubble is at the zero mark on the Caster side of the tool. For a digital gauge, set zero once the wheels are turned and the tool has been leveled.

4) Turn the steering wheel to the left so that the right front wheel is turned past straight ahead and ends up left of straight ahead by 20 degrees.

Turn the wheel past zero degrees on the turn plates and go past that to twenty degrees left turn. Level and read the gauge and that is the amount of caster in the RF wheel. Repeat in reverse order for the left front wheel.

5) Again, level the gauge and then note the location of the bubble on the scale of the manual gauge and record the amount of caster in the Right Front wheel. For a digital gauge, just read the display and that is the caster amount for the RF wheel.

6) While the wheels are still turned left 20 degrees, remove the Caster gauge and place it onto the Left Front wheel hub.

7) Level the gauge and set the bubble on the Caster gauge to zero, or zero the digital gauge.

8) Turn the steering wheel to the right past straight ahead until the LF wheel has turned 20 degrees to the right of straight ahead.

9) Level the gauge and read bubble scale or display to know how much caster is in the LF wheel.

Adjusting Caster - To adjust the amount of Caster in each wheel, you will need to move the upper or lower ball joint fore or aft. To increase the amount of positive caster, move the upper ball joint towards the rear of the car, or the bottom ball joint towards the front. Some cars have slots cut into the upper chassis mounts or upper control arm shaft for this purpose.

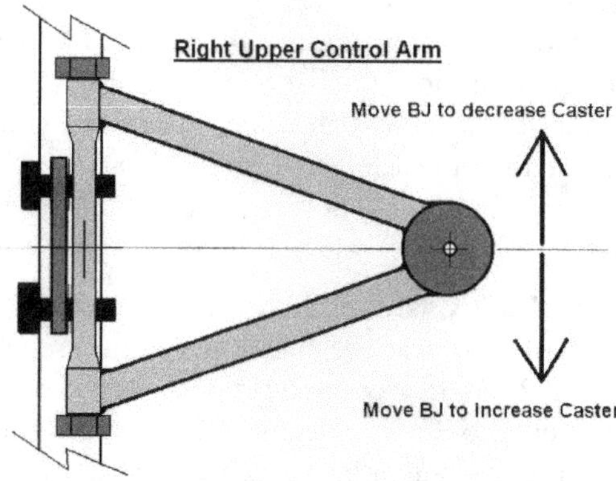

We introduce caster into our front suspension system by moving the upper ball joint forward or backward. We can also move the lower ball joint fore or aft to accomplish the same thing. It is more common and many times easier to move only the upper ball joint.

If you have permanently attached vertical mounting plates that the upper control arms are attached to, then you can vary the amount of shim spacing for each of the bolts that attach the control arm to the chassis. Wider spacing at the front bolt (control arm shaft inside of the mounting plate) will move the upper ball joint to the front creating less Caster at that wheel and so on. This is not the preferred method though.

You can move camber adjustment washer shims back and forth from each mounting bolt, or slide the control arm shaft if it is slotted like this one is, to adjust the amount of caster in each wheel. This will move the upper Ball Joint forward (to reduce positive Caster) or to the rear (to add positive Caster).

If you are not using slotted control arm shafts, once you have established the exact Caster amounts for each wheel using the above method, you should order an upper control arm that has the ball joint mount offset to give the correct amount of caster at each wheel. That way, you can use the same camber shim spacing for each mounting bolt to connect the upper control arm shaft to the chassis, which is much safer.

How Much Caster Split - Normal Caster splits for most short track asphalt applications are around 2 to 4 degrees of difference, left to right. The Left Front caster might be 1-2 degrees and the RF caster might be 3-5 degrees. The left front caster could also be (-) 1-2 degrees and the right camber a (+) 1-4 degrees. The higher the banking angle of the race track, the less Caster split you will need because less steering effort is needed due to the banking.

Also, the tighter the turn radius, the more Caster split that is needed. Driver preference plays a big role in getting the Caster split right for your application. Some drivers don't want any amount of caster split so they can feel the steering better.

Exam - In The Context Of This Lesson:

Caster Settings Are Relative To?

1) The track banking angle

2) The track turn radius

3) The type of race car

4) Driver preference

5) All of the above

Caster Is Present When?

1) The upper ball joint is behind the lower ball joint

2) The upper ball joint is ahead of the lower ball joint

3) The upper ball joint is closer to the centerline than the lower BJ

4) 1 and 2

To Measure Caster We Turn The Wheel?

1) Lock to lock

2) The same as when we race

3) Twenty degrees left and right

4) Until the level bubble moves

Road Racing Cars Might Need?

1) More caster than circle track cars

2) Less caster than circle track cars

3) Zero caster split

4) The same caster on both sides

5) All of the above

To Change Caster We Do What?

1) Move the upper ball joint forward for less positive caster

2) Move the upper ball joint rearward for more positive caster

3) Turn the steering more or less

4) 1 and 2

Racecar Technology – Level Two

Lesson Five – Setting Camber

The basics of changing the camber in your wheels in a AA-ram suspension is pretty straight forward and simple. You didn't take this Course to learn the basics, you took it to go beyond the basics and come to know more and gain information not readily available elsewhere.

We make camber changes by moving the upper ball joint, in most cases, in or out from the center of the chassis. This changes the angle of the wheel and tire to the racing surface. And we do that by either moving the upper control arm in or out, or lengthening or shortening the arm itself if it is designed that way. There, that is the basics. But there is much more to know and understand.

When the camber changes, so do other settings. Here we will learn how to change cambers and how to understand what that involves. There is more than meets the eye.

In this Lesson, we will tell you about the different ways camber changes have an effect on other chassis setup parameters, and how we can optimize camber and for what reason. Of the two most critical necessities for optimum race car performance, the first being loading on the tires, contact patch size is the other and camber dictates what our ultimate contact patch shape and size will be. Therefore, understanding all that is related to camber design becomes a very important part of chassis design and setup.

To read your camber, we can use a simple bubble vial gauge where we read directly the camber by noting where the bubble is located on the scale. Make sure your gauge is level side to side before taking a reading. We need to note what our static cambers are as well as the dynamic cambers, or the camber when the car is at the mid-turn attitude.

As easier way to read cambers is to use a digital camber gauge. Again, be sure to level the gauge and then just read the number. If a negative sign shows, then the camber is negative, and if the sign is not there, it is a positive camber. We can even use a digital angle finder set against the side of the tire, but we need to be careful of the lettering and how that might influence the angle we are reading.

Why Change? – The reasons why we might want to change cambers include the following. We might see where the tire temperatures are not ideal in our view for the type of car, type of tire and design of race track we are running at. We also might want to experiment with camber to see if the tire could use more or less to provide us with better traction. The tire temperatures might look good to us, but hey, what the heck, let's see if more is better.

Tire temperatures are not really the final determining factor in judging whether we have the best camber in our tires. In some suspension designs and especially tire designs, the tire can roll through camber changes such that temperatures rise across the tread, but the tire is never at the ideal camber. The temperatures lie to us sometimes. I've personally seen this happen.

An example of this is the NE pavement modifieds, being the SK division as well as the Tour modifieds. Some teams run very little right upper control arm angle, close to 5 degrees or so. With that angle, the tire runs the straightaways on the inside edge, through entry the transition on the middle of the tire and the mid-turn on the outside. When the team checks the tire temperatures, they might look OK. But the tire never has the best tire contact patch.

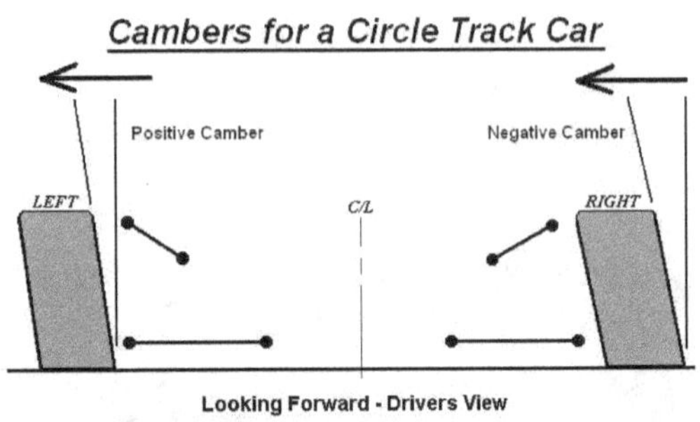

Basic camber knowledge tells us that on a circle track, turning left, we need positive camber in the left front tire and negative camber in the right front tire. When the chassis travels in dive and roll, these cambers will change. How much depends on the control arm angles, especially the upper angles.

You can get a clue this is happening by simply observing the car going through mid-turn. When your car goes out, look at the tires and see if they look like they have the camber you think they need on the race track through the turns. Visuals can tell us a lot. A straight up tire does not have the camber it needs in order to provide the most tire contact patch.

Ride Height Changes – When we decide we need to change the cambers, we need to know what else we are changing when we do that. A quick camber change on pit road does not just represent the "one change at a time" edict racers have followed for many years. It goes, only make one change at a time so you can better judge what that change did. Make more than one change and you will end up guessing which, or were both, the cause for the change in handling.

Changing camber will change the weight distribution on the tire you are making the camber change to as well as the other three tires. Here is how that works.

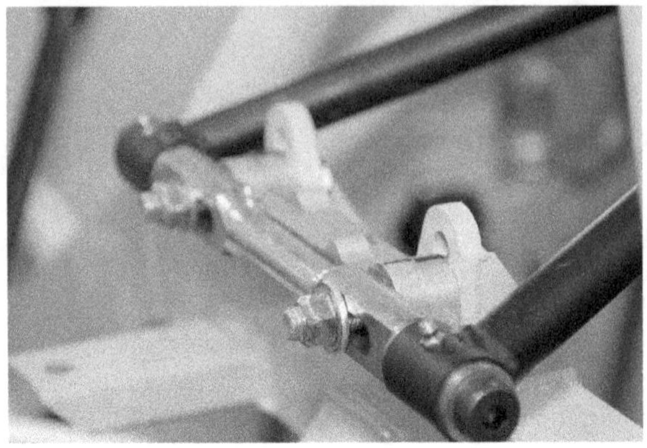

We simply add or subtract spacers between the control arm shaft and the upper control arm mounting plate to change the cambers, right? When we do that, we are changing the chassis height at that corner and also changing the corner weights on all four corners. We need to understand how this happens and adjust the ride height to compensate.

For this design for upper control arm, we adjust the length of the links to change the camber of the wheel and tire. We need to be careful when doing this so that we don't inadvertently change the caster settings. This is not the ideal design for upper control arms by a long shot.

On the left front, the spindle inclination, or angle of the ball joints from a front view, is usually less than at the right front. Typical left side king pin angles are 5-7 degrees for circle track cars. This puts a lot of the tire outside the intersecting line that passes through the ball joint centers.

As such, when we add positive camber to the left front, we essentially jack up the chassis. In other words, we load the LF and RR more so which unloads the RF and LR corners. This de-wedges the car (reduces cross weight) and can make the car looser, or oversteer for those who know that language.

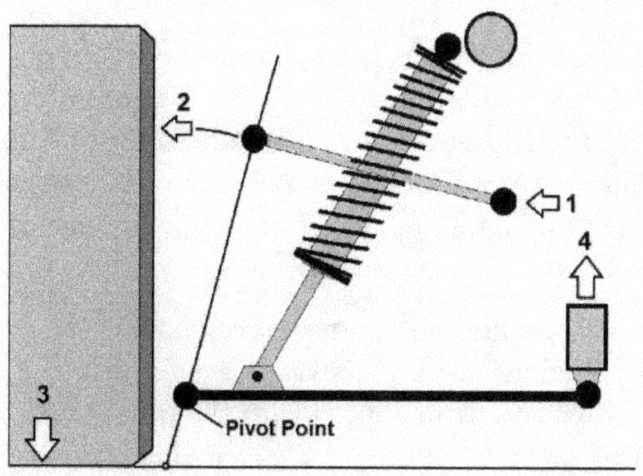

On the left side, we can see here how the intersection of a line through the centers of the ball joints, upper and lower, intersects with the ground inside the tire contact patch. Not all suspensions are setup exactly this way, this is just a demonstration of approximately how this works. If we add positive camber by moving 1, 2 moves in the same direction. Then as 2 rotates around the Pivot Point, 3 moves down lifting 4 the chassis up. We would need to adjust the spring length to bring the chassis back to normal ride height. On the track, we cannot do this, so there is load change happening with chassis travel.

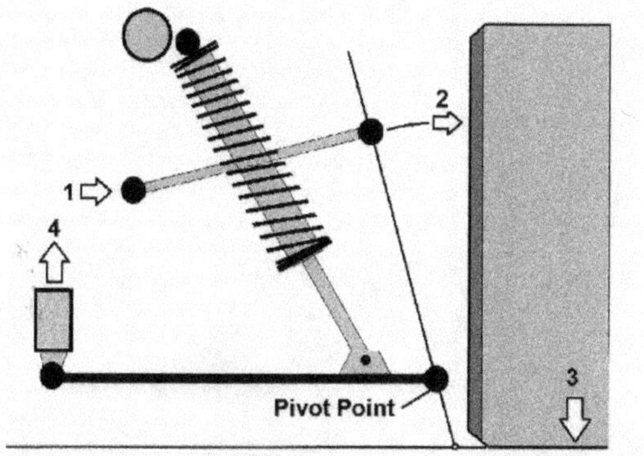

On the right side, what we are seeing is that as we reduce negative camber, or move the wheel towards positive camber, we are also raising the chassis and adding cross weight percent if we don't adjust the spring length to compensate for the chassis height change.

If the car turned better after the change, we might think that the added camber did the trick when in reality it was the drop in cross weight that made the car turn better by making it looser.

At the RF corner, if we reduce negative camber, we add to the wedge, or cross weight, to the car because that tire moves down with the change. Since that spindle has a higher degree of inclination, the intersecting point falls closer to the tires center of contact patch. So, the effect is less than at the LF for the same degree of camber change.

If the LF addition of positive camber de-wedges the car and reduction of negative camber at the RF adds to the wedge in the car, then they might cancel each other out, right? Not really because for each degree of change in camber, the LF changes the wedge more than the RF does. The net is still a looser car, but how much is dependent on the amount of camber change and the overall design of the suspension.

If we add positive camber to the LF and also add negative camber to the RF, we have really loosened up the car because both of those actions took wedge out of the car. The car is turning better now, but not necessarily because the camber change caused the tires to work harder.

The bottom line here is this, when you change the camber in one wheel, bring the ride height back to normal, or what it was before the change. Then you will not have changed the cross weight percent and loading on the four tires. At least not statically. Remember, make just one change at a time.

Bump Setup Camber Change – With the advent of the bump setups in circle track and other forms of racing, camber change of any significant amount while on the bumps is a thing of the past. Even so, there is still a lot of camber change going on from when the car is on the grid to when it is going through the turns. That camber change can cause other setup changes we might not easily see.

The tires provide better grip when the cambers are not changing and will stay very close to ideal due to very little vertical chassis movement. But what about the transition onto the bumps? What happens during that process? Few teams consider what happens then. The associated movement can cause other problems.

As we have discussed above for a more conventional car, when we change the cambers, we are also changing the corner heights and along with that, the distribution of loading on the four tires. If the chassis travels some three or three and a half inches down onto the bumps, where does the wedge, or cross weight, go to then? It has to change from where it was statically based on what we just learned.

Weight change due to the jacking effect of camber change (and that is essentially what it is, the tire jacks up, or down depending on the change) also changes the loading on the four tires.

We already know that the extreme travel associated with bump setups, from ride height to on-track height will also most times pre-load the sway bar. This adds cross weight to the setup. If the camber change from all of that travel changes the cross weight too, then where do we end up? Good question.

Finding Loaded Cross Weight – The major point made here in this Lesson is that camber change will change the loading on the tires and we have to know how it does that in order to plan for it in our preparation of the race car.

If we know how much the cross weight will change due to chassis travel, then we can adjust our static cross weight percent with the car at ride height. Say we gain 2% of cross weight at full travel, then all we need to do is set 2% less cross weight in the car at normal ride height from what we know we will need.

If we put the car onto the bumps with the springs removed and the sway bar preload set, we can note what the scales read. Then we can calculate the gain or loss in cross weight and therefore know how much the weight distribution changes as the car goes down.

When a race car is down on the bumps, bump springs in this case, there will be very little vertical movement and therefore, very little camber change. Depending on the control arm angles, we could even see some change when on the bumps that we can further eliminate and we will cover that part of the design in Race Car Technology Level Three.

To eliminate the change in weight distribution, you can adjust the bump spacing between it and the shock body to bring the cross weight percent back to what you want while on the bumps. That will compensate for the sway bar and the camber change that happens due to the high amount of travel.

One way to know the amount of change associated just with the sway bar is to push the car down onto the bumps with the bar set to neutral at ride height and see how much the bar is loaded while down on the bumps. You can then adjust your sway bar pre-load with the car at ride height so that it will not add or subtract cross weight as the car travels onto the bumps.

Summary - A large part of knowing your settings isn't just knowing what is happening when the car is at ride height, but more so knowing what happens when we make changes and when the car travels on the race track. This becomes a part of the art of chassis setup and one of the primary reasons why this school exists. We want you to become better at the art of race car setups.

Exam - In The Context Of This Lesson:

Cambers Are Mostly Set By Moving?

1) The upper control arm fore and aft

2) The upper ball joints

3) The lower ball joints

4) The lower control arm fore and aft

Cambers Are Read Using What?

1) A bubble and vial gauge

2) A digital camber gauge

3) A digital angle finder

4) All of the above

Why Do We Make Camber Changes?

1) To make the car turn better

2) To create a better heat pattern on the tire

3) To create a larger contact patch

4) All of the above

Adding Positive Camber To The Left Front Will?

1) Lower that corner of the car

2) Add load to that corner of the car

3) Raise that corner of the car

4) Take away load from that corner of the car

5) 2 and 3

Adding Negative Camber To The Right Front Will?

1) Lower that corner of the car

2) Add load to that corner of the car

3) Raise that corner of the car

4) Take away load from that corner of the car

5) 1 and 4

Bump Setups Help The Tire Why?

1) The center of gravity is lower

2) There is better aero efficiency

3) There is very little camber change on the bumps

4) The heat on the tire is ideal

As The Front End Travels Down On A Circle Track Late Model, The Sway Bar?

1) Provides more grip

2) Reduces the cross weight or wedge

3) Increases the cross weight or wedge

4) Provides more bite off the corner

Racecar Technology – Level Two
Lesson Six – Bump Steer

Part of the process of designing our cars for proper alignment is minimizing bump steer. Not only do we need to measure the steer as the wheel travels in a AA-arm suspension, we need to see what happens with our bump steer when the wheels are turned and what else may be affected when we make changes to our suspension components.

There is bump steer in a solid axle suspension as well, but here we are only discussing bump steer in a double A-arm, or what we might refer to as a AA-arm suspension. In Lessons 11 and 12 we will address rear solid axle bump steer.

Bump steer is one of the most basic geometry functions in the front end of the race car. There are some interesting things to know about how bump steer occurs and how to measure it. We will also discuss ways we can interfere with the perfect bump steer settings by making chassis setup and geometry changes.

Certain setups have evolved to where your bump steer might not be what you think it is. When you make changes to your Anti-dive or Moment Center location, you may be changing your bump steer characteristics. So, when you change to a new setup, you might be introducing BS and not know it.

The greater amount of travel associated with the soft spring setups may be out of the range in which you last checked your BS. As a result, the car may begin to behave erratically and cause the driver to be uncomfortable with the new setups because you now have excess bump steer.

We will use the term, near zero bump steer simply because most cars will never be able to have absolute zero bump steer. If you can get your bump to under 0.020 inches of toe change in the total amount of vertical travel, you will have a good design where the driver will not be able to feel the movement.

That amounts to a little over 1/64 inch of toe change. If you are inclined to want a small amount of bump in some direction for one or both front wheels, that's OK as long as you know what it is.

Basics of Bump Steer – As the front wheels move up and down, we want both wheels to stay pointed in the direction they were to begin with. It is most important for the wheels to have minimal bump when we are negotiating the turns. There are certain elements of the construction of the front end components that will reduce or eliminate bumps steer.

The components that make up the suspension parts for bump steer are the upper and lower control arms and the tie rods that connect the spindle to the steering mechanism, being either a rack and pinion or drag link system.

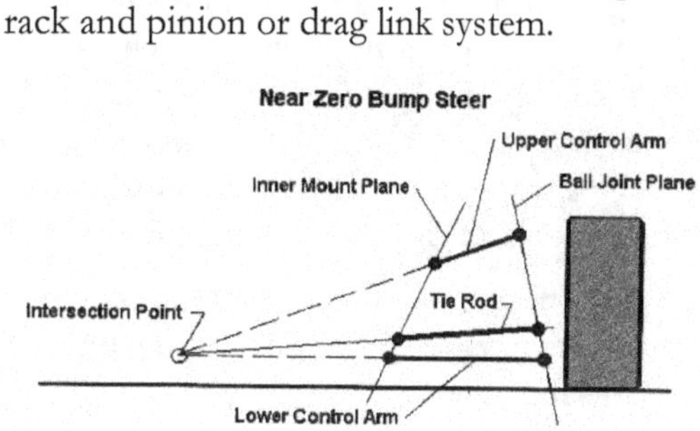

In order for your car to have near zero bump steer, two conditions need to be met. 1) A line through the tie rod needs to point to and intersect with the Instant Center and, 2) its length need to be specific to planes created by the inner mounts and the ball joints.

The direction of the upper and lower control arms caused by a line extending through the center of rotation of the ball joints and inner mounts of each arm, will intersect at a point we call the Instant Center (IC). This is one of the components used to determine the Moment Center location too. In order to have near zero bump steer, the intended goal, we need to setup the car so that a line through the centers of rotation of the tie rod ends points to, and intersects with, the IC formed by the control arms on that side. This is one of two criteria for near zero BS.

The second thing we need is for the tie rod to be a specific length. That length must be equal to the distance formed by the intersection of the line extending through the centers of rotation of the tie rod ends with, 1) a vertical line extending through the upper and lower ball joints, and 2) the plane that is formed by the inner chassis mounts of the upper and lower control arms.

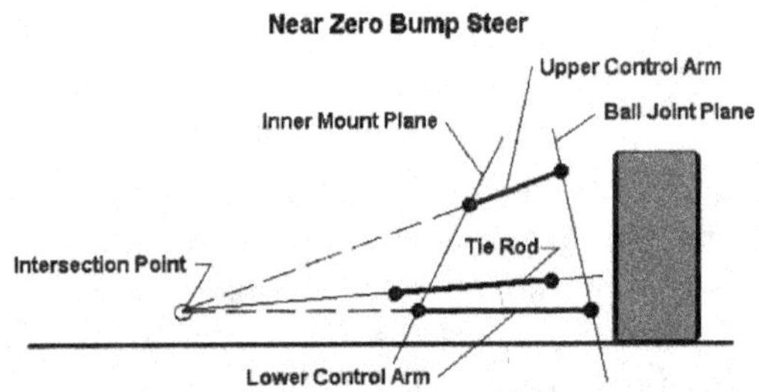

The tie rod length requirement does not mean the tie rod needs to be positioned laterally exactly between the chassis mounting plane and the ball joint plane, it can be offset as long as it is still pointed at the Instant Center.

This can get a little complicated because although the ball joints do form a single line, the chassis mounts form a twisted plane because of the difference in distance from the centerline of the front and rear mounts on the lower control arms, and sometimes with the upper control arms.

So, the inner tie rod intersection point is where the tie rod line intersects the plane of the inner mounts and the outer line intersection point is where it intersects the line through the ball joints. A three-dimensional geometry program can simulate this very well, but most of us don't have the luxury of owning and knowing how to operate one of those. So, we must go through the process of physically measuring the BS in our cars.

The other thing to know is that the tie rod does not need to lie within the two planes, but the length must be that length. Most of the time, the tie rod is located closer to the centerline than it would be if it were to fit inside the intersection points. That's OK because our length is still correct and we should have near zero bump steer.

What Creates Bump Steer – When the tie rod is not aligned with the IC and/or the length is wrong for the system, we have BS. As the wheel moves vertically, the wheel will either steer left or right. We will refer to the direction from a driver's perspective only, in this discussion.

In a front steer car, meaning the tie rods are in front of the spindle pins, if the tie rod were pointed so the tie rod line passes below the IC, then the wheel will bump in (towards the centerline of the car) as the wheel travels up. The wheel will bump out when the wheel travels down. If the tie rod line passes above the IC, then we will have bump out as the wheel travels up and bump in when the wheel travels down.

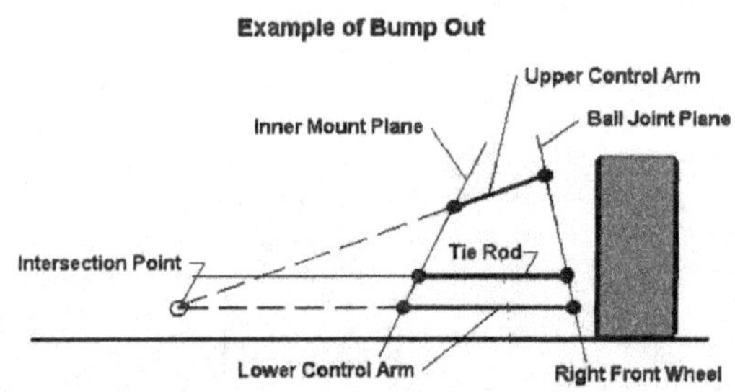

If the tie rod is aligned to be pointing above the Instant Center, the wheel will bump out when moving up and bump in when moving down. The arc it follows will not be correct in relation to the arc of the ball joints.

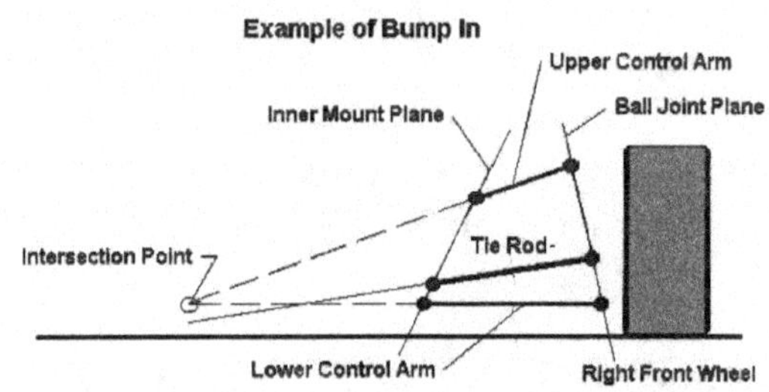

If the tie rod is aligned where it points below the Instant Center, then the wheel will bump in when traveling up and bump out when traveling down due to the difference in arc path to the ball joints.

If the tie rod were too short, we would have bump steer in when the wheel travels in both directions from the static ride height position. If it were too long, then the wheel would bump out as the wheel traveled in both directions from ride height.

These indicators can tell us if we have either a tie rod alignment problem or a tie rod length problem. In some cases, both may be present and that causes a very erratic motion of the wheel. To determine which of these might be present, record the amount of toe change per inch for several inches of travel in both directions from static ride height and note the tendencies. You might have perfect alignment of the tire rod to the IC and a tie rod that is wrong for length. This could be due to a poorly designed drag link or the wrong width rack and pinion steering unit.

If the tie rod is the correct length, and the wheel bumps in or out, we will know how to reposition the ends of the tie rod to make it intersect the IC based on what we discussed above.

Design Changes That Affect Bump Steer – We could buy a car that was near perfect for BS and then make design changes that would unknowingly change our BS. When racers and manufacturers started to install spindles that were designed for rack and pinion systems into a drag link system some years ago, they inadvertently changed the bump steer characteristics, along with the Ackermann geometry.

With the drag link system, the outer tie rod end should be closer to the centerline of the car than the lower ball joint by using angled (from a top view) steering arms. This design feature cancelled out the natural tendency for the system to cause toe changes as the car was steered.

When the new, "rack" spindles were installed, with their straight ahead steering arms, the length of the tie rods changed necessarily. This created a bump out situation as the wheel traveled up and down. New drag links with the inner tie rod ends placed further out were needed so that the tie rod would remain the correct length to eliminate the adverse BS those spindles created.

If we make changes to the front end geometry to improve our Moment Center location, we might accidentally change the BS characteristics at the same time. For example, when we install extended lower ball joints to take angle out of the lower control arms, we change the angle of the lower arm and move the IC height. The tie rod may not now intersect with the IC and we have now introduced BS. When making changes to the arm angles, we need to re-align the tie rod so that it stays pointed towards the IC.

In this example, the race team added spacers to the lower ball joint to change the angle of the lower control arm for better geometry. The outer end of the tie rod shown closest to you has not been moved and is no longer pointed at the IC. This system probably now has a considerable amount of bump steer. Remember that when you make changes to the geometry of your AA-arm suspension, you need to re-check your bump steer and make corrections.

Anti's Affect Bump Steer – In both dirt and asphalt racing, anti and pro-dive is used in various degrees. If we make changes to enhance those effects, we could be causing changes to our BS. This is because with anti-dive for example, when the right wheel travels up, the upper ball joint moves towards the rear of the car and this rotates the spindle from a side view, counterclockwise. This rotation moves the outer tie rod end upwards and changes the angle of the tie rod. Now it no longer points towards the IC.

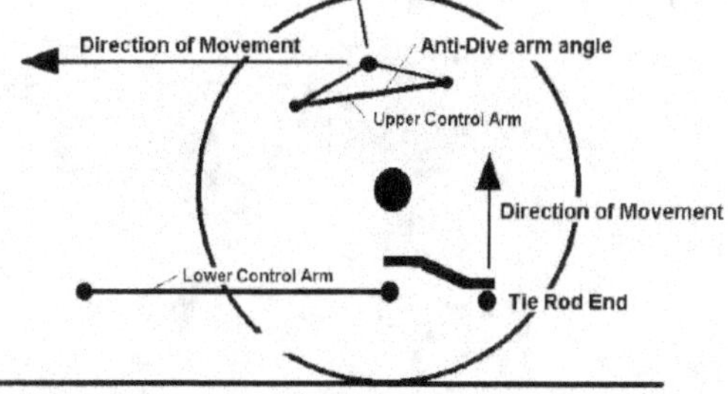

With Anti-dive, as the wheel moves up, the upper ball joint moves rearward, and in some cases the lower ball joint moves to the front. This causes the spindle to rotate and with it the steering arm. The outer tie rod end will move up changing the front view tie rod angle. This changes the alignment with the Instant Center and will change the bump steer characteristics.

Where we had near zero BS before with no anti-dive, we now have BS when we introduce anti-dive. With pro-dive, we see a similar affect, the tie rod end moves down with vertical travel and again the tie rod is miss-aligned with the IC. If you originally checked your BS and found it acceptable and then experimented with Anti's, and didn't recheck your BS, you could, and probably do, have a problem.

Steering Affects Bump Steer – When we steer our front wheels, we change the angles of our tie rods due to caster, camber and degree of spindle on both sides. The tie rod ends travel in an arc that is not parallel to the ground.

This changes the outer tie rod end height and therefore the BS. It is for this reason that we recommend doing your BS with the wheels both straight ahead and then again with the wheels turned equal to the mid-turn steering angle you steer to at the track you will run.

Measuring Bump Steer, Some Tips – We can measure our BS using several different types of equipment. There is the double dial caliper system, the single dial system and the laser system. Each one will tell us if the wheels steer when they are moved in bump (moving up) or in rebound (moving down).

The most common tool is the bump steer gauge. It consists of a flat plate bolted to the hub and a stand that holds dial indicators. It comes in two configurations, the double dial indicator type with a stationary stand and the single dial type with a swinging stand. With the single dial type, when the wheel moves vertically, the stand follows the plate. With the double dial type, the stand is stationary and the two dial pins are always moving one way or the other.

The most common way to measure bump steer is to support the car on jack stands and using a jack, move the spindle up and down in the range of motion it will have during a racing lap. If your wheel travel is 3-4 inches, moving the wheel one or two inches up and down will not simulate the conditions. For most asphalt circle track racing, both front wheels travel up in relation to the chassis. For dirt cars, the left front may well travel down considerably.

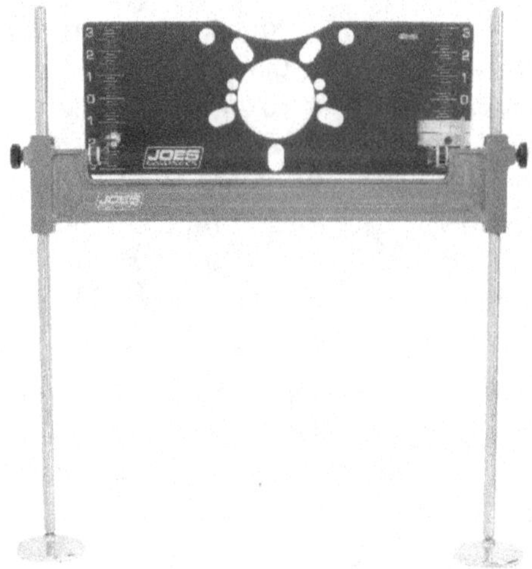

This is a bump steer gauge that uses one dial indicator. It moves in and out with the wheel movement and uses a roller follower on one side and a dial indicator on the other. If the wheel steers when moved vertically, then the dial will move in or out, depending on the way the wheel is steering.

Using a laser system mounted to the wheel hub, you need to place targets in front of and to the rear of the wheel at equal spacing. Preferably, measure the tire diameter and multiply by 3, 4 or 5 so your multiplier used to divide into the readings is an even number. Example: for a 28 inch diameter tire, go 84, 112 or 140 inches out to each target. The farther you go, the more accurate the readings will be.

You can check your bump steer with a dedicated gauge available from many suppliers, or a laser system like this one from DRP Performance Products. This one uses a remote laser that projects a laser line onto a fixture mounted to the wheel hub. Whichever one you choose, make sure you understand the process and how bump changes when other components are adjusted.

If the double dial system has zero BS, then both dials will move together the same amount. If the front dial moves farther as the wheel moves vertically, then we have bump-in at that wheel. If it moves less, then we have bump-out. Be sure to count the number of turns each dial makes when moving the spindle vertically. Subtract the readings to find the BS amount related to the distance the wheel has moved. We usually refer to BS as decimal inches of bump per whole inches of travel.

The single dial indicator gauge is a little different and one I personally like. Using a swing stand, it has one dial that rides on one side of the plate and a roller that rides on the other side. As the wheel moves vertically, the stand follows it in and out. If the wheel has zero BS, then the roller and the dial shaft will move together the same amount and the dial will not change its reading.

If the dial does move, it is recording the total amount of bump steer. Since the roller is stationary, the dial records the movement between itself and the plate, or the total distance between them. With the double dial, when you subtract the two readings, you have the total amount of bump steer.

At the right front, if the dial is on the front of the plate, movement causing a greater number means the wheel is bumping out. If the reading is getting smaller, or the dial shaft is moving out, then we have bump in.

We can also use a laser alignment system to check for bump steer. When using a laser systems, the laser is mounted on the hub or wheel and a targets is placed ahead and behind the wheel center the same distance. This way, any difference in movement of the laser on the two targets, as the wheel moves vertically, will be divided by the distance to the target from the center of the wheel divided by the diameter of the tire.

So, if a tire where 28 inches in diameter (an 88 inch circumference tire), and the targets were 112 inches away, we would divide the difference in movement of the laser front to rear by four (112 / 28 = 4). If the wheel were bumping 0.030 inch, the differential readings on the targets would be 0.120 inch or about an eighth of an inch.

Remember too that when we move the wheel up or down, it will most times move in towards the centerline of the car, or move out. That is why we take a difference in movement on the targets, not the movement itself of the laser lines. If the both move the same distance, then we have near zero bump.

Gauge Ratio – It is important to consider that most bump steer gauges are not the width of the tire diameter for practical purposes. So, we must translate the readings from the width of the tool to the tire diameter if we want to record the bump as being equal to toe.

We can see how the width of the dial indicators is much less than the tire diameter we use for measuring toe. To translate the numbers for Bump into toe, we need to multiply the results by the ratio of the gauge width to the tire diameter. Divide the gauge width into the wheel diameter. For example, 0.030 inch of change on the bump steer gauge could be as much as 0.050 inch of toe change.

This is easy, just divide the measurement between the dials, or the dial and the roller depending on the type, into the tire diameter and use that number to multiply times the bump reading to see what the bump is in toe equivalent, which is something we all more readily understand.

How To Measure Bump Steer - With the car at ride height, measure your shock lengths and write those dimensions down. We want to reference those measurements when we move the wheel in the range of motion it will experience on the race track.

Now you can put the car on jack stands at a height closer to ride height but not too high and remove the front wheels. Remove the coil over, or shock and spring in a "stock" spring car. Check the suspension for freedom of movement by moving it up and down. If the motion feels rough or sticks at points in the travel, you might have dry ball joints or corroded bushings at the chassis mounts. Correct these problems before you proceed.

Start with the right front wheel. Place a jack under the lower control arm and bring the wheel up to your ride height shock length. This is where we will begin to take measurements. For this exercise, we are using a standard bump steer gauge setup that uses a plate mounted to the hub and either one or two dial gauges.

Point the wheels straight ahead and mount your bump steer plate onto the wheel and rotate it so that it is level with the floor and then move your floor fixture up to the plate. The dial indicators will move away, or out, so be sure to bring the fixture towards the wheel enough so that the indicators can move out without bottoming out.

Note the position of the dial shafts on the plate. We will be noting how far the wheel moves vertically and recording measurements at one inch increments as we go along. So, move the dials to the zero line or some line on the plate. Now zero each dial.

Jack the wheel up one inch slowly noting how far each dial moves in relation to each other in a two dial system. Each dial will go out and the indicator will go backwards. So, you will need to be aware of the distance the dials move. If one dial is outrunning the other considerably, things are really bad in the bump steer department.

If the rear (one closer to the rear of the car) moves out faster than the front one, then you have bump out. The reverse is true for bump in. If the movement is more than fifty or sixty thousandths in that first inch, you need to stop and look for the problem.

A wheel that bumps out a lot either has a longer than necessary tie rod or the angle of the tie rod is wrong and the outer end is low and/or the inner end is high. Remember that the tie rod must point to the instant center created by extending the planes of the upper and lower control arms. Bump in is the reverse of the above.

If you can look at the tie rods from a front view looking roughly parallel to the ground plane, you can visualize where the tie rod is pointed and where the instant center probably is located. If it looks like the tie rod will never intersect the IC, start making changes to the height of the ends to get it pointed correctly. Once it looks like it is more so inline, we can proceed.

Let's assume we are getting little difference in movement of the two dials. Record the dial readings at the first inch of travel. Then jack the wheel another inch and record those readings. Now note how much the two differ in total movement. This is your average bump steer per inch of wheel travel in two inches. We are ideally looking for numbers under 10 thousandths of an inch.

Many modern race cars travel the front wheels much more than in the past. For the right front wheel on dirt and asphalt, it is now normal to see three to four inches of travel. So, we need to now bump our wheel up another two inches for those cars, recording one inch at a time.

Your readings may indicate many things when going this far with bump. What you had for bump in the first two inches may move back toward zero bump as you go higher, or it may continue to bump in the same direction even more. If you have less than 20 thousandths in the first two inches and it goes back towards zero, you are probably good to go.

If the bump steer continues to increase, note which way it is going. The back dial moving farther out and reading less than the front dial indicates bump out. Again, adjust your tie rod end heights moving the inner mount down and/or the outer mount up. Again, the reverse is true for bump in.

For readings on the left with a two dial system, the same holds true as to the movement of the front and rear dials. The rear dial moving out more and reading less equals bump out. We need to think out the direction of the movement of the wheel for our car. For dirt cars, the left wheel moves vertically up and down. So, we typically bump that wheel up and down two inches each way to read bump steer.

For asphalt cars, the left front will, or should, always move up in bump. For bump setups, that movement might nearly equal the right front movement upwards of 3 to 4 inches. So, take into account your type of setup and how much each front wheel moves when checking your bump steer.

For single dial bump steer gauges, you will only be reading the one dial. The upright follows the plate as the wheel moves in towards the centerline of the car when it is bumped and moving upwards. This is admittedly easier to use because you only need to record the movement of one dial.

In bump, moving up, on the right side, with zero bump steer, the dial should not move at all. With the dial mounted on the floor fixture towards the rear of the car and the roller to the front, if the dial is moving out and the reading is showing less, then you have bump out. The front of the tire is moving out and the back is moving in.

For bump in, the reverse is true, the dial will indicate an increase in reading since the shaft is moving in. The reading on the dial for each inch of vertical movement is the bump steer amount.

When checking the left wheel, the gauge will be at the front and the roller at the rear. Now your movement for bump out is reversed, and the readings will increase and the dial will move in. The reverse is true for bump in.

Summary – Once we understand all of the things that affect bump steer, we will know when we need to re-measure the car after a chassis or setup change so we can maintain near zero bump steer. If you make Moment Center changes or Anti's changes or spindle changes, or even change setups from conventional to bump, re-measure your BS.

If you have only measured bump at static ride height with the wheel pointed straight ahead, maybe it's time to re-measure the bump at mid-turn configuration. Correcting bump steer when the wheels are turned could make a difference in how the car feels to the driver as the car moves vertically on corner entry and when going over those ruts on a rough dirt track.

Exam - In The Context Of This Lesson:

Bump Steer Happens With/When?

1) Turning the steering wheel

2) Vertical movement of the chassis causes the wheels to steer

3) Steering away from bumps on the race track

4) When we make changes to geometry or setups

5) 2 and 4

Bump Steer Can Occur When?

1) The tie rod is not pointed at the Instant Center

2) The tie rod is too short

3) The tie rod is too long

4) All of the above

Anti's Affect Bump Steer By?

1) Changing the control arm angles

2) Changing upper control arm angle

3) Moving the outer end of the tie rod vertically

4) Restricting movement of the wheel

With Double Dial Bump Steer Gauge?

1) Both dials move together

2) Only one dial moves

3) We can read the bump amount using one dial

4) The readings equal the toe steer amount

With Single Dial Bump Steer Gauge?

1) Both the dial and roller move together

2) We only read the one dial

3) The readings do not equal the toe steer amount

4) All of the above

Racecar Technology – Level Two

Lesson Seven – Ackermann Settings

Ackermann, as described in RCT Level One, is a very destructive effect in race cars. We need very little of the Ackermann effect as a part of our steering system. In this Lesson, we will tell you how to check for Ackermann and how to adjust your system to eliminate excess Ackermann.

You can easily check your steering system for Ackermann. Use a string method or laser system. In this Lesson we will tell you how to measure and how to fix excess Ackermann.

When we have Ackermann effect present in our steering design, it means that the amount of toe-out increases as the steering wheel is turned. With Reverse Ackermann, the toe is reduced. There are different static settings for front end toe that are dependent on the radius of the turns, the banking angle, and the type of tire used.

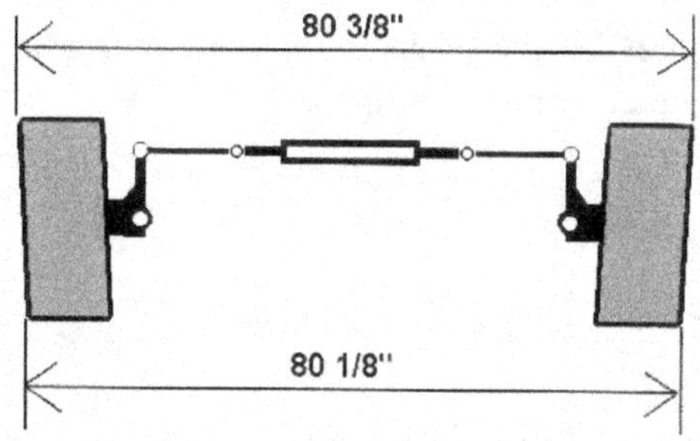

Toe is when the distance between the front edge of the tire is more or less than the rear edge. Toe-in is when the front is less, and toe-out is when the front is more than the rear. This car has ¼" of toe out. For short track racing, we almost always use toe-out. Ackermann and Reverse Ackermann are caused when we turn the steering wheel and the system gains or loses toe. For most types of racing, we need very little or zero toe change, or Ackermann, when we steer the car.

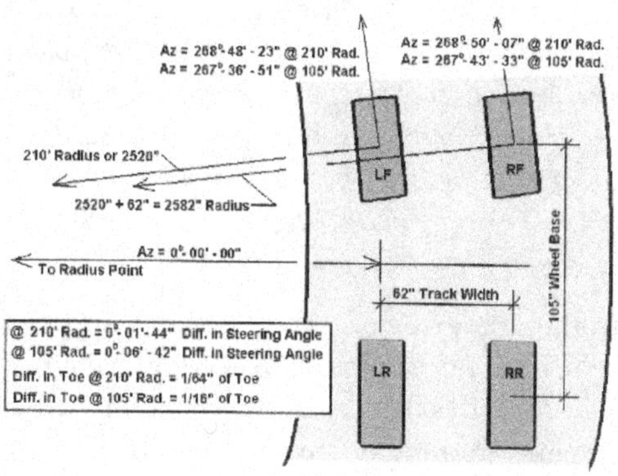

This chart represents calculations done with a very sophisticated coordinate geometry software program. We see where we need very little difference on direction of the left wheel verses the right wheel in order for the tires to track along their individual radii. We are talking about tenths of degrees of difference, not whole degrees.

Checking For Ackermann – To check to see if, and how much, Ackermann you have, you can use either simple string or if you have it, a laser system. The process is the same, just the tools are different.

Aim your wheels straight ahead in the shop, or anywhere for that matter. String along the sidewall of your tire making sure not to go over any lettering or other extrusions. Go out ten feet in front of the axle and place a piece of tape on the floor under the string line.

You will carefully lay the string so that it just touches the edge of the sidewall in the front part of the tire while having someone hold the string wrapped around the side of the rear part of the sidewall. Bring the string down to the tape and make a mark with a felt tip pen.

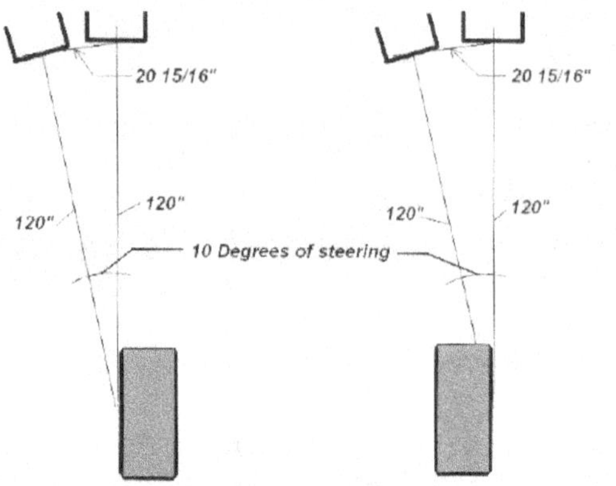

The easiest method for checking for Ackermann is by use of strings pulled across the tire sidewalls or a laser system attached to the wheel hub. With strings, we simply pull the string along the sides of the tire bulge where there are no lettering or other protrusions. We make marks out ten feet from the axle at straight ahead and again when the wheels are turned. The difference in gap between the marks should be zero if you have aero Ackermann.

Go to the other side of the car and repeat the above for the left front tire. When you have two marks, one for each side, turn the steering wheel to the same amount as you would at the track going through mid-turn. Then repeat the string method for both sides.

For road racing cars, you should turn the wheel both ways the same amount to check for Ackermann in both steering directions. You system might not be symmetric, especially if your car is a converted circle track car.

Now you have two sets of marks, one set for the right wheel and one for the left. Measure the distance between the straight ahead mark and the "turned" mark for each wheel. If you measure the same on each side, you have zero Ackermann. The direction of the tires is exactly the same when the wheel are turned as when they are straight ahead.

If you measure a half inch more for the left wheel than the right wheel, you have gained about one 1/8 inch of toe over what you started with. This is the most you'll ever need and might be good for very short ¼ mile ovals. For longer, faster tracks, you'll want less.

When using a laser, you will do the very same procedure, except you'll be using the laser instead of a string. Attach the laser to the hub and point it at a target ten feet in front of the axle, or down at tape on the floor. You'll make marks and measure between them.

If you are measuring a race car that turns left and right, do this same procedure with the wheels turned right also. These cars could include dirt cars and road racing cars. You might be surprised to learn that the Ackermann may be different each way as you turn the wheel.

Correcting For Ackermann – If you find that you have too much Ackermann, or if you have Reverse Ackermann where you lose toe (i.e. the left measurement is less than the right side measurement) and you want to correct it, here are some methods for doing that.

If there is excess angle in the tie rods from a top view (the inner mounts rearward of the outer mounts) there will be Ackermann present. Reducing this angle reduces the Ackermann and fixes it for turning both left and right, like you would do in a road racing car or a dirt car on an oval track.

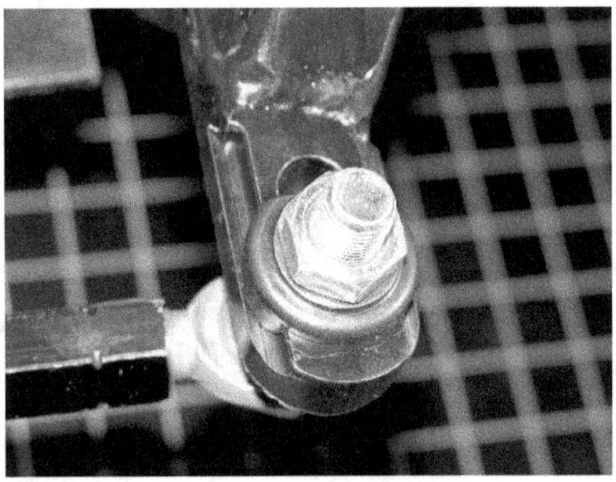

Most modern race cars that are designed with the rack and pinion steering systems use slotted steering arms so that teams can adjust the Ackermann in their cars. For most applications, the tie rod end bolt will be to the very front of this slot. Be sure to check your Ackermann using the methods described to be sure of what you have.

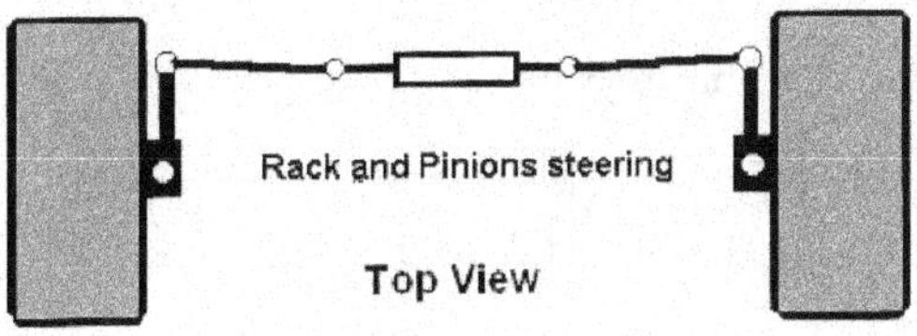

The rack and pinion steering systems can be adjusted for Ackermann by moving the rack (at the center) forward or rearward to change the top view angle of the tie rod. The greater the tie rod angle, the more Ackermann we have in the system.

For rack and pinion designed steering, one way to adjust for Ackermann is to move the rack forward or aft. To reduce Ackermann, move it forward and to increase it, move it rearward as in the case of Reverse Ackermann. Make sure both steering arms are the same length when making this change.

Another way to adjust Ackermann in a rack system is to change the length of both steering arms at the spindle. Be sure to change both arms the same amount and make sure they are the same length.

For drag link steering systems, we suggest a similar fix. Instead of moving the rack, we move the part of the drag link where the tie rods are attached forward if there is too much top view tie rod angle. This is not easy to do and usually involves installing a new drag link designed to place the inner tie rod mounts more forward. As with the rack system, we can (if possible in the design) change the length of the steering arms, moving both the same amount.

Some teams have taken it upon themselves to re-design the drag link, but we don't recommend bending or cutting and welding steering parts or doing something that might be unsafe. It is up to you to make sure any changes you make to your car are safe.

In past years, race cars fitted with the drag link steering systems used spindles that had steering arms with two holes. This was intended to be used to adjust the speed, or quickness, of steering by either using the front holes, or the back ones. Teams that experimented by using the back hole on the left and the front hole on the right learned very quickly that this produces a large amount of Ackermann and the car refused to turn after that change. Please don't do this. Small changes in the length of the steering arms have a pronounced affect.

Again, the above fixes are perfect for road racing cars and dirt cars where you would be steering in both directions and need to have the same toe settings whichever way you steer. But for race cars that only turn the steering wheel left, there is an easier way to go about this.

For a circle track car that always turns left, we can usually lengthen the left steering arm and that reduces the Ackermann effect. For stock type of cars with cast spindles, you'll need to cut and add material and weld back together. We would never recommend doing something that would be unsafe and you will need to determine if any changes you make are safe.

For most modern late model cars, the spindles are made with slotted steering arms, sometimes on both sides. You could lengthen the left arm and/or shorten the right arm length to reduce Ackermann. Remember that small changes in the length of the steering arms make a big change in Ackermann.

Summary – Ackermann was a design feature for early automobiles. It helped the wheels to track along the turning arcs in very tight turns having very short radii so they didn't create ruts in gravel driveways. In most types of racing, we have much larger radii for our turns and therefore, we don't need the Ackermann effect.

We have shown through calculations that a very small amount of Ackermann, less than an 1/8th inch for most tight turns, is all we need. This represents a fraction of a degree of steering, not whole degrees as some race teams think.

Exam - In The Context Of This Lesson:

Ackermann Is Present When?

1) The wheel steers when it bumps

2) Only one wheel turns

3) We gain or lose toe when the wheels are turned

4) Our steering arms are too short

We Can Correct and Reduce Ackermann By?

1) Moving the steering rack fore or aft

2) Changing the lengths of the steering arms

3) Moving the drag link forward or backwards

4) All of the above

For Cars That Turn Both Ways, We Reduce Ackermann How?

1) Shortening the left steering arm

2) Lengthening the right steering arm

3) Make both steering arms the same length

4) Moving the rack or drag length fore or aft

Lesson Eight – Alignment Front To Rear

In this Lesson, we will explain the concept of the Alignment of all four wheels and how to accomplish that. This Lesson will involve the initial alignment at static ride height and not dynamic alignment as the chassis moves which is a subject we will speak about later on in RCT Level Two. The overall thought presented here is that the four wheels need to be pointed in the same direction and lined up relative to the type of racing you do.

At this point in the RCT Course, Level Two, we need to align our wheels and tire contact patches. Many handling problems can be traced back to poor alignment issues. In this Lesson, we will tell you how to align your wheels and in what order using different methods.

Of all of the setup parameters, including moment center location, alignment ranks at the very top of the list. Why? Because when your alignment is wrong, no other setup parameters or changes will be able to compensate for it. The alignment package has a tremendous influence on the way a race car handles.

Alignment is a first priority in setting up any race car. It really doesn't take long and the benefits are as great, or greater, than any other setup task you will perform. In this Lesson we will explain race car alignment, why it is necessary and how to do it. For this exercise, we will be aligning a race car with a AA-arm front suspension with a solid axle rear suspension.

For formula race cars that are designed with the AA-arm front and rear suspension, the alignment will be for toe only since those cars are symmetrical and the tires and wheels are equidistant from the centerline.

The elements of proper alignment of the subject race car are as follows:

Toe Settings – For the purpose of this Lesson, we will set the car with zero toe, meaning that the front, and rear tires are pointing straight ahead. This will facilitate some of the procedures we will undertake. We will provide a separate Lesson on proper racing Toe settings later on in this Course.

Front to Rear Tracking – The tire contact patches must track along the same path and parallel to the centerline of the car for all race cars. In most cases for circle track racing where the cars turn left, we need to line up the right side tire contact patches. Our final alignment will show the right side patches are in line with one another along with the rear end, or rear axle, being perpendicular to that line.

For road racing cars and formula type cars, we will place the wheels equal distance off the centerline of the chassis for the front and rear wheels and point them parallel to the centerline for alignment purposes. In many cases, the track width (distance between the centers of the tires at the contact patch for opposing wheels) is different front to rear. We will set the toe for both ends of the car later on after the alignment has been completed.

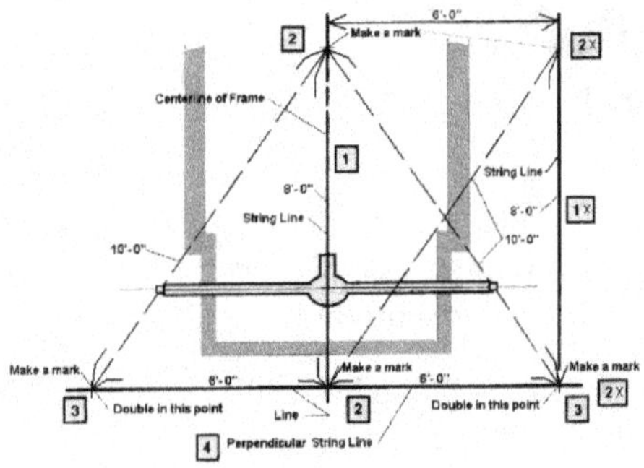

In the past it was common to string a car by making marks on the floor. This is a good method for checking rear end square and can be used to line up the right side tires for circle track cars. It is a little slower to do than the raised string method but can be more accurate due to the process of dropping points down to the floor with a plumb bob and then making direct measurements to the rear end. This sketch shows how to setup a parallel line outside the chassis as well as a line that is ninety degrees off of centerline to assist in rear alignment.

Rear Alignment – The direction, in relation to the chassis, that the solid axle rear end, or each of the wheels in a AA-arm rear suspension, are pointed can dictate how a car will behave in the turns. For example, on turn entry, if the rear wheels are pointed to the outside of the turn from the chassis centerline, no amount of setup tuning will prevent the car from being loose, or oversteering in road racing terminology.

That looseness will stay with the car throughout the turn, especially ruining the turn exit. If the rear end is pointed to the inside of the turn from the centerline of the chassis, the car will be tight, or understeer, through the middle and off the corner.

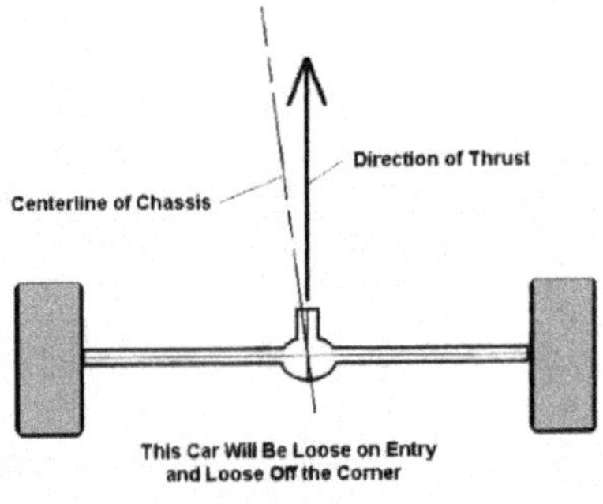

In this example, the rear end is pointed to the right of the centerline of the car. This will help free up a tight car in a left turn, but will produce a severe loose condition under throttle application off that same turn. The loss of bite off the corner will hurt the lap times more so than what might be gained from freeing the car up for corner entry, which is usually the reason why teams do this.

There should never be any reason to miss-align the rear wheels at static ride height. They should always be perpendicular to the centerline of the chassis and centered on the centerline of the chassis for road racing and formula cars, and with the outside tire contact patches lined up for circle track racing. This does not include rear steer, a subject we will get into later on in this Level Two course.

Measuring The Car - There used to be only one reliable way to align a race car and that was by using a string and either measuring to the tires at hub height or at the floor by creating right triangles on the floor to measure from. That is still a viable way to do it and necessary for the lower budget race teams.

A quicker and more accurate way to align the car in all areas is by the use of a laser system. The key to maintaining accuracy in a laser system is to be able to check the tool to make sure the beam is truly tracking parallel to the wheel and perpendicular to the mounting device. The actual laser must also be tested to be sure it is pointing inline with its base structure.

The True Track Laser System sends a single laser beam to the ground and to targets placed at the front and rear of the car. The system can check wheel base, alignment, Ackermann, toe, etc. It can also be checked for accuracy because we always need to check to see if the beam is pointed parallel to the body of the laser. We do this by finding the direction of the beam by its showing on a plate ahead of the car. Then we remove the laser, turn it 180 degrees and remount it to the hub fixture. We then rotate it around to again point to the target to see if it again hits the same spot on the target. If not, the beam is not pointing parallel to the laser body. This means we cannot rely on the laser for accuracy in doing any kind of alignment.

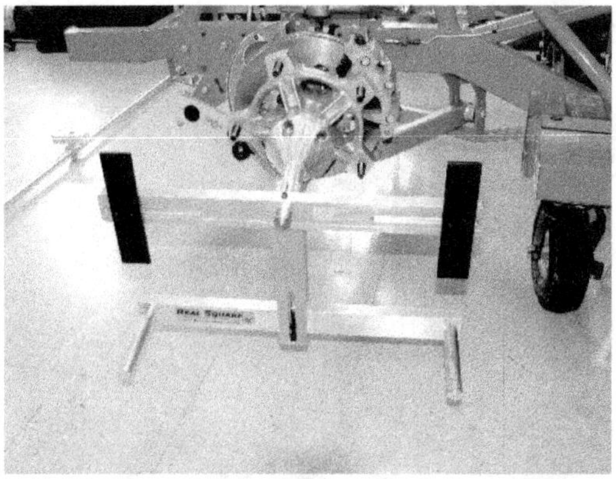

The Real Square™ Alignment Systems by DRP Performance uses scaled bars attached to a fixture that bolts directly to the hub. After the lasers are setup and aligned with the chassis, the wheel alignment and toe can be read on the scaled bars. Ackermann can also be checked with this system along with other parameters. This system too must be checked for accuracy of the direction the lasers are pointed. When the lasers are mounted to a bar, the measurement between the laser beam can be checked at different distances from the car. If the measurements are the same, then the beams are parallel and we can rely on the system.

The units I have used in the past are the True Laser Track, Gale Force Laser and the Real Square™ Alignment Systems. Each of these systems allow for checking the accuracy of the laser beams. This must be done each and every time we use the tool to check alignment.

The laser systems attach to the wheel hubs (models for both Wide 5 and 5 on 5 hubs) and are pointed at targets or measuring scales. For the "Analog" method, the nylon string we used was from the local hardware store, or you can use heavy monofilament fishing line.

We will go through the process using both methods so you can see how to perform each step. Note that the order in which we will proceed through the alignment process is important. Performing your alignment out of order can upset portions that have already been done.

Step 1 - Wheel Run-out Check – Check both the front wheels and the rear wheels for Run-out. This means that as the wheel rotates, the outer edge of the tire will wobble slightly. We must compensate for this slight distortion by finding the extreme high spot at a point equal in height to the hub height. We can simply use a jack stand to hold the tape steady and rotate the tire noting the distance from the stand. Once we locate the high point, we mark it with an arrow and then rotate the tire (be it front or rear) so the arrow is at the top pointing straight up.

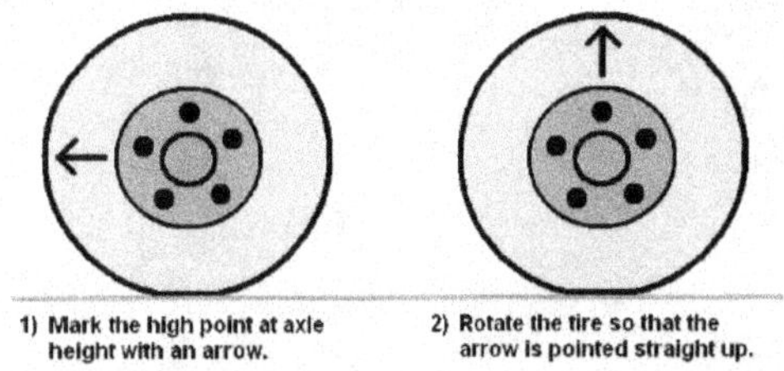

You need to check for run-out in the wheels before you start your alignment process. We tell you how in the Lesson. Move the high point to the top when you are finished. Now, check the toe at the rear end for all suspensions. Use toe plates or toe bars for the "analog" method and the laser systems for a more accurate assessment. Even small amounts of toe-in or toe-out are not acceptable when doing alignment. So, you might need to straighten any straight axle rear end before proceeding. When checking toe, be careful how each person holds the plates so that the measurements are consistent. Do the measurement several times to ensure accuracy.

When using the laser systems, attach the laser to the hubs remembering to thoroughly clean the surface of the hub and make sure there are no protruding threads from the bolt holes. It is recommended that you go over the hub surface with a flat file to eliminate any bulges or protruding edges of metal that would cause the laser to not be aligned properly.

Follow the manufacturer's recommendations for setting up the laser systems. Remember that the accuracy of the measurement is directly related to how closely you follow the directions and how carefully you read the system. There is a logical progression to alignment and each company has put a lot of thought into the methods. The end results, if properly applied, will be the same.

Step 2 - System to Frame Setup – Once the rear wheels have been checked for zero toe with both pointing straight ahead, square the laser system to the frame making the laser line parallel to the centerline of the car. Most car builders will align at least one frame rail parallel to the intended centerline of the chassis for circle track cars. It may be located on the right side or as the weight box on the left side next to the driver.

The offset distance of the laser beam to the frame can be taken by using a laser or by the use of a string. The string is ideal for a low budget team, but the laser systems are much more accurate and do more alignment tasks in a shorter amount of time.

You can also align the laser system to be parallel to the centers between the front and rear clip rails closest to the main frame rails for perimeter and road racing cars, or those where the dimensions from the centerline to the wheels are equal for both sides of the car. Some circle track cars are built this way and all road racing and formula cars are built this way.

For the analog method, this setup can also be done using strings. We will setup a "box" with strings on each side and at the front and rear. We use plumb bobs to make marks on the floor off the outsides of the frame rails or the centers of the front and rear suspensions. For Perimeter cars, or ones with no offset in the chassis, we can split the measurement between the frame rails to find the centerline of the chassis.

For offset chassis, we should be able to use a straight rail, or we could again split the front and rear frame rails. Measuring between the marks on each side, we can split the distance in half and place a mark at the halfway point which will represent the centerline of the car.

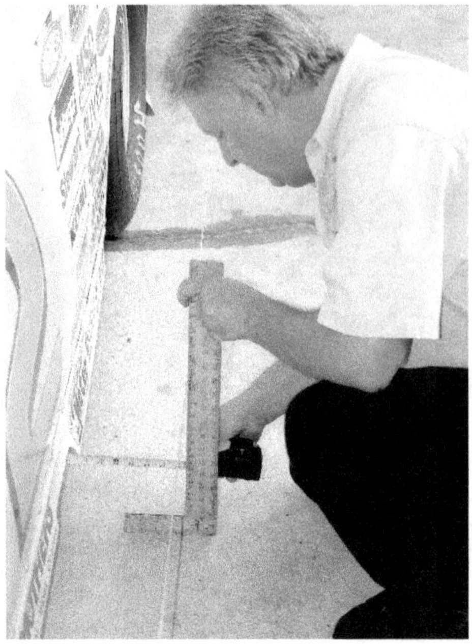

When using the string method, make sure your tape is level and not bent or held at an angle other than level. A tight string is always straight, but we can introduce errors in our measurements if we are not careful. Try not to eye ball too high above the string or your reading of the measurement will be in error.

Step 3 - Center the Steering Box – Center the front steering system. This is done by turning the steering wheel lock to lock and back half the number of turns from full lock in either direction. Once mid-rack (or mid-box with drag link steering) is found, lock the steering shaft against the frame rail with two vise-grip type of pliers applied in opposite directions. This will become obvious when you try to do it.

We want to make sure the steering is centered and then we can adjust the wheels to be pointed straight ahead. Once the steering box has been set to center, adjust each tie rod length so that the right and left wheels are pointing straight ahead. With the laser systems, this can be done quickly and accurately.

With the string method, run a string down each side of the car at hub height and parallel to the centerline you have established. Then, simply measure to the side walls of each front tire, or to the rim, with the tape measure and make both front and rear measurements equal by adjusting the tie rod lengths. We can set our race toe later on after we have aligned the car.

Step 4 - Rear Tire Contact Patch Alignment – Once the front wheels have been adjusted to point straight ahead and parallel to the centerline, we need to align the rear tire contact patches to the front tire contact patches. We do this by using our laser systems as described in the user manuals for each system.

If using hub mounted lasers, mount the laser to the front wheels only at this time. When moving the rear end to adjust the rear wheels to proper placement of the tire contact patches off the centerline, the rear end may steer. If the lasers were mounted to the rear wheels, then they would change direction as we move the rear wheels laterally.

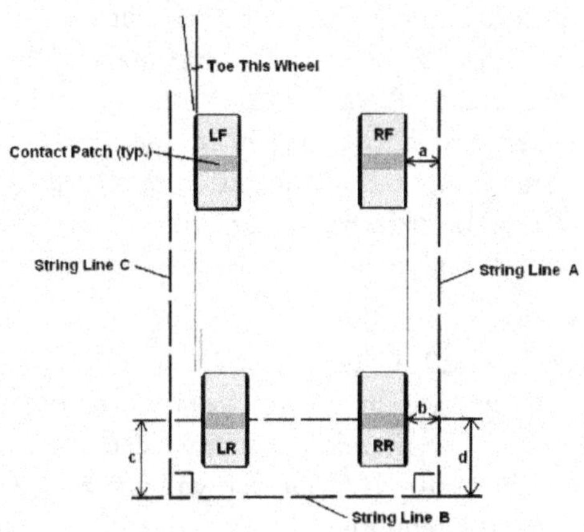

This is what we should end up with after aligning the rear to the right side tire contract patches for a circle track race car. Measurements "a" and "b" should be equal at the contract patch, as well as dimensions "c" and "d". We don't need to be concerned with the alignment of the left side tires for most circle track applications. There is much less loading on those tires through the turns and less influence on the handling of the car due to miss-alignment. For road racing cars with solid axle rear suspensions and formula type cars, we place the tire contact patches equidistant from the centerline of the car, or from the offset lines running outside the car.

Camber Offset Chart

Deg. of Camber	Tire Circumference 80"	82"	84"	85"	86"	87"	88"
2.0	0.444	0.455	0.466	0.472	0.478	0.483	0.489
2.5	0.555	0.569	0.583	0.590	0.597	0.604	0.611
3.0	0.666	0.683	0.700	0.708	0.716	0.725	0.733
3.5	0.777	0.797	0.816	0.826	0.836	0.845	0.855
4.0	0.888	0.910	0.933	0.944	0.955	0.966	0.977
4.5	0.999	1.024	1.048	1.061	1.074	1.086	1.099
5.0	1.110	1.137	1.165	1.179	1.193	1.207	1.221

The tire contact patch at the ground is offset from where the contact patch is located at the hub height. We need to know what this offset amount is when the wheel has either positive or negative camber. The chart shows how far the contact patch is offset from vertical due to wheel camber from hub height. If the front camber is in the negative direction (the top is closer to the centerline of the chassis), we subtract that offset distance from any measurement we make to the rear axle tire sidewall at hub height to align the rear to the front. For example, if we measure 10 inches to the front sidewall at hub height and the negative camber offset is 0.5", then when we measure to the rear tire sidewall (assuming it has zero camber) at hub height, we subtract the 0.5" from 10" to measure 9.5" to the rear tire sidewall so that the front and rear tire contact patches at the ground will then line up. If we just measured 10" to each front and rear sidewall, then the tire contact patches would not line up because of the front wheel camber.

To align the rear tires to the front tires for solid axle rear suspensions, adjust the panhard, J-bar, or Watts links, length so that the right side tire contract patches are: 1) the outside (the turns) contact patches are in-line for circle track cars, or 2) each of the two rear contact patches are equidistant from the centerline of the chassis for road racing cars.

NOTE: The rear wheels may also have camber set into them. Look at the chart to estimate how much to compensate for the cambers. Find the difference in the compensation amount and the offset read at the wheels to find how far from the string the wheels need to be in order to line up the tire contact patches. Negative camber settings will require subtraction of the camber offset from the hub height offset. Positive camber settings will require adding the camber offset amount to the hub height offset.

As we adjust the rear end side to side in a solid axle rear suspension, there is a possibility that the rear end may not be square to the centerline and will change as the rear end moves laterally. That is why we do the rear wheel front to rear alignment first. Now that we have aligned the contact patches the way we want, we will now need to check to make sure the rear end is perpendicular to the chassis centerline for the solid axle suspension and if not, adjust it.

Step 5 - Rear End Squaring Alignment – Once the front to rear tire contact patches alignment has been completed, we can then square the rear end. The rear wheels should always be set parallel to the centerline of the car for most types of racing. For road racing and formula type cars, there is a desired amount of toe-in usually designed into the alignment of the rear wheels. This toe-in can now be set.

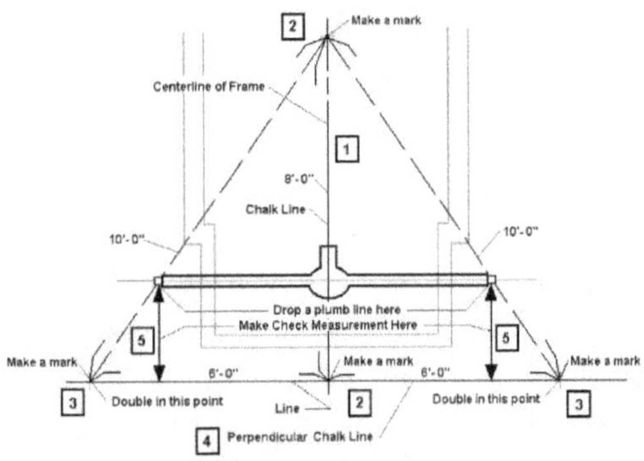

Here we see the layout for the process of setting up to square the rear end. Note that the lines formed by the intersecting marks are right triangles. We use 6-8-10 foot sided triangles. Take your time and measure very accurately. Re-measure if you feel that everything is not right once you pull the rear string line across the outer points of the triangles. A line over points 3 should align with point 2 on the rear perpendicular line.

For solid axle rear suspensions, we simply set the rear axle to be perpendicular, or at ninety degrees, to the centerline of the car. We do this by creating a line that is perpendicular to the centerline we have already established. Using a simple 3-4-5 right triangle with the lengths double to 6-8-10 feet, we can measure using the centerline to establish the perpendicular line that will be used to measure to the rear end. We simply make a mark on the centerline under the car and measure 8 feet back down the centerline to the rear and then make a second mark, the rear point being placed well behind the rear end.

We then set an outer point by measuring from the rear point 6 feet out and 10 feet from the front centerline point to establish a point on each side. A line between these two outside points forms a line that is ninety degrees off of the centerline that we can then measure from to the rear hubs to square the rear end.

Now all we do is drop a plumb bob line down off the front or back of the hub on each side and measure to the rear line. Those two measurements must be the same. If not, adjust the trailing arm lengths so that they are equal. Now you have squared the rear end and it is aligned properly.

Summary - Now we should have all of our wheels pointed straight ahead and parallel to the centerline of the chassis. And, our contact patches should be aligned so that either the front and rear tire contact patches are equidistant from the centerline for road racing and formula cars, or for solid axle circle track cars, with the outside contact patches aligned, or offset if so desired.

When we know that the wheels are pointed where we intend for them to be and the tire contact patches are the desired distance off the centerline, we can then know that we can proceed with the other setup settings that will not be affected by miss-alignment of the wheels.

Exam - In The Context Of This Lesson:

If Our Alignment Is Wrong, We Can Fix It How?

1) By using different spring rates

2) Through shock developement

3) By correcting the alignment issue

4) By changing the weight distribution

For Circle Track Racing, Where Do The Rear Wheels Point?

1) Into the turns

2) Straight ahead on entry and through the mid-turn

3) To the right of centerline

4) To the left of centerline

The Proper Tool To Use To Align The Race Car Is?

1) The string, it never lies

2) A hub mounted laser system

3) A remote mounted laser system

4) All of the above

When Using The String Method, What Is The First Thing We Do?

1) Find the high point or runout in the tires

2) Center the steering

3) Point the front wheels straight ahead

4) Square the rear end

When Doing Alignment, We Set The Front Toe To?

1) An eighth inch out

2) An eighth inch in

3) To zero toe

4) What works for our race track

Rear Solid Axle Alignment Is Always?

1) Perpendicular to centerline

2) Lining up the right side contact patches for circle track

3) Centering on the centerline for road racing

4) All of the above

Lesson Nine – Alignment Rear with Roll Steer

Now that we have the wheels aligned and the tire contact patches where we want them, we need to address the alignment of the wheels as the chassis dives and rolls through the turns. There may well be a different alignment once the chassis moves. For cars with AA-arm suspensions, front only or front and rear, the steer for that system would result strictly from bump steer, which we already talked about in this Course.

The steer and alignment that we want to concentrate on in this Lesson is what results from chassis motion that commonly exists in the straight axle suspension systems. That is what this discussion is all about. Most of these types of suspensions, with their various attachment designs, will steer as the chassis dives and rolls through the turns.

Rear steer is mostly desired in dirt track racing. Here we see two different angles for each of these cars. The 45 car has much more rear steer (notice the LR tire moved forward in the wheel well) than the 0 car. The rear steer is caused by the angle of the links that hold the rear end in the car and the movement of the chassis pushing and pulling on the links to in-turn push and pull the rear end. Note also that the greater rear steer causes a larger angle of the chassis to the direction the car is traveling. The 45 car is "hanging it out" much more than the 0 car. There is a reason why dirt teams do this.

It is very important to know how this happens and how we can dictate when and how much our rear end steers as we enter and drive through the turns. The presence and amount of rear steer needed is very dependent on what type of race car it is.

For asphalt race cars competing on circle tracks or even road racing, we neither need nor want very much rear steer at all. The rear of these straight axle cars are very sensitive to the direction the rear tires are pointed. Even if we desire and incorporate rear steer, a little goes a long way.

For dirt cars, it has been the trend to cause a lot of rear steer for various reasons. We'll get into those cars designs and see where we can cause steer that might be beneficial to performance. Let's now discuss how rear steer happens and how we can design our cars to steer, or not.

What Is Rear Steer? - Rear steer in a race car is a geometric effect that is mostly caused by suspension movement. Under the right conditions in measured amounts, RS can be beneficial and enhance performance. Under the wrong conditions, it can ruin your handling. We need to have a solid understanding of what produces RS and what effect RS has on the handling in our cars in order to have the best performance.

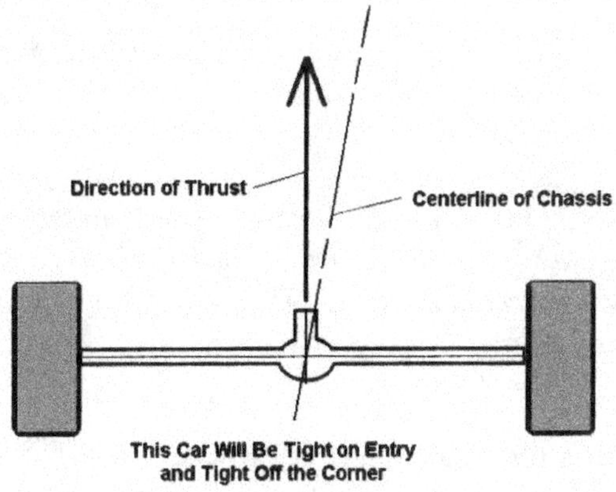

With the rear end pointed to the left of centerline, the steering will cause the rear of the car to want to run under the front end causing a very tight condition, especially under acceleration. With the rear end pointed to the right of centerline, the car will be freed up going in, through the middle and possibly loose off the corners with the rear end wanting to run around to the right of the front.

When we have rear suspension movement, as the rear corners of the car move and the control arms that locate the rear end fore and aft move, the components for each side can cause the wheels to either move forward or to the rear.

A rear end that is steered to be pointed to the left of centerline will cause the thrust angle to be left of centerline and that makes the car tighter on entry and tighter on exit under acceleration. A rear end that is steered to the right of centerline will cause the thrust angle to be pointed to the right of centerline and that makes the car loose on entry and loose on exit under acceleration.

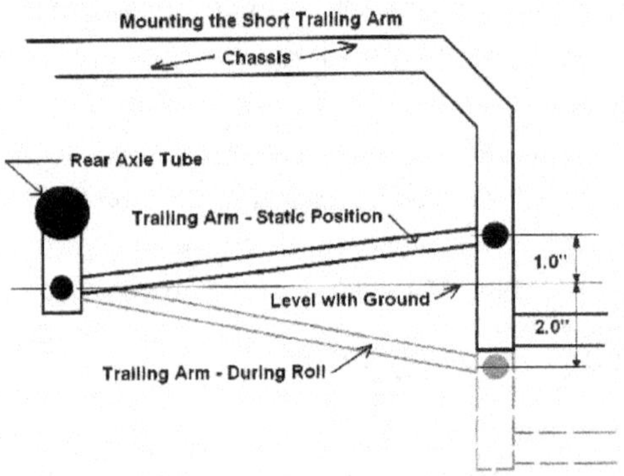

The three link rear suspension system can produce rear steer in both directions. As an example, let's say the right rear of the chassis moved down 3.0" through the turns and the front mounting point was 1.0" higher than the rear mounting point. Then as the chassis moves down on the right side 1.0", the RR wheel will be moved back as the front mount approaches the height of the rear mounting point. As the front mounting point continues to move down, the RR wheel will be pulled forward and return to the original fore and aft position as the front mount goes 1.0" down from level. Normally, to produce a small amount of rear steer to the left on asphalt, we mount the front pivot point 1/3 of the total travel distance higher than the rear pivot. In this case, the chassis traveling another 1.0" to make up the total of 3.0" of travel continues to pull the rear end forward a slight amount to create the rear steer to the left.

The asphalt racing surface provides a lot of traction, even on those flat "slick" tracks. Because there is very little slip of the tires on asphalt, the range of useable rear steer is very small. There are six predominant rear suspension systems used in stock car racing and all of them produce some amount of rear steer. They are:

The Three Link System– This system has two trailing arms mounted near the rear tires under the axle tubes and one third link, usually mounted on top of the rear differential. The top, or third, link controls rear end wrap-up. The trailing arms can be permanently mounted from a top view to be parallel to the centerline of the car or angled with the front mounts closer to centerline. They can also be designed to be adjustable to create different angles from a side view.

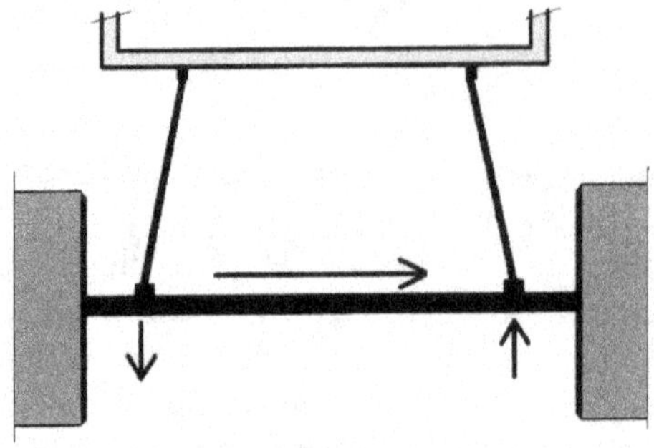

Here we see the lower links in a three link suspension. The front mounts are attached to the chassis to be closer to the centerline of the chassis than the rear mounts. With this design, if the rear end were to move laterally, there would be rear steer. If the rear end moved to the right, the side would move forward and the left side would move rearward. This causes rear steer to the left which would tighten a car turning left. If the rear end moved left, the opposite would occur, and the effect might loosen the car. This is important to think about with a panhard bar system. Which way does the bar push the rear end as the chassis travels?

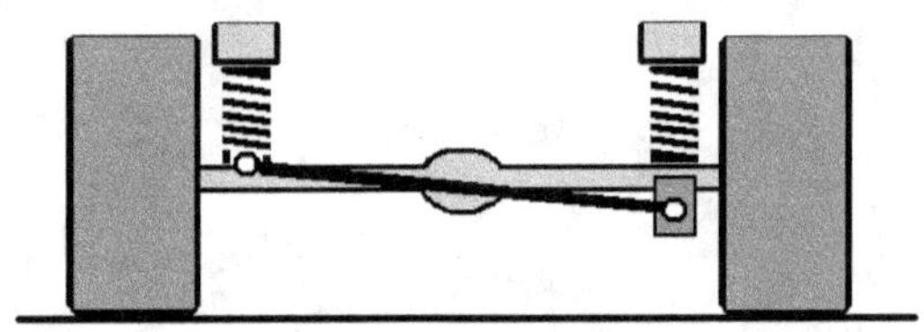

It is a result of panhard bar angle, and the motion of the chassis in roll, that can either pull the rear end to the right or push it to the left. If the panhard bar were mounted with the right end (chassis mount) lower the left side that is mounted to the rear end axle housing, then with chassis movement down resulting from chassis roll, the rear end would be pulled to the right. If the bottom two links in a three link system were like the top view angles in the previous illustration shows, then the rear end would move into steer to the left. If on the other hand, the panhard bar were mounted with the right end higher than the left end in this same car, then the motion of the chassis would push the rear end to the left and that would produce rear steer to the right. This would act to loosen the car, but also cause an angle to where aero flat plate effect might just play into causing a force to the inside of the turn helping to go faster through mid-turn. Many of the modern asphalt late model circle track cars have mostly flat sides that would help create this effect.

RS in this system is caused by chassis movement which can produce several secondary effects. Usually, the right rear corner of the chassis moves more than the left side, and on most flat to medium banked tracks, the left rear moves very little.

On most asphalt 3-link cars, the RR trailing arm mostly controls RS due to body roll. We usually need to position the angle of the trailing arm so that the front mount is higher than the rear mount by roughly one-half to one-third the distance that the front mount will move down during cornering. This compensates for the movement of the front end of the link and in reality pushes that side of the rear end back first and then pulls it back to where it was resulting in very little rear steer.

The Truck Arm System – The truck arm system has been adapted from the design for a 1964 Chevy pickup trucks and is used on many Late Model Stock cars as well as the three premier divisions of Nascar, being the Camping World Trucks, the Xfinity cars and the Cup cars. These systems only steer to the left and have a limited amount of steer. The roll of the chassis and the movement of the panhard bar are the two components that influence the amount of steer in these systems.

As far as geometry related to rear steer is concerned, this is a good system for asphalt cars. The amount of rear steer due to body roll is regulated by the height of the front mounts of the arms which are always mounted lower than the rear point of rotation which is the axle. Rear steer amounts due to the panhard bar vertical movement are regulated by the angle of the bar.

The Standard Dirt 4-Bar System – The 4-bar suspension is highly adjustable and can be made to steer in either direction. The rule about never steering the rear end to the right on an asphalt car does not apply to a dirt car. There are times when we definitely want the rear to steer to the right.

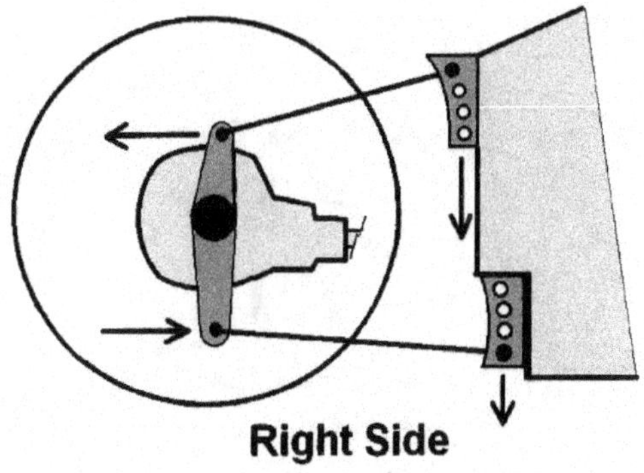

If the links on a 4-bar car are set in certain predetermined holes, the movement of the top and bottom mounts at the rear end will compensate for fore and aft movement that results in zero rear steer. Most of the existing rear steer comes from movement at the left side of the chassis and the mounts located there.

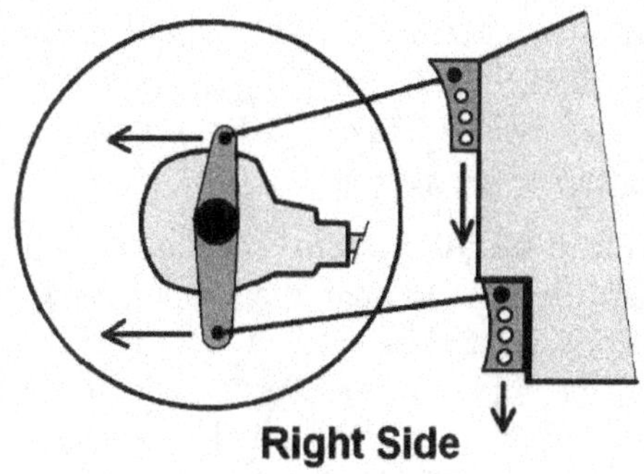

Here we can see how, on the right side of a four-link system, mounting the forward ends of the links in the top holes causes movement of the right side of the rear end back when the chassis moves down. On the left side, the opposite happens when that side of the chassis moves up and the rear end moves forward.

To determine the amount and direction of rear steer in this system, we must know the movement of the corners of the car. On dirt cars, this can be very different than what we see on asphalt cars.

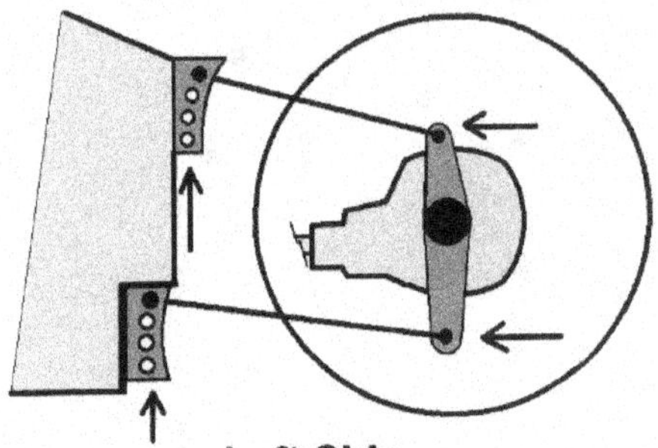

The way a four-bar car steers so much is by the use of high angles in the links on the left side of the chassis. When these links forward ends are mounted higher than the axle mounts, the motion of the chassis rolling up going through the turns plus the force of the thrust of acceleration pushing the chassis up cause the left rear wheel to move forward to create a lot of rear steer. This pushed the rear of the car out to the right by necessity in order for the rear wheels to line up with the direction of travel. The flat plate aero created by this angle of the body helps push the car to the inside of the turns and helps it go faster, especially on dry slick tracks that have low grip.

The Metric 4-Link System – The metric four link is a widely used system that comes on some models of stock automobiles. It uses four links as the name implies that are not parallel to the centerline of the car. The top links are angled from a top view with the front pivots wider than the rear pivots. The lower links are angled from a top view with the front pivots narrower than the rear pivots.

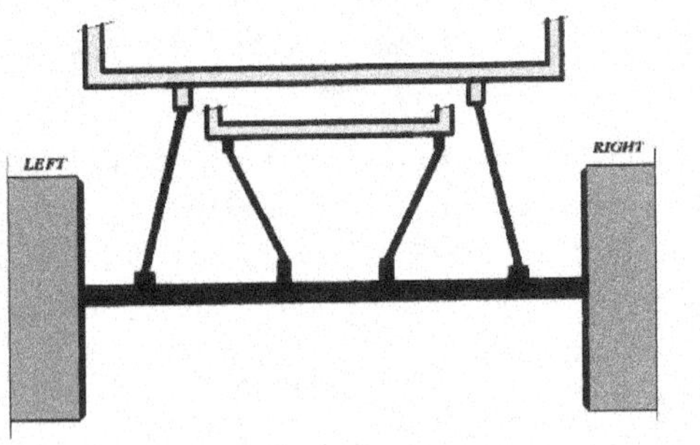

The Metric 4-link suspension used on many models of street stock and some trucks has two links above the rear end and two links below the rear end. They are angled from a top view to prevent the rear end from moving side to side as the chassis experiences lateral loading through the turns. These rear suspensions have a traditionally high rear moment center. The rear steer is dependent on the ride heights in the rear which determines the link angles.

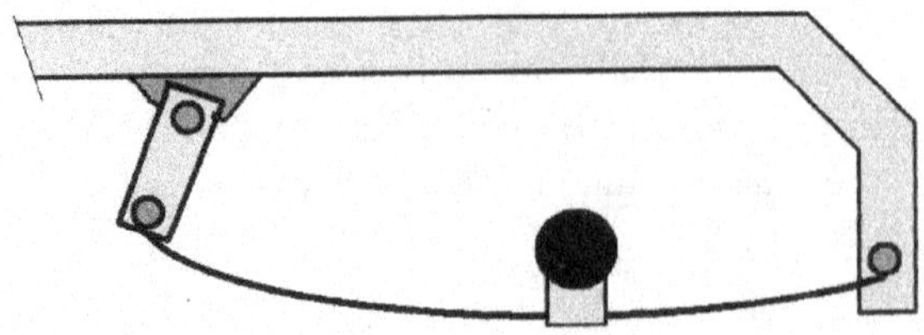

We can see how the front and rear chassis mounts in a leaf spring suspension act somewhat like the links in a three link system. The rear mount swings in a shackle and has no effect on rear steer, or movement of the rear end. But the front mount can move the rear end on that side fore or aft depending on if it is higher or lower than the center of the axle. If higher, it would push the axle back as the chassis moves down. If lower, it would pull the axle forward as the chassis moves down.

With this system, the rear end stays located by virtue of the opposing angles of the upper and lower links. There may be rear steer inherent in this system depending on the ride heights because the original car was designed for stock ride heights. When we race a stock car, we usually change the ride heights. Under most current rules, there is no adjustment for amounts of rear steer with these systems.

Leaf Springs Systems – The leaf spring rear suspension system locates the rear end fore and aft as well as laterally using only the leaf springs. There can be a small amount of rear steer as the chassis rolls and squats, but it is both minimal and mostly fixed as far as adjustability. The advantage of this system is that it keeps the rear end squared up and the thrust under acceleration more straight ahead, if that is what is needed.

The height of the front eye of the leaf spring in relation to the axle centerline determines the amount of rear steer. Spacing the leaf different heights from the axle tube can be a way to change the height relationship of the front eye and the axle, which will change the amount of rear steer for this system.

The Z-link System – The Z-link rear suspension, or swing arm as it is sometimes called, is another system used on dirt cars. Compared to the 4-bar cars, it has more limited adjustment for rear steer and historically has worked well on the tighter and more highly banked race tracks because the direction the rear end is pointed is more straight ahead.

Some manufacturers have added multiple mounting points on the front and rear chassis mounts. This helps make the rear steer characteristics more adjustable for the changing conditions.

A Z-link suspension system uses a link extending from the under the rear end forward to the chassis and one from the top of the rear end rearward to a mount on the chassis forming a Z pattern. Most designs use many mounting holes that enable the team to adjust for the amount of rear steer in those cars. You can set your angles and then move the chassis to determine how much the axle is pulled forward, or pushed backwards. This helps to determine how much rear steer is in a certain system. There is no right or wrong in these designs, but experimentation at different race tracks helps teams find the right combination that will work for both the track conditions as well as what the driver feels comfortable with.

Measuring Your Rear Steer – To measure the rear steer in your car, first know the shock travels at each corner of the rear. Then you can recreate the rear attitude at mid-turn. You can use a tape measure for simple measurements or a laser system for more detailed analysis. With the tape, simply measure, at ride height, from the rear wheels towards the front of the car level with the ground to a point. This can be a mark on a tape near the front wheel. For the laser systems, follow the manufacturer's instructions.

Move each of the rear wheels in any increment you choose ending up in the position they are at when the car is at mid-turn. and measure the fore and aft distances at each increment. If the movement on the right and left are in opposite directions, add them to determine the amount of rear steer. If they are in the same direction, subtract them. You might be surprised at how much rear steer is going on.

Why Do We Want Rear Steer? – Dirt cars need all of the aero help they can get. There is an aero effect called Flat Plate aero. That means, if a flat plate were traveling through air at an angle, it produces a force. If the front of the plate were lower than the rear of the plate, then that force would be directed down. The air being pushed out of the way exerts a force on the plate in that direction the plate is pointed in.

If the flat plate were mounted, or positioned vertically, then the force would be exerted laterally. If the front of the flat plate were positioned left of the rear, then the force would be directed to the left. A cars body can be made to create an aero force in a lateral direction.

A modern late model dirt cars body is flat on the left side. That side acts as a flat plate and when driven through the air at an angle with the front left of the rear, it exerts a force to the left. If that car turns left, then the aero force helps it resist lateral forces that try to run it off the track in the direction of the lateral force.

So, why do we need rear steer in a dirt late model? It's not related to helping the car turn in or help it point off the corners like some people think, it is to help provide the lateral aero force pushing to the inside of the turn needed to resist the lateral force trying to push the car to the outside of the turn.

Remember too that there is a limit to how much angle you can use to create the aero sideforce. Too much angle and the effect is diminished and the sideways attitude creates a lot of drag slowing the car down.

On asphalt, we can utilize rear steer in smaller amounts than what is used on dirt for a similar reason. If the rear were hung out a bit, there would also be a lateral aero force helping to keep the car on the track. Everything helps and if that force can help us go a little faster, even better.

But along with that steer to the outside, we need to be able to get off the corner and steer to the outside of the turn is detrimental to getting up off the turns under power. In the next Lesson, we'll tell you how to have your rear steer and get off the corner with rear steer under power.

Summary – Rear steer design is something we need to know about and utilize, but telling you exactly how much is like telling you exactly how far to go into the corner before braking. Every track is different and every car is somewhat different, even in the same class.

So, in this Lesson, we wanted to provide enough information about how to create rear steer and why so that you can make an intelligent decision as to how much would be good for you. The rear steer we get from chassis dive and roll is not the entire picture of rear steer. It is only the beginning. In our next Lesson we will describe another type of rear steer that can be utilized along with the dive and roll rear steer.

Exam - In The Context Of This Lesson:

Rear Steer Is Defined As The?

1) Rear direction of the tires when we turn

2) Change in direction of the rear tires when the chassis moves

3) Alignment at static ride height

4) Alignment at dynamic ride height

For Asphalt Race Cars, We Need?

1) A lot of rear steer to make the car tight

2) Rear steer to the outside of the turns

3) Little or no steer at all

4) The rear wheels to steer with the front wheels

For Dirt Race Cars, We Need?

1) Rear steer to the outside of the turns

2) More steer for slick tracks

3) Less steer for tight tracks

4) All of the above

In A Three Or Four Link Rear Suspension With Straight Axle, The Rear Will?

1) Never move through dive and roll

2) Will always steer through dive and roll

3) Will always steer to the right

4) Will always steer to the left

For Dirt Race Cars, We Use A Lot Of Steer Why?

1) To make the car tighter

2) To make the car looser

3) To make the car run at an angle for better aero

4) To help turn the car

Truck Arm Systems Always Steer?

1) To the left

2) To the right

3) Very little at all

4) Quite a bit

Steer In A Leaf Spring System Is Determined By?

1) Height of the rear spring eye

2) Height of the front eye

3) Angle of center of axle to the center of the front eye

4) Angle of rear eye to front eye

We Use What To Measure Rear Steer?

1) The shock travels

2) Chassis roll angle

3) Chassis dive

4) All of the above

Racecar Technology – Level Two

Lesson Ten – Rear Steer Under Power

In the last Lesson we learned how to utilize rear steer that comes from chassis dive and roll. We talked about the straight axle suspension system exclusively because that is the only system we can use to create significant rear steer. The AA-arm suspension systems don't work well to produce rear steer, and we wouldn't want them to. So again, this discussion involves only the straight axle rear suspension related to rear steer, and now as it is created while under power.

Some classes of dirt cars use quite a bit of rear steer to provide some advantages for getting around the corners. The car is not loose, as in the rear tires are sliding, they are steered this way on purpose.

To refresh, rear steer can occur while the chassis dives and rolls through the turns. We can use this steer in various ways depending on the type of race car and the needs of the team. The other form of rear steer happens while we are applying power when accelerating off the corners. We use the torque of the engine and the thrust of acceleration to create rear steer independently of the rear steer we get from dive and roll.

In fact, the RS we get from acceleration can counteract the rear steer we got from dive and roll. That is an interesting concept and one we will explore further. In my opinion, very few race teams know how to properly utilize RS. Of all chassis related performance areas, Power RS is second in importance being just below mid-turn dynamic balance.

We use mid-turn dynamic balance to make sure the loading on the tires is correct and what the car wants, and we use RS from dive and roll to help the car stay on the track. But at some point we need to accelerate off the corners and the RS we have to get us through the middle of the corner can be detrimental to having good bite off the corner. In other words, we need the opposite RS for corner exit.

How Does Power Induced Rear Steer Happen? – Three things are happening when we accelerate that can be used to create RS under power. One, we have weight transfer from the front to the rear. As this weight transfers, it puts more load on the rear springs and they can be made to compress causing the rear of the race car to move downward. This front to rear weight transfer only happens while under acceleration and therefore doesn't affect the mid-turn, steady state handling.

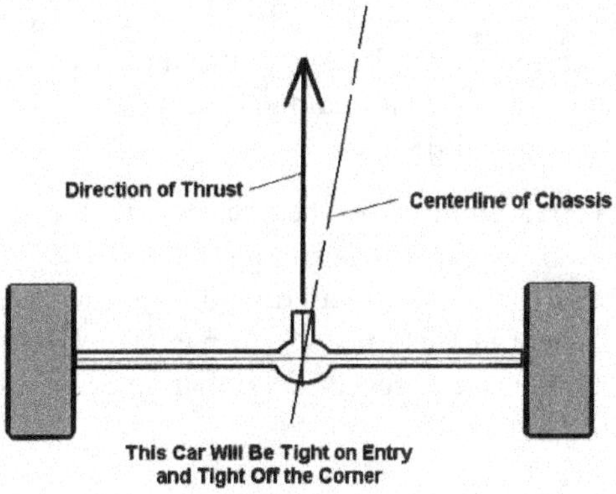

A race car steered this way, with the rear tires pointed to the left of centerline, will always be tight. The front must try to steer to compensate for the rear of the car moving left. This is almost never desirable.

The second thing that happens is that we have the torque of the motor providing thrust to push the car faster, and so there are uses for that thrust to create rear steer only under acceleration. And the third thing that occurs while under acceleration is that engine torque is also trying to twist or rotate the rear end as the pinion tries to climb the ring gear in the rear end. The rotation is clockwise when viewed looking at the left side tire and we can use the rotation to create rear steer again only while under acceleration off the corners.

OK, all of that sounds pretty cool because what we do to the setup and design of the race car for better performance off the corner must not upset or interfere with the mid-turn handling. Remember, our number one goal is to create the ideal mid-turn dynamic balanced setup. Anything we do to the chassis to improve, or solve problems with exit performance that changes the mid-turn performance takes us backwards in overall performance.

Now that we have three components we can use to enhance corner exit performance, without interfering with the mid-turn performance, let's study each one, one at a time. These effects can be mixed or used individually. As with any performance adding technology, more is not necessarily better. There are limits to how far we can go with our use of RS.

Weight Transfer Utilization – The use of weight transfer to provide better bite off the corners has long been in use. Older circle track setups often used softer right rear (outer rear springs) springs than the left rear (inside) spring. When the weight transferred and the rear of the car compressed, the left rear with the stiffer spring compressed less and gained more of the weight that had transferred.

This increased the cross weight percent and tightened the car when exiting off the corners. As setups became more modern and teams started running stiffer RR (outside) springs, the effect was diminished or reversed, so that option became un-useable.

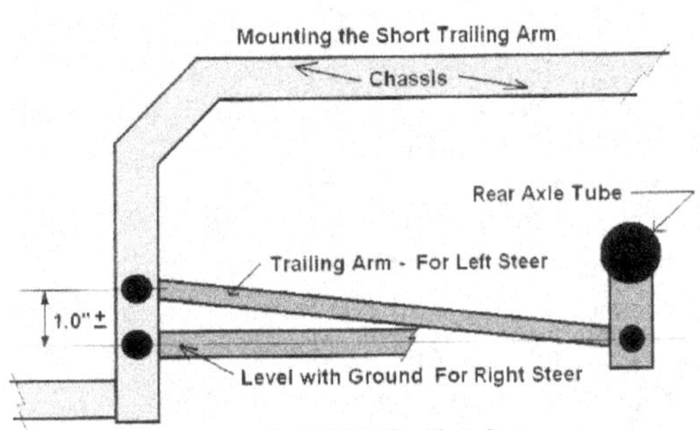

When we add angle to the left trailing arm in a three link suspension system, when the car squats under acceleration, the angle will drive the left rear back causing rear steer to the left for a left turning race car. We can vary the amount of rear steer by changing the trailing arm link angle.

Later on, teams started utilizing the rear link angles to help create RS. As the car squats from the result of added loading from the weight transfer, adding angle to the left rear (inside) link, with the front mount higher than the rear mount, on a three link suspension caused the left (inside) wheel to be pushed back. This created RS to the left (inside) which definitely will create more grip in the rear tires.

Along with that effect, installing a shock in the LR corner of the chassis that is stiffer in compression settings than the one in the RR corner will also load the LR tire more, but only while the shock was moving as the weight was transferred onto the spring. Once the shock stopped moving, the effect goes away whereas the RS stays as long as the car is accelerating.

This RS effect is significant because on most flat and medium banked race tracks, with modern circle track cars and even with road racing cars, the inside rear corner does not move vertically much from where it is statically at ride height. Therefore, any angle we put in the link to the corner has no affect until that corner moves and that movement is mostly while under acceleration.

Thrust Created Rear Steer – As we talked about above, when the car accelerates, there is thrust created that causes the rear end to push against the links and the chassis. This thrust is significant. To accelerate a 2800 pound car from 60+/- MPH up to 120 MPH takes a lot of thrust. If we measured it in pounds of thrust, it would come in at around 1,500 to 2,000 pounds total. If we divide that between the two rear tires like we should, each tire would be pushing at 750 to 1,000 pounds. That is a lot.

One way to create RS is to install compressible links in a three or four link suspension so that when the car accelerates and creates thrust, the link will compress and that side of the rear end will move forward creating RS. This rear steer almost always points the rear end to the inside of the turn to create more angle of attack for the rear tires. This creates more grip, or side force enabling the driver to use more throttle without the rear tires slipping out.

This concept can be used in reverse as well. With some stock classes, a compressible link or mount can be made to work both ways, compressing while under acceleration and extending while under braking. Both produce RS, the first pointing the rear end to the inside of the turn, the second pointing the rear end to the outside of the turn helping the car to turn.

I've personally seen this used in a stock class at a tight, flat turn race track. That car was dominant and no other car was doing anything like that. For those of you who are racing in stock classes, or know someone who is, this little gem might just give you a huge edge if done right.

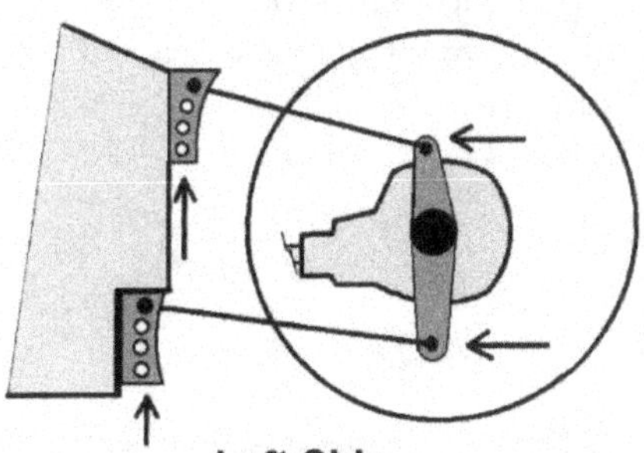

When the left side links on a four-link suspension are in the top holes, as the car accelerates and the rear wheel drive forward, the angle of the links pushes the chassis up at the left rear and forces the right front corner of the car down. This is a desirable attitude for maximizing the aero efficiency of the car.

Torque Induced Rear Steer – The other effect that can be used to create RS only under acceleration is from the torque of the motor trying to rotate the rear end. As the pinion gear in the rear end pushes against the ring gear, it tries to climb the ring gear and as a result pushes the rear end housing and tries to rotate it. This rotation is clockwise when looking at the left rear tire.

The three link rear suspension, and the lift arm and torque arm suspensions all create AS. On the three link, the link is pulled under acceleration, and the lift and torque links forward ends are lifted. When we allow the rear end to rotate a small number of degrees, we can use that rotation to create RS.

We do that by installing a compressible spring or rubber type of device in the third link or lift arm that would compress and allow the rear end to rotate. When it does rotate, steer is created by staggering the height of the side links and the distance from those mounts to the center of the axle.

When the rear end moves in rotation, the axle moves back because it pivots on the bottom link mounts, but if the left (inside) link mount was lower than the right (outside) link mount, then the left side of the axle would move farther back creating rear steer to the left, or towards the inside of the turn.

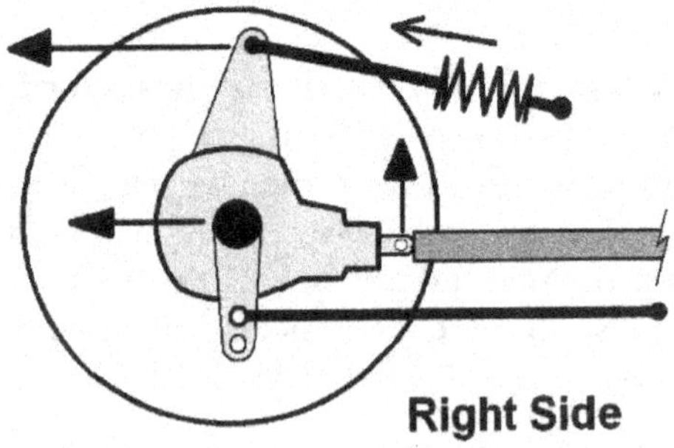

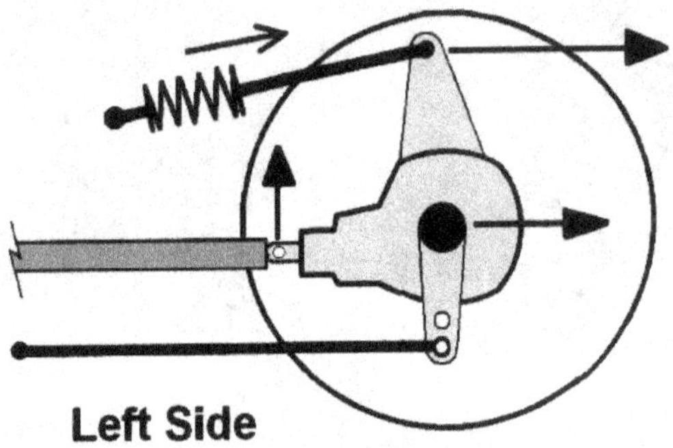

This type of rear steer is utilized mostly on asphalt race cars and only with circle track cars because it only works when accelerating off of a turn in one direction. It would work the opposite way, or the rear steer would be the other way, if accelerating off a turn to the right making the car have less grip rather than more.

The actual amount of steering, or degrees of rear end change in direction, is very small, but then again, asphalt race cars are very sensitive to any amount of rear steer in any suspension system. In practice, we usually see where the left rear tire will move back 0.050" to 0.075". The angle created is about one tenth of a degree. To put the numbers into real tape measure dimensions, the left rear tire is moving a little less than, or more than, one sixteenth of an inch.

Roll Steer Plus Power Rear Steer – We talked about RS created by chassis dive and roll in the last Lesson, and now we can talk about using both RS types to enhance mid-turn and exit performance. Let's start with asphalt cars because they are a little different in needs from dirt cars.

On asphalt, we might need a slight amount of rear steer to the right (inside of the turn) to allow the car to rotate into and around the corner and lessen the amount the driver has to turn the steering wheel. This is, or could be, a two edged sword so to speak.

If the RS to the right did indeed help us into and through the middle of the turn, then surely it would make us loose off the corner because the thrust angle would be right of centerline and that would want to take the rear of the car towards the outside of the turn. We would quickly learn this is not a good thing if we didn't lose the car in the process.

What if we could loosen the car into the corner and through the middle, but tighten it off the corner by using RS to the left (inside) to counter or reverse the initial RS to the right? From what we learned in the bulk of this Lesson, yes we can do that. And winning teams are doing that right now, we just don't hear much about it, until now.

For dirt cars, especially the late models and modified classes, rear steer is being used to create aero side forces, as we have explained, to enhance the lateral force working against centrifugal forces trying to push us off the track in the turns. That is all well and good, but when we accelerate off the corners, we don't need as much aero help anymore since we are going more straight and not turning.

It is at this point that power induced RS can be utilized to bring the rear end back to pointing more inline with the centerline of the car. This too is a two edged sword because with those cars, the driver is sometimes on the throttle through mid-turn and the RS would be working back and forth and possibly destabilize the car.

Only the team can determine if and when RS either way would work to enhance performance. Our job in this school is to teach you methods and give you ideas, not setup your car. There are too many variables involved to say it has to be one way or another.

Exam - In The Context Of This Lesson:

Rear Steer Under Power Can?

1) Add to the mid-turn rear steer

2) Subtract from mid-turn rear steer

3) Be designed to be independent of mid-turn rear steer

4) All of the above

One Way To Create Power Rear Steer Is?

1) Using the squat of the rear from weight transfer

2) Using the acceleration force driving the car forward

3) Using the torque of the motor

4) All of the above

In A Three Link Suspension, More LR Angle Will?

1) Cause rear steer to the right under acceleration

2) Cause rear steer to the left under acceleration

3) Help keep the rear end straight

4) Make the car loose off the corners

In A Three Link Suspension, More Third Link Angle Will?

1) Cause more squat and rear steer

2) Cause less squat and rear steer

3) Reduce weight transfer

4) Increase weight transfer

In A Four Link Dirt Suspension, More Left Side Link Angles Will?

1) Cause more rear steer under power

2) Help produce aero side force

3) Load the LR tire under acceleration

4) All of the above

Power Rear Steer Should?

1) Be used to correct corner exit problems

2) Always add to mid-turn performance

3) Take away from mid-turn performance

4) Be used to correct mid-turn handling

Racecar Technology – Level Two

Lesson Eleven – Toe Front and Rear

This Lesson is a short and concise review of what Toe is in a suspension system and some theory on why we have it and how to set Toe. Basically, Toe is a feature of a AA-arm suspension that adds stability to the car. If is defined as: the miss-alignment of two tires/wheels on a suspension system. Perfect alignment is what we know as both wheels being pointed in the exact same direction. With Toe, we cause them to point in slightly different directions.

To further define Toe, Toe-In is when the front of the tires are closer together than the rear of the tires. Toe-Out is the opposite. Normal front Toe-In for most applications is from 1/16" (0.063") to 3/16" (0.188") of Toe. Some teams have actually used ¼" (0.250") of toe for very short ¼ mile tracks, but based on calculations, that is too much in our opinion.

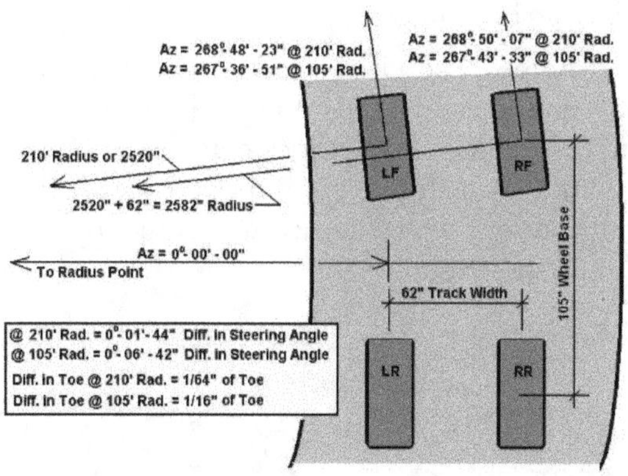

These calculations show that we need very little Toe-Out in the front wheels in order for each wheel to track correctly along its individual arc. The different arc radii create the need for small amounts of Toe.

We need Toe in our race cars to stabilize a AA-arm suspension, be it front or rear. In a straight axle suspension, we usually do not use any toe. The amount of toe is similar to the amount of Ackermann we learned about. We need very little. For most double A-arm applications, we need from 1/16 inch to ¼ inch of Toe-out for most circle tracks and road courses. For the rear AA-arm suspension, we usually need Toe-in, say from 0.020 to 0.035 inch of toe depending on the race track.

How Much Toe Do I Need - The general rule is that you need less front Toe-Out for: 1) larger radius turns, 2) Higher banked tracks, 3) very fast tracks with large radius turns. As for rear Toe-In For a AA-arm suspension, much smaller amounts are used. These range from 0.020" to 0.035" in most cases.

It has been learned through the years that this miss-alignment does stablilize the car. And the amount and direction of the Toe is dependent on the type of suspension. A race car with rear wheel drive and a AA-arm front suspension usually uses Toe-Out in the front. For a rear AA-arm suspension such as on a prototype or formula car, the Toe is usually set to Toe-In.

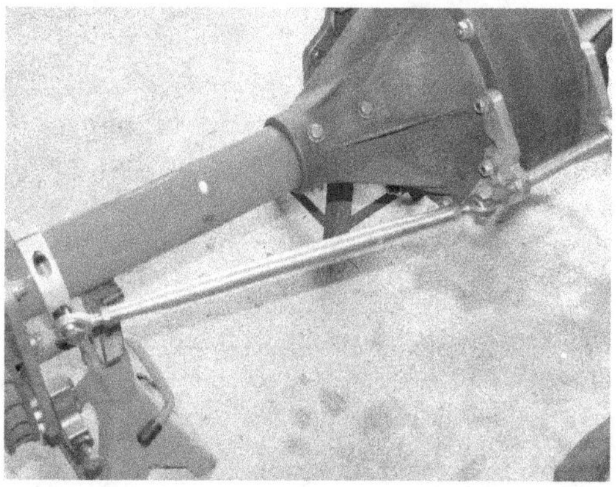

Like we said, for a straight axle rear suspension, we need very little, or no, toe. This rear end has a device that can be used to straighten the rear end and/or help it to stay straight under acceleration by use of braces coming off the center section.

Proper toe settings at the front and the rear are very important. A set of tires that are not toed correctly will create a lot of drag, much like applying the brakes and sometimes more efficiently than the actual brakes. The standard of the circle track industry is to toe the front out 1/16" to 3/16". The rear wheels are to be straight ahead and parallel to each other, having no toe at all.

We can use lasers to measure toe by placing a target at a distance in front of the wheel hub and another one to the rear at the same distance. If we put the targets far enough apart, then the toe will be magnified and make it easier to set exact toe amounts. We commonly use toe plates, but we must be careful to read the tape measure accurately.

Some theory exists that the tire contact patch deflection dictates the toeing of one or more of the rear wheels to compensate for this, but it is hard to justify and can probably be best defined as a crutch used to help solve some other setup miscue.

We can set the toe using one of three different methods. We can use simple toe plates that rest against the tire sidewalls. These have two slots, one at the front and one at the rear about an inch off the ground. We run a tape measure across the rear between the slots and remember that number. Then we measure at the front between the slots and subtract to find the Toe. There can be errors in this method from tire/wheel runout, sidewall protrusions like lettering, etc. and sloppy handling of the toe plates. This procedure should be done several times until the toe reading is consistent.

We can also use strings, but we need to makes sure the strings are parallel on each side of the car. And, the strings must be small diameter, preferably heavy fishing line or similar. We have used this method on prototype cars for setting the rear toe.

Finally, we can use lasers. This method could be the most accurate because we can place the laser far to the front of the wheels where the Toe amount is multiplied. If we set the targets four times the diameter of the tire, then we can easily dial in smaller amounts of Toe.

Again, we need to make sure the lasers are pointing true. We can do this by setting plates at the front and rear of the wheels at the same distance from the center of the wheel. These targets should be equal distant from each other at each end when there is zero Toe. Set zero toe with the toe plates, a much more simple task than measuring some amount of Toe and then check the laser to see if the beams are running parallel.

If we are using a four times multiplier, for an 85 inch tire, the diameter of the tire is 27.056". Four times that would be 108.22 inches, or simply 108 and ¼ inch. A Toe of 1/8" would measure a full half inch at the target. It is much easier to see and read, and more accurate than using a tape measure on a toe plate.

How To Set Toe:

Step 1 - Wheel Run-out Check – Check both the front wheels and the rear wheels for Run-out. This means that as the wheel rotates, the outer edge of the tire will wobble slightly. We must compensate for this slight distortion by finding the extreme high spot at a point equal in height to the hub height.

We can simply use a jack stand to hold the tape steady and rotate the tire noting the distance from the stand. Once we locate the high point, we mark it with an arrow and then rotate the tire (be it front or rear) so the arrow is at the top pointing straight up.

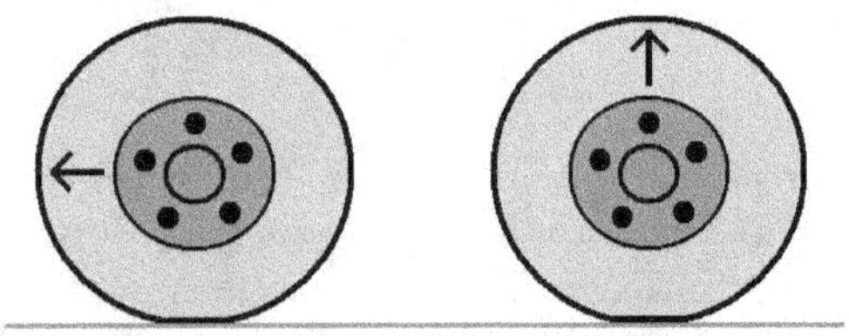

Before we set the toe in our race car using toe plates that will rest against the tire sidewalls, we need to find the high point in the sidewall of the tire. There is almost always a high point in a tire and if we place a tape measure or end of a screw driver for instance against the tire sidewall while we rotate the tire, we can visually see the high point. We then mark that point and position it to the top. That way the high point will not point our toe plate at some angle from perpendicular to the axle line and parallel to the wheel.

Step 2 - Center the Steering Box – Center the front steering rack. This is done by turning the steering wheel lock to lock and back half the number of turns from full lock in either direction. Once mid-rack (or mid-box with drag link steering) is found, lock the steering shaft against the frame rail with two vise-grip type of pliers.

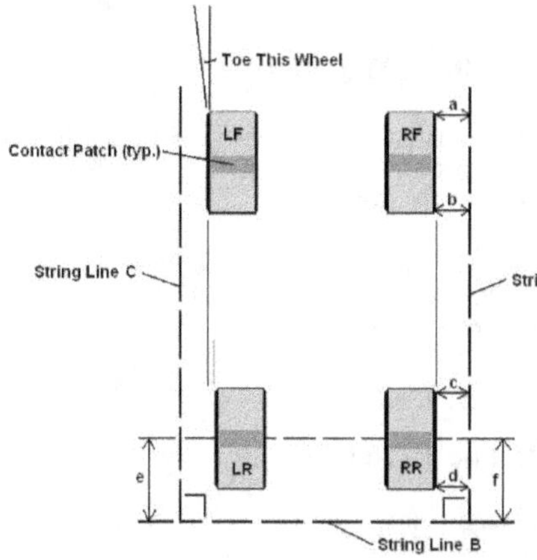

Another critical point to make sure of when setting toe is to center the steering rack or box, then adjust the tie rods so that both wheels are pointed straight ahead. Then we can add toe-out to the front wheels or toe-in to the rear wheels in a AA-arm suspension system. We can add all of the toe to the left front, right front, or split the toe and add to both the front wheels in a AA-arm front suspension. For the rear AA-arm suspensions, we always split the total amount of toe between the two wheels.

We want to make sure the steering is centered and the wheels are pointed straight ahead. Once the steering box has been set to center, adjust each tie rod length so that the right and left wheels are pointing straight ahead. With the laser systems, this is done quickly and accurately.

Step 3 - Setting Front Toe – It is time to set the static toe at the front wheels. We can use toe plates or strings to measure the Toe at the wheel and/or tire sidewall. Remember to be careful, be accurate and do the measurements several times to be sure of the numbers.

If you use lasers, you need to check that the laser is pointed parallel to the wheel and tire. We told you how to do that above. Then just read the measurement from straight ahead and divide by the multiplier of the distance from the center of the wheel.

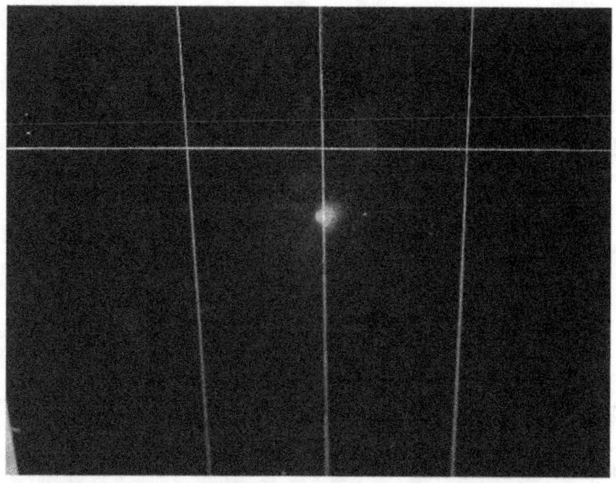

In this view, we can see how accurately we can read the toe amounts from straight ahead, being the line the laser is lighting up. When the plates are a distance from the wheel hub by say four times the tire diameter, then an 1/8 inch of toe becomes a distance of ½ inch on the plate.

Step 4 – We need to decide how we are going to set the Toe. There are three different ways to do this and it really depends on your preference and the type of race car.

For circle track cars turning only one way, here are some theories. You can either only adjust the left side tie rod to set your toe, leaving the right side alone, or half the toe amount to be split between each side. For 1/8' Toe, the left wheel would be set 1/16" out and the right wheel would be set 1/16" out.

If only setting one wheel for the total Toe you want, we may want to set the left wheel only as opposed to the right wheel because with most circle track cars, the caster split causes the car to pull to the left and driver to steer slightly right when going down the straightaway. Setting the Toe in the left wheel only adds to this pull and the driver won't really notice the effect.

Another way to look at this is to set the Toe with only the right wheel. This way, the steering wants to center left of straight ahead due to the Toe being set this way and that opposed the caster split pull to the right. The driver does not need to steer as much right as before and the car might track much straighter.

After you have set the Toe, roll the car back and forth and recheck the toe setting. Remember that for toe-out, the front measurement on the toe plate will always be more than the rear measurement. If you are using a string or a remote laser to check your toe-out, the opposite is true. The front measurement from the laser/string to the wheel would be less than the rear measurement.

Remember to re-check the Toe after an on-track incident. Small collisions can bend the tie rods and change the Toe. Re-think the Ackermann settings because Ackermann will change the Toe in the car as the wheels are turned.

Exam - In The Context Of This Lesson:

Toe Settings In A Race Car Are?

1) Where the wheels are pointed different from in-line with centerline

2) Always with the wheels pointed in

3) Always with the wheels pointed out

4) Not necessary in a straight axle system

Toe-In In Race Car Design Is?

1) Good for stabilizing front suspensions

2) Helps stabilize rear AA-arm suspensions

3) Where the front of the tires are wider than the rear

4) Important to use in a straight axle system

Toe-Out In Race Car Design Is?

1) Good for stabilizing front suspensions

2) Helps stabilize rear AA-arm suspensions

3) Where the front of the tires are narrower than the rear

4) Important to use in a straight axle system

Before We Set Toe, We Always?

1) Center the steering

2) Put the car at ride height

3) Find the run-out in the tires

4) All of the above

The Use Of Toe Plates Is?

1) As accurate as lasers

2) More accurate than lasers

3) Less accurate than lasers

4) The only way to measure toe

The Correct Way To Set Toe Is?

1) Right side out, Left side straight ahead

2) Left side out, Right side straight ahead

3) Both sides our equally

4) Any of the above

Racecar Technology – Level Two

Lesson Twelve – Driveline Alignment

Drive Line Alignment is one of the basic design features of the car and it can and must be done initially when building, or re-building, a race car in order to save a lot of work later on moving things around. The basic parts of the alignment are the transmission output shaft, the drive shaft itself and the pinion shaft at the rear end on a straight axle rear.

Most of this drive line technology had initially been developed for production cars, but racing applications require a closer look and slightly different approach to this information. The advantages of this knowledge are in the area of reduced power loss and increased component life. To begin with, we need to know what to call the various components related to our driveshaft.

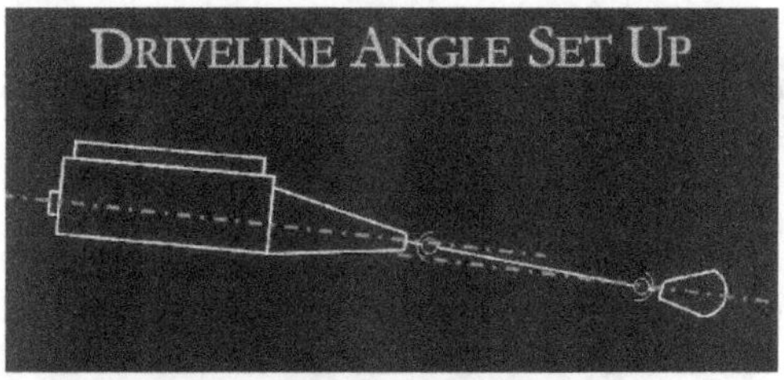

Proper driveline alignment is when the engine/transmission alignment is parallel to the pinion shaft alignment. If those two are not inline, then there will be an angle created between the transmission and pinion shafts. This angle should be minimal in a racing application.

Drive Line Terminology – The component names for the driveline parts are as follows:

Slip Yoke – this is the yoke at the front of the driveshaft that goes into the transmission and can slip to take up the slack of chassis movement.

Weld Yoke – is the yoke at each end of the drive shaft that attaches to the pinion yoke at the rear end and the Slip yoke at the front. These are welded to the shaft tubing.

Universal Joint Kit or U-Joint (UJ) – is the actual part that forms the rotational connection between the driveshaft and the transmission and rear end.

Tubing – is the tubing between the weld yokes and is usually steel but can be aluminum or carbon fiber.

Pinion Yoke – is the yoke that is attached to the pinion shaft at the rear end and accepts the U-Joint.

Drive Line Angles – Drive line angle is when the drive shaft tubing is not inline with the transmission output shaft and/or the pinion shaft at the rear end. When the UJ operates with any amount of drive line angles, this creates a problem. The bearings speed up and slow down twice per revolution of the driveshaft. This causes an oscillation in the power train. The more angle that we have, the higher the peaks of oscillation we see and therefore the greater chance of vibration.

Drive line angles are a cause of vibration and power loss. If your race car must have driveline angles from a design standpoint, the angle of the drive shaft to both the transmission output shaft and the pinion shaft should be equal and also opposite. The angle should also be kept to a minimum when at all possible. To get to "equal and opposite", the transmission shaft and the pinion shaft must be parallel.

Drive shaft angles are not only measured from a side view, but also from a top view. Some offset late model circle track and road racing cars can have as much as a 1 ½ inch displacement of the rear of the drive shaft from the front. That equals almost 2 degrees of drive shaft angle at both the transmission and the pinion. So, we can align the drive shaft, from a side view, to zero angle and still have 2 degrees of drive shaft angle present overall.

Strictly for racing applications, near Zero Drive Shaft Angles are encouraged. Based on recent studies and testing, a race car with zero drive shaft angles at the transmission and pinion will not harm the u-joint bearings. The natural vibrations that are produced by the race car will cause the u-joint bearings to rotate enough to stay lubricated and not flat-spot.

In the real world, the drive shaft will never really maintain a zero angle configuration through the diving and rolling of the chassis as we lap the track. Starting at zero means that we will stay very close to zero at all times. If the race car will be at some attitude other than it is at normal ride height, then we must align the driveline with the chassis at the position it will be while we are racing. The teams need to align the drive line for near zero angles with the car placed at the on-track race attitude.

NOTE: For overall consideration in drive line angle design, number one is to keep the driveline angles as low as possible and keep the angles equal and opposite at each end of the drive shaft.

Controlling Driveline Vibration – The sources of vibration in our racing driveline, in order of importance – most critical at the top, are:

1) Run-out (similar to wheel runout)

2) Improper Drive line Angles (not equal and opposite)

3) Looseness in the fit of any of the parts (commonly in the U-joints)

4) Unbalanced parts (drive shaft)

5) Component Deflection (shaft wobbles at high speeds)

6) Reaching Critical Speed (a speed too great for the diameter of the tubing)

We will examine some of these causes and see how we can cure many of them just by using better parts that are designed for the racing environment.

If your race car must have drive line angles from a design standpoint, the angle of the drive shaft to both the transmission output shaft and the pinion shaft should be equal and also opposite.

Yoke Design – The only attachment design for holding the U-joints that is suitable for all out racing is the strap design. The U-bolt attachment that is commonly used on passenger cars and trucks is no longer considered acceptable for several reasons: 1) it has less strength than the strap design, 2) it may distort the bearing caps if over-torqued, and 3) it grips the cap at three points whereas the strap design grips it in four places which translates to less distortion of the caps.

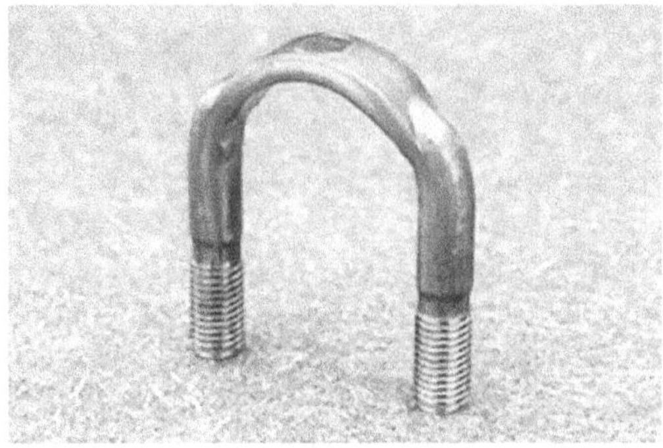

For racing we should always use parts designed for racing applications. This stock type of U-joint strap is not to be used in racing. There are stronger yokes and straps that will endure the high stress caused by racing.

U-Joint Kit Designs – There are basically two designs of U-joints available. The popular standard zerk designs is used for production vehicles and has, as is obvious, a grease zerk fitting for lubrication. Inherent in this design are hollow shafts that provide the means for the grease to reach the bearings. This hollow design also makes the part weaker than if it were solid.

In this speedway application, note the heavy yoke and thick strap holding the U-joint in place. And note the large diameter drive shaft tubing. The higher the RPM of the drive shaft, the larger the tubing needs to be in order to prevent vibrations and failure.

The other design is called the Sealed Design, or Solid U-joint and has no grease fitting and therefore a solid core. This unit has precision seals that keep the lubricant with the bearings while also sealing out dirt. The sealed design is therefore much stronger.

Proper lubrication of the sealed U-joint is simple but can be overdone. We always want to coat the bearings, but not excessively. We also need to fill the trunnion cavity with grease, but not overfill. If too much grease is applied, then when we attach the caps, the pressure from the grease trying to escape will literally blow out the seals and ruin them.

Adjusting Run-out at the Pinion Yoke – A major cause of driveline vibrations is when the pinion yoke has a measurable degree of run-out. We can test the run-out after the installation of the U-joint in the pinion yoke by using a dial indicator that is attached to the rear end housing.

Runout in the yokes can cause vibrations that can damage your driveline system. This tool is used to check for runout in professional applications. You can use a dial indicator mounted to the rear housing and measuring to the end caps. If those are the same distance off of the centerline of the yoke, then the driveshaft is centered on the yoke shaft.

We measure at both sides of the u-joint and then if the offset is different, we must remove the pinion yoke and rotate it on the pinion shaft to find a position that will index more correctly. This is a very important and necessary step in reducing driveline vibration.

Failure Mode – Fatigue Crack Propagation – Drive line components can fail due to a number of reasons. Cracks in the welded yoke or drive shaft will accelerate the failure of the drive line system.

Cracks can occur from the stress of an already weak part or upon construction of the parts due to the heat of welding. It is recommended to weld any drive line part using the TIG welding technique instead of MIG welding.

Balancing the Drive Assembly – The only true way to balance a drive shaft is through the use of a Two Plane balancer, or one that simultaneously balances both ends and has the capability to Cross Talk between ends. This allows the equipment to derive a dynamic model of the forces of the imbalance to determine the force vectors involved and formulate a solution that takes into account both ends influence in the overall balance of the shaft.

A simple automotive drive shaft balancer is fine for production automobiles, but for racing, with the very high RPM we experience, we need more precision.

Snap Ring Failure – A common failure in our racing driveline is when the retainer snap rings come out of the yoke. There is a simple and easy way to reduce this occurrence by applying a spot of epoxy to the ring. This prevents the ring from collapsing and falling out of the ring groove.

In this photo, you can see where this team placed a small dab of epoxy in between the ends of the snap ring so they cannot move and allow the ring to come out.

How To Align A Drive Line System:

Here is how we can determine what we have, what to do and how to fix it. You will need to put your car up on blocks or boxes so that you can get underneath of it safely. Be sure to block the tires from moving while doing this process. Once the car is secure, we will need to take several measurements.

You need to align the drive shaft with the car at the attitude it will be while racing. Note the raised left rear corner and compressed right front corner. When this team measured their driveline angles at this attitude, the car needed to be re-aligned. Because the engine was angled down to the rear, the pinion shaft needed to be aligned up to the front to make it parallel to the engine/transmission centerline. When this was tested, the car became vibration free.

The attitude of the car needs to be the same as it is when running around the race track. So, if you are running on bumps, put the front of the car down to where it would be while on the bumps. For a dirt modified, you could jack up the front on the left side and down on the right just like it would be during the race. The left rear would also be hiked up again like it will race.

Also consider the change in pinion angle if you run a lift arm or pull-bar. Those pieces of equipment will change the pinion angle as they move under acceleration. When power is applied to the drive train, any miss-alignment will cause the most damage. You need to measure your pinion with the rear end at the position it will be under full acceleration.

How And What To Measure – First measure your transmission output shaft angle. There is an easy way to do that. Your engine is in line with this part. The valve covers, unless they are of a strange design, will be parallel to the block deck, which is parallel to the driveshaft which is parallel to the transmission shaft. So, just lay an angle finder on the edge of the valve cover, read the angle and write down the number. Also note which way it is inclined, down to the rear or up.

Next crawl under the car and lay your angle finder on top of the driveshaft and measure the driveshaft angle and note the inclination. Then measure the pinion angle and also note the inclination. To get the pinion angle, you might have to use a straight edge placed against the flat part of the flange with the angle finder against that.

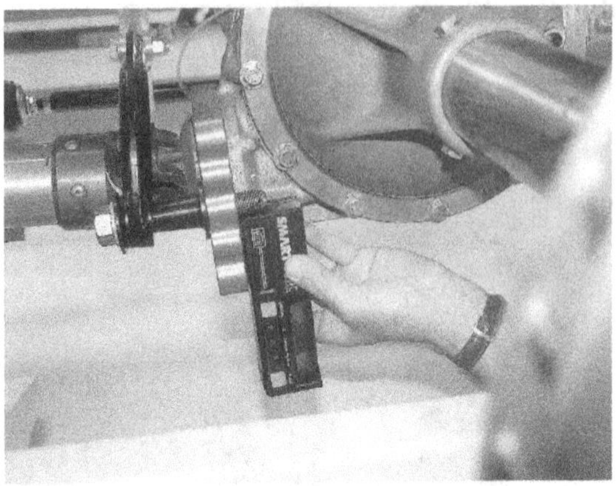

The first measurements you take are of the pinion shaft, ninety degrees off of this measurement shown, and of the engine/transmission shaft angle. These two should be equal and in the opposite direction (Example: engine up to the rear/pinion down to the front) for proper driveline alignment. We make changes to the engine/transmission and/or pinion angle to make them line up.

Driveline angles can occur from any view. We have just measured the angles from a side view, but what about the top view? Now we need to look at that scenario.

To measure top view angularity, we need to do some lateral measurements to the center of the tranny shaft and the center of the pinion shaft. If you have a straight rail car and know that the rail is parallel to the centerline of the chassis, this will be easier. If not, you need to setup a string line that is parallel to the centerline.

Measure from the centerline or straight rail to each of the above mentioned shafts using plum bobs or similar devices to make sure the measurements are correct. We need to determine if the two shafts are in line, or if not, how far they are out of line.

For most offset late models, I have found the two to be out of line by 1.0 to 1.5 inches. For a 40 inch long driveshaft that is out of line by 1.5 inches, the driveshaft to pinion and driveshaft to tranny shaft angles would be 2.15 degrees. That is plenty of pinion angle and almost too much by today's standards. But the good thing is that they will be equal and opposite if the rear end is at 90 degrees to the chassis centerline and the engine and transmission is also parallel to the centerline.

After we have lined up the engine/transmission and pinion shafts, we can then measure the drive shaft to see what the difference in angle is. This is out driveline angle and it should be minimal. If it is over 2.0 degrees, then we need to make changes to the engine angle to reduce the driveline angle. We then re-adjust the pinion angle to match the new engine angle.

If those conditions are met, then you would not need any side view angles in either the pinion or tranny. So, you could place the tranny shaft, driveshaft and pinion shaft in line with no side view angular difference. That is because the U-joints don't really know which direction they are mis-aligned, only that there are equal and opposite angles in their relationship to the driveshaft.

How To Adjust The Angles – Now that we know what the angles are, let's see how we go about making changes. Let's assume that from a top view, the tranny is in line with the pinion. So, we only need to work with the side view angles.

For an example, we have an engine that is downhill to the rear by 6 degrees and the pinion going downhill to the front by 4 degrees. We can reduce the engine angle, but it is impossible to run it uphill to the rear. So we could re-shimmed the motor mounts we get to only 2 degrees downhill to the rear.

Next, it is fairly easy to change the pinion angle because we have a three link rear suspension in our sample car. We then rotate the pinion by extending the length of the third link until it is pointed uphill to the front to an angle that is equal and opposite of the transmission angle, or the same 2 degrees.

If you have designs different than the solid axle three link suspension, you'll need to get creative when changing the pinion angle. For cars with leaf spring or truck arm rear suspensions, there are wedges made to varying degrees you can use to place between the spring or truck arm and the pad on the axle tube. This will rotate the rear end and pinion to change the angle to the drive shaft.

Some designs like this lift arm in a dirt late model, promote rotation of the rear end upon acceleration. The forces that drive the car forward also force the arm up at the front to soften the application of power and to redistribute the loads on the rear tires. We need to compensate for this movement and rotation of the rear end when we set our pinion angle.

For stock four link rear suspensions, you'll need to get even more creative. Since the alignment of the driveline is so important, it is not out of the question to cut and re-weld the suspension links to change the lengths in order to rotate the rear end and change the pinion angle. Just do it in a safe manner. Make good welds or have a professional do it.

Summary – The important things to remember are to use drive line parts that are made specifically for racing when possible. Make sure you install the components correctly to reduce drive line vibrations and parts failures.

In high speed and high horsepower applications, teams use specially made pinion yokes that are much stronger than stock units. And, they are carefully balanced as indicated by the drill marks on the side that removed weight in order to provide balance.

Check the alignment of your system and correct any miss-alignment. The drive line must be aligned when the chassis is at the attitude that it will race at, not at normal ride height. Allow for the deflection of pull bars, lift arms, or other rear suspension parts that move under acceleration. Always use a drive shaft that it is up to the task for the intended RPM range you will be running in your type of racing. Periodically check driveline parts for cracks or for signs of a bent or dented shaft.

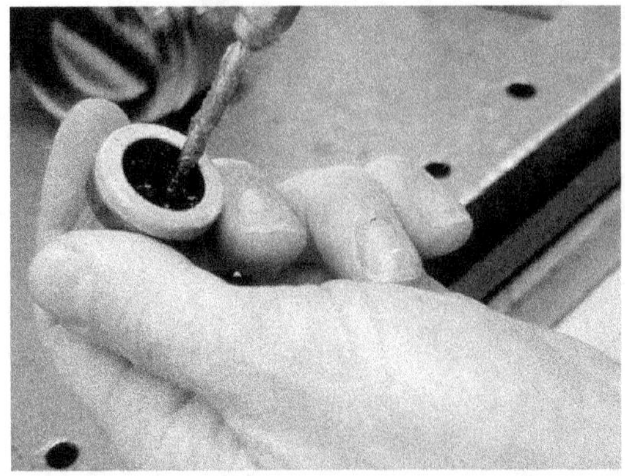

Lubrication of the U-joint caps must be maintained on a regular basis. Racing U-joints don't have the grease fittings like the OEM U-joints do. Also, be careful to provide enough lubricant but not too much or when the caps are installed, the excess lubricant will blow out the seals.

Exam - In The Context Of This Lesson:

Proper Driveline Angles Are When?

1) The pinion is angled 2 degrees down to the front

2) The transmission shaft and pinion shaft are parallel

3) The drive shaft angles at the transmission and pinion are equal and opposite

4) 2 and 3

Driveline Angles Occur When?

1) From a side view only

2) From a top view only

3) From either a top view or side view

4) When the transmission, driveshaft and pinion line up

Driveline Vibrations Can Be Caused By?

1) Runout in the pinion yoke

2) A bent driveshaft

3) A miss-aligned driveline

4) All of the above

Driveline Angles Are Critical When?

1) The car is at ride height

2) The car is under accelerating

3) The car is decelerating

4) Never

Top View Miss-alignment Is?

1) Never seen in a race car

2) The best way to achieve driveline alignment

3) Most often seen in circle track race cars

4) A way to achieve driveline alignment

5) 3 and 4

We Can Adjust Driveline Pinion Angles By Using What Method?

1) Adjusting the length of the third link

2) Adjusting the height of the front of a lift arm

3) Using angled wedges for leaf spring or truck arm suspensions

4) All of the above

The Pinion Angle Must Always Be?

1) Pointed down to the front

2) Pointed up to the front

3) Made level to the ground

4) Any angle that is equal and opposite of the transmission angle

Racecar Technology – Level Two

Lesson Thirteen – Spring Selection for Balance

In RCT Level One, we spoke about what springs are, what they do and a little about how the chassis reacts to spring rates. This Level Two course is all about going to the race car. We are setting up the race car and working with the springs. So, we need to know a little about what our goals are and how springs can help us achieve those goals.

In earlier Lessons we have alluded to the concept and idea of a balanced setup. The term "balance" is used frequently today in broadcast of motorsports events from Nascar racing to Formula One. So, most everyone has accepted the notion that cars that have perfected the dynamic balance part of their engineering are better cars.

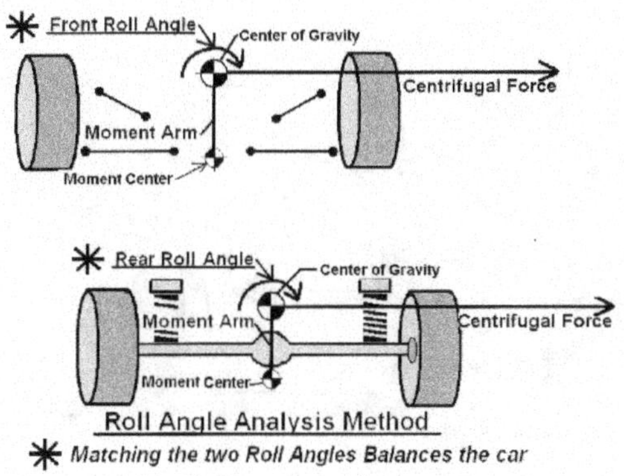

The concept of a Balanced Setup involves critical elements that we will define and explain in this Lesson. In the most basic terms, it is the front and rear being made to work together to put the maximum and ideal loading on the four tires. This method is a refinement of the older concept of Roll Axis / Roll Stiffness optimization. What RA/RS lacked was the inclusion of influences that affect the "roll stiffness" such as spring split which we will discuss here.

Now we will present what we hope is a complete explanation as to what constitutes a balanced setup. Just to differentiate, we are not talking about a neutral handling car, we are referring to a dynamic balance where the two suspension systems are in harmony and working together.

The balanced setup has noticeable positive result in race car performance. That is why we are so adamant about getting everyone to fully understand the concept. The following is an excellent definition describing the results of balance and was written by championship racer Larry Bendele. It is a classic and it deserves repeating and so here it is.

"A balanced setup for asphalt has the following characteristics: A perfect handling car allows the driver to have full confidence that he can drive it into the turn as hard as he wants without the slightest worry if it is going to spin out or head for the outside wall. For corner entry, the steering wheel movement is made one time, and doesn't need to be corrected when you approach the corner apex.

"Just prior to the center of the turn, the car rotates slightly (points) without getting loose and begs the driver to give it full throttle. You are able to give it full throttle at the corner apex and keep the hammer down, and the car is willing to run the lower groove on corner exit. However, you can reduce front tire drag (and accelerate more) by purposely steering the car to exit near the outside wall.

"It does the above for 90% of the laps of the feature and you are able to win by several car lengths. Late in the race, you think about taking it easy because of your big lead, however, the car is handling so perfectly and comfortably that it wouldn't be as much fun to drive if you backed it down a notch. So, you smoke the competition."

What Larry describes is what winning race car drivers have sometimes experienced. What we will do in this lesson is teach you how the springs have a major influence in making that happen. In RCT Level Three, we will be setting up real race cars and showing you exactly how springs are used in real situations. For now, we need to come to grips with the process of realizing perfect setup balance.

New Methodology - In the mid 1990's a method was developed that represented an advancement related to vehicle dynamic modeling. It involved treating the vehicle as two separate masses, front and rear, each with its own separate suspension system, weights, etc.

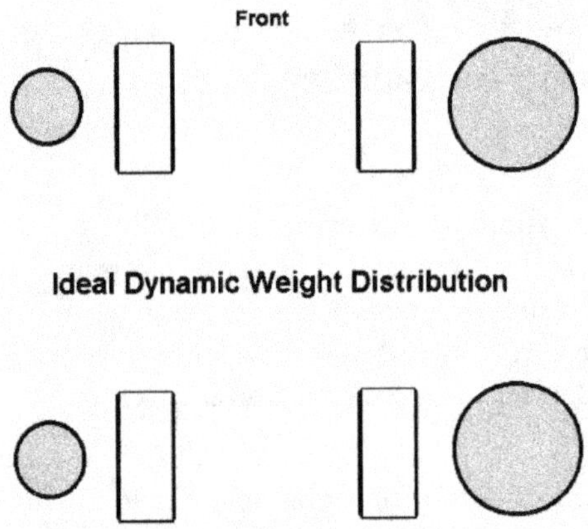

When we have achieved perfect dynamic balance, our weight distribution after weight transfer has taken place will be as indicated in this sketch. The size of the circles represent loading, the larger the circle the more loading there is on the tire. This is for a car turning left. There will be equal loading on the outside tires, front to rear, and equal loading on the inside tires, front to rear. This provides the same grip level for each axle and also represents a perfectly balanced race car.

This method made a lot of practical sense because in a stock car we have two "axles", each supporting the weight of each end of the car, with each resisting the lateral forces created by cornering. In physics, we are taught that everything is fluid, or somewhat flexible, and dividing the car into separate halves allows us to analyze each one independently to determine its desire.

There are several critical reasons why a balanced setup is essential to optimum chassis performance. If a cars setup is balanced, then we can accurately predict load transfer. An unbalanced setup redistributes the loads on the four tires in a very unpredictable way. If we cannot match the desires of each end of the car, we then cannot accurately predict the exact amount of load transfer and ultimately how much load ends up on each tire at mid-turn.

We need to have the two sets of tires at each end of the car doing equal amounts of work at mid-turn in order to have a true balance. In most racing series and types of race cars, we will almost never have all four tires doing equal work (having equal loading on each tire) under most current weight rules. If we can setup the car so that after the load transfers in the corners, we have equal working pairs of tires, then we will have a truly balanced setup.

We will also have less (almost non-existent) chassis flex with a balanced setup. Compliance, or flexing of the chassis, cannot occur if we remove the forces that cause this to happen. When the two suspension systems are unbalanced, they are trying to roll to different angles and that causes the chassis to twist.

More importantly, a balanced setup is much more forgiving when the track conditions change or the driver runs different grooves. The speed of the car does not fall of as much, either, as the race goes on and the tires wear. It is the tendency of the balanced setup to maintain a neutral handling characteristic and retained speed after a long run and that helps win races.

The AA-arm Suspension - A double A-arm suspension, the type used on the front of most circle track race cars and also on the rear of formula type race cars, has a moment center that represents the bottom of the moment arm. The top of the moment arm is the overall center of gravity of the sprung mass of the car for both the front and rear suspensions and their moment arms.

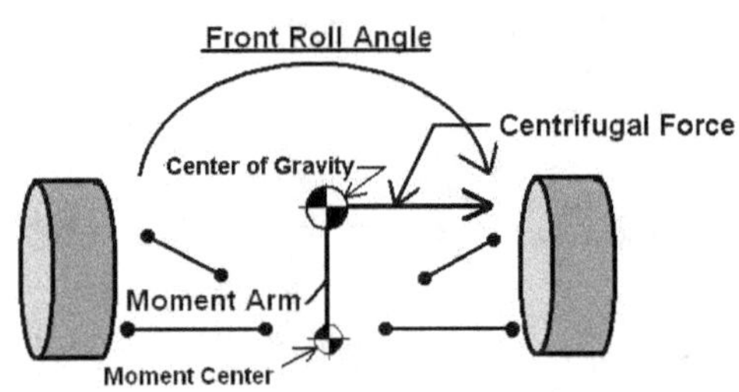

This sketch represents a AA-arm suspension and how the lateral force acts on the CG. The top of the moment arm is the CG and the bottom is the MC. This force acting through the moment arm tries to roll the chassis. If the front and rear suspensions are trying to roll to the exact same angle, then we have a perfectly balanced setup and we can then calculate the loading on the four tires accurately.

When the center of gravity/center of mass tries to continue in a straight line as we turn the corner, a lateral force (Centrifugal Force) is exerted on the chassis and that force is resisted by the tires and acts on the center of gravity and resisted by the moment center. If you stick a shovel blade firmly into the ground and then pull on the end of the handle, your arms represent the lateral G-force, the end of the shovel handle is the center of gravity, and the blade at the ground is the moment center.

The Rear Suspension Dynamics - The rear suspension is a much different system than the front and we look at it differently. At the front with the double A-arm suspensions, the spring base is felt at the wheels. But a rear solid axle system (and this relates also to a front straight axle car) has a spring base felt on top of the actual springs.

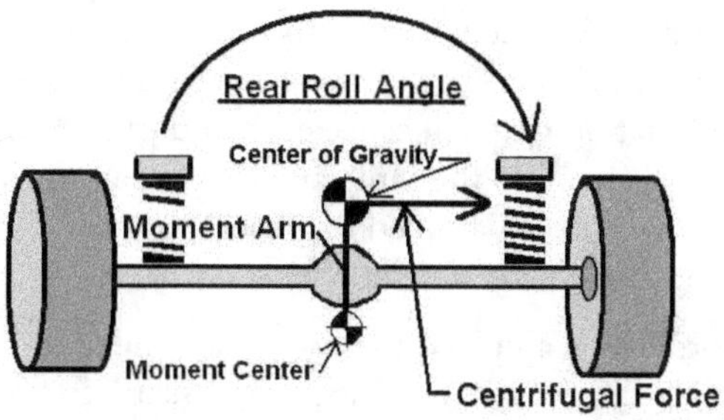

For a solid axle suspension, the chassis rides on the top of the springs. The spring base is the distance between the top of the two springs for this system, unlike the AA-arm systems where the spring rates are translated out to the tire to create a wheel rate. There is no wheel rate for a solid axle system for the calculation of the roll angle and roll resistance. In this system, the moment arm is like with the AA-arm system, it is the distance between the CG and the MC.

This dynamic model is not a new concept, but was developed and published some sixty years ago. It is also described and illustrated in the popular book by the late race engineer Mr. Carroll Smith called Tune to Win. So, the width of the top of the springs represents the Spring Base.

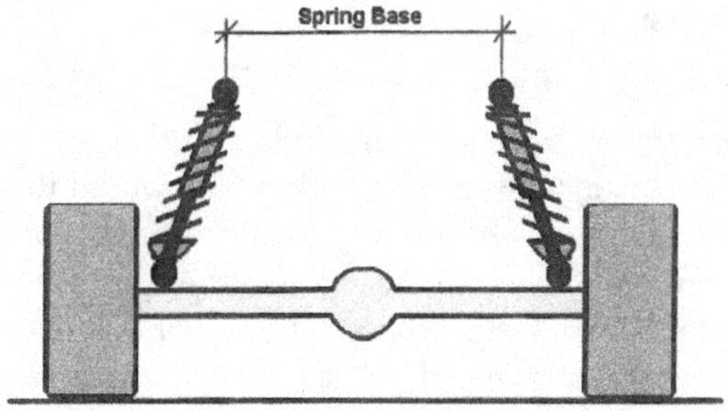

For a solid axle suspension, as we have discussed, the spring base is the distance between the top of the springs. Even with the springs being angled with the bottom wider than the top, we still consider the spring base at the top where the chassis rests on the springs. The angle of the springs will still influence the rate of the spring "felt" by the chassis, but that ultimate rate will be felt at the top. This is an important concept to understand because with some race cars, it is possible to move the top of the springs in or out from the centerline of the chassis. This can increase or decrease the spring base width and influence the roll angle.

The rear suspension has a center of gravity of the sprung weight of the car that represents the top of the moment arm just like at the front. The bottom of the moment arm is the moment center created by the lateral locating device known to us by the terms Panhard/J-bar, metric 4-links, leaf springs, or Watts link.

These four devices comprise the majority of lateral restraint systems used for straight axle suspensions in race cars. Each restraint system has its own moment center that is the bottom of the rear moment arm.

Two Forces at Work – The lateral force is not the only force acting on a race car here on Earth. Again, the moment arm is much like a pry-bar or shovel handle. The CG is equivalent to the end of the bar we hold on to, and the roll centers are the opposite end, which is the object we are trying to move. The longer the bar (moment arm) the more leverage we have and the easier we can move the object, or in this case, roll the car. Now here is what else is happening.

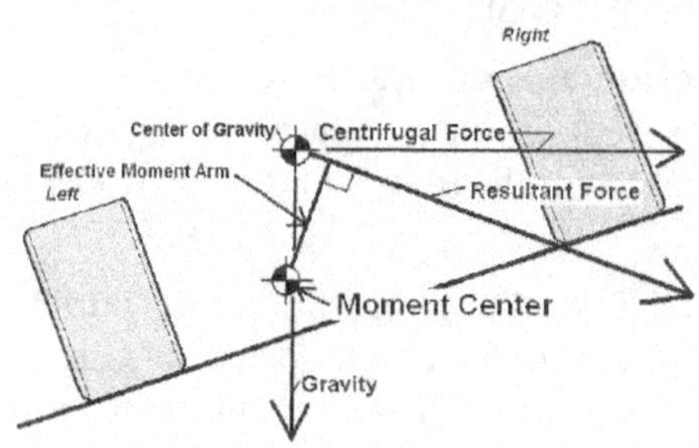

In every race car, there are two forces acting on the chassis. There is the ever present Gravity and then there is the lateral force we call Centrifugal force which is acting parallel to the path of the car going through the turns. In almost every case, this path is parallel to the earths surface, or to be more specific, perpendicular to a line running from the CG to the center of the earth. In physics, when two forces act on a body, they combine into a Resultant Force which has a magnitude and direction.

In a race car we really have two forces at work being applied to the center of gravity. One is the lateral force of cornering we described as the centrifugal force, and the other is Gravity. We cannot forget about this basic force that is acting on the chassis at all times, even when at rest. The two forces combine into one resultant force for which we can calculate a magnitude and direction.

By looking at the direction of the resultant force in the sketch, we can see the true picture of how a combination of the two forces will react in the front suspension through the Center of Gravity and be resisted by the moment center. The Effective Moment Arm length or Combined Force Vector is the result of the direction of the combined force and the location of the moment center.

Effective Moment Arm - The Moment Arm length is the vertical distance between the CG height and the MC height. The amount of lateral force and the length of the effective moment arm are contributing factors that help determine exactly what angle the front of our car desires roll to. If the front end of our car was not rigidly connected to the rear it would roll to a certain angle and a predictable amount of load would transfer from the left front tire onto the right front tire. This is the Front Roll Angle.

But we know that the two ends of the race car are connected with a rigid chassis that resists twisting. The two ends of the car are connected and unless we know for sure that their desires are matched, then we cannot accurately predict the load transfer at each end of the car.

Primary Influences On Roll Angles – The following are the primary influences that affect the magnitude of the front and rear roll angles. They are:

1) The weight of the sprung mass of the portion of the car supported by the suspension (scale weights at each end minus the weight of the un-sprung components),

2) The height of the sprung mass Center of Gravity,

3) The height of the Moment Centers front and rear,

4) The magnitude of the lateral force (measured as G-force),

5) The overall spring stiffness as well as side to side spring split (i.e., right rear stiffer than the left rear spring, etc.) at each end,

6) The width of the spring base, which is the distance between the centers of the top of the springs for a solid axle suspension and the track width of the tires for a AA-arm suspension,

7) The stiffness of the sway bar (if used) which has an effect of anti-roll and must be taken into account. The larger the diameter of the bar in this system, the more resistance there is to roll,

8) And the Track Banking angle.

Now that we know what influences the roll desires of each suspension system, the key to creating a balanced setup is to change spring rates, sway bar, etc. so that we can find the combination that will cause each end of the car to WANT to roll to the same angle. We can detect when our car is balanced by observing tire temperatures as well as tire wear.

In order to set up our cars, we would need to be able to change components, if necessary, to balance the desired roll angles of the front and rear of the car. These changeable components include the spring rates, spring split, the moment center heights front and rear, and if used, the sway bar sizes and arm lengths.

A Balanced Setup – A dynamically balanced setup is where both ends are working together and WANTING to roll to the exact same angle. Some teams will experiment and sometimes find a really good setup and they might not know how they got there or why it worked so well. Often the performance could not be duplicated when they built, bought or otherwise acquired a new car because of differences in moment arm locations or other differences that affect the roll angles.

The Balanced Setup concept is one that means we should work to match the front to rear desires until our setup is balanced. The "desire' can be thought of as the roll angles that each end of the car would want to achieve if that end were independent and not attached to the other end.

For a front AA-arm / rear solid axle car, the front roll angle is not easily adjustable within the range of spring stiffness we usually want to work within. So, to balance the setup in these cars, we need to work with the rear of the car and make changes there to match what the front wants to do.

Our choices for adjustments include making spring changes for both stiffness and spring split for circle track cars, raising or lowering the rear moment center, and that is it. So, spring split is the primary way to get close to the balance we need and the panhard bar is the fine tuning device to use once we get close.

For road racing cars with AA-arm front and rear suspensions, we work with overall spring stiffness combined with the anti-roll bar stiffness. We can also tune the MC height for each system and we will get into more on that in RCT Level Three.

How to Select the Right Springs - The desires of both ends of the car can be matched through the use of many combinations of spring rates, spring split amounts, and MC heights. So, we need to make a decision as to how soft or stiff our spring rates need to be for the track we will be running on. The general rule is to stiffen the spring rates for high banked and faster tracks and soften them for lower banked tracks with less speed and traction.

For road racing cars, we run equal spring rates on each side of the car at each end. What we call spring split, or running different rate springs on each side does not apply to road racing cars for the most part. So, for those cars, spring stiffness, MC height and sway bar rates front and rear, are the deciding factors in determining the roll stiffness and ultimately the roll angles for the front and rear.

Circle Track Race Cars - Spring split at the front can enhance the transitional parts of the track. On flatter tracks, a softer RF spring than the LF spring, can improve corner entry. This is because as we brake, a softer RF spring can start the chassis rolling in the same direction that it will roll to in the middle of the turns. On medium and higher banked tracks, there is no need to run a softer RF spring and we would either even them up or run a stiffer RF spring on tracks banked 14 degrees or higher.

The same is basically true for the rear of the car. A softer right rear spring for lower banked tracks for cars using the Metric four link works where improved bite off the corner is usually needed. In addition, those cars inherently have a very high rear Moment Center and this form of spring split will help create a rear roll angle to match the front.

In modern day circle track racing, it is common to run a stiffer RR spring rate to try and match the low front roll angle created by using stiff bump stops or bump springs. The force needed to do that in the RR can be pre-loaded into a softer rate of spring so that the RR corner will compress more on exit. We'll describe in detail how that happens and how to do it in RCT Level Three.

On the higher banked tracks, we can run the same rate for the rear spring, or use a higher RR spring rate than the LR spring rate to create an anti-roll effect. To help reduce the roll tendencies for the Bump setups where the front has a low desire to roll when riding on the bumps, a much stiffer RR spring rate will help reduce most of the roll tendencies in the rear. Or as we have stated, a softer spring rate pre-loaded to a sufficient force to resist roll.

Whichever type of race track you run and whether you run dirt or asphalt, attention must be paid to the relationship between the front suspensions desires and the rear suspension desires. When both ends are close to wanting to roll to the same angle in the turns (more easily referred to as desired roll angle), then the more work all four tires will be doing and the faster and more consistent your car will be.

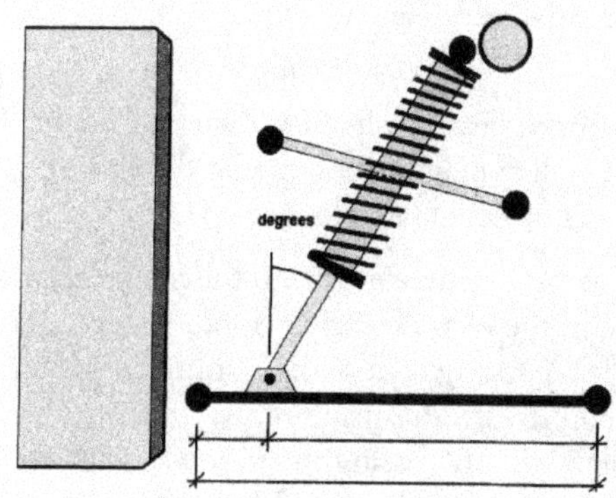

The front spring rate is translated out to the wheel for a Wheel Rate in this coil-over spring system. We need to find the motion ratio and spring angle to find the Wheel Rate for this application in order to find the roll angle.

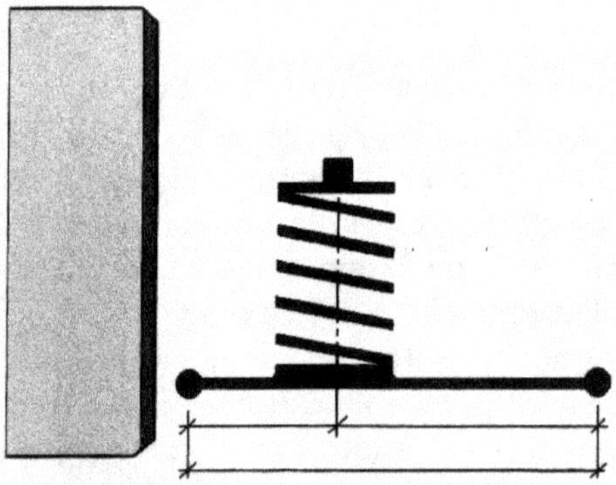

In a way similar to the coil-over system, the stock type of big spring suspension has a motion ratio that must be used to find the wheel rate. The wheel rates and the track width (spring base) are determining factors in finding the roll angle.

What Affects Roll Angles – The following are general rules concerning roll angles and can be considered when wanting to know how to create larger or smaller roll angles.

Spring Rates – A higher overall average spring rate at the front or rear results in a lower roll angle. A lower overall average spring rate at the front or rear results in a higher roll angle.

Spring Split – A spring split with the inside spring softer than the outside spring results in a reduced roll angle. A spring split with the outside spring softer than the inside spring results in a greater roll angle. And, increasing the inside spring rate results in a greater roll angle. Decreasing the inside spring rate results in a reduced roll angle. Conversely, decreasing the outside spring rate results in a greater roll angle and increasing the outside spring rate (or spring force) results in a reduced roll angle.

Spring Base – In a solid axle suspension, a wider spring base results in a reduced roll angle. A narrower spring base creates a greater roll angle. Remember that the chassis feels the spring base at the top of the springs, not at the bottom. It basically rides on the tops of the springs.

Moment Center – A higher moment center at the front or rear will result in a reduced roll angle. The moment arm becomes shorter and has less overturning moment as a result. Conversely, a lower roll center at the front or rear results in a greater roll angle.

Sway Bars – Changing to a larger diameter sway bar, thicker wall, shorter length sway bar, or shorter sway bar arms, results in a reduced roll angle. Conversely, changing to a smaller sway bar, thinner wall, longer length sway bar, or longer sway bar arms, creates a greater roll angle.

Lateral G-force – A higher lateral G-force results in a greater roll angle, and a lower lateral G-force results in a reduced roll angle.

Center of Gravity Height – A higher center of gravity results in a greater roll angle. A lower center of gravity creates a lower roll angle.

Track Banking Angle – If the G-forces remain the same, the following apply. A higher banking angle results in a reduced roll angle. A lower banking angle results in a greater roll angle. However, a higher banking angle usually results in the race car being able to achieve a higher speed and higher Lateral G-force and therefore must be considered in the influences on roll angle.

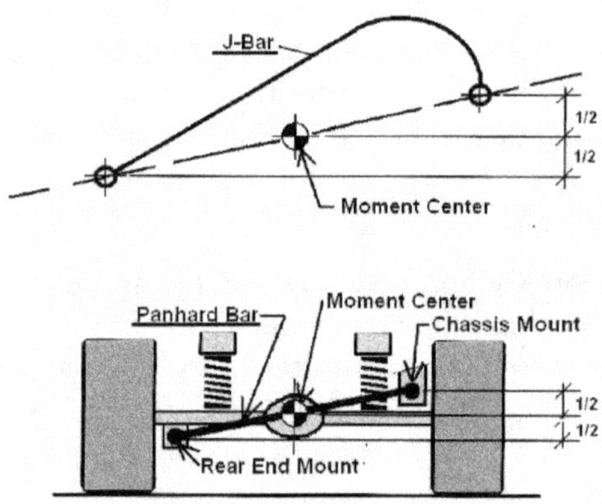

The rear moment center height is the average height of the center of the bolts connecting the ends of the panhard, or J-bar. When we raise or lower the ends of the bar, we raise and lower the rear moment center affecting the roll angle of that end of the car. As the car rolls through the turns, if the outside end of the panhard bar is mounted to the chassis, it will move down causing the rear MC to also move down. If the car squats on exit, then the chassis end of the panhard bar will move down and with it the rear MC. These are all things to take into account when working out our setups.

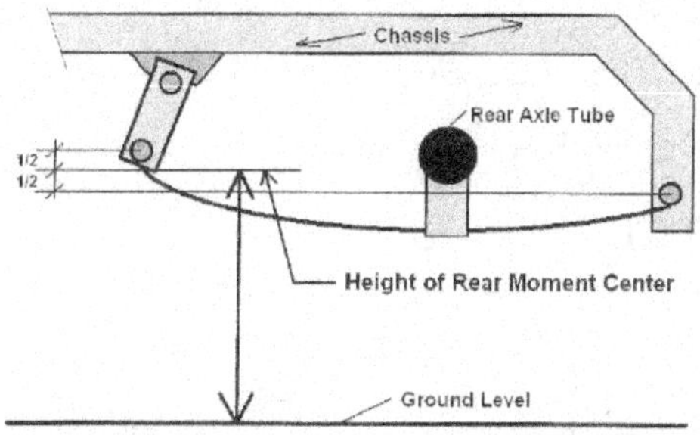

The height of the rear MC for a leaf spring suspension is the average height of the two mounting bolts on the leaf spring. As the chassis rolls through the turns, these points move down on the outside leaf just like the chassis mount of a panhard bar does lowering the rear MC.

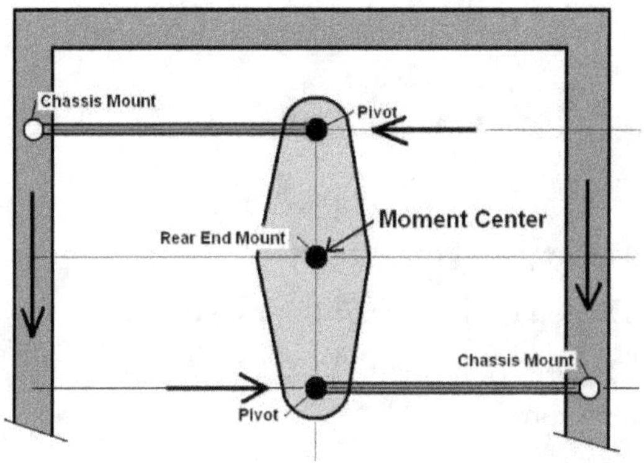

The Watts link system, used mostly with road racing cars using solid axle rear suspensions, is a system that maintains the moment center height through chassis roll. The point of resistance to lateral forces is at the vertical center of the rear end and mounted to the rear end. The other notable feature of this link is that it does not influence the chassis to move side to side as the car dives and rolls. There is no rear steer due to lateral movement of the chassis with this system.

How To Tell - Tire temperatures and tire wear tell us a lot about how our car is working. Equal front to rear tire temperature averages indicate a neutral car, though that car might not be well balanced. Equal temperatures and/or tire wear on each side of the car is an indicator of a balanced setup. When the LF and LR tires are doing equal work, and the right side tires are doing equal work, you will have found that balanced setup.

Just try to remember which components affect the roll tendencies of each end, and to what extent, and making those adjustments will be quicker and more accurate in your search for that winning setup.

Aero Balance Influence – Although this discussion is all about mechanical balance and how we go about achieving that, I need to put in a short mention of aero balance. If we truly achieve mechanical balance and our wheel loads are ideal, then where does aero downforce fit in?

If we end up with a lot of front aero downforce and very little rear downforce, then obviously the car will have more grip on the front tires due to having more loading. We need to be careful when we analyze our aero package that we don't overwhelm one end by creating too much aero downforce on the other end. We'll talk much more about that in RCT Level Three.

Summary - All race teams need to know certain basic information about their cars so that the foundation of a balanced setup can be laid. Learn to recognize the tendencies that make a car's suspension want to roll more or less and try to match the two ends of the car for a more balanced setup.

Know your front geometry settings, especially the Moment Center height and camber change characteristics. Be prepared to adjust the weight distribution (cross weight distribution) as you make changes in the direction of a balanced setup.

In RCT Level Three, we will be using this information to setup an actual race car. We will demonstrate how we can perfect corner entry and exit performance without changing the roll angles in the suspensions. What you have read and learned in this Lesson is among the most important pieces of race car technology information you will ever learn.

Exam - In The Context Of This Lesson:

Dynamic Balance Is When?

1) The car is neutral in handling

2) The front and rear weights are the same

3) The two suspensions are working together

4) The inside and outside tire loadings are equal

Ideal Roll Angle Is When?

1) The chassis does not roll

2) The front rolls more than the rear

3) The front rolls less than the rear

4) Any roll angle where the front and rear are equal

Roll Axis/Roll Stiffness and Roll Angle Methods Are The Same?

1) True

2) False

Matched Roll Angles Create What?

1) Ideal weight distribution

2) A faster turn speed

3) More overall grip

4) Better tire wear

5) All of the above

The Bottom Of The Moment Arm Is?

1) The ground

2) The bottom of the springs

3) The moment center height

4) The tire contact patches

The Top Of The Moment Arm Is?

1) The top of the springs

2) The upper ball joints

3) The upper chassis mounts

4) The center of gravity of the sprung mass

Name The Two Forces Acting On The Race Car?

1) The jacking force

2) The overturning moment

3) The force of gravity

4) Centrifugal lateral force

5) 3 and 4

Spring Base Is Measured At?

1) The bottom of the springs

2) The height of the springs

3) The top of the springs for a solid axle

4) The tire contact patches for a AA-arm suspension

5) 3 and 4

Which One Is A Primary Influence On Roll Angles?

1) The Center of Gravity height

2) The spring stiffness

3) The spring base width

4) The lateral G-force magnitude

5) All of the above

Which Will Increase The Roll Angle?

1) Raising the moment center

2) Lowering the CG

3) Softening the RR spring rate

4) Softening the LR spring rate

Racecar Technology – Level Two

Lesson Fourteen – Sway / Anti Roll Bars

A sway bar, or what is sometimes called an anti-roll bar is a device that resists chassis roll. The device can be constructed in many different ways and some of the components may not be "bars" at all. They just have to produce resistance to the rolling action of the chassis.

In this Lesson, we will tell you how to rate your sway bar the accurate way. We won't rely on published data or simple calculators and we will tell you why you shouldn't do that. Once we know our installed sway bar rate, we can then use that information to setup the car. We will be doing that in the RCT Level Three course.

As we prepare our test car for finding the true rate of the sway bars at both the wheel and at the ride spring placement, we install solid links in place of the spring/shock combo. This Method Number One is both easy to do and very effective and accurate. We also can measure the actual force the sway bar exerts on the suspension at mid-turn using the Gale Force Sway Bar force measuring tool.

As a chassis rolls, the inside (of the turns) springs decompress and the outside springs compress as the car rolls towards the outside of the turn. This is an over-simplification of what occurs, but gets you thinking in the right direction. In reality, the outside springs compress more than the inside springs in most cases and that results in chassis roll.

Chassis roll in and of itself is not a bad thing, but the trend in racing is to limit the amount of roll in a race vehicle. Of course, some forms of racing need the cars to roll in order to facilitate weight transfer and the chassis compliance that is needed. The racing I refer to is dirt racing.

Dirt cars just roll more than other forms of racing and most dirt cars do not even have a sway bar installed. So, we can assume from this point on that we are talking about asphalt race cars including circle track and road racing sedan and formula types of cars. First let me give you a little information about sway bars.

On the formula type of race car, the systems are necessarily made very compact and the sway bars designs can be varied and somewhat odd. With the links attached to rockers that join the sway bar and the spring running at odd angles from a top view and attached to bendable blade arms at the ends of the bar, we can count many variables and non-linear functions as the car rolls. Calculating the influence of the sway bar becomes almost impossible without the wheel rate measurement method we describe here. The two gray bars are the links from the sway bar arms to the pushrod rockers attached to the springs. The sway bar is to the right and the springs are to the left.

The blades on this prototype cars sway bar can be turned to provide varying amounts of resistance to bending. This combines with the twisting of the tubing to produce many combinations of sway bar rate. We would need to test each setting of the blades, different blade thicknesses and different sized sway bars and note those rates for future use.

Items That Affect The Rate of The Sway Bar – There are several things that affect the rate of a sway bar. The bar itself may have a stiffness, but it does not have a rate. When we attach an arm to it, we can then install it and when the car rolls, it twists in resistance to the motion of roll and then has a rate. We need to know the installed rate of the sway bar related to wheel rate and/or ride spring rate. More on that later.

So, the sway bar resists chassis roll and with some of the more popular setups, minimizing and/or eliminating chassis roll is a desired effect helped along by using sway bars. The following are the seven most common variables that affect the amount of work the bar does and there could be more in some applications.

- The length of the portion of the bar that will twist affects its rate. The longer this portion is, the softer the rate.
- The outer diameter of the bar affects its rate. The larger the diameter, the stiffer it will be.
- Hollow bars are less stiff than solid bars, so the wall thickness (or hole size) affects the rate. Thinner walls are softer rates.
- The hardness of the steel used for the bar affects the rate. There is a modulus of elasticity for each type of steel that relates to the mixture of metals and the hardening process used in the manufacturing of the steel. This is the part we cannot absolutely know because it is almost never published.
- The sway bar arm length affects the rate. The longer the arm, the softer the bar rate.
- The arm material can affect the rate. If the arm bends under load, the rate will be softer.
- The installation ratio affects how the bar rate relates to the wheel rate and/or ride spring rate.

It is easy to see that calculating a sway bar rate and knowing how much work it does is very difficult at best. That is why we should measure the rate and force as it is when the bar is installed in the car with all of the motion ratios, arm lengths, and other variable taken into consideration. There just is no other way to accurately do it and be able to use the information for the actual setup.

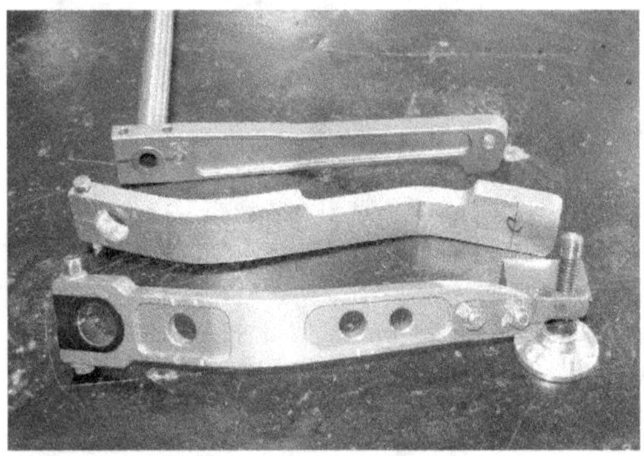

In the stock car world, sway bars can sometimes be larger for some uses. If the type of racing requires the use of stiffer bar rates, then stiffer arms are needed to connect the bar to the lower control arms. The old-style aluminum arms are just not strong enough and will bend, greatly influencing the overall rate and influence of the bar. And, the rate will not be linear. We proved that in tests done almost twenty years ago. The steel arm in the rear of the photo is best because it flexes very little.

It is important to know your sway bar rates in order to translate that information to the setup you desire. There are problems associated with looking up the rate in a table or other publication or doing a calculation because of several factors that affect the working rate of the bar. All of the variables are taken into consideration when we rate the bar after it is installed in the race car.

There are several methods that take into account all of the variables to provide the true working rate of any sway bar. The first one has been used successfully in both circle track cars as well as formula prototype race cars where the sway bar system is very complicated.

Method Number One - Giving credit where credit is due, I came across this method when it was presented by James Hakewill on his website. With this method, your team can go through the process of measuring the rates of all of the combinations of sway bar sizes and settings for the more complicated sway bars.

The method works for any type and installation of sway bar from stock car types to formula types including Formula One and Indy cars. The method rates the sway bar system and provides a wheel rate of the bar. The wheel rate can be used for many different setup functions. The following is a detailed account of how you can rate your sway bar. This simple yet effective method takes into account all of the variables and any we might not have mentioned. Here is how it works.

Overall Principle of the Method – What we are going to do is cause the sway bar assembly to support the weight of one corner of the car. Once we do that, we can accurately calculate the sway bar rate as a wheel rate and then through the motion ratio, as if it were a ride spring so we can know how our sway bar affects our car.

We will use the Right Front corner to rate the bar in this super late model circle track car. The process is the same for other types of race car. Sitting at ride height, the RF tire supports a portion of the total weight of the car. If we can cause the sway bar to act as a spring holding up that corner, then there will be a deflection of the wheel that we can measure and convert to a spring rate at the wheel in pounds per inch.

Once we find the wheel rate of the bar, we will convert that rate back to the ride spring mount to get the rate as a ride spring. From there we can use that information to help setup the car. I think that now you can see where the sway bar rated at the end of the arm is of little use in the real world.

This sway bar arm attachment to the sway bar itself is easily adjustable. For our testing, we set the bar neutral to find just the bar rate. Then we apply the number of turns of pre-load we would use on the race track to find the sway bar force that combines with the spring and bump force at mid-turn. The sway bar force is added to the spring and bump forces to match the force needed to support the right front suspension at mid-turn.

You will need to install solid links at all four corners of the car in place of the shocks. Make sure your link is strong enough to support the weight of that corner. Using opposite threaded heims helps us to adjust the lengths easily. We will read the loading on the tire over the scales at normal ride height and also the un-sprung weight of the wheel, control arms, etc.

Before taking any measurements or weighing the car, lock the steering shaft with the wheels pointed straight ahead. Movement of the steering and wheels will negatively affect the results. We used opposing locking pliers to hold the steering shaft.

You will need to support the rear of the car at one point in the middle between the two rear wheels. We used a section of large angle iron placed on the cross bar behind the rear end. Remove the rear wheels so that there is room for movement. Then lower the rear to the normal ride height plus any offset such at the height of the scales in this case.

When measuring the static loading on the RF scale with the solid link installed, make sure the bar arm is loose. Then, before getting ready to support the car with the sway bar, snug the left arm up against the lower control arm, but do not pre-load the bar. Then you can find the sway bar wheel rate. If you are measuring mid-turn force, go ahead and pre-load the sway bar just as you would for racing because the pre-load will add force.

Once we have caused the sway bar to support the weight of the RF corner, we need to place thicknesses of wood under the tire in order to shim the tire up so that the chassis ride height remains at the previous static ride height. Remember that when we weigh the un-sprung components, we need to add the weight of those wood shims to the un-sprung weight since they will be a part of the wheel assembly when we supported the car by the sway bar.

When taking your measurements, take your time. Get down at right angles to the tape and look at it several times to be sure you get the correct reading. Read the tape to 1/32nd of an inch for best results. A small error can add up to a lot of sway bar rate difference.

When rating the sway bar, it is a good time to check the connection between the end of the sway bar arm and the control arm. This link needs to be perpendicular to the sway bar arm and the control arm.

Getting Started – We start off by taking the following steps in preparation to take the measurements we will need to determine the sway bar wheel rate. This routine is fairly easy and fun to do. It may take some time, but the results will last forever.

• Install the desired bar and arms and leave the connections loose for now. For stock car installations, make sure that the link attaching the bar to the right control arm is perpendicular to both the sway bar arm and the lower control arm. This is very important and can affect the motion of the bar causing it to bind and be artificially stiff.

• Remove the springs and support both of the front corners of the car with solid links in place of the shocks. Adjust the length of the links to be the exact same length as the shock/springs were at normal ride height. Also make sure that the material used for the links is strong enough to support the weight of the car. We usually use steel tubing with nuts welded onto the ends and heims attached with opposite threading so we can easily adjust the length.

• Support the rear of the car at one point at the center of the frame on a jack stand, or similar, and use a piece of angle iron placed on top of the stand. Set the height to be the same height for the chassis as it is at normal ride height. You will need to remove the rear wheels to provide room for chassis movement.

• Place the front wheels onto the scales and record the RF weight reading.

• Raise the frame at the RF with a jack and remove the link. With the chassis still supported, lower the RF wheel/tire onto the scale and record the weight of the wheel assembly. This is the RF un-sprung weight. Move the assembly up and back onto the scales several times to take the friction out of the pivots. Also weigh each individual shim you will use later on to space under the tire. When we know which shims we need to use, we can add the weight of those shims to the un-sprung weight.

• While the link is out and the load is off the RF wheel, we can measure the shock motion ratio. Measure the shock length (the distance from center of bolt to center of bolt of the mounting bolts for the shock with the tire resting on the scale pad with no load on it.

• Place a thickness of wood or other material under the RF tire. Re-measure the shock length. Divide the amount the "shock" moved by the thickness of the material placed under the tire. Example: if the material was 2.0 inches thick and the shock moved 1.5 inches, then our motion ratio would be 0.750 (1.5/2.0 = 0.750). We will use this number later on.

• Remove the block, put the link back in place, remove the jack and let the front tires again rest on the scales. Snug up the sway bar arm being careful not to pre-load the bar. Record the RF ride height from the frame to the floor. We will need to keep this ride height the same throughout the process.

• Record the weight on the RF scale pad to make sure it is the same as before. Now jack up the RF corner carefully and remove the link. Let the RF corner back down onto the scale so that the sway bar is supporting that corners weight. The frame will be lower and we need to return it to the previous ride height.

• Place the shims under the RF tire so that the chassis will be back to the same original ride height. This is so that no load transfer takes place from chassis roll. The RF scale weight will now include the weight of the shims.

• Now that we know the weight of the shims, subtract the combined RF un-sprung weight, plus the weight of the shim, from the total RF weight to find the RF Sprung Weight that the bar is now supporting. Record that number.

• Measure the new, compressed "shock length". Find the difference between the new shock length and the original static shock length and divide that number by the RF shock motion ratio (0.75) that we found before. This tells us how far the wheel moved vertically. Record that number.

• Divide the sprung weight of the RF corner by the amount the wheel moved to find the Sway Bar Wheel Rate. Example: if our RF sprung weight were 600 pounds and our wheel moved 2.0 inch, then the sway bar wheel rate would be 300 lb./in (600 / 2.0 = 300).

Finding The Bar Rate - We now have the wheel rate of the installed sway bar assembly. This is not the last step in the process. We need to know the bar rate as if it were a ride spring. Then we can add that rate to the ride spring rate to determine total ride spring rate. Here is how we do that.

• We already know the motion ratio of the wheel to the ride spring. In this example it is 0.750. The spring moves 0.750 the amount the wheel moves. So, we need to square that number. 0.750 times 0.750 = 0.5625. Our dynamic motion ratio is 0.5625.

Instructors Note: To find spring rate from the wheel rate, or the wheel rate from the ride spring rate, we use the motion ratio squared. This is the law of motion ratios as a part of the study of dynamics and forces.

• To find the sway bar rate as a ride spring rate, we divide the Sway Bar Wheel Rate by the sway bar motion ratio squared (0.5625). Example: if we have a sway bar wheel rate of 300 lb./in. and a sway bar dynamic motion ratio of 0.5625, we calculate the sway bar rate as a ride spring rate to be, 300 / 0.5625 = 533 lb./in. The work the sway bar does is like installing a 533 pounds per inch stiffer ride spring.

What This Tells Us – Once we find the bar rate, we can relate that to our setup. You might be using a bar that is much stiffer, or softer, than needed for your intended use. In any routine, we can only expect the results to be as accurate as the effort we put into it. We need to make sure all of the steps are done correctly and that no other influences affect the results.

Make sure you zero your scales before taking any readings and lock the steering shaft so the wheels don't turn. This could affect the results. Also take your measurements a couple of times to make sure you read the tape correctly. Take your time in all phases of the process and keep good notes.

Method Number Two – This next method will let you rate the sway bar as a force in the same way we rate a spring as to force. Once we know the travel of the wheel and spring in our system, on our race track at mid-turn, we can measure the force the sway bar is adding to the spring/bump combination. Here are the steps.

This shows the Gale Force sway bar force measuring tool installed on this modified at the right front. With this tool, we can move the suspension up to the same travel as it shows at mid-turn and then measure the force of the sway bar only. We have removed the ride spring and placed the tool in place of the ride spring. The force in pounds is the exact work the sway bar is doing. If we know the ride spring force we will need for this car, we would set the ride spring to the total needed force minus the sway bar force.

• We first use a Gale Force Sway Bar force measuring tool. Leave the sway bar arm loose for now.

• Mount the GF rig into where the ride spring/shock would mount, adjust it to the correct length equal to normal ride height, and then zero the reading.

• Now set the sway bar links to where they are for on-track racing. If the sway bar is set to neutral, or preloaded some amount of turns, we just duplicate where it is for racing.

• Now jack up the wheel using the GF rig and the included hydraulic pump until we reach a shock length equal to what we read for on-track movement at mid-turn. We can then read the force on the digital readout to know what force the sway bar is adding to the ride spring and bump forces to support that corner of the car at mid-turn.

What This Is Used For – If we know that the force needed at the ride spring, to hold up the RF corner of the car at midturn is 2400 pounds, then if we know the sway bar force, we can set our ride spring, plus any bump device, to a force that is 2400 pounds minus the sway bar force. This number will be the net force the ride spring plus bumps will need to produce to support this corner of the car at mid-turn.

If the sway bar force were 533 pounds like the first example, then our spring/bump combo force should be 2400 minus 533 = 1,867 pounds. Now we can accurately set our ride spring height and bump spacing to generate 1,867 pounds of force at the mid-turn travel amount.

The overall use for these forces and how we calculate the required force will all be covered in RCT Level Three in much more detail. This is a very important and necessary part of race car setup and applies to all race cars.

Old School Method – The old school methods we have used in the past to rate a sway bar involved using either a scale to read the rate at the end of the sway bar arm, or a formula to calculate the rate at the end of the arm. Both of these methods provide a number that is hard to relate to our setup, but could be used to sort out different sway bars and arm combinations.

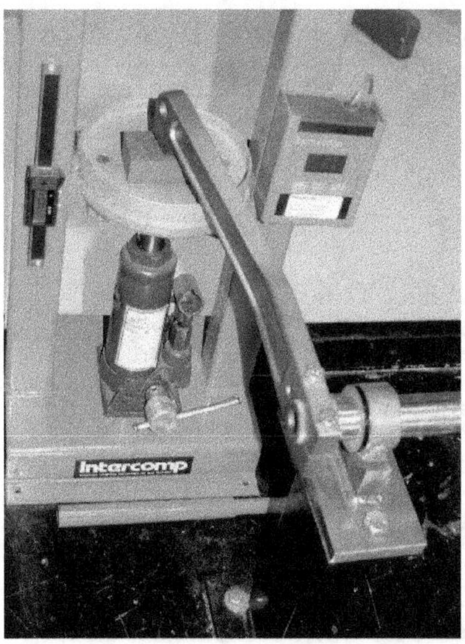

In years past, we may have rated the sway bar using a scale pad or spring rate like in this illustration. This tells us exactly what the rate of the bar is at the end of the arm in pounds per inch. But it does not tell us how much work it is doing in combination with the springs at the motion ratio our springs perform at.

Shear Modulus Numbers for Types of Metals

Metal	Shear Modulus
Cold Rolled Steel	11.5×10^6
Nickel Steel	11.0×10^6
Iconel (nickel-chrome-iron alloy	11.5×10^6
Molybdenum Steel Compound Almost all ultra-high strength steels contain molybdenum in amounts from 0.25 to 8%.	17.1×10^6
Stainless Steel	10.5×10^6
Carbon Steel	11.5×10^6
Titanium	6.5×10^6

This is the chart showing different Modulus numbers for different types of steel. We can easily see where a formula might not give us the correct information due to the many different modulus numbers and the fact we don't know exactly which metal our sway bars are made of.

Formula for Determining Sway Bar Rate

$$\text{Rate} = \frac{SM \times (D^4 - d^4) \times Pi}{L \times A^2 \times 32}$$

Where:
SM = Shear Modulus
D = Diameter of the Sway Bar in inches and decimals
d = Inside diameter of a hollow sway bar
L = Effective Length of the sway bar in inches and decimals
A = Effective Arm length in inches and decimals
Pi = 3.1416
Rate = Bar rate in pounds per inch of arm movement

Here is the formula for finding the rate of a sway bar or torsion bar and that rate is the rate at the end of the arm. Note again that the Shear Modulus is the number that is hard to be sure of, and that makes using a formula problematic.

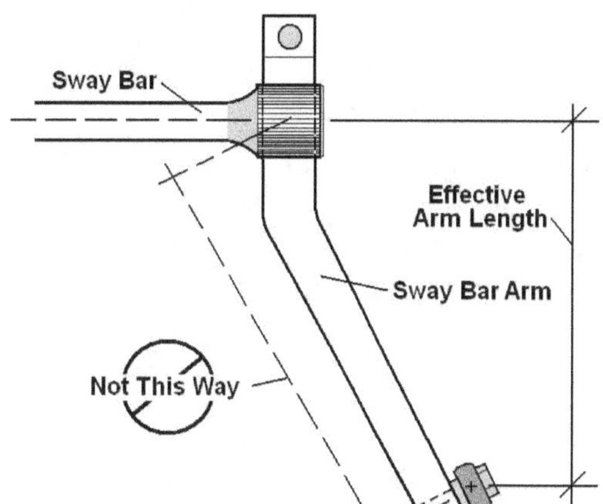

If you decide to calculate your sway bar rate, you need to measure the arm lengths at a distance that is at ninety degrees off of the sway bar, not at an angle. This sketch shows the correct way to measure.

Conclusion – What you find for rates for your sway bars might be what you expected or not. This is why we go through this type of exercise. We need to know every influence that affects our setups. If our sway bar force exceeds half the overall force we need at mid-turn, then maybe we need a softer rate of sway bar.

In any case, you now know how hard your sway bar is working and you can make some decisions about what size bar to use. Having this type of information is the key to being able to properly plan out your overall setup.

Exam - In The Context Of This Lesson:

A Sway Bar or Anti-Roll Bar Does What?

1) Keep the car from rolling

2) Helps to roll the car

3) Resists chassis roll

4) Keeps the chassis balanced

What Affects The Rate Of A Sway Bar?

1) The diameter of the bar

2) The length of the arms

3) The length of the bar

4) The modulus (hardness) of the metal

5) All of the above

We Can Accurately Rate A Sway Bar By What Method?

1) Using a formula

2) Using a chart from the manufacturer

3) Making the sway bar support a known weight

4) Measuring the sway bar force as wheel rate or spring rate

5) 3 and 4

A Sway Bar Assist The Ride Springs By?

1) Adding spring rate to the suspension

2) Providing more wheel rate

3) Providing roll stiffness support

4) All of the above

To Determine The Correct Mid-turn Ride Spring Force, We?

1) Add the sway bar force to the wheel rate

2) Add the sway bar force to the mid-turn force

3) Subtract the sway bar force from the mid-turn force

4) Add the sway bar rate to the wheel rate

Racecar Technology – Level Two

Lesson Fifteen – Shock Selection Matching Spring Rates

In previous Lessons we have learned what shocks do and cannot do. That is important to understand if we are going to be successful with developing a shock program for a race car. Some race teams continually try to make the car do things with their shocks that could be better served with other settings and components.

Here we will be setting up a race car with shocks based on what each shock will be working with, and those are forces. All components that contribute to the generation of force in a race car suspension must be controlled by the shocks while the chassis and suspension is in motion. In motion. To re-state, a shock does nothing if it is not in motion.

We experimented with Dick Andersons super late model car some years ago to find the proper front shock rates to run with the bump springs we installed. We matched the shocks to the spring rates on the front and back of the car and had a very fast machine. A lot has changed in the layout of both the spring rates and the shocks since then. What we present in this Lesson is inline with the current thinking. The shocks on this car at mid-turn is not influenced by the rates of the shocks because they are not in motion. This is called a steady-state condition because neither suspension is moving.

If we expect the shocks to transform or simply control our suspension, then we can only expect the shocks to contribute when the suspension is in motion. We need to think out time when the chassis moves, how it moves and at what part of the track it is moving.

The other major consideration is: how does the control of movement of one shock affect the movement and control of the other three shocks on the car? It is this interrelationship of the four shocks and springs that will ultimately determine where the loads on the tires go during the transition periods, or those times when the suspension is moving.

On exit, we have weight transfer and some suspension movement. In this case, with the suspension movement, shocks can help us distribute the loading on the four tires differently to enhance our transitional handling.

Based on the above statements, we can now declare that **shocks control the distribution of loading on the four tires while the chassis is moving**. Movement can be defined as: 1) chassis dive on corner entry from braking and weight transfer from the rear to the front, 2) chassis squat on corner exit under acceleration from weight transfer from the front to the rear, and 3) chassis roll throughout the corner from lateral centrifugal force that rolls the chassis and transfers weight from the inside tires to the outside tires.

Working With Shocks At The Track – Everything we do to tune the car with shocks should be based on the knowledge we have learned above. We will get into specifics about each corner of the car and each of the transition segments of the race track. We will study the motions for each transition and how those motions can be used to enhance our setup and improve the performance of the race car through the segments.

There are many different types of setup these days and racers are trying to do things that the laws of physics just won't allow. You don't have to be an engineer to understand the problems that can come as a result of this. With just a basic understanding of how the dynamics of the race car work, we can all choose our shock rates correctly to complement our setups.

In the modern world of short track racing for both dirt and asphalt competition, shocks have become one of the most important tuning tools we have. They can complement the current setups, especially the radical bump setups if applied correctly.

The following information is useful whether for teams who are running either the more conventional setups or the more radical bump setup ones. Shocks affect the speed of movement and the load distribution at the four corners of the car as we transition from high speed to minimum mid-turn, and then back up too high speed again.

One of the most basic and important things to remember is this. Shocks do not affect the handling of the car if they are not in motion. To say that you can tune your mid-turn, steady state handling with shock changes is wrong. Any competent shock expert will tell you that.

Shocks Work With Springs - Controlling wheel movement would be much easier if all we had to work with was the shocks. But in reality, our race cars are supported by a set of springs and other components that produce forces. Basically, we always want to match our shock rates to the suspension forces of: 1) the ride springs, 2) the sway bar, and 3) the bump devise you might be using. In all situations, the shocks rebound rate will always need to be greater than the compression rate for any racing shock because the springs help resist compression and promotes rebound.

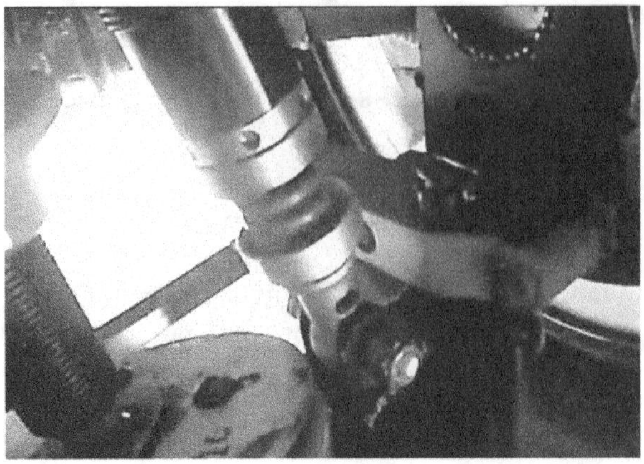

This image taken from a video shows an early version of the bump spring in action. The shock rates need to be high enough in rebound control to manage this high spring rate, in this case 1200 ppi.

https://www.youtube.com/watch?v=u9clWQpps6Q

This video shows a late model circle track car on-track with a 1200 ppi RF bump spring installed. This is a fairly bumpy track and we can see that the motion is enough to almost coil bind the spring. We could have used either a stiffer rebound shock or a stiffer bump spring.

https://www.youtube.com/watch?v=RlvO7tDJm4Q

In this video, we ran the same car with a stiffer 1500 ppi bump spring and the motion was reduced. Modern bump springs are available in much higher spring rates that would further limit the motion of the suspension on bumpy tracks.

As we install stiffer springs, we would naturally need to increase the rebound resistance and decrease the compression resistance. Stiffer springs would include adding bumps to one or more corners of the car. Each bump device has a spring rate in the range of motion it is operating in. Bump springs have a constant rate that is just added to the equivalent ride spring rate.

Let's expand on this thought. A stiffer spring that supports the same load, or generates the same force, will rebound faster than a softer spring supporting the same load. So, even though you are not generating more force to carry the intended load, your rebound speed will increase with the spring that has a stiffer spring rate.

Conventional Vs. Bump Forces – The actual force needed to support the car at all points on the race track is no different between the conventional setups and ones what use bump stops or bump springs. The natural forces of gravity, centrifugal (lateral) force and mechanical downforce due to track banking are the very same if the turn speeds remain the same.

The difference between the setups is the dynamic spring rates, or stiffness of spring the suspension is using during the trip through the turns. A conventional setup using 250 ppi (pounds per inch) coil-over springs will rebound with less force than a bump setup running on 1500 ppi bump springs or bump stops that rate in that range.

The shock rates we need to use to control the conventional setup is much different and softer than the ones we will use to control the much higher bump setups spring rates. It is all about controlling the speed of movement during times of load transfer. Obviously a stiffer spring will rebound faster when X amount of weight is put on it, or taken off of it than a softer spring with the same X amount of weight change.

Soft Spring Setups Are Not – In a soft spring setup running on bumps, the setup is not really soft. The ride springs are soft to allow the car to be pushed down onto the bump devices and then the suspension becomes very stiff with a very high spring rate.

So, we may say we have a soft spring setup because we install 150 ppi ride springs at the front in our coil-over circle track race car, but when we add the stiff bumps to that, we end up with 1200-1500 ppi or more spring rates. That is why I avoid saying, soft spring setups when referring to bumps setups.

Therefore, with those setups, keeping with the idea that the shocks must control the installed spring rate, we must run shocks with a rebound rate upwards of 1200-1500 pounds at 3-5 inches per second of speed.

"Tie-Down" Shock Terminology – This brings us to an important discussion. The use of the term "tie-down" has been around for some time to describe shocks that are high in rebound resistance. The idea initially was that if we install these high rebound shocks, we can tie that corner of the car down and keep the tire more in contact with the track surface. This way of thinking is completely wrong.

First off, a tire is not attached to the track surface, it is free to rise and fall according to what the other three corners of the car are doing at that instant on the track. If our setup wants that corner to move vertically because load has shifted off of it, then the load will come off regardless of how much we "tie down" with shocks. You just won't see much suspension movement.

So, we might be fooled into thinking that it indeed did work. No, it did not. A suspension can lose or gain loading without moving. Movement in and of itself is no indication of tire loading. If enough load comes off that corner, or the setup is unbalanced enough, the tire might well come off the track surface even without suspension movement.

The unloading of the tire will occur on the inside front or inside rear corners in most cases. So, we cannot tie the left front tire down and we cannot tie the left rear corner down, although I've seen some teams try. Lack of movement of the suspension does not mean that load remains on the tire.

Entry Tuning – If we split the front shock compression rates with a RF shock using a stiffer compression rate than the LF shock, then, while the suspension is in motion due to load being transferred to the front on entry, the RF suspension will move slower than the LF suspension. Additional load will be transferred onto the RF and LR tires causing a momentary increase in the corresponding diagonal weight percentage in the car.

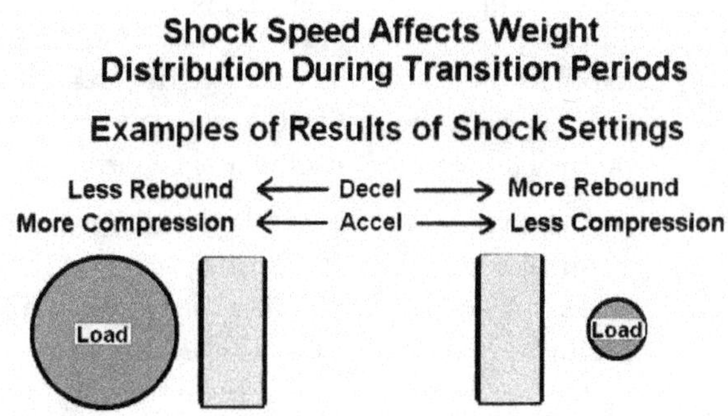

In this example, for the rear tires, the larger circle represents more tire loading. Under deceleration when the car is braking and weight is transferred from the rear to the front, less rebound in the LR shock leaves more retained load on the LR tire and conversely, more rebound in the RR shock removes more weight from that tire. Under acceleration, more compression resistance in the LR shock causes the LR tire to support more of the transferred weight over the RR tire and conversely, less compression in the RR shock causes less load to be transferred onto that tire.

Shock Speed Affects Weight Distribution During Transition Periods

Examples of Results of Shock Settings

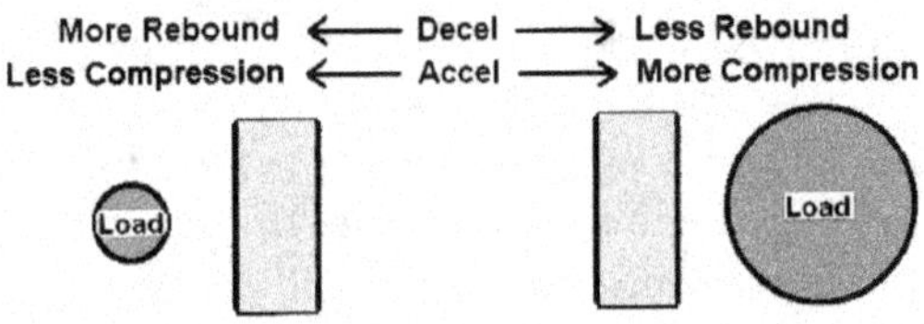

This shows the opposite results for rear tires while under acceleration and deceleration. On Decel, more rebound in the LR shock pulls more load off that tire and less rebound causes the RR to retain more loading on Decel. On Acceleration, more compression in the RR shock will load that corner with more of the weight that is being transferred.

The affect is more pronounced with conventional setups that might be used now days in the classes that are restricted from using bumps, or the stock classes. That is because those suspensions move much more than ones running on bumps. But it can still affect the bump setup cars to a certain degree.

It is important to note here that during the transitions, the load transfers almost immediately when a force is presented to enact that transfer such as applying the brakes. As we apply brakes entering the corner, the load transfer to the front happens quickly. If on entry we transfer 300 pounds from the rear to the front, the 300 pounds goes to the front in an instant.

The distribution of that 300 pounds between the two front tires, while the suspension is in motion and assuming a new attitude, will depend entirely on differences in stiffness of the suspension systems at all four corners. Stiffness is defined as the resistance to movement influenced by the shocks and springs.

The result of all of the above is this. The slower moving (or stiffer) corner will momentarily retain more of the transferred load while the suspension is in motion traveling to a new attitude. This added loading also affects the diagonal corner and loads it as well.

Cross weight in circle track racing is defined as the percentage derived from the combined RF and LR weight divided by the total vehicle weight. If the cross weight percent increases, then the car will be tighter on entry. This is exactly why it has been said that a stiffer RF shock will speed up load transfer to that corner, although we won't be using that terminology here.

If we install a LF shock that is stiffer than the RF shock, and/or run a stiffer LF spring than on the RF, then we can effectively reduce the cross weight in the car on entry while the suspension is in transition by loading the opposite diagonal, the LF and RR. As one diagonal goes up in percentage of supported weight the other goes down.

Modern bump setups use much lower front compression settings due to the high spring rate of the bumps. We can still utilize this method by creating a compression split. The effect will be less than with a conventional setup, but still somewhat effective.

Once the car has decelerated enough to make the turn, and the shocks are firmly on the bumps, the load distribution will equalize. Therefore, the advantage of shock compression split will be nullified. If the shock rebound settings are high enough, the shock will never leave the bump and no amount of compression split will work because there will be very little shock movement.

We can see where with conventional setups, we can probably utilize differences in shock compression rates where with the bump setups, the restricted movement nullifies the effect. We are better served working with the rear shocks only for the bumps type of setups.

Exit Tuning Using Split Valve Shocks – Corner exit performance can be improved by utilizing the shocks. This is done by either splitting the compression settings in the rear shocks and/or utilizing the rebound settings in the front shocks. In a car with equal rear spring rates, a stiffer compression setting in the LR shock than in the RR shock will load the LR and RF corners as weight transfers onto the springs.

Load is transferred to the rear under acceleration and while the rear suspension is in motion, this split will tighten the car by increasing the cross weight percent. But, this event is very short lived. Because as we stated, weight transfers very quickly, so the effect of shocks happens just as quickly and goes away just as quickly.

For all setups, a shock with a stiffer rebound rate at the LF corner can help accomplish the same effect by causing a slower movement of that suspension and a more rapid transfer of load off of that corner and the diagonal RR corner, which will in-turn increase the percentage of load supported by the RF and LR tires.

With the bump setups, the rear compression settings can really help a car that is otherwise limited in adjustments. If the RR spring is somewhat stiffer than the LR spring, and this is very common with bump setups, there will be a loosening effect on acceleration when load transfers to the rear. The stiffer RR spring causes load to be placed on the RR and LF corners reducing the cross weight percent.

By installing a LR shock with a stiffer compression rate, and a RR shock with a softer compression rate, you can nullify the negative effect of the stiffer RR spring. This serves to equalize the unequal resistance to compression due to the dissimilar spring rates and helps keep the car tight on exit. This again is a short lived event.

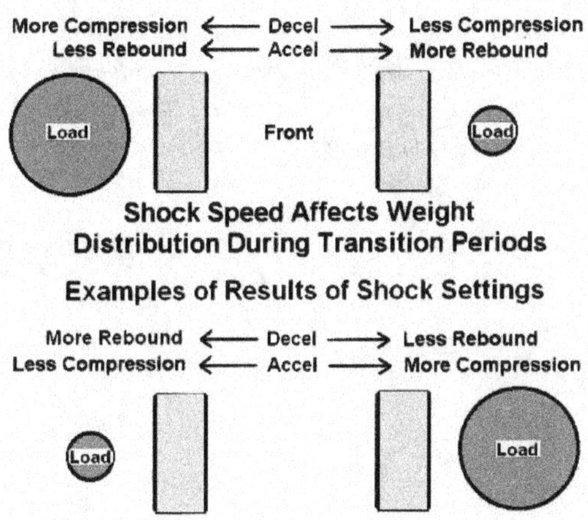

Here we can see the results of shock rate split for both the front and rear shocks and how the corners react. To load the RF and RR tires for either Accel or Decel, we follow along with the suggestions.

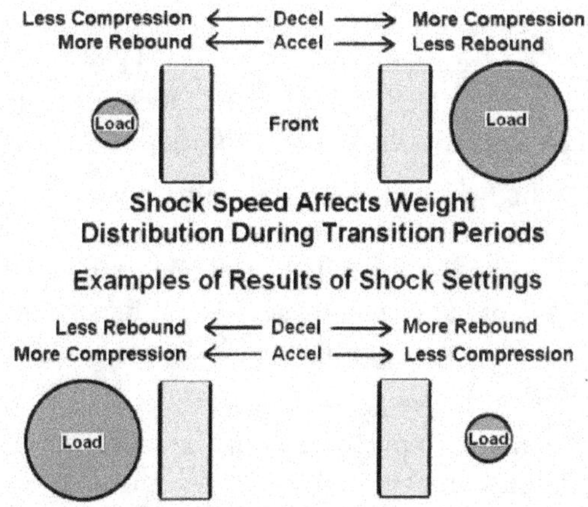

To load the RF and LR tires for Accel or Decel, we follow the suggestions here to see the entire picture of how the four tires are affected by shock rate differences side to side.

Putting All Of This to Use – In order to utilize the configurations we have discussed above, we must be able to use a range of different rates of shocks in order to find the right combination for a race car at a particular race track for a particular setup. For a team that races at only one track, the process is fairly simple.

You would need to experiment to find the fastest set of shocks and ones that suit the driver's style and stick to those staying within the boundaries of physics. For teams that travel to different tracks, some changes might be necessary if the setup (read as spring rates) needs to change and/or the track layout is different from track to track.

Most shock experts agree with the following basics:

1) The shock package should be softer overall when racing on dirt and when the track is flatter when on asphalt for the conventional setups.

2) Get your basic setup close to being balanced before trying to tune with shocks. Shocks cannot solve basic handling balance problems that occur at mid-turn when the shocks are not in motion.

3) Higher banked tracks require a higher overall rate of shocks and springs as opposed to flat tracks. This is because of the higher speeds and the extreme amount of suspension forces.

4) Shocks that are mounted farther from the ball joint should be stiffer than if they were mounted close to the ball joint. That is because with each

inch of travel of the wheel, the shock mounted farther away will move at a slower speed which means less resistance in both rebound and compression. This is also true of shocks mounted at high angles to the direction of motion.

5) Of the two transitions, tune entry performance first with your shocks. If there are no entry problems, make small changes if you want to experiment to see if entry speeds can be improved. Entry problems include a tight or a loose car.

By far the worst problem would be the loose-in condition. This can involve an alignment problem, but far too many times it is as we have discussed, a inside rear shock that is too stiff in rebound. For road racing cars, this could include both rear shocks.

6) Tune exit performance last. If there are no exit problems, don't make any significant changes. Exit problems can include a car that pushes (under steers) under acceleration or one that goes loose (over steers) under power. Be sure that you do not have a Tight / Loose condition where the car is basically tight in the middle and goes loose just past mid-turn. This is fixed with spring rate and/or moment center adjustments, etc.

7) On dirt race tracks, reduce rebound settings on the left side and decrease the compression rates on the right side for dry slick surfaces to promote more chassis movement. This helps to maintain grip as the car goes through the transitional phases of entry and exit.

8) For the bump setups on asphalt, the whole shock package must be much different than when running conventional or soft conventional setups. The bump spring rates (either bump stops, or bump springs) will be very high and so the shock rebound rates must match those high rates in order to control the ride spring and bump device.

Using a bump that is rated in the 1000 pound per inch spring rate range will need a shock that is rated at around 1000 ppi at 3 - 5 inches of movement per second. Usually the low speed rate of the shock will be comparatively high too and we often see a "nose" rate of between 500 and 800 ppi or more at less than one inch per second speed of movement.

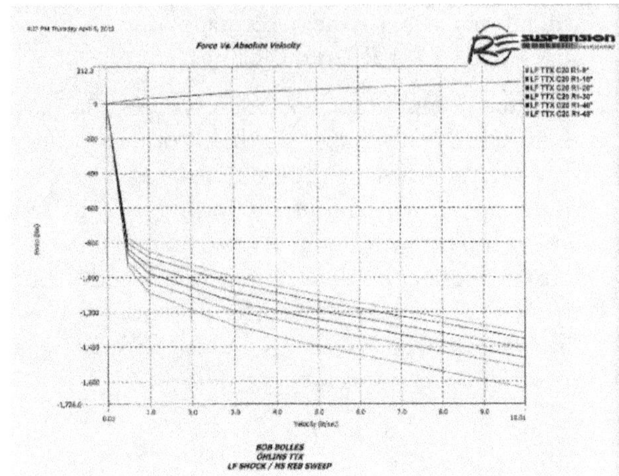

These are typical shock rebound and compression rates for a Left Front shock when running on very stiff bump setups. Note that at 3.0 inches per second of Velocity, we are in the range of adjustment of between 1,000 pounds up to about 1280 pounds. The rate builds quickly from motionless up to 0.5 inches per second of speed. It then builds more slowly up to 10.0 inches per second of velocity. Each line represents the force of the shock at an adjustment of the shock bleed valve. The bottom line on the rebound side represents a closed bleed valve. The other five lines represent openings of the bleed valve from 10 "clicks" to 48 "clicks".

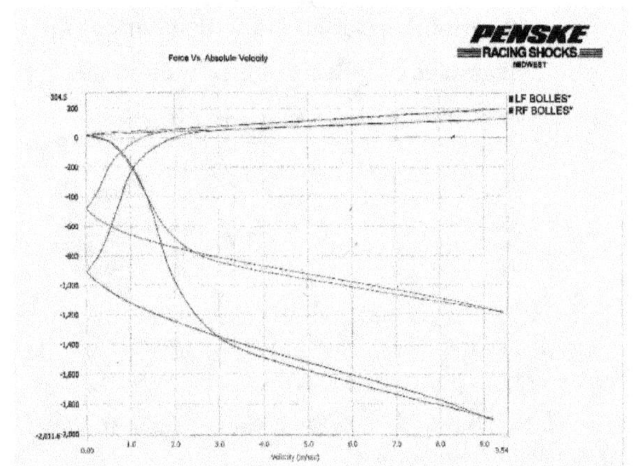

In advanced tuning for both the LF and RF shocks on a car running bumps setups, there is very little compression (lines above the zero force mark) as the bump provides plenty of compression resistance, so the shock doesn't have to do much work in compression. In rebound (lines below the zero force mark) we have what is called a "nose" rate. The red line, or RF shock, has about 500 pounds of force resistance to movement before the shock even starts to move. Then the force builds up to about 700+ pounds of force at 3.0 ips of velocity. The LF shock has about 900 pounds of "nose" force at zero movement and over 1,300 pounds of force at 3.0 ips velocity. We generally control the LF spring and bump to their rates and control the RF bump to about half of their rates for circle track cars turning left.

How To Make Changes – Racing shocks are either adjustable or not. The adjustable shocks might have rebound adjustments only, or both rebound and compression adjustments. For non-adjustable shocks, you would then need to physically change the valving in the shock, or have someone else do it. If the shocks are not re-valvable, then you would need to acquire shocks with a different value.

Every race team needs to have a shock dynamometer or send their shocks to someone who does, to be rated. The team must know the force of resistance of each shock on the car represented by a shock graph generated by a shock dyno. If you do not have this information, then you are guessing at the amount of work and the forces your shocks produce. No part on a race car can afford to be guessed at.

Summary – The suggestions provided here are representative of trends that can enhance your handling package. Before any of this can work, the setup must be balanced, the steering characteristics must be ideal and the car must be aligned properly. If not, you will probably chase the setup and experience a lot of frustration and expense. Shocks are to be used as a tuning device only after all of the other settings on the car have been perfected.

Shock tuning is the last thing to experiment with in order to try to increase your race cars performance, but it is nonetheless a necessary step in finding the ideal total handling package. That said, before you setup your car and chose your shocks, evaluate what you will need as far as shock rates that will match the spring and bump rates you will run.

When we experiment with the bump setups, we would do well to consult with the shock experts so that we can match our shocks to the spring rates and overall forces we will be working with. Most racing shock companies have technicians who are very familiar with those setups and can advise as to the best rates to use to match those bumps.

Because there are so many different setups, we cannot suggest specific rates of shock to use. This Lesson is intended to help you think out your changes as you evaluate your setup and try to make improvements. In RCT Level Three, we will be working with actual race cars and developing shock packages to help those setups.

Exam - In The Context Of This Lesson:

Shocks Control What?

1) Speed of suspension movement

2) Suspension forces

3) The rate of the spring

4) Transitional handling

5) All of the above

A Higher Rebound LR Shock Than The RR Shock Does What?

1) Loads the LF and RR corners on acceleration

2) Loads the LF and RR corners on deceleration

3) Loads the RF and LR corners on acceleration

4) Loads the RF and LR corners on deceleration

A Higher Rebound LF Shock Than The RF Shock Does What?

1) Loads the LF and RR corners on acceleration

2) Loads the LF and RR corners on deceleration

3) Loads the RF and LR corners on acceleration

4) Loads the RF and LR corners on deceleration

A Higher Compression LR Shock Than The RR Shock Does What?

1) Loads the LF and RR corners on acceleration

2) Loads the LF and RR corners on deceleration

3) Loads the RF and LR corners on acceleration

4) Loads the RF and LR corners on deceleration

A Higher Compression LF Shock Than The RF Shock Does What?

1) Loads the LF and RR corners on acceleration

2) Loads the LF and RR corners on deceleration

3) Loads the RF and LR corners on acceleration

4) Loads the RF and LR corners on deceleration

To Control A Stiff Spring, We Need To?

1) Have more compression control and less rebound

2) Have more rebound control and less compression

3) Make the rebound and compression the same

4) Run more compression than rebound control

To Control A 150 ppi Ride Spring Plus A 1350 ppi Bump Spring,

We Need What Shock Rebound Rate?

1) 1500 pounds of force at 10.0 ips velocity

2) A nose force of 1350 pounds

3) 150 pounds of force at 3.0 ips velocity

4) 1500 pounds of force at 3.0 ips velocity

Lesson Sixteen – Bump Setups Cause and Effect

Race car suspensions and setups have gone through many changes in recent years. Ever since the onset of the "Big Bar Soft Spring" era, racers have found themselves trying to go softer and softer with their suspensions to achieve faster turn speeds. The trend started in Cup racing when those teams started running coil-bind setups intended to lower the race car on the track to improve the aero effects.

Now that the short track industry and even the road racing community including prototypes and formula cars have now started to use bumps in their setups, it is time to explain how to utilize bump technology to go faster.

The stock car industry has taken to setups utilizing bump devices such as bump stops and bump springs. The reasons why include a lower front end to produce a lower center of gravity and better aerodynamic properties, and a consistent ride height which reduces camber change, something tires like. To get this attitude, the ride springs need to be soft, the bumps need to be stiff and the shocks need to be of a sufficient rebound rate to control the stiff bumps.

Every expert will agree that bump setups offer the following three distinct advantages over conventional setups. They are:

1) *Lower center of gravity.* This is important because the weight that is lower is at the front in stock and circle track cars, and the engine is a big part of the weight up front. Lower that chunk of metal and you've really lowered the CG by quite a bit. A lower CG transfers less weight in the turns and more weight ends up on the left, or inside, tire providing more equally loaded front tires and more grip.

2) *Better aerodynamics.* A lower front end and valance cuts the air that would go under the car and provides more low-pressure area under the front bodywork for more downforce. Also, it provides more angle of the hood area above the engine and this increases the effect of flat plate aero and offers more downforce.

3) *Reduced Camber Change.* Finally, a race car on bumps moves vertically very little and therefore there is very little camber change in the tires. Race car tires don't like camber change. The elimination of camber change could be the most important advantage in using bump setups.

Now in our current situation in racing, bumps are being used on almost every corner of the car. Certain advantages can be had by using a bump device for increasing the spring rate and/or loading on one or more corners of the race car.

There are several types of bump devices used in racing today. The first ones on the market were loosely called bump stops because they stopped the soft spring suspensions from moving too far and were used to prevent the car from bottoming out on the track. Even road racing prototypes used bump stops on high banked and fast tracks like Daytona.

A variation on the bump stops were the bevel washers that provided a consistent spring rate, but the available travel was not much more than the bump stops. Then the bump springs hit the market and a whole new and exciting way to bump became available.

The bump stops were hard on suspensions when run on rough tracks, and the bevel washers bottomed out. The bump springs offered considerably more travel and a consistent spring rate. Here are detailed explanations and applications of each.

This is one type of bump stop made out of fairly soft foam material. These are available in stiffer materials for different rates and properties. They can be combined with other bump stop designs. These are installed on the shock shaft and are compressed between the bottom of the shock body and a retainer cup at the bottom of the shock shaft.

Bump Stops - Bump Stops are one of the best methods of achieving the three bump advantages. When using bump stops, you install a stack of properly selected bump stops on one or both front shock shafts. You then select a front spring and sway bar package that will give you the best on track body attitude. Bump stops work best when used on a smooth track surface.

The key is knowing and achieving the proper wheel load in critical parts of the corner to balance the chassis handling. We do this by selecting different hardness ratings in the bump stops themselves, and by adding spacers or packers to alter the timing and engagement points of the shock body to the bump stop stack.

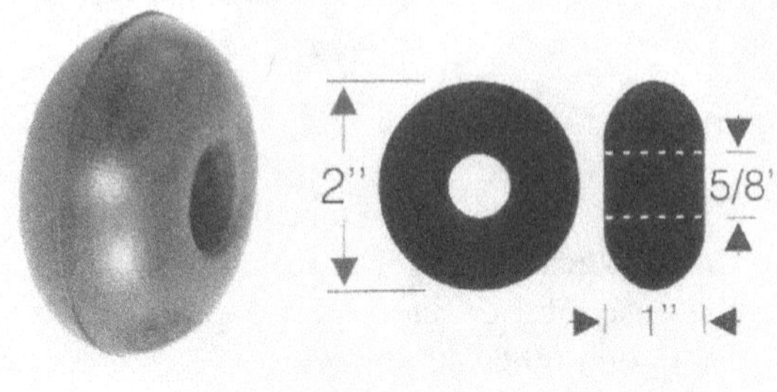

This is another type of bump stop and is a stiffer spring rate and can be stacked to provide a softer rate over using just one.

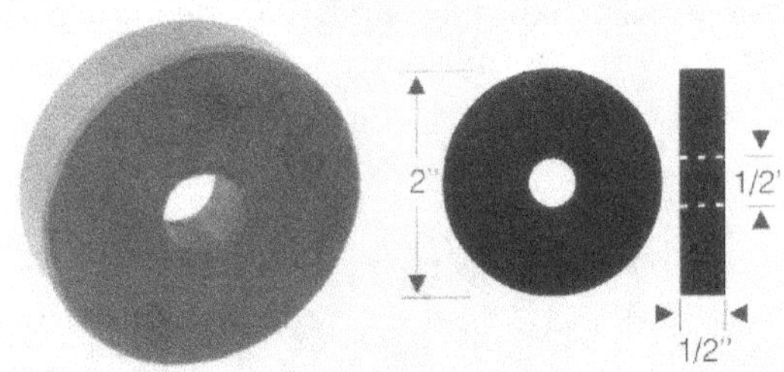

Here is yet another type of bump stop referred to as a "hockey puck" due to its resemblance to that. It too can be stacked and is available in different densities for a different spring rate. All bump stops are rising rate springs. That is, the spring rate rises as the bump stop is compressed.

This in effect creates a specific amount of wheel load for each wheel, based on a given amount of suspension travel to achieve that optimum body attitude and optimum distribution of wheel loads at the same time. We have force measurement rigs to apply the shock travel to the coil-over with the bump stops and then read the force it is under.

Here is an example: A traditional balanced setup with a 300 ppi RF ride spring, produces 900 lbs of force at 3" of shock travel to support the weight of the RF corner of the car at ride height. If this spring traveled an additional 3 ½ inches, the spring force would equal 1950 lbs. The problem is getting the spring to compress early enough to provide the benefits that staying on the bumps provides.

Alternately a balanced bump stop set up might utilize a 150 ppi RF Spring that is preloaded to provide the same 900 lbs of force needed at ride height. Then when the braking and corner loads are added to that spring, it easily moves down because of its soft rate and is stopped by the bump stop. Once it engages the bump stop stack at the same 3 ½" of shock travel from ride height, the force goes up immediately to the required 1800 lbs or more of total force to hold the car up against the turning forces. As you can see both setups are now achieving the about same wheel load at max travel keeping the balance very similar, but one very important difference exists.

The bump stop setup with softer springs reaches its optimum load sooner, and maintains it for a much longer time frame through the corner. With the right shocks installed, the shock will stay on the bump stop most of the way around the track. This has several major advantages.

First, as we discussed earlier, the camber will be at its optimum settings for maximum grip much earlier in corner entry, through center and remain longer through corner exit. The overall center of gravity of the car will also remain lower throughout the entire lap, also enhancing every part of the corner for more overall speed. Lastly, by using softer springs in the front, the front end will stay down longer, and will therefore offer better aero downforce throughout the turns from initial turn-in to corner exit.

The simplicity of a good bump system and ease of fine adjustments make bump stops the number one choice for a suspension package wherever rules allow. There are tools available now that help you determine the forces your bump package produces and then you can fine tune the loads on the bumps to maximize the loading on the four tires.

Drivers first impressions of a well-balanced bump setup are usually that they have much more feel in the front of the car. The cars usually respond very quickly to small inputs in steering, and less overall steering input is usually required.

The very beginning of the trend to create a lower front ride height came from using coil bind where when the softer ride springs compressed, then ran out of travel and then the springs coils contacted each other. This prevented the car from bottoming out on the race track, but the suspension now had no spring rate except for the tires spring rate. The suspension was basically solid at this point. There had to be a better way, and bump stops was the next step.

Initial experimentation with softer springs involved letting the spring go into what we call coil bind. The spring ran out of travel and the coils contacted each other and that was what kept the car from bottoming out. It was a harsh way to go and did provide the benefits, but the ride was harsh to say the least.

Bump stops have an advantage over coil-binding because they do compress somewhat and the ride improved when teams started using bump stops. With bump stops you can fine tune the weight distribution and cross weight settings much more easily without changing any critical dynamic chassis heights or stop heights. Bump stops can also provide much more compliance than the coil-bind setup making it a better choice when rough track surfaces are a concern.

When we use bump stops, we need to know the shock travel for mid-turn so we can duplicate that travel on a spring force rig. Once we know the force, we can adjust the force to what we need, no more and no less.

A well-balanced bump stop setup using soft springs is much closer in relative wheel rate to our older traditional setups than many racers realize. The key word used here is balanced. We talked about balance before and the need to keep the two ends of the car working together. With the bump setups, this is still the primary goal and balance provides the platform for proper load distribution through the turns.

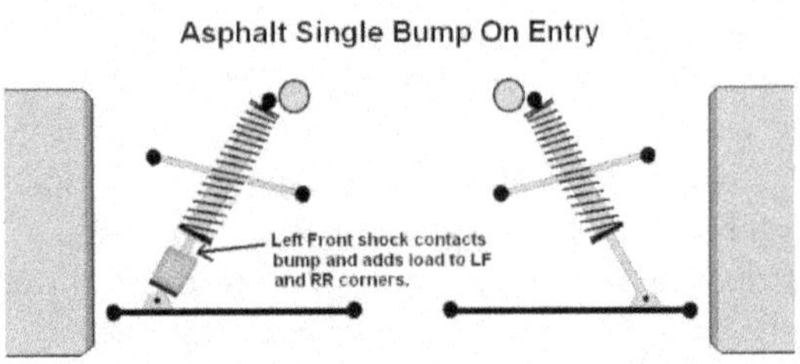

For many asphalt circle track race teams, only one bump is used on the car and it is mostly run on the LF corner. This forces a lot of loading onto the LF tire and may overload that tire, rather than providing a balance of loading on the two front tires.

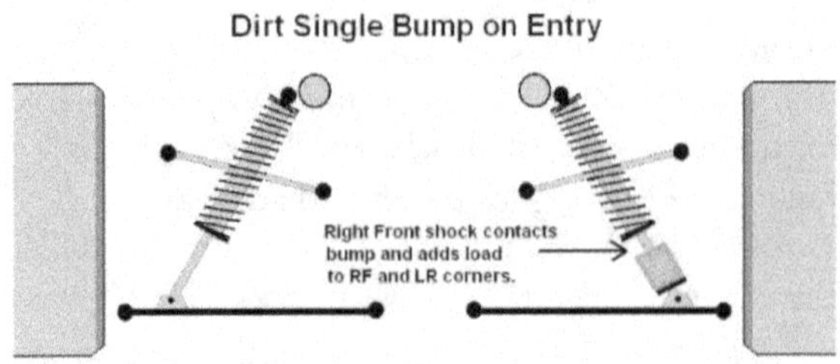

For dirt circle track late models, the bump is often used on the RF corner. This loads the RF and LR under braking and cornering. For slick tracks, it is an advantage to load two tires heavily in order to cut through the slick layer of sand to get down to the hard clay and to get more grip.

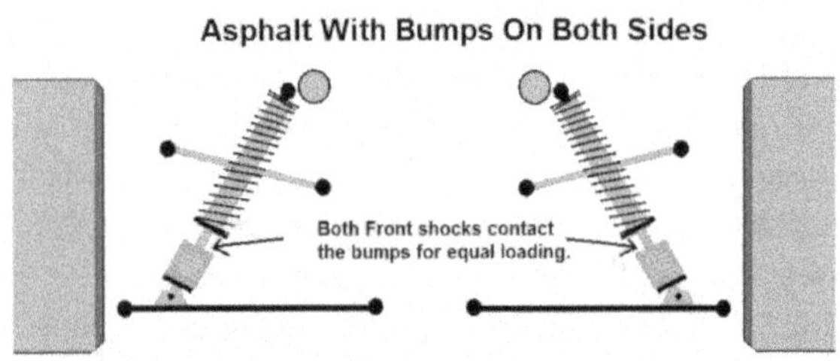

A better way to setup an asphalt circle track race car, or a road racing stock car, is to install bump stops or bump springs on both the front shocks. This balances out the forces and the setup providing better load distribution and a more balanced setup. The method we talked about for setting the cross weight on the bumps cannot be done when you only have one corner on the bumps. We can then see where there will be a drastic change in the weight distribution when the chassis transitions from ride height off the bump onto one bump.

Bump Springs - Things are changing fast in the bump arena. A couple of years ago, someone introduced a commercial bump spring. The concept of bump springs was not new and racers have been experimenting for many years with using various springs such as valve springs to use as additions to the softer ride springs. But the bump spring had never been commercially available.

This is the first commercially available bump spring produced and sold by Chassis R&D. It was an adaptation of a flat coil spring called a "die spring" that was normally used in industrial die stamping machines. It proved that using a spring with a consistent spring rate could be a viable alternative to bump stops. It provided more travel than a bump stop, a consistent spring rate throughout its range of travel, and it was durable unlike many bump stops that deteriorate quickly. The available travel for this spring is 0.75 inches.

Now we see where most major racing spring companies offer their version of a bump springs. Here we will explain what a bump spring is, what it does and how it may be superior in some cases to other bump devices.

Traditional bump stops are non-linear in their spring rate. This means that the farther they are compressed, the more spring rate they develop. This tends to make your setup tricky to balance and sometimes unpredictable. The traditional bumps also have very little travel available to absorb bumps and holes in a rough race track.

With the Bump Spring, we have a consistent spring rate at all positions. We also have much more travel available to smooth out the rough tracks we sometimes run on. The traditional bump stops usually allow around a quarter inch of movement and go to a very high spring rate approaching maximum travel. Most bump springs have three-quarters to one full inch of travel before coil bind and the spring rate never changes.

Bump springs are available in many different spring rates and heights. This collection of the first commercial bump spring was available in rates of 1,000, 1,200 and 1,500 ppi rates. Now, all major racing spring companies offer their version of the bump spring in many different rates and configurations.

So, we can see where the bump spring has characteristics that are better than the other bump devices due to the increased available travel and the consistency in spring rate. The fact that most of the racing spring companies have developed their own line of bump springs speaks to the fact that they may be better for many applications for teams choosing bump setups.

In reality, by using bump springs, we are actually going back to yesteryear and the stiff spring setups from the 1970's to the 1990's. The only change is that now we can run the car much lower and take advantage of the improved aero, a lower center of gravity and the reduction of camber change.

Setting Cross Weight For Bump Setups – Once the shocks are firmly on the bump springs, the load distribution will change very little since there is little movement of the shock through the turns. So, if you have not pre-set the cross weight percent with the car on the bumps, it may change dramatically from what was set at ride height when the car is off of the bump springs.

The following is how you should set your weight distribution when running bump setups. Failure to do this can result in a significant change in weight distribution from when the car is at ride height to when it contacts the bumps.

This shows how bottom of the shock comes in contact with the bumps spring. This shock would need a screw mount for its top mount in order to adjust its spacing, or the timing when the shock contacts the spring on entry and through the middle of the turns. Other applications use spacers, or what are called packers, to adjust the point in shock travel where the shock body contacts the spring.

You can set your cross weight both on and off the bump stops or springs. The procedure is as follows:

1) Begin by setting the correct cross weight percent on the scales at normal ride height;

2) Now remove the ride springs and allow the car to rest on the bump springs or bump stops with the bumps supporting the weight of the front end;

3) Bring the corner weights back to what you had at normal ride height by adjusting only the amount of spacing, or packers, between the front shocks and the bump springs or bump stops. If you have height adjustable shock mounts, just make changes to the shock heights to get your static cross weight number. Then just reinstall your ride springs and go racing.

Today's bump spring is of a round coil bar design just like the ride springs. They are designed to be compact and have significant travel for a bump device. And they are offered in a wide range of rates for any application and loading required.

This way, your car will not experience a change in load distribution when transitioning from ride springs onto the bump springs. This seems to be a very common problem when using any type of bumps.

Load Changes Due To Sway Bar Loading – Another effect caused by the bump setups is the loading of the sway bar when the front travels to the bumps. On many cars, if you were to set your sway bar to neutral at ride height, when the car is down on the bump springs, the sway bar will load by several turns because the two sway bar arms don't move the same distance. The right sway bar arm travels more than the left sway bar arm and that loads, or twists, the bar.

This definitely causes a change in the load distribution at the four wheels. When using larger sway bars, the change is much greater. What you can do is set your bar to neutral at normal ride height and then put a few crew members on the front of the car and push it down onto the bumps. While on the bumps, check to see how many turns it takes to return the bar to a neutral setting with no pre-loaded.

Once you know how many turns are involved, just set to neutral at ride height and then back off that number of turns. Or, if you are using a smaller sway bar, say 1.25 inches or less, you can just leave it preloaded when you set your cross weight on the bumps as we described above. That way, you can run pre-load on the sway bar for bite off the corner while still maintaining your correct cross weight percentage.

Another important thing to remember about the sway bar is that it provides added spring rate to the suspension. It acts, in roll, like the car has a stiffer left and right side ride spring. When the car rolls, the sway bar resists the rolling motion and uses its spring rate to do that. So, if we know what force we need in the suspension to support the car at mid-turn, we need to take into consideration the sway bar force that is working in addition to the spring and bump forces. If we don't consider the sway bar force, we may overload that corner of the car with too much ride spring force.

Rear Spring Rates For Bump Setups – Initially when teams started running front end bump setups, the only way to match the front and rear roll angles, and balance the setup, was to run a spring split in the rear with the RR spring much stiffer than the LR spring. Many teams overdid this split and tried to run much more split than was necessary.

The problem with running this spring split is that on corner exit, the weight transfers onto the rear springs and the LR spring compresses more than the RR spring. This loads the LF and RR corners and loosens the setup by reducing the cross weight percent.

To fix this, the teams noted the shock travels using the stiffer RR spring and then recorded the force that amount of travel produced. Then they installed a much softer RR spring that was preloaded some amount to help provide the force needed at mid-turn to support the car and provide the balance needed.

Then when the car accelerated, and weight transferred to the rear springs, both rear springs compressed about the same amount and the weight distribution remained unchanged. Some teams are even experimenting with a RR spring rate that is actually softer than the LR spring rate, but with a preload that works at mid-turn. Then as the weight transfers from acceleration, the RR compresses more than the LR corner and the cross weight will then increase providing more bite off the corners.

Entry Tuning – Teams soon learn with bump setups, that they need a much stiffer front shock setting for the rebound of the shocks. They may see this trend of running stiffer rebound shocks as the way to go for all of their shocks. Remember that we set the rate of the shocks in relation to the stiffness of the springs they will be required to control. For the most part, the rear springs are not overly stiff for bump setups.

The left rear corner is usually sprung fairly soft and less than what was used for the older conventional setups. If the spring used in the LR is softer than what was run for conventional setups, then the shock rate in rebound that we use for that corner must necessarily be softer.

Since our spring rate for the LR corner is averaging around 150 ppi with bump setups, our shock rebound only needs to be in that range at three inches per second. Most left rear shocks are valved to a higher rebound setting than that because back in earlier days using conventional setups, the LR spring rate was higher. If we use a shock that is too high in rebound, then on entry to the corner, the LR spring will be slow to extend and weight will be lifted off of that tire and the car will become loose.

This is a very common problem and easily solved by just reducing the rebound in the LR shock. Many crew chiefs I have talked to tell me that the LR shock does not need very much rebound, only enough to control the spring that is installed.

Putting All Of This to Use – In order to utilize the bump configurations we have discussed here, we must be able to use a range of different rates of shocks in order to find the right combination for our car at a particular race track for a particular setup. For a team that races at only one track, the process is fairly simple.

For the bump spring setups on asphalt, the whole shock package is much different than when running conventional or soft conventional setups. The bump spring rates will be very high and the shock rebound rates must match those high rates in order to control the bump device.

A bump spring that is in the 1500 pound per inch range will need a shock that is rated from 1200 to 1500 ppi at around 3 inches per second of movement. Usually the ultra-low speed rate of the shock will be comparatively high too and we often see a "nose" rate of between 300 and 600 ppi or more at close to zero speed of movement.

Summary – The use of Bump devices has revolutionized setups for circle track cars. For road racing and formula cars, bump springs and bump stops provide a convenient way to keep the chassis from over-traveling on fast and high banked tracks, or from high aero downforce. For those cars, the bumps help to maintain the aero platform and provides consistency.

As for the circle track cars, the setups are becoming much easier to balance, the cars are faster, and the racing is better. What used to be normal in performance differences in a field of race cars is now much less. Race teams have to work much harder to fine tune their setups in order to become better than the competition. In RCT Level Three we will be setting up race cars on bump devices and telling you how to properly do that.

Exam - In The Context Of This Lesson:

Bump Setups Provide An Improvement For Which?

1) Aero downforce

2) Camber change reduction

3) A lower center of gravity

4) All of the above

Bump Stops Have The Following Characteristic That Bump Springs Don't?

1) A consistent spring rate

2) A lower spring rate

3) A progressive spring rate

4) More colors available

Bump Springs Have The Following Characteristic That Bump Stops Don't?

1) A consistent spring rate

2) A wide range of spring rates

3) More spring travel before coil-bind

4) All of the above

A Circle Track Sway Bars Can Do What When The Chassis Dives?

1) A load itself

2) Unload itself

3) Change the cars weight distribution

4) 1 and 3

The Bump Device Gap To The Shock Body Must Be Timed So That?

1) The force will be correct

2) It will not coil bind in the turns

3) The weight distribution stays consistent

4) It will load the right front corner first

Is It Considered Better To Run Bumps On One Corner or Two On Asphalt?

1) One bump

2) Two, one on each side

Racecar Technology – Level Two

Lesson Seventeen – Anti Dive and Anti Squat

This Lesson on Anti Dive and Anti-Squat is simple and straight forward. We discussed the basics of these Anti's in RCT Level One. What we now have to discuss is how we actually use these Anti's in today's racing environment. Things have changed.

First off, we will talk about Anti-Dive. This was an effect that provided an important function years ago with all race cars, whereas today, the use of this effect is limited by necessity, or sometimes out of lack of necessity. For stock class cars and race cars using more conventional setups, Anti-Dive is still a viable setup option.

On entry to the corner, a race car brakes and if not restrained, will dive at the front end. This dive can be mitigated and reduced by the effect of Anti-dive. Anti-dive uses the braking force to restrict the movement of the chassis in a downward direction. In this Lesson we will tell you how this is done.

Anti-Dive Technology

What Is Anti-Dive? – Anti-dive is a mechanical function that utilizes the force of braking to impede the movement of a AA-arm suspension, in most cases the front suspension. For rear AA-arm suspensions, there is an effect of this function that limits droop, or rise, in the rear suspension when braking, but that is very advanced and does not apply to 99% of modern race cars.

So, we have a mechanical function that only works while we are braking. The force created by the brake pads gripping the rotor wants to rotate the brake caliper which is bolted to the spindle is called the torque force. By inclusion, the spindle is also trying to be rotated.

It can't because it is firmly held in position by the two ball joints. But, there still exists a powerful force again that is applied to the spindle while the brakes are applied and the car is being slowed from a very high speed. We can measure, or calculate this force by knowing the torque force created by braking.

There is also the force applied to the chassis and wheel that is the longitudinal force acting through the center of gravity. As the car is braking and the car is slowing, the components still want to move at the same speed, so a high force is created. This force "pushes" on the front chassis mounting points for the control arms and also pushes on the balljoints. The force ultimately acts on, and is centered on, the axle centerline, or spindle pin.

With Anti-dive design, the control arms that attach to the chassis are mounted at an angle from a side view. The upper control arm is mounted so that the rear mount is lower than the front mount. The angle varies, but usually we see between 1 and 6 degrees with 2.5 to 5.0 being the most commonly used angles in stock cars.

In this photo we see two cars entering the turn. The outside car is much higher under braking than the inside car. It is possible that the inside car has less anti-dive than the outside car. With the bump setups (both of these cars are running bumps), you don't need much if any at all of Anti-dive. We want to get onto the bumps quickly and Anti-dive restricts the movement.

Most teams split the rates of Anti-dive between the front upper control arm angles in a circle track car that turn only one way. The outside, or right side in the US, upper control arm Anti-dive angle is around 3-5 degrees and the left upper control arm Anti-dive angle is from 1-3 degrees.

The lower arm usually has much less angle and that angle is created by mounting the rear chassis mount higher than the forward chassis mount. These angles are anywhere from 1 to maybe 3 degrees. Now let's analyze what happens with the ball joints in relation to these control arm angles.

If the upper arm is angled the way we explained, then as the spindle/wheel moves up, the position of the upper ball joint moves to the rear. Just imagine rotating the upper arm to vertical, the ball joint would be well behind and to the rear of where it started out if we measured its location from a forward baseline. This motion moves in opposition to the braking torque and in the opposite direction of the longitudinal forces we talked about created by braking.

As for the lower arm, as the lower ball joint moves up with the spindle/wheel, due to the angle, that ball joint will move forward from its original position. It moves much less than the upper ball joint because the arm itself is longer and it has much less sideview angle. This movement is in the opposite direction as the torque movement, and in the same direction as the longitudinal forces.

To sum up the two movements for the upper and lower ball joint, in dive, the spindle rotates counter clockwise when viewed facing the right spindle and clockwise when facing the left spindle as the spindle moves upwards. The motion is in relation to the chassis when the chassis dives.

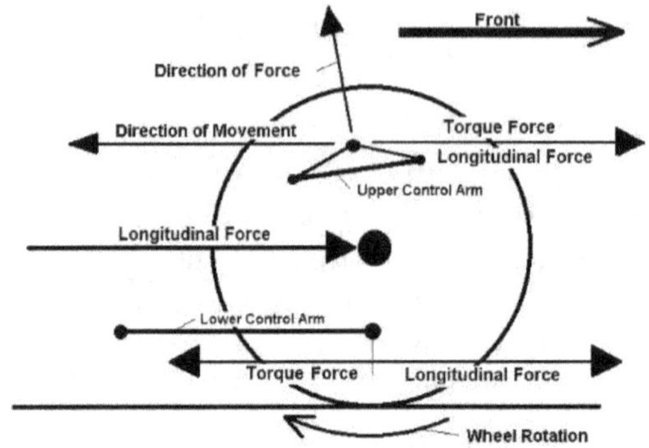

As this sketch illustrates, as the braking force is applied, the wheel and brake rotor is grabbed by the brake caliper that is mounted to the spindle. This causes the spindle to want to rotate in the direction of the wheel. If the chassis dives, its motion will move the upper ball joint to the rear against the rotation of the wheel. These opposing motions are what cause the Anti-dive effect.

So, chassis dive will rotate the spindle when we have Anti-dive design in the cars suspension. But as we stated before, the braking force wants to rotate the spindle in the opposite direction, or clockwise when facing the right front wheel and counter clockwise when facing the left front wheel.

These two opposing motions cause Anti-dive in some degree and the amount depends on two factors:

1) the amount of braking force being applied, and

2) the amount of Anti-dive design angles in the control arms.

If we brake less and have less angle in the control arms, we will have less Anti-dive effect, and visa versa.

To provide the upper control arm angle needed to produce Anti-dive, offset slugs are used for mounting the arms to the chassis. We would put in a higher hole in the front mount and a lower hole in the rear mount.

Here is what the mounts look like on the car. This car has no Anti-dive as noted by the level control arm mounts. If this car did have Anti-dive, then the front mount would use a slug with the hole higher off the frame rail.

Here is a closeup showing the slug and mount. The slug, if offset say ¼ inch and with the hole down from center, could be flipped and that puts the hole up ¼ inch providing a half inch of difference if the rear hole was in the middle of the slot. A half inch is about 5 degrees of control arm angle from a side view.

Why Use Anti-dive? - In earlier periods in racing, and even today with the more stock applications and conventional setups, Anti-dive limits dive in the front end under braking. If we didn't have Anti-dive, then when the car moved down in the front, the cambers of the front tires would change quickly and to a large degree.

Race car tires don't like camber change. We have stated that fact throughout our lessons in the RCT courses. If we can limit the camber change in the tires in a AA-arm suspension, we can create a larger and more consistent contact patch and experience more grip.

So, the race teams that run more conventional setups and the stock classes that by rule must run conventional setups still to this day need Anti-dive. But what about the race cars utilizing bump setups? Good question.

When we run our race car on bumps, the shocks go onto the bumps fairly quickly when we brake into the corner and don't come off the bumps until well past mid-turn. If the shocks are on the bumps, the vertical motion is limited by the very high spring rate of the bump device, whether we are braking or not.

So, if there is very little vertical motion of the chassis while on the bumps, then there will also be very little camber change because camber change only happens with vertical chassis motion and chassis roll, which is also limited with bump setups.

Getting to the point here about bumps setups, if the chassis moves very little, then we don't need to further restrict this motion by having Anti-dive designed into the suspension, do we? No. In fact, we may well want the chassis to dive quickly onto the bumps on entry braking and eliminating Anti-dive would help accomplish that.

As we setup our race cars, we need to think out what we need for effects like Anti-dive. Are we running bump setups? Are we running softer front springs where we might need more help limiting chassis dive? Can and should we control dive through shock settings in addition to the Anti-dive? These are all to be taken into consideration when planning out your setup and the settings that go along with that.

What Is Pro-dive? – Some teams have experimented with what is called Pro-dive. This is an effect that contrary to Anti-dive promotes the dive of the suspension while the car is under braking. In this scenario, as the brake force is applied, the angle of the control arms will cause the ball joints to want to move in the same direction as the spindle under braking and helps, or speeds up, the process of dive, again only while under braking.

This author personally does not use nor recommend Pro-dive as a way to promote chassis dive. On entry, the teams that do use Pro-dive suggest that this will move the suspension onto the bumps faster. It may, but in the process, it tends to pull up on the spindle/wheel before actually causing the chassis to dive. This motion lifts load off of the LF tire causing it to lose grip.

With the proper shock rates for compression and the correct soft ride spring, you will have little need for Pro-dive in your front geometry setup. In fact, with those same bump setups, you probably won't need Anti-dive either.

A Final Note About Anti-dive – From our lessons on AA-arm geometry, we learned that the elimination of bump steer is partly dependent on the angle of the tie rod. If the spindle rotates while it is moving vertically, and we have designed Anti-dive into our suspension system, then then the angle of the tie rod will necessarily change if the chassis does dive. We will no longer have near zero bump steer after the chassis dives and we would then need to adjust our settings to compensate for the chassis dive influence on the tie rod angle.

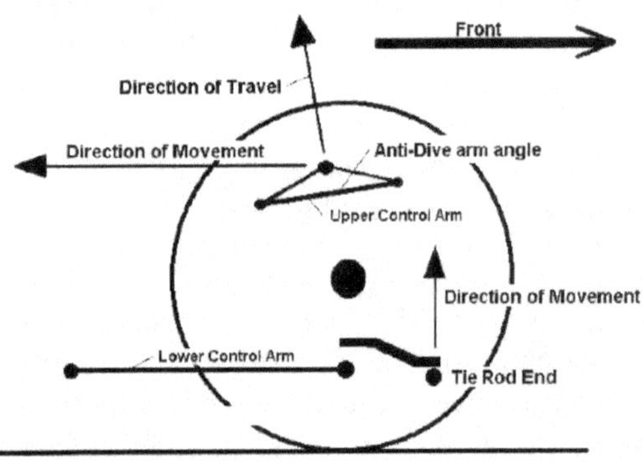

When we introduce Anti-dive into our suspension system, as the chassis dives, which it will eventually do after we are off the brakes and entering mid-turn, the spindle will rotate from the movement of the ball joints and this rotation also moves the end of the tie rod up. This changes the angle of the tie rod and we might not have near zero bump steer as a result. This has to be re-checked after we change our Anti-dive settings.

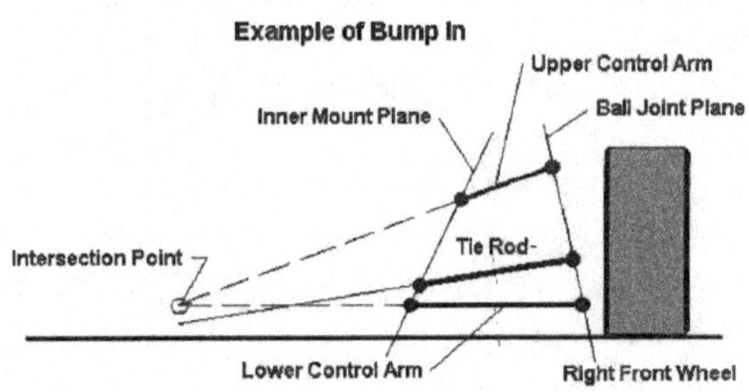

Here is where we end up after chassis dive when we have Anti-dive present in the system if we don't adjust the bump steer settings. The outer tie rod end at the spindle will move up as the chassis dives and this points the tie rod at an angle that falls below the Instant Center.

Anti-Squat Technology

What Is Anti-squat? – Anti-squat is a mechanical effect that helps the rear suspension resist squatting when the car is accelerating. Squat comes from the load, or weight, transfer from the front tires onto the rear tires when the car is accelerating.

On corner exit, we might want to incorporate some Anti-Squat, or Pro-Squat. The car will squat due to the weight transfer that takes place when the car accelerates. This weight can be captured by the Anti-squat links and redistributed between the rear tires for better weight distribution and more grip.

This effect is very similar to weight transfer during cornering, but it exists at 90 degrees from lateral weight transfer. The same elements are a part of longitudinal weight transfer as are a part of creating lateral weight transfer.

Anti-Squat uses the sideview angles of the upper and lower control arms, or suspension links, to create forces that oppose the movement, or squatting of the rear of the car due to weight transfer during acceleration.

The two forces that assist Anti-Squat are the torque of the motor and the acceleration force pushing the car forward to a greater speed. Both of these forces act on the suspension links. The upper link(s) and the lower links both contribute to Anti-Squat, but we will only concentrate on the upper link for this discussion.

It has been thought that using Anti-squat enhances and adds to the loading on the rear tires when the car is accelerating. That is not exactly true. Like lateral weight transfer, there is a defined amount of weight transfer that happens for the conditions. We cannot change that number unless we change the conditions.

The "conditions" we refer to are:

1) the Center of Gravity height,

2) the weight of the sprung mass of the car,

3) the wheel base length, and

4) the accelerating force, or longitudinal G-force.

As you can see, none of these are easily altered, especially by manipulating the rear geometry in our race cars.

What we can do with Anti-squat is redistribute, some, or all, of the weight that is transferred to the rear and shared among the two rear tires. That is, we can cause more of the weight that has transferred to be placed upon one tire over the other. This can help us move towards more equally loaded tires in the rear and this we have learned provides more overall grip for any pair of tires.

A 100% of Anti-squat condition means that the car will not squat at all from weight transfer due to acceleration. All of the transferred weight is supported by the Anti-squat mechanisms and does not find its way onto the ride springs. We'll tell you how that happens next.

More than 100% Anti-squat will actually lift the rear of the car upon acceleration and lift weight off of the rear springs. This is an extreme example, but some teams have experimented with this effect. It is not advised as a viable alternative to less than 100% Anti-squat.

How Is Anti-squat Created? – Anti-squat utilizes the force of the torque of the motor that is accelerating the race car. This torque is acted through the driveshaft, pinion gear and into the ring gear in the rear end. As the car accelerates, the pinion gear tries to climb the ring gear in the rear end. This force is also trying to rotate the rear end in a direction that pushes up on the pinion. From a side view looking at the left rear wheel, the rear end is trying to rotate clockwise.

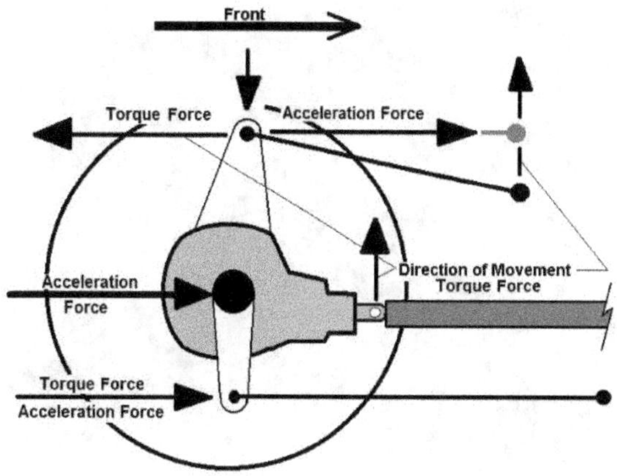

This shows how under acceleration, the pinion at the end of the driveshaft wants to move up as the pinion gear tries to climb the ring gear inside the rear end. That motion produces a torque force pulling on the third link on top of the rear end. That pulling motion produces a downward force on the rear end at the rear of the link and an upward force at the front of the link that resists squat, hence the term Anti-squat.

There is also an Acceleration Thrust force on the upper link mounts that is opposite of the torque force. These two forces must be reduced to a sum of forces to determine the ultimate lifting, or Anti-Squat force.

The solid axle rear suspension has a system of links that locate the rear end fore and aft from moving when braking and while under acceleration. There is also a mechanism attached to the rear end that is used for preventing the rear end from rotating under those same conditions. In a three or four link system, this part of the mechanism could be either a third link, a lift arm, a lift bar, or torque arm. They basically all do the same thing and that is to prevent rear end from rotating. But these links can also act to distribute the transferred weight differently.

The Third Link - So, we have a third link made of a piece of tubing with heim joints at each end that is mounted to the rear end on the top with brackets. The front of the link is mounted to the chassis and the height of the front mounting bolt is usually lower than the height of the rear mounting bolt. It is this angle of the third link that produces the Anti-squat effect.

We can adjust the height of the third link at the rear end mount. A lower position, keeping the same link angle, produces more Anti-squat force. It is the distance between the upper and lower control links that determines how much force is produced. We can also raise the lower links to shorten the moment arm.

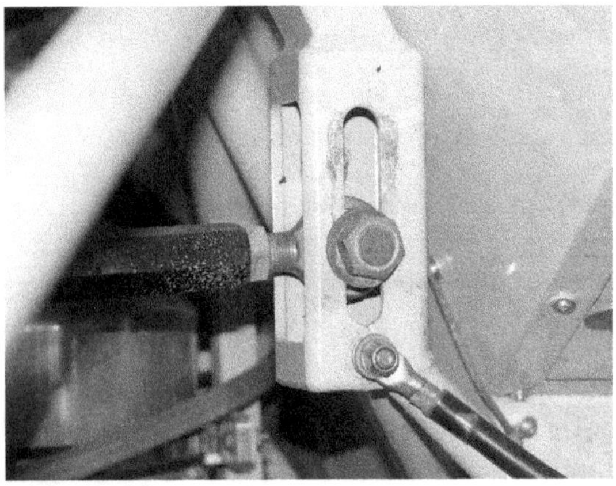

We can adjust the angle of the third link by lowering the front mount shown here, or raising the rear mount, or both. There is a limit to how much angle we can put into the link for Anti-squat purposes. This is because on entry, the reverse action takes place and there is a tendency for the rear to unload on entry because the Anti-squat mechanism restricts the rear from unloading when the weight is transferred to the front.

When the race car accelerates, the rear end pulls on the third link with a lot of force. This pulling force is directed at ninety degrees from a line running through the center of the mounting bolt through the center of the axles. Since the front mount is lower and the angle of the third link is not parallel to the ground, it wants to become parallel to the earth and this exerts both a lifting force on the front link and also a downward force on the rear end through those mounts.

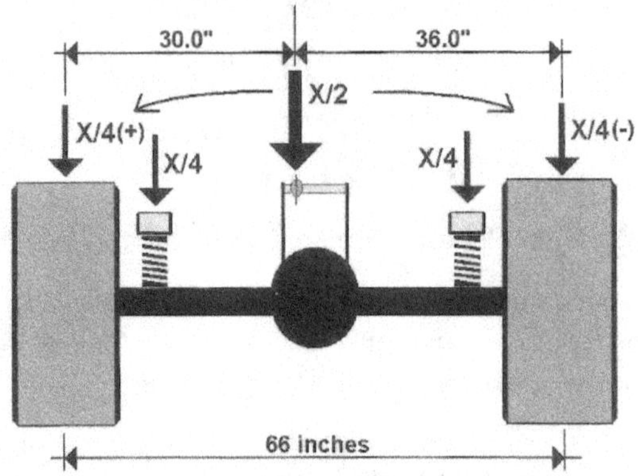

The Anti-squat forces created by the third link can be used to place some of the weight that has been transferred from front to rear under acceleration. If we move the upper link closer to the left tire, then that tire will receive more of the transferred weight than the right tire.

For a normal late model racing engine putting out 600 ft/lbs of torque, we can expect a lifting force of about 200 pounds at the chassis for a third link angled at 15 degrees. If 300 pounds transferred to the rear under acceleration, then the chassis would only squat 1/3 of the amount it would have if there were no Anti-squat. A full 2/3 of the squat loading would be absorbed by the third link angle.

Now comes the best part. That 2/3 of the transferred weight, or about 200 pounds, now is placed onto the rear end instead of onto the ride springs. The center of this force is positioned at the center of where the rear of the link is located side to side on the rear end. If that link were equidistant from each of the rear tires, then 100 pounds of load, or half of the total, would be placed on each of the tires. If this 200 pounds of load were transferred to the springs instead, it might be distributed differently depending on the spring rates of the ride springs.

Now, if we move that third link mount towards one of the rear tires, then that tire would have more of the transferred percentage of weight added to it. Let's say we move the third link 6 inches to the left on a rear end with a 66 inch rear track width. Then the percentage of LR added weight of the 200 pounds of available transferred weight would be 118 pounds instead of the 100 pounds we added when the link was centered between the rear tires. Not only is the 18 pounds added to the LR tire, 18 pounds is taken off of the RR tire. This represents a difference in loading of 36 pound for the two rear tires from what it was before. That 36 pounds equals 1.4 percent of the total vehicle weight.

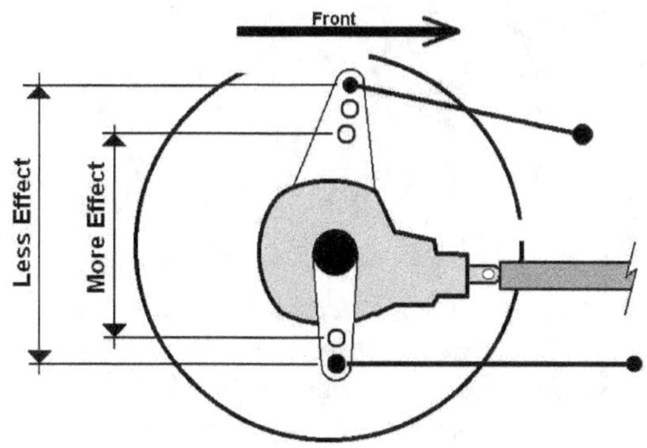

If we lower the upper link in a three link suspension and/or raise the lower link, the distance between the upper and lower mounts becomes shorter. This shorter distance provides more Anti-squat force for more effect. The opposite produces less effect.

The Lift Arm, Torque Arm and Lift Bar – These devices are designed to prevent the rotation of the rear end while the race car is under acceleration as a first priority. Secondly, they act much like the third link and transfer some of the acceleration weight transfer onto the rear end.

These devices have torque softening springs and shocks to help absorb the application of torque when the race car is throttled up on corner exit. But those don't affect the Anti-squat properties. What does affect those properties and how much Anti-squat is present is the length of the arm from the centerline of the axle to the front chassis mount.

This example of a Lift Arm is used on a dirt late model. Some asphalt late models are now experimenting with lift arms too. Note that the rear of the arm is mounted to the rear end and has no lateral adjustment to move the force towards on tire or the other. The position the coil-over is mounted can be adjusted to make the arm longer or shorter.

If we shorten this distance, then the greater the force of the torque that is applied as Anti-squat. If we lengthen the arm, then less force is applied to the chassis and rear end. Lift arms and such are not easily adjustable for lateral location on the rear end like a third link is. Therefore, where the load absorbed by the lift arm, etc. is applied is mostly fixed.

What Does Anti-squat Do? - Anti-squat transfers some or all of the weight that has transferred from the front tires to the rear tires off of the springs and onto the links attached to the rear end. The law of physics state that for every reaction there is an equal and opposite reaction. This means the Anti-squat effect cannot add load onto the rear tires. The total loading that would normally be there if we didn't have Anti-squat remains the same.

This is understandable because where would the added load come from? In the study of dynamics, we cannot "receive" additional load without "taking" from somewhere else the same amount of load. So, we receive the loading on the rear end from the action of Anti-squat and that same load is taken from the rear ride springs where it would have ended up if we didn't have the Anti-squat effect.

Let's look at it another way. If we had zero Anti-squat effect, the car would transfer a certain load or weight onto the rear springs as we accelerated and they would compress to carry that added load and the car would squat. If we apply 100% Anti-squat meaning the effect would prevent the car from squatting at all upon acceleration, then the springs would not have compressed and therefore do not carry any more load. The load the springs would have carried is now applied to the rear end through the Anti-squat device because it has to go somewhere.

Adjusting The Amount/Location Of Anti-squat - There are several ways to adjust the magnitude of the Anti-squat vertical force that is applied to the rear end through the device. One is by changing the angle of the third link and the other is by changing the distance that the upper link is mounted from the lower links. The greater the distance between upper and lower rear mounts, the less mechanical force the system has and therefore the less Anti-squat effect.

This third link has a spring that compresses when the car accelerates. The motion that help produce Anti-squat is the same force that pulls on the spring. This cushions the movement, but does not restrict the Anti-squat effect.

Adding angle to the third link will increase the Anti-squat effect but this has its limits because the effect is reversed when we decelerate and brake into the corners. Instead of transferring load onto the rear tires, the reverse means that we reduce load on the rear tires when entering the turns and that can make the car very loose.

Usually an 8 to 10 degree angle is sufficient to cause the desired effect while not being excessive. If more Anti-squat effect is needed, we can lower the third link and/or raise the lower trailing arms while maintaining the angles of all of the links.

This close-up shows how with the third link, there is plenty of room to move the rear heim of the link to the side. Moving it more to the left, and closer to the left rear tire, will cause more loading of the left rear tire from the weight that transfers while under acceleration.

As we have stated, with Lift Arms, etc., we can adjust the length of the arm to change the amount of Anti-squat we have. The shorter the arm, the more force we will have for Anti-squat.

Summation - Anti-dive is useful for setups that do not utilize bumps. These include stock classes of race cars, road racing cars, and those with more conventional setups like late model circle track cars.

In other Lessons we discuss how to geometrically steer the rear end by utilizing the squatting action of the rear of the car. For this to happen, we need for the car to squat to a certain degree. If we ran 100% Anti-squat, then there would be no squat, and therefore no motion that would have the desired effect for rear steer.

Anti-squat, as we have said, does not produce more loading on the rear tires under acceleration. What it does is restrict the rear from squatting and can be used to re-distribute the weight that has transferred due to acceleration. It is not the magic bullet so to speak, but it can be used to our advantage if setup correctly. We have given you some ideas on how to do that.

Exam - In The Context Of This Lesson:

Anti-dive Works When?

1) When the race car is under acceleration

2) As the car goes through mid-turn

3) While the car is braking

4) As a product of roll center location

With Anti-dive, Under Braking The Ball Joints Move?

1) Opposite of the braking force

2) In the same direction as the braking force

3) Towards the inside of the turns

4) Towards the outside of the turns

Advantages Of Anti-dive Include Which Of The Following?

1) Less camber change

2) Reduced over-travel of the front end on braking

3) Better grip on corner entry

4) All of the above

Anti-dive Should Not Be Used For?

1) Bump setups

2) Stock class setups

3) Conventional setups

4) Road racing setups

Pro-dive Should Only Be Used For?

1) The right front corner

2) The left front corner

3) The right rear corner

4) The left rear corner

Anti-squat Cannot Not Do What?

1) Reduce chassis movement when accelerating

2) Redistribute the loading on the rear tires when accelerating

3) Reduce the loading on the ride springs under acceleration

4) Add loading to the rear tires under acceleration

To Increase The Anti-squat Effect, We Do What?

1) Lengthen the lift arm

2) Raise the third link

3) Increase the third link angle

4) Lengthen the third link

100% Anti-squat Does What?

1) Reduce chassis movement to one-half

2) Puts all of the rear loading onto the rear end

3) Provides 100% of rear grip potential

4) Eliminates all rear movement under acceleration

Moving The Third Link Left Does What?

1) Increases the effect of anti-squat

2) Puts more loading onto the RR tire under acceleration

3) Puts more loading onto the LR tire under acceleration

4) Loosens the car coming off the corner

To Enable Rear Steer, We Need?

1) The most anti-squat we can get

2) Less than 100% of anti-squat

3) A higher third link angle

4) A lower third link location

Lesson Eighteen – Ride Heights

Every race car from dirt cars to Formula One cars have established ride heights. These heights provide the basis for every vertical and angular measurement we make throughout the life of the race car. Changes in ride height that go unnoticed will change control arm and suspension link angles and cause chaos.

We can either create and note forever our ride heights, or we can go by the original manufacturers suggestions. In any event, we first off need to establish what ride heights we will use and then return the car to these established ride heights whenever we change a setting or part.

We can measure our ride heights to the floor from raised stands. These stands are the same height at each corner of the car and allow us to move around under the car to take measurements of control arm angles, and other geometric functions.

Ride Height Definition – I guess it would be a good idea to define just what constitutes a ride height. For most cars, we would choose a point on the chassis near each tire that we can easily get to in order to measure from the ground to the frame. We can re-create these heights later on using other measurements in places that may be easier to get to.

These heights may be dictated by the rules you are governed by, or what makes sense in your class of racing. There are many ways to measure ride height. For formula cars that have no easy place to measure from, the height can be dictated by the length of the spring/shock mounts. These are usually on top of the chassis and easier to get to.

For those cars, once the correct suspension angles are achieved, and the correct heights of other components that matter are achieved, the team will simply measure the shock/spring length and then go back to that measurement in the future. That measurement of the shock/spring length relates directly to the chassis height.

For most circle track cars and road racing cars, the heights are taken from the chassis to the ground. Like the formula cars, we can translate these measurements to other suspension parts to make life easier later on, like when at the track.

Ride Height Considerations – The most important thing to remember about measuring ride heights is to make sure the surface on which you are measuring is flat. I didn't say level because we can get away with a small amount of slope. The spots the tires rest on must be on the same plane or our ride heights will change once we run the car on the relatively flat plane of the race track. If your floor is not level, then the slope amount must be under an inch in the length of the wheel base and a half inch in the span of the track width of the tire centers.

To get a surface for our four tire contact patches that is on the same plane, we use a level across the spots at each end of the car. We use spacers under the low spots to bring them up so that each pair of contact spots are on the same plane, or level to each other. The floor can have a rake angle as long as it is not excessive. Up to a one inch difference from level in the length of the wheel base, or up to a half inch difference from level in the track width is acceptable.

To make sure your tire spots are on the same plane, you can just place a level across the centers of each tire location at each end of the car. So, the front tires should be at the same angle as the rear tires. I didn't say level here either, but the difference must be under a half inch of height difference from level over the track width, or distance between the tire contact patch centers.

The best thing to use is a six foot level obtainable from any hardware store for under $50. Or, you can use a length of tubing, square metal, stiff piece of wood, or whatever works to use to span the distance and measure that with a smart, or bubble, level. Shim the low end until the level is, well, level. Then use those same shims at the other end. To create a "same plane" condition for the four tires, find the lowest tire location and place plates/spacers under that tire, or scale if you are using scales, until the tool is level. Then your four tires will be on the same plane.

Example: Let's say the front level is established and the shims are under a half inch. Then we move to the rear and we find that we must raise one end of the tool to get to level again. Simply place plates of thin metal, or even layers of paper or plywood under that end of the measuring medium until the level shows it is level. Use those same spacers under that tire, or if scaling the car, on top of the scale for that corner every time you setup the car or change the weight distribution.

Spring Changes – We will cover this section in more than one Lesson, but it helps to drive the point home sometimes. When making a spring change to a new rate of spring, or a new length, the ride height will change. If you are installing a softer rate of spring in one corner of the car, that corner will be lower unless you adjust the spring loading back to where it was by use of an adjuster.

When we change spring rates on one corner of the car, the ride height will change. So, we need to adjust the spring height so that our ride height will return to the original ride height. Installing a spring with less rate will cause the ride height to be less. We then adjust the spring height up to compensate. Here the crew member is using the ring adjuster on a coil-over shock/spring combo.

The adjuster can be a ring on a threaded shock body, a large threaded bolt placed on top of the spring mount, or other means, but in any event, something must be adjusted in order to get the ride height back to what it was before the spring change.

For big springs that are used in stock class applications, we can use an adjustable spacer to achieve the ride heights we desire. If the class you run in does not allow adjusters, these can be welded solid after you find your desired ride height. Then you will be legal. Just remember that if your change spring rates, your ride heights will no longer be what you had originally.

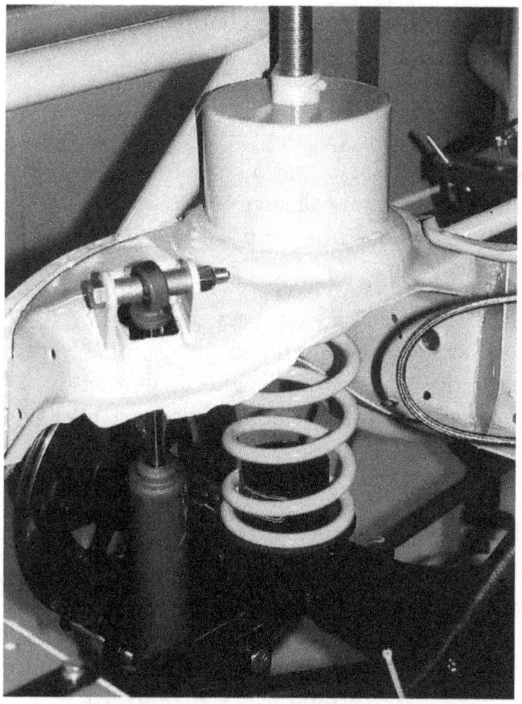

For this application, the car builder installed screw jacks over the springs to easily adjust the ride heights. These are typically used on the front of stock clip (stock front frame rails) that use stock sized big springs.

If you want to make the job of changing a spring easier, just change one spring at a time. Then you will only need to make one ride height adjustment at a time. Once you are done with that corner of the car, go to the other corner or corners you want to make a spring change for. But do them one at a time.

What Height Do I Need - For circle track cars, tradition has it that there is usually a rake, or difference in ride heights among the four corners of the car. The Left Front is usually the lowest corner with the Right Front next followed by the Left Rear (sometimes the same as the RF) and then the Right Rear being the highest ride height.

With the modern setups, these "norms" may not be so normal anymore. Teams and manufacturers car re-thinking the ride height layout to accommodate the newer low attitude setups and making the ride heights so that the car has less rake and is more level, like it will be on the race track.

For road racing cars, there can be a rake, but the corners at each end of the car are usually the same height. So, the front would be the lowest ride heights on those cars. Much of the decision for ride height on road racing cars is connected to rules governing wing heights, or front valance heights, etc.

The road racing cars might have a flat bottom where they need to maintain what is called a platform. The platform is the angle and height of the flatter bottom of these cars to the racing surface. For aerodynamic purposes, the dimensional heights of the platform are critical. Small changes to the ride heights can cause large gains or losses in the aero downforce that are created by the platform.

Tools To Measure Ride Height – There are various tools racers use to measure ride height. These range from a simple tape measure to sliding blocks that fit between the frame and the ground. Once you establish your ride heights at the shop, you can recreate them at the track by recording the shock lengths and making sure they are still the same after the trip to the track if, and only if, the spot you park your car creates the plane we used at the shop.

NOTE: The spot you park your car at the track is seldom flat or to where the four tires are on the same plane. Therefore, you will need to either shim the spots where the tires rest, or record new ride heights that you can use to maintain the geometry of your car. These won't match the ride heights you established at the race shop, but will maintain the heights once the car hits the track.

Even if the tires rest on uneven spots that are not on the same plane, when you make spring changes, if you keep the same "track" ride heights, your overall geometry and static tire loading will not change when the car is on the "same plane" track surface.

Ways To Record Ride Height – There are several, or multiple, ways you can record ride height. You can just measure from the corners of the frame to the ground, and this is by far the most common. You can also measure the shock length at each corner, or from the bottom of the wheel rim up to a point on the body for full bodied race cars.

Once you have established your chassis ride heights by measuring from the floor to the frame, you can record other dimensions to replicate those ride heights. These alternate methods of measuring are many times much easier to get to and perform. One way is to measure from the bottom of the wheel rim to the fender well.

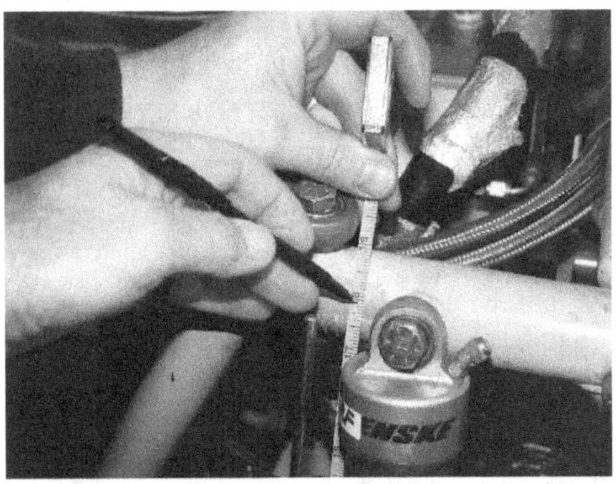

Another way to replicate ride height measurements is to measure from the ball joint up to the frame tubing or measure the shock length. Either of these will place the chassis at the same attitude that the ride heights created initially.

For open wheel cars, you can measure from the top of the lower ball joint up to a point on the chassis. Whichever way you choose, be sure to record those measurements so you can access them later on. Some teams just write them on a piece of tape attached to the body or frame so anyone can read them easily.

Other Uses For Ride Heights – We have already covered some uses for ride height. Just making sure our ride height don't change when making spring changes is the most common use. But we can use them for other things.

If we make careful and precise measurements of the corners of the car, then after a hard hit, we can remeasure those corners to see if the frame is bent. A bent frame will alter our component angles and change our setup.

If we have a need to make many measurements, such as for geometry purposes, and we need to get under the car, we can add a constant offset measurement to the ride heights at all four corners to raise the car, but maintain the rake as if the car were on the ground.

After we make the measurements, we can then just subtract the vertical offset to record the component heights as if the car were on the ground. If say, our ride heights were LF = 4.0", RF = 4.25", RR = 4.5" and LR = 4.25", and we add 10.0" to raise the car, what would our new heights be? They would be LF = 14.0", RF = 14.25", RR = 14.5" and LR = 14.25".

Alternate Ride Heights – Sometimes we want to take measurements on the car that will relate to how it is situated on the race track at, say, mid-turn. This could be called the dynamic ride heights because they are caused by the dynamic forces as the car rides through the turns.

We not only need to know our static ride heights, we also want to know our dynamic ride heights. These are the heights the chassis is at when the car is at mid-turn and all of the forces are on the car. We can record the dynamic heights using data acquisition and position sensors mounted to the shock, or by other means, but the goal is to know how far each corner of the car travels at mid-turn so we can duplicate that travel in the shop.

If we know the shock travels from reading the travel indicators on the shock, then we will know how much shorter our shock mounts are at dynamic ride heights. If we use adjustable solid links to replace the shocks, we can adjust them to the new shock length, place them in place where the shocks are normally mounted, and then place the car on the ground, or better yet, onto raised stands.

Simple shock travel indicators are mostly used in many different forms of auto racing to record the maximum travel of the springs, and with those, the suspension. We need to be careful when relying on those measurements. Maximum travel might also include the dive the front suspension experiences during heavy braking, or the squat the rear experiences during acceleration off the corners. The team can make a run reaching full turn speed without heavy braking or hard acceleration to get these travels to represent just the chassis movement related to the mid-turn forces.

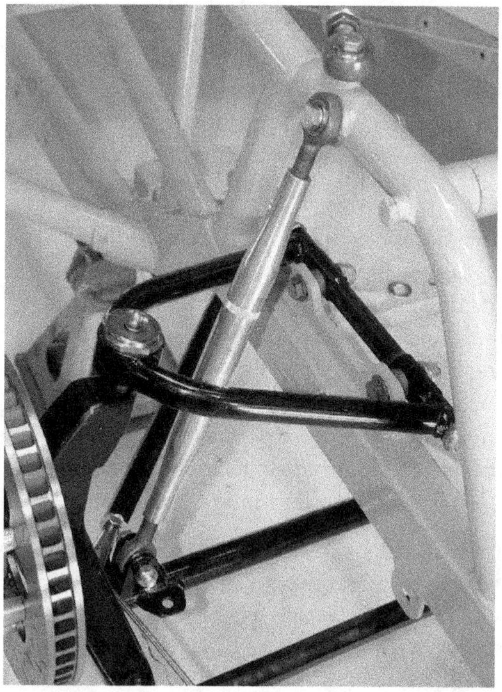

Once we know the shock lengths from dynamic shock travel, we can install links that are the same measurement so that the chassis will be at dynamic ride height. Once there, we can measure front AA-arm angles, link angles in the rear, wheel cambers, clearances, and other important features. To be able to measure tire loading while at dynamic ride height, we need to use a pull-down rig that actually loads the chassis and springs. Then we can record the dynamic loads seen on the scales.

Pulling down the chassis to dynamic ride heights will tell us how our control arm angles are positioned, what our driveline angles are, and what our front wheel directions are. All of these might change under loading as opposed to just placing the components at that attitude with no loading.

We might have many needs to put the car at dynamic ride height. Some of those include: measuring control arm angles, measuring wheel direction, or steer, measuring frame clearance to the racing surface, measuring driveline alignment, etc.

We now have available what are called Pulldown Rigs. These will pull the chassis down to the ride heights that the car has traveled to at mid-turn. Then if the car is on scales, the team can read the loading on the tires, as well as geometric measurements that reflect where the components are under the stress of dynamic loading. For instance, the load on the chassis and control arms from loading might alter the toe readings.

Summary – It is important to establish and record ride heights for your race car. You can take the measurements in a lot of different ways, so choose the one that best suits the design for your race car.

Make sure the surface where you will be setting ride heights in your race shop where the four tire contact patches will sit will all be in the same plane. And the overall tilt of that plane should be no more than one inch in the wheel base length and a half inch in the width of the track, or tire width.

Exam - In The Context Of This Lesson Three:

What Is A Ride Height?

1) The maximum height the car will achieve on the track

2) The distance from the frame to the ground

3) The distance from the body to the ground

4) 2 and 3

We Maintain A Defined Ride Height Because?

1) The suspension angles will always be consistent

2) The static loading on the tires will stay consistent

3) Aero downforce will be predictable

4) All of the above

The Surface We Measure Ride Heights To Must Be?

1) Flat

2) Level

3) On the same plane

4) The same as at the track

We Can Maintain Our Ride Heights By Measuring What?

1) The distance from the floor to the corners of the chassis

2) The shock lengths

3) The ball joint stud to the frame

4) All of the above

When We Get To The Track, We?

1) Re-set our ride heights to what they were at the shop

2) Level the spots under the tires

3) Record new ride heights without changing tire loading

4) 2 or 3

When We Change A Spring, We?

1) Make sure we don't change the spring adjuster

2) Return to the original ride heights

3) Return to the original corner weights

4) 2 and 3

Racecar Technology – Level Two

Lesson Nineteen – Tire Selection and How To Read

Everything we do with race car setup all comes down to one thing, making the tires do as much work as possible. In the very first introduction to Race Car Technology we said that our goal is to: "Maximize the amount of traction that is available from the four tires on a race car, any race car, and that will make you as fast as you can be, all other things being equal. Everything we present will ultimately lead to optimization of the race cars tires grip and we use of that grip to go faster."

The art of the selection of race tires and the preparation and maintenance are critical to your cars performance. Everything we do for chassis setup is directly related to helping the tires work harder. This Lesson will provide you with some ideas for choosing and preparing your race tires.

In order to maximize the grip in our tires, we need to know how to select those tires, how to prepare the tires and how to read them once they are run on the race track. Remember, the goal is to produce the ideal loading among the four tires and create the largest contact patch possible. Then our race car will be as fast as possible through the turns.

Every race team should select a tire specialist or a small team of specialists to choose, prepare and maintain the teams race tires. Success depends on it. That teams needs to keep good notes and go over those notes often. The following will help those guys with their duties.

Tools Of The Tire Trade – The tools that are considered absolutely necessary are:

1) a quality, large dial pressure gauge,

2) a small, flat tape measure,

3) tire chalk or suitable marker,

4) extra stems, bleeders, valve cores, etc.,

5) a note book you can use to record your tire sizes and pressures,

6) and a nitrogen tank to use to pressurize your race tires.

Other not so necessary tools, but ones that can help you, include a durometer to test the hardness of your tires, a tread depth gauge, purging equipment to remove most of the atmospheric air, that contains moisture, from your tires and wrapping material to keep your tires from drying out when not in use.

You can also invest in tire covers for use at the track when the hot sun is on one side of the car heating those tires when the other side is in the shade. Some teams also carry tire brushes to wipe off the sand or pebbles before measuring the sizes so that the numbers come out correctly.

Use a large diameter tire pressure gauge so that you can read, and estimate values between the whole pound marks. We need to work with half pounds of tire pressures in many cases.

Tire temperatures tell us a lot and help us to apply the correct tire pressures as well as cambers. When you take your temperatures, poke the probe into the rubber at a 45 degree angle and push it close to the core. The surface temperature cools very quickly and won't help us understand the true heat the tires generate.

Measure your tire diameter around the middle just off centerline to avoid the seam. Measure the tire un-inflated first to get relative sizes, then with race pressures plus 10 pounds. The chalk marks on new tires are the sizes of the tires inflated to the same pressures plus some amount beyond race pressures.

How To Select New Tires – The overall theory on tire selection deals with a few critical musts. They are:

1) The tires must be as fresh as possible,

2) They must be equal in age,

3) And for circle track race cars, they must be sized correctly so that you can mount the correct stagger on the car.

Most race tracks will provide a stack of race tires for each division. It is up to the team to dig into that stack and find the correct tires that will work well for your race car. It is the age, sizes and grouping of the tires that matters the most.

For Circle Track Asphalt Applications - When you go to the tire shack at your track, you are looking for sizes and dates. Let's deal with the sizes first. The manufacturer measures all new tires that have been inflated a few pounds above the highest race pressures. They write these sizes on the tire in chalk. These are referred to as chalk marks.

Since all tires, left and right, are inflated to the same pressure, naturally, when you mount your tires and inflate them to different left and right side pressures, the left sides will be smaller due to the lower pressures. So, choose a size difference that is much smaller than your required stagger.

Measure all of the tires in their deflated state. Measure the tires around the middle just off center to avoid the seam. Keep the same sizes for the right side tires and keep the same size for the left side tires. The stagger you are looking for with the deflated tires is about half what you need on the car when inflated to race pressures.

So, if you choose an 85 inch tire circumference for the right side tires, then if you need 2 ½ inches of stagger, you would choose 83 ¾ inch tires for the left sides, because the inflated stagger will be twice as much. This number may differ for different tires at different race tracks, and only repeated selection will determine the amount of difference in stagger between the deflated and inflated tires that you will end up using.

The date codes are molded into the tires and printed on the sticker. It is critical to match the date codes for each side of the car. That means, the right side and left side tires don't necessarily have to have the same date code, and chances are, they won't. But the RF and RR tires, as well as the LF and LR tires must have the same date codes or the car will not be balanced and the handling will suffer.

This will take some work and sorting through the various tires that are mixed and un-matched lying around your tire shack, but the top teams make the effort to get this right. It should be done as soon as possible and it should take as much time as is necessary to get the right tires, period. Get this wrong and you'll be out to lunch for as long as you have those miss-matched tires on the car.

Storing your race tires in a plastic wrap helps keep them fresh and delays the evaporation of important chemicals that make the tire soft and compliant. This is true for normal or treated tires.

A hard tire does not adhere well to a rough surface because the contact patch does not conform to the entire racing surface area. The total square inches of rubber that meets the track is reduced when the tire is harder. Older tires, even if they have very few laps on them, will be harder. The chemicals that make the compound pliable evaporate over time. Sun and heat accelerate the process. If legal, a team can apply commercial chemicals made for racing applications to the tire to make them more pliable.

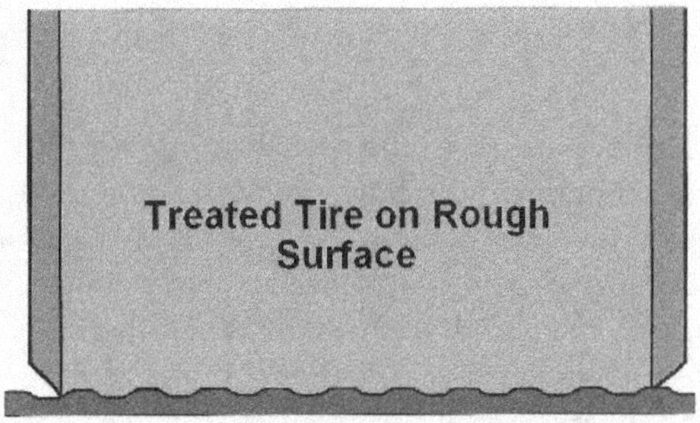

Tires have more grip by being newer, or by being treated. By treating the tires, the surface is changed chemically so that it is more compliant and "fills" the areas around the high spots in the track surface. A treated tire would provide a performance benefit nearly as much if the track was very smooth.

Road Racing Tire Selection – Most road racing series and sanctions choose the tires for the teams. If you have a choice, then go by the date codes and choose all the same dates for your four tires. If you are choosing multiple sets of tires, make sure to practice on the older dated tires and save the newer dated tires for the race. This last statement is true for all racing types.

If there is a difference in tire diameter sizes and all of the tires you use are the same width, choose the same size for each end of the car if you cannot get the same size for all four tires. Put the smaller diameter size on the front when mounting the tires.

If your type of racing uses different width tires for the front and rear, choose you tires so that the date codes are the same (might be hard to match) or as close to the same date as possible. For instance, if you choose different dated tires for the front and rear in two sets, and the two sets are reversed (older fronts on one set, newer fronts on the other), then the handling between these two sets of tires will be vastly different.

How To Read The Sticker – A new race tires has a sticker on it that tells certain important information. Learn how to read your tires sticker. If you need to, have a talk with the tires representative at the track, or call the company. They should be glad to help you with this information. The information includes the date the tire was manufactured, the group or batch it was made with and other information about the tire compound codes, etc.

The stickers on new tires can tell us a lot about the age of the tire. They help us select tires that are the same relative age, which is important maybe more so than the age. If we cannot find left side tires that have the same dates as the right sides, at least match dates for front to rear tires on the same side. That way we will maintain the balance of the car.

For a Hoosier race tire, the sticker has information you need. The top number is the model number and is usually five digits. It might say 10615. The next number is the tire size. This will look like 27.0/10.0-15. Under that number is the compound number and it might look like F40.

Then there is a bar code and just under that is a date code and sequence number that no one can decipher. So, you need to try to find tires with the same sticker date number, and that might be very hard to do because a production run of a certain tire might last several days at the factory.

What is the same, for the same batch is the date code embedded in the tire sidewall. This code will be the same for all of the tires coming out of that mold in that production run regardless of the date and sequence. When that run is finished, they will change that code for the next run which usually involves producing tires with a different tire compound and/or core design.

Mounting Your Tires – Make sure you mount the tires on the correct rims. Some teams will run different wheel offsets for each side of the car. Make sure the imbedded date code for each tire is on the left side, or infield side of the rim for both sides for circle track cars. So, on the left sides, you'll be able to see the code when mounted on the car and on the right sides, you won't.

Don't allow the use of any liquid like bead soap, or use very little if needed. Liquid that gets into the inside of the tire will expand when heated and influence the tire pressures when they heat up on the track.

Purge the tires with nitrogen by inflating and deflating two or more times. Inflate your tires to about ten pounds above the race pressures and measure them. Then deflate them to race pressures and let them sit. Once they have stabilized, then re-measure them at the race pressures.

This is the point in time where you create your sets of tires from a group of eight or more. The sets will consist of tires at each end of the car that will have the correct stagger for your track. If a tire needs to "grow", you can inflate to 10 pounds above race pressure and then let it sit for a while and that may increase the size slightly.

This tire has been marked for the corner of the car, the set number, the track is has been run on, and the date. In chalk or other less permanent style, the size has been marked as well. The size can change with pressure changes and/or ambient temperature changes. That is why we mark them with less permanent markers so we can change the number.

Mark the tires as to the corner they belong on, the set number, the track it was run at and the date. In your notebook record each tires size and pressure. As you run the tire, record the number of laps run also. Race tires, like any other tire, will degrade in performance the more they are run and the older they are.

Safety Note: Never inflate the tires more than 10 pounds above race pressures. Do not ever use air hose ends that can be clamped onto the valve stem. This is a recipe for disaster. If you become distracted, the tire can over-inflate by accident and explode with the power of a hand grenade. Race team members have died using these types of clamp hose ends.

Never inflate your tires with an air hose that clamps onto the valve stem. This is a very dangerous practice and if left unattended, the tire can overinflate and explode with the force of a hand grenade. Racers have been injured and killed doing this.

Choosing Stagger – Even when we know our desired stagger amounts, we can mount the tires such that changes can be made to improve the car due to unexpected stagger changes. Once the tires get hot, the gas expands and the bias tires will stretch. If we put the correct stagger in the rear, we can set the front stagger so that we can swap tires on each side of the car to effect the rear stagger if one of the rear tires changes its size unpredictably after being run.

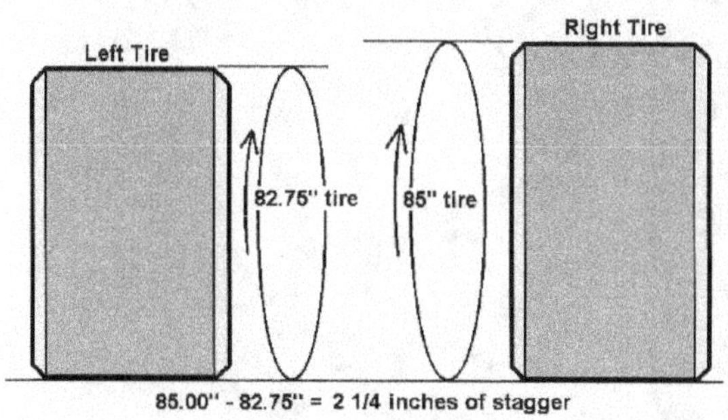

Stagger is the difference in circumference of two tires on the same end of the car used for circle track race cars. The example is for a car that turns left. Measure the stagger around each tire near the middle of the tread. Subtract the two measurements to find Stagger or "Roll Out" as it is sometimes called. Always inflate to race pressures before measuring.

To find stagger the easy way, follow this simple method and never subtract tire sizes again. Measure the larger tire first. Then remember the number.

Wrap the tape measure around the smaller tire and note where the larger tire stagger number falls near the end of the tape. Read the measurement from the end and you have the stagger.

Suppose we need a rear stagger amount of 2 ½ inches. We mount a RR tire that is 86.0 inches in circumference and a LR tire that is 83.50 inches around. A common occurrence is when one of the rear tires will grow excessively causing an increase or decrease in stagger. If we mount the same stagger on the front, but at a larger size, we can swap either left side or right side tires to correct the problem.

The tire stagger must match the race track and groove radius we will be racing on. We do not want to correct handling problems with excess or deficient stagger. For every track there is an optimum stagger for the rear tires and if we use a locked spool rear differential the need for correct stagger is even greater.

The spool will need a stagger that is an average of the radii that the tire will experience in the turns. Very few driving lines result in a single constant radius. The driving line through a turn is more like a parabola, or constantly changing radius with the smallest radius in the middle portion of the turns.

A Detroit Locker rear differential unlocks going into the turns and locks back up upon acceleration. For this type of rear end, we need to match the stagger to the radius of the last third of the turn. This may be less stagger than what we might have needed at the very tightest portion of the mid-turn.

Pressures For New Tires – The following is for circle track cars mostly, but much of this information is relevant for road racing tires too. Starting pressures for new, cold tires will be lower than race pressures. The left sides should be about 4-5 pounds lower and the right sides 7-8 pounds lower. Run the tires for one practice session and then adjust the pressures and check the stagger. Make changes based on the tire temperatures and hot pressures as necessary.

If you are allowed to run bleeders, don't run them in practice so you can see what the pressures are doing. This can identify a handling problem that would otherwise go un-noticed, such as a growing RR tire due to being loose, etc.

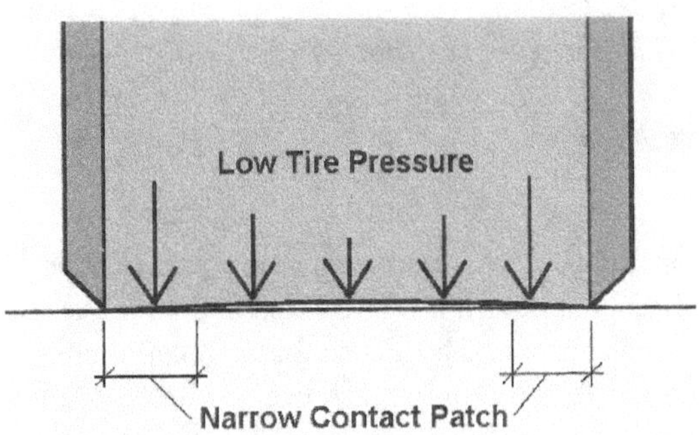

It's a the fact that the whole area of the contact patch is not exerting the maximum pressure on the racing surface, and that is a the problem. This tire will have less grip than one that is properly inflated. Here we see low pressure where the center of the contact patch is not providing maximum loading due to the pressure being insufficient to press in onto the racing surface.

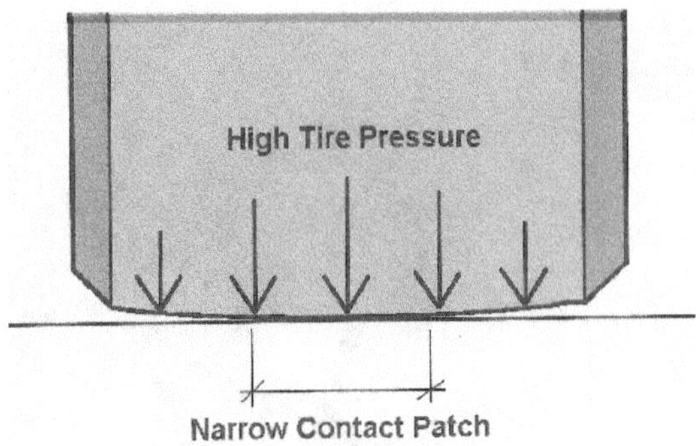

Using excessively high tire pressures will also cause the tire to have a smaller contact patch area and therefore reduced traction. It is not that all of the width of the tire is not on the ground. Here we see excessive pressure puts most of the load at the center of the contact patch instead of distributing it across the entire tread width.

Tire Temperatures – It is very important to take tire temperatures when evaluating your cars setup. The temperatures can tell us how well each tire is working in conjunction with the other three tires. We set camber and tire pressures based on how our tire temperatures look.

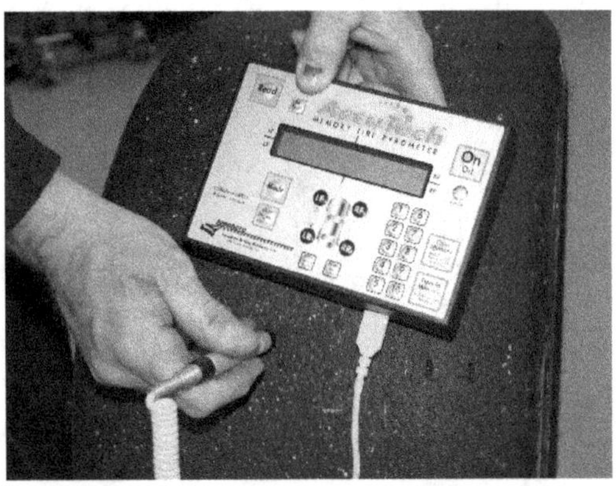

This shows a tire temperature gauge while the crewmember is taking a reading. Take the outer edge temperatures in from the outside edge of the tire approximately 1 to 1 ½ inches first and then the middle and then the inside edge. Allow enough time for the reading to settle before recording each number.

The proper way to take tire temperatures is to read the temps in three places across the face of the tire with a tire temperature probe. This instrument has a needle like end that is inserted into the rubber to get the inside, or core temperature. It should be inserted at about a 45 degree angle into the rubber. The temperatures for the three positions should be taken near the inner and outer edge of the tire about one inch in, and at the middle of the tire.

Give the instrument time to react to the temperature of the tire. Once the reading stabilizes, record the amount and move on to the next reading. Don't be in too much of a hurry to do this because a lot of decisions will be made based on the temperature readings. They must be accurate. Follow the manufacturers recommendations for which temperatures to take first, second, etc.

During the practice sessions, it is very important to measure and record the tire temperatures. These will tell you a lot about tire pressures and cambers. Most of the time the tire temperature readings will not be the same at the inside edge (towards the inside of the turns) as they are at the outside edge. But the middle temperature should be the average of the two side temperatures.

If the middle temperature of the tire is cooler than the average of the three tempertures, then the tire is underinflated and you need to add pressure. If it is higher than the average, reduce the pressure. Once you have found the hot pressures the tire likes, record those and maintain those hot pressures from now on.

What Temperatures Mean – When you look at your tire temperatures, note if a tire has more average heat than the other tire on that same side of the car. This can be an indication of a car that does not have the proper dynamic setup balance. It can also indicate a car that has corner exit problems that cause one tire to heat up from spinning.

If the RR is hotter than the RF, we tend to think the car is loose. But it may be possible that the car is tight/loose meaning it is tight in the middle and as the driver turns the steering wheel more to get the car turned. At this point, the front end quickly develops more traction and overcomes the tight condition taking the car into a loose condition.

This happens so quickly, the driver will believe the car is loose, when in fact it is tight. When you correct the tight condition to where the driver does not need to steer excessively, then the loose off condition, along with the hotter RR tire, will go away.

If the RF is hotter than the RR, it is a simple case of being very tight and you probably need to think about making spring rate changes and/or roll center and weight distribution changes. A simple way to tell if the car is tight is to note the position of the driver's hands when slowly arcing the turns. Then note again the position of the hands when the car is at speed. If the driver turns more at speed, you have a tight setup.

Running and Evaluating the Tires – When we run on a set of tires at race speeds, we need to look at three things to determine how well the tires are working. We can look at the tire Temperatures, tire Pressures, and tire Wear. The last two are fairly easy to read for both asphalt and dirt teams, but the first may be difficult for dirt racers to get due to the way heat bleeds off the tire after a run. None the less, if possible, get all three to help understand all that the tire is doing.

The temperatures tell us how well the tire footprint is working. Don't make major changes to your tire pressures or front cambers based on the tire temps. until you have had a chance to run the car at full speed for more than five laps. We need to get maximum temperatures in order to see the true picture of what the tires are doing.

We have learned a lot about how the tire temperatures can be an indication of ideal cambers at the front of the car. What we know now is that the ideal camber that produces the maximum tire footprint for a front tire will show the inside (towards the radius of the turn) temperature of the tire to be 15 to 20 degrees or more hotter than the outside.

If the temperatures are more even, then the cambers probably need to change. Having the correct camber is necessary in order for the tire to generate the largest foot print and give the most traction. This should be taken care of right away as we practice with the car. Do not make handling changes until the tires have the correct pressures and cambers.

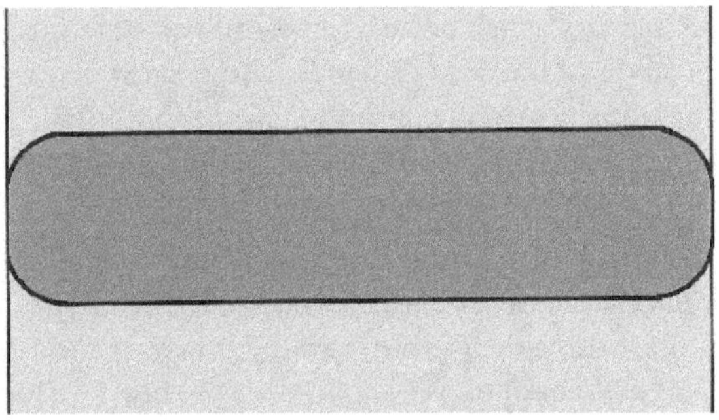

A front tire contact patch can be maximized by adjusting the camber of the tire. We want to adjust the front cambers using tire temperature and/ or wear patterns to tell if the camber can be improved. We then need to maximize the pressure distribution on the contact patch to create the largest contact patch possible. This sketch shows how our contract patch pressure distribution would look if the tire temperatures were even across the face of the tire. This is not ideal for the front tires on a circle track car.

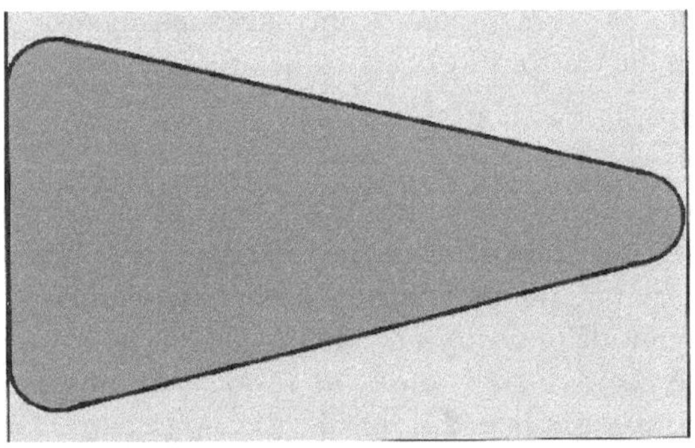

If we introduce more camber to where the inside edge temperatures are higher by more than 15 degrees from the outside edge, and reduce the pressures, then the contract patch would look somewhat like this. The area of this contact footprint is larger than when we had less camber and uniform tire temperatures across the contact patch. Therefore, it has been proven that this cambered pattern will have more grip.

The tire wear also tells us a lot about how a tire is working. If we measure the depth of the tire grooves or wear slots across the face of the tire, we can see if the cambers and pressures are correct. This is especially useful for situations where taking tire temperatures is not practical.

Wear pattern are especially useful for dirt applications where taking tire temperatures is not practical. We should look at a tire after having run enough laps that wear can be measured. Wear on a cambered wheel can tell us a lot about how our selected camber works with a particular setup. Also, if we are wearing out the outsides on any of the four tires, the pressures are probably too low. Excess wear on the middle of the tire tells us the pressure is too high.

Evaluating the following charts will help you see how race teams use tire temperatures and pressures to correct tire problems to create the maximum tire contact patch area.

Chart 1 — **Session #3**

Left Front

Loc 12	Loc 11	Loc 10
146	154	150
Avereage Temp 150.00		
Pr Cold	Pr Hot	Pr Gain
20	23	3

Right Front

Loc 1	Loc 2	Loc 3
211	204	207
Average Temp 207.33		
Pr Cold	Pr Hot	Pr Gain
24	29	5

Front Average = 178.67

Left Rear

Loc 9	Loc 8	Loc 7
162	168	158
Average Temp 162.67		
Pr Cold	Pr Hot	Pr Gain
20	25	5

Right Rear

Loc 4	Loc 5	Loc 6
216	214	210
Average Temp 213.33		
Pr Cold	Pr Hot	Pr Gain
24	32	8

Rear Average = 188.00

In this chart, we have the third practice session. This is relevant because it may take several sessions to get the car up to speed and to heat the tires. Right away we see where the left side tires show a higher temperature on the middle reading. We need to reduce the pressures for the LR and LR tires. We always work with tire pressures first.

We see from the average tire temperatures that the RR tire is 6 degrees hotter than the RF tire. And the LF tire is a full 12 degrees cooler than the LR tire. This car exhibits characteristics of being what is termed, Tight/Loose. The LF is not doing as much work as the LR tire and this indicates a tight setup. Then as the car snaps loose on exit, the RR tire heats up causing the high RR temperatures. This car needs setup help right away.

Chart 2 — **Session #4**

Left Front

Loc 12	Loc 11	Loc 10
148	152	157
Avereage Temp 152.33		
Pr Cold	Pr Hot	Pr Gain
18	22	4

Right Front

Loc 1	Loc 2	Loc 3
212	208	204
Average Temp 208.00		
Pr Cold	Pr Hot	Pr Gain
26	32	6

Front Average = 180.17

Left Rear

Loc 9	Loc 8	Loc 7
164	162	159
Average Temp 161.67		
Pr Cold	Pr Hot	Pr Gain
18	23	5

Right Rear

Loc 4	Loc 5	Loc 6
214	212	209
Average Temp 211.67		
Pr Cold	Pr Hot	Pr Gain
26	34	8

Rear Average = 186.67

In Session #4, the tire pressures for the left side tires have been corrected by dropping them 2 psi, but we still need to work to balance the setup so the LF tire will do more work and then gain heat to match the LR tire average temperatures.

Chart 3

Session #5

Left Front

Loc 12	Loc 11	Loc 10
154	153	152
Avereage Temp 153.00		
Pr Cold	Pr Hot	Pr Gain
18	21	3

Right Front

Loc 1	Loc 2	Loc 3
209	208	207
Average Temp 208.00		
Pr Cold	Pr Hot	Pr Gain
26	32	6

Front Average = 180.50

Left Rear

Loc 9	Loc 8	Loc 7
164	162	159
Average Temp 161.67		
Pr Cold	Pr Hot	Pr Gain
18	23	5

Right Rear

Loc 4	Loc 5	Loc 6
214	212	209
Average Temp 211.67		
Pr Cold	Pr Hot	Pr Gain
26	33	7

Rear Average = 186.67

In this Session #5, changes were made that caused more problems. The team took positive camber out of the LF tire and then also took negative camber out of the RF tire. If anything, they should have put more positive camber into the LF tire and more negative camber into the RF tire. And, nothing was done to correct the balance problem as we see that the left side tire temperatures are now still 8 degrees different. This is a tight car based on the rear averages being 6 degrees hotter than the front averages.

Chart 4

Session #6

Left Front

Loc 12	Loc 11	Loc 10
170	163	155
Avereage Temp 162.66		
Pr Cold	Pr Hot	Pr Gain
17	21	4

Right Front

Loc 1	Loc 2	Loc 3
214	204	193
Average Temp 203.66		
Pr Cold	Pr Hot	Pr Gain
25	32	7

Front Average = 183.16

Left Rear

Loc 9	Loc 8	Loc 7
164	163	159
Average Temp 162.00		
Pr Cold	Pr Hot	Pr Gain
17	21	4

Right Rear

Loc 4	Loc 5	Loc 6
204	202	201
Average Temp 203.00		
Pr Cold	Pr Hot	Pr Gain
25	32	7

Rear Average = 182.50

Now we're looking pretty good on the tire pressures and tire temperatures. We have corrected the front cambers to where the LF is now hotter at the inside edge and so is the RF tire. It takes this kind of temperature difference from outside to inside to produce a larger contact patch.

The average temperatures are now almost even on each side and the front to rear averages are almost the same as well. The tire pressures front to rear are also equal on each side of the car. This car is about ready to race.

Making Changes – We want to react to any tire problems quickly, whether they are front camber issues, tire pressures or stagger. As we test and race, setup adjustments will not be correct if the tires do not have what they want to in order to provide the ideal tire contact patch.

Adjust the pressures first, and then camber as you run the car. After we are up to speed and have run several sets of laps, see can evaluate where we ended up with stagger. We are mostly concerned with the rear stagger, but make a note if a front tire grows more than normal. This may indicate a bad tire or one that is being overused. The tire temperatures will increase with the tire growth if the tire is working too hard.

Make all changes quickly after noting a problem. Correct a stagger problem right away too, so you won't chase the setup to correct a loose or tight off condition made worse by incorrect stagger.

What is obvious is that new tires are better than old ones. Tires degrade with age, even when they have never been run. Some of the chemicals that make a tire compliant will evaporate and the tires will become harder, or less compliant with age.

Don't chase a quicker car that is running fresher tires. If your lap times are down 3 or 4 tenths from a car that has newer tires on it, verses your tires that have 100 laps on them, chances are your setup is better. Once you mount your stickers (new tires), your lap times may improve 5 or 6 tenths and you will probably out qualify the other car.

Qualifying Tips – You need to start your qualifying laps with more pressure than if you were starting the race or going out for the first practice. There will be less time for the pressure to build when you are only running a few laps. So, a general rule is to set the left side pressures 2-3 pounds lower than race pressures and the right side pressures 4-5 pounds lower than race pressures.

In practice, many teams will put on sticker tires and run a simulated qualifying run. If you will quickly measure the pressures right after this run, you be able to see if the tires got up to race pressures and if not, make an adjustment in your notes for future starting pressures for qualifying.

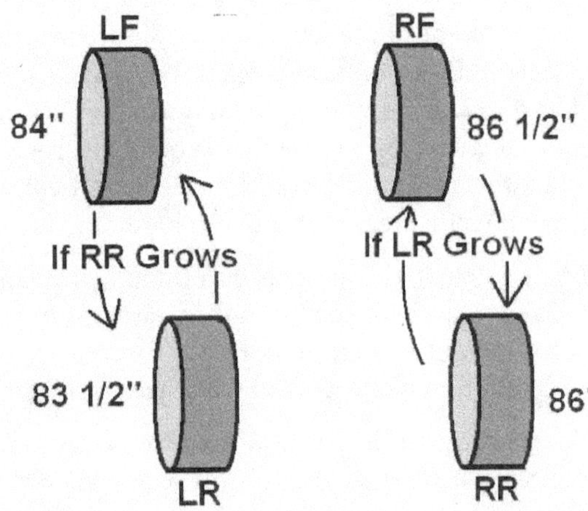

There is a way to select and group tires so that if stagger changes are needed, you only need to swap left side or right side tires. In this example, we put the larger diameter tires on the front. If your stagger closes up, you can swap the right side tires and put the larger RF tire on the RR to bring the stagger back to where you intended it to be. And, if the rear stagger increases, then you can switch the larger left front tire to the left rear to reduce the stagger. The front stagger is not so critical.

Race Preparation – To prepare for the race, you need to make longer runs in practice and evaluate the pressure gain. A run of 15 to20 laps will probably tell us what we need to know. Start out with the tires cold, make the run and then take tire pressures and temperatures immediately after the car returns to the pits.

The amount that each tires grows in pressure will determine the starting pressures for the race. If you inflate the right side tires 5 pounds below your intended race pressure of say 30 psi and they come in after the long run at 32 psi, 2 pounds above our target race pressure, then we need to start out with a lower cold temperature.

The amount of difference should be a percent of the excess gain. So, this tire gained 7 pounds. The percent gain was 32 psi divided by 25 psi, or 1.28. If we divide the 30 psi desired race pressure by 1.28, we get 23.44 psi as a starting pressure, or about 23.5 psi on the pressure gauge.

For longer races, you might want to adjust the starting pressures down a pound or two. For shorter races where you anticipate a lot of cautions and shorter runs, you might want to add a pound or two to the starting pressures.

Afterwards What to Do – Once a set of tires has been used for practice or a race, mark how many laps are on the tires and note unusual tire growth or tire temperatures on the problem tire. If the car ran loose all night in the race, chances are the RR tire was getting very hot. Noting that may save problems if this set is used on the car for practice later on and the car begins acting funny.

Store the tires in a cool and dark place. If convenient, cover the tires to prevent exposure to the air. Used tires can always be used in practice or testing providing we understand that they will provide less performance than newer tires.

Dirt Tire Management - Tire management for dirt cars is somewhat more critical and more technical than for asphalt cars. The reasons why are equally complicated, but there are some basic rules to follow that will teach you where to start. The rest is just practice and evaluation. The tire guy on a dirt team must know how to read a track and what to do to the tire to maximize the grip.

For dirt tires, tread depth is possibly the best way to evaluate camber settings. If the inside wears excessively, then there may be too much camber. Many teams are experimenting with running more camber to create more contact patch, just like the asphalt race teams are doing.

In this photo taken by the author, we see a very high degree of camber in the right front tire for the number 71 car. In keeping with the theory that more camber can produce a larger contact patch, this proved this out. This car started on the pole and ran up front all day. The car to his left has much less camber and did not run as well.

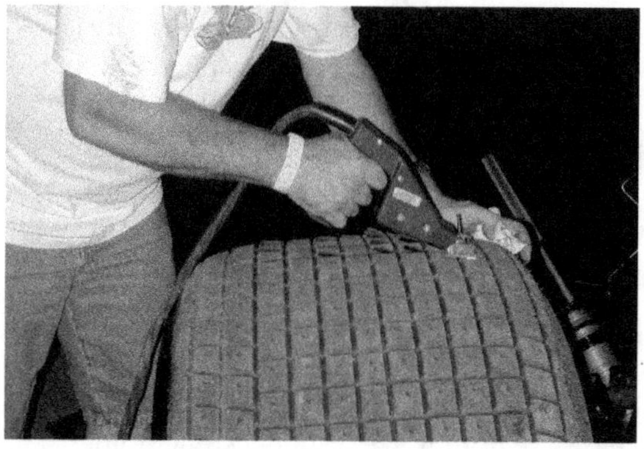

This is an electric tire grooving and siping tool that uses heat to help cut through the rubber. The tire is grooved so that there will be more edges in the tread and more grooves help cool the tire. There will be accelerated wear when grooving and/ or siping, but the team can judge the wear factor when evaluating how much to groove the tire.

Dirt racers need to pick the right tire compound, but there is another important step to getting the most out of a dirt tire. Dirt racers often have to alter the design of a tread pattern to best work with the conditions at a particular track, at that time of the day and with their driver's style or habits. No one can tell you exactly what you will need that will work best for your situation, but there is some basic information you can use to be able to decide what to do as you encounter different racing conditions.

We have to consider the abrasion of the track, how wet or dry the track is and if the track contains rocks or other debris that will tear up the tire. All of these conditions must be considered when deciding on what kind of groove to use, how deep to cut and how many grooves can be cut out without causing the early demise of the tire. Along with these factors dirt racers must consider the amount of heat a track puts in their tires. In addition to enhancing traction, grooving helps the tire dissipate heat and can be used to help control tire temperatures.

Grooving and Siping For Dirt Cars - Tires used on dirt cars will undergo some special preparations prior ever hitting the track. The one preparation process is grooving tires. The main job of a grooving tool is to place a greater number of cutting edges into the tire to aid in gaining grip from the dirt surfaces it is racing on and to help cool the tire.

There are three basic shapes used in grooving: 1) Square, 2) V grooves, and 3) Sipe cuts. Square grooves are the same width through its entire depth. V grooves start out wide at the top and taper to nothing at their bottom. Sipes are thin slits cut by installing the blade upside down in the holder and using the separate ends of the blade to cut slices in the tire.

All three types of grooves can be used in various depths depending on the conditions and the length of the race. The V groove is often used when the track is expected to need more tread contact later in the race. As the tire wears, the grooves become smaller or disappear completely. Square grooves can be used the same way but the extra width at the bottom of the groove could provide enough leverage on an abrasive track to tear the tire if the track becomes tighter.

Siping is usually meant to make the tread more pliable and does not produce the edges the Square or V grooves do. Siping also helps the tread maintain a more consistent wear that helps keep the tire working uniformly.

Grooving the shoulders can be helpful if you plan to run the high line or on a cushion and need to be moving some dirt. Grooves cut into the shoulders help clean away some of the loose dirt to get at moisture beneath it. Sometimes the outer row of blocks is also grooved on the right rear to work with the shoulders.

On some of the harder, natural rubber tires teams sometimes sipe the shoulders. This can really help when you are rolling the tire under when running lower tire pressures on a very slick track.

The sipes can help prevent the shoulder area of the tire from glazing over and losing traction. The shoulder is as important a part of the tire as anything else and if you are going to be running on it due to low air pressure or because you are running against the cushion you need to make the best use of it.

The angle at which grooves are cut determines how much of the edges are exposed to the track when the car is in various degrees of slide. Dirt race cars seldom if ever run in a straight line. When the track is tight we keep the grooves pretty straight because the car is being driven more straight ahead. But when we spend a lot of time with the car sideways on a slick track, we put more angle into the grooves.

How much of an angle is dependent on the driver's style and experience is the only way to determine the best angle for your situation. The idea is to keep the maximum amount of the tires edges facing the direction the tire is actually traveling.

The grooving process is something that any racer can learn and there are a good number of tools on the market that make it a very simple process. They range from a simple sharp razor blade type of tool to heated knifes. The heated knifes, or grooving irons, are the logical choice for this process. They are faster and they give you a greater level of control.

Most dirt racers are using blades in the range from .125, .187 and .250 of an inch in width. The power required to run most grooving irons is no greater than what would be required to run a 3/8" electric drill motor. So, you will not have to use a large capacity generator to run a grooving iron at the track.

Any time a groove or sipe is cut into a tire it accelerates wear. The trick is to balance the benefit of grooving with the increased wear. One of the reasons we sipe a tire is to prevent the tread surface from glazing over and becoming slick. Sipes help the tire to wear and to keep it working throughout the race.

Learning to recognize the amount of wear you can expect from a track not only helps you choose the best compound tire for the night but is important information for deciding what pattern and depth of grooving to use. The object is to maintain the highest level of traction throughout the race without using up all of the tread with five laps to go.

This is a new Hoosier dirt tire before any grooving or siping has been done. Note the square blocks, These blocks will be cut into smaller squares by grooving if needed. Those smaller squares can be cut by siping to create more edges in addition to the grooves.

These blocks have been grooved and cut into four smaller blocks. This tire has been used and exhibits a lot of wear to the point that the grooves cut into the tire are almost gone. Note also that the middle 2/3 of the tire is worn, but the outer parts are almost like new and still have the molding remnants. This tire could have possibly been run with less air pressure.

This is a closeup of a tire that has been recently grooved. Note the cut across each of these two blocks. These cuts are beneficial for a slick track and will provide more edges to grip the track plus provide more cooling.

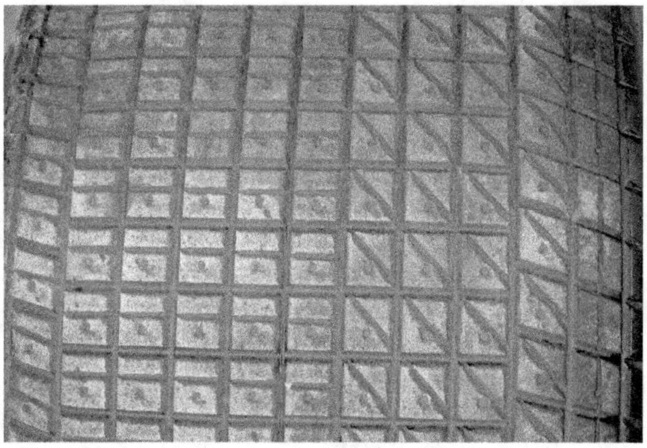

The angles of the grooves can be any angle relative to the direction of the tires path. They can be straight across like those in the left middle section as opposed to the right third with diagonal grooves.

Note Keeping - The first step in proper tire management is to keep detailed notes. This is your first line of information gathering. You can purchase a multitude of pre-printed setup sheets from a variety of racing vendors. In fact, many of them are free. The issue is not so much the format but the fact that you are keeping notes.

You most likely will develop a format that works for you over time, but in the beginning you will need to keep track of a few things from a tire management perspective that may include but are not limited to: 1) Brand and compound number of each tire. All tires have a S/N, or a manufacturing date, or other unique identifier molded into the tire that will help you single them out

2) Tire age, meaning the number of laps the tires have on them. You should be able to document the number of laps each tire in your inventory may have accumulated to date.

3) Which corner the tire was run on,

4) Tire and wheel width size,

5) Air pressures (cold and hot),

6) Circumference of the tire, pre and post run,

7) Tire grooving for dirt tires (size of grove, pattern and depth),

8) General condition of the tire pre and post race, as in are the edges sharp, is the tire feathered or blistered, and

9) Tire temperatures recorded after each run.

Creating a note sheet with each of the parameters you are tracking in a "fill in the blank" type of format will make this data much easier to keep track of and much quicker to do. The easier and quicker it is to develop or execute, the more likely you can get the crew to accomplish the task.

Make this task as easy as possible. It will become important to designate a person on your team who always has this responsibility, or you can rotate this job between several team members from race to race to further develop your team. These note pages need to go into a master notebook that you will be able to refer to for future setup information.

When you have developed a good note keeping routine, you can refer to the data and it can help you to understand just how the tires are performing. It also helps you to keep a better level of control over the number of tires you keep in your inventory. You will know which ones to sell and which ones to keep and know just when new tires are required.

Search For Information - Asking questions or searching online are ways we can glean information. Your first resource is to go to the manufacturer's website for information, compare this information to your notes and use this information to further improve your individual race tired data bank.

The information would include, required optimum wheel widths for a given tire and the recommended maximum air pressure settings. Find out what the recommended tire temps are for the best performance and tire life. On the same thought learn what temps are too high and inversely which temps are too low. The majority of the tire manufactures have web sites that explain and answer all of these questions.

Goodyear and Hoosier both have great web sites for the dissemination of this type of information. Your next resource should be the dealers who are selling you tires. They want you to do well with their tires, that way you will buy more as the need arises.

Summary – Like we have stated, all that we do with the setup of the car is for the benefit of the tires. Tires provide the final ingredient for performance. Tire loading and tire contact patch optimization equals high performance for any race car. Race tire optimization is track dependent in many cases. What works for one track may not work for another.

A lot has been learned in the recent past about how to optimize the tire contact patch area. It is known that a tire with a larger contact patch, with the same loading, will provide more grip. The shape of the contact patch itself will ultimately determine the amount of grip.

To learn more about how to manage your tires at your track, or with your series, look at what the other teams around you are doing. Develop a rapport with your fellow racers so that you can ask and learn. Share information you might have learned in order to gain their trust. And talk to the tire pros at the track and at the factory.

Exam - In The Context Of This Lesson:

The Most Useful Tools Used For Working With Race Tires Include?

1) A large dial pressure gauge

2) A narrow and flat tape measure

3) Nitrogen gas

4) All of the above

When We Take Tire Temperatures, We?

1) Record the surface temperature

2) Check the temperature at the middle of the tire

3) Probe deep into the tire at three points

4) Can use an infrared digital thermometer

Race Tires Are Measured How?

1) Around the middle of the tread

2) At normal racing pressures

3) With a narrow and flat tape measure

4) All of the above

Race Tire Stagger Is When?

1) The cambers are different side to side

2) When the front tire circumference is larger than at the rear

3) The inside tire circumference is smaller than the outside circumference.

4) All of the above

If The Middle Of The Tire Is Cooler Than The Average Tire Temperature, Then?

1) The pressure is too high

2) There is not enough camber

3) The pressure is too low

4) There is too much camber

If The Rear Tire Averages Are Hotter Than The Front Tires?

1) The setup is tight

2) The car might be tight/loose

3) The car is loose

4) 2 and 3

Dirt Tire Grooves And Sipes Are Cut Into The Tire For?

1) Increasing the number of edges for more grip on dry slick tracks

2) To cool the tire

3) To better clean the tread

4) All of the above

Cutting V-grooves Into Dirt Tires Is Useful When?

1) On tight tracks that will go slick

2) On slick tracks that will go tighter

3) For dry slick dusty tracks

4) For tacky tracks

Racecar Technology – Level Two

Lesson Twenty – Summation RCT Level Two

Congratulations, you have completed the Race Car Technology course, Level Two. This final Lesson will remind you about all that we have learned in this course. We outline all that we have accomplished in taking you to the race car. In Race Car Technology Level Two we have taken all of those parts and pieces we described in Level One and applied them to an actual race car. And, we presented all of that important information in a logical order, just the way you would approach an actual race car and prepare it for competition.

Before we do anything to setup our race car, we first check out all of the parts to make sure they are able to hold up to the rigors of racing. We told you how to go through the entire car to make sure it is track ready and safe.

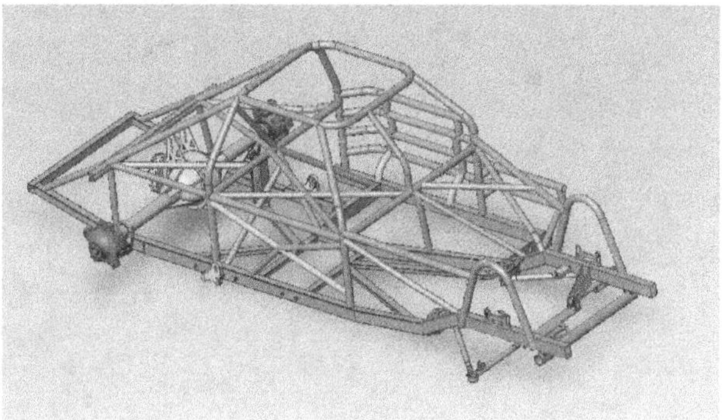

After learning all about the parts and pieces of the race car, we now, in RCT Level Three work on the actual race car using all of those parts. We go from one end to the other carefully explaining how to setup and use all of those systems.

We first outlined all of the goals of Level Two. We proceeded to go through all of the systems on a race car and help you understand how to work with the race car to accomplish the stated goals.

We ten taught you how to go through the car to make sure it is track worthy. We taught you how to set your ride heights, distribute your weights and what those weights should be.

In the geometry sections, we discussed the settings and considerations, caster, camber, bump steer, Ackermann and alignment. In the alignment sections, we went over front and rear alignment, rear steer, toe and the driveline alignment.

Once our car is aligned correctly, we moved on to the spring selection in and how to arrange our spring rates in order to achieve dynamic balance. As a part of the springs choices, the sway bars play a part and act as springs. Then we described how we use shocks to control those springs.

We then set the ride heights and established the corner weights we will need to go racing.

In advanced setups, we described how to setup the race car using bump technology in todays racing. Bumps stops and bump springs are being used in almost every form of motor racing today.

The next item we think about for our setups is the Anti-dive and Anti-squat settings. We fully explained how those are designed into the suspension and when, and when not, to use the Anti's.

We went through the settings for caster, camber, bump steer, etc. to get the chassis setup for the type of racing we will be doing with each specific type of race car.

We told you how to check Ackermann and when it is, and when it is not, appropriate, and in what amounts.

The alignment of the wheels is the next step we covered and explained in detail how to properly align a race car for each type.

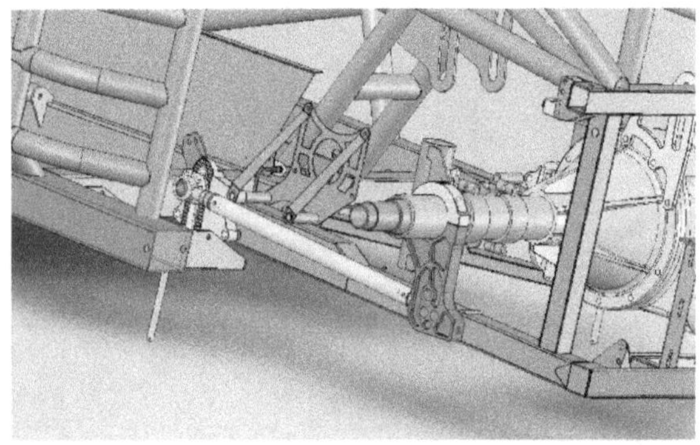

Next came the explanation of how to use rear steer to enhance your drive off the corner and other uses for an advanced understanding of the technology of rear steer.

You were able to gain valuable knowledge about spring selection, roll stiffness and roll angle dynamics as well as how to control those springs using shocks.

Another current trend and one that is used in almost every form of modern racing is the bump devices. The bump stop, or bump springs, add to the spring rate of the ride spring and we told you how that works and how to adjust your shock package to work with the bump setups.

Now we have to go to the track and actually test and race our race car. We outlined how to read and select your race tires. After all, the tire is the final point of contact between your race car and the race track. All that we do involves making those four tires work as hard as they possibly can.

The final thing to go over before we actually take our race car to the track is tires. We describe how to choose and mount the tires. We also tell you how to read the tire temperatures and pressures so that you can get the most grip out of your racing tires.

And once we get all of the systems set correctly and we have our tires mounted, we are ready to test all of that. We tell you how to conduct your test sessions to get the most out of the race car. We outline what to adjust for what is happening with the car, in an order that makes sense.

When we have done all that we need to do to setup our race car, the day comes when we take the car to the race track for the first time. We tell you how to weigh the car and how to conduct a test session to get the most out of the experience.

Now that you have all of this information, you should be well on your way to setting up your own race car. In RCT Level Three, we continue with our Lessons by describing advanced setup technology and current routines and processes that have never been presented before, anywhere.

Now it is time to run the car. We outline what to evaluate once the car is on-track and how to make changes to the setup for dynamic balance and proper load distribution. We tell you how you can make your car faster and how you will know when it is working properly.

Level Three actually goes on to explain the engineering behind setting up real race cars. We take you through the process in detail describing what we are doing and how exactly to do it on an actual race car. There are methods and knowledge that has only recently been discovered and applied to race cars, and in Level Three, we explain and present all of that knowledge. We look forward to your continued education in race car technology in the ORS continuing education series of courses.

Hopefully with all of this newfound knowledge, your team can extract the most performance out of your race car that is possible. Then maybe you'll be able to visit the winners circle and take home a trophy.

Exam - In The Context Of This Lesson:

The Primary Goal For Setup In A Race Car?

1) Make it easier to drive
2) Increase grip
3) Balance the two suspension systems
4) Create less chassis roll
5) 2 and 3

Why Should I Go Over The Whole Car If It Was Running When I Bought It?

1) To make sure it is correct for the rules you race with
2) Because it might have the wrong parts
3) It is your responsibility to make sure it is safe
4) It probably does not have the correct setup to be competitive
5) All of the above

We Maintain A Defined Ride Height Because?

1) The suspension angles will always be consistent
2) The static loading on the tires will stay consistent
3) Aero downforce will be predictable
4) All of the above

The Following Can Change The Weight Distribution?

1) Sway bar pre-load
2) Front end travel
3) Travel onto the bump stops or springs
4) Camber Changes
5) All of the above

We Can Change The MC Location By?

1) Raising or lowering the chassis mounts
2) Using taller or shorter ball joints
3) Changing the lower control arm angles
4) Changing the upper control arm angles
5) All of the above

Caster Settings Are Relative To?

1) The track banking angle
2) The track turn radius
3) The type of race car
4) Driver preference
5) All of the above

Why Do We Make Camber Changes?

1) To make the car turn better
2) To create a better heat pattern on the tire
3) To create a larger contact patch
4) All of the above

Ackermann Is Present When?

1) The wheel steers when it bumps
2) Only one wheel turns
3) We gain or lose toe when the wheels are turned
4) Our steering arms are too short

One Way To Create Power Rear Steer Is?

1) Using the squat of the rear from weight transfer
2) Using the acceleration force driving the car forward
3) Using the torque of the motor
4) All of the above

Proper Driveline Angles Are When?

1) The pinion is angled 2 degrees down to the front

2) The transmission shaft and pinion shaft are parallel

3) The drive shaft angles at the transmission and pinion are equal and opposite

4) 2 and 3

Matched Roll Angles Create What?

1) Ideal weight distribution

2) A faster turn speed

3) More overall grip

4) Better tire wear

5) All of the above

Shocks Control What?

1) Speed of suspension movement

2) Suspension forces

3) The rate of the spring

4) Transitional handling

5) All of the above

Bump Springs Have The Following Characteristic That Bump Stops Don't?

1) A consistent spring rate

2) A wide range of spring rates

3) More spring travel before coil-bind

4) All of the above

Race Tires Are Measured How?

1) Around the middle of the tread

2) At normal racing pressures

3) With a narrow and flat tape measure

4) All of the above

The First Part Of The Track We Tune For Is?

1) Turn entry

2) Mid-turn

3) Turn exit

4) Straightaway speeds

Race Car Technology – Level Three

Lesson One – The Goals Of RCT Level Three

This is Race Car Technology – Level Three. It has the latest information for setting up a race car you will find anywhere. This course will cut no corners. In this course, we will present information and techniques that will vastly improve the programs for most race teams. For the others, it will provide performance you never thought possible, even for current championship winning teams.

The goals for designing our race cars are basically the same be it a formula road racing car or a late model circle track car. Performance gains in the slower portions of the track cause all of the speeds to increase around the track including on the straights.

A True Statement: 99% of the crew chiefs, race engineers, or whomever, who are in charge of setting up a race car do not know how, and do not have the tools, to setup a race car without trial and error. 15% of these are decent at tuning their race car setup to the point of winning. The other 85% are just along for the ride.

The proof is this. How many teams at the local level win during the season, 2, 3, 4? Out of 15 to 25 entries? That is anywhere from 8 to 20% +/- of the teams that show up that can win and even then, they don't win all of the time.

What about in the most expensive form of motor racing where hundreds of millions of dollars are spent each year and where the most educated engineers and technical chiefs reside? On a good race day, only three Formula One cars out of twenty-four are competitive and in a position to win for a 12.5% performance rate. Most of the time it is two cars for an 8% performance rate.

Explain to me how, with all of those resources, can the bottom 25% of the cars be a full two to three seconds slower than the leader on the same tires? A tenth of a second for an average F1 race track is about 15 feet in one lap. If you are one tenth faster than another car, you will gain 15 feet per lap. Two seconds is 20 tenths. 20 times 15 is a whopping 300 feet PER LAP that the fastest car gains, or moves ahead of, the back 25% of the cars, on the same tires.

If you follow F1 and watch closely, those same bottom 25% cars actually have very quick straightaway speeds. The gains that the leaders have are through the slower turn portions of the race track where aero downforce is lowest because the speeds, something aero needs, are lacking. A great deal of the gains are with enhanced mechanical grip and few of these teams have maximized mechanical grip.

If all of that educational power and knowledge gained through higher education was correct and perfect throughout that industry, we wouldn't see this much difference in performance. Obviously, most of these engineers lack the knowledge and the tools to get the job done.

Another True Statement: Almost every team that wins does not get the most performance out of the car. How do I know this? By experience. Just because a team wins does not mean that car is as fast as it could be, it's just faster than the competition, in that race, at that race track.

Major track championships and major series championships have been won by setups that are not the best they can be. This is because the person in charge of the setups didn't have the knowledge and/or the tools to be able to setup the race car to be the best it could be.

It's not the desire of the crew chief or engineer, nor the effort that keeps them from setting up the race car as good as it can be, it is the lack of knowledge and tools that holds most crew chiefs back. The education for motorsports engineering must be lacking in today's racing world.

In this course, we will give you the knowledge and the tools to get the job done.

What We Will Learn In Level Three - At this RCT Level Three point in time, we would hope that you have completed Levels One and Two. In those Lessons, we learned about all of the parts and pieces of a race car, and we also learned how to adjust and position all of those parts so that they can work together and not hinder the actual race day setup. So, now we come to the actual setup part and how the winning teams perform their magic. We will setup actual race cars.

I say it is magic, but it is not really magic at all. It is knowing what to do, in what order to give the car what it wants, and that is a setup that takes advantage of aerodynamics, mechanical balance and tire contact patch optimization.

When we get the car as low as possible giving it a low center of gravity and better aero, and when we arrange the spring rates and moment centers so that the load distribution on the four tires is ideal, and when we create the optimum geometry so that the tire contact patch is as large as possible, we will then have a winning race car. All the driver has to do after that is drive it to victory.

In this Level Three Course, we will show you how to do all of the above, exactly like every winning team does it, or at least how they should. Even winning teams get some of this wrong. But how do they win, you might say? They win because everyone else they are racing against isn't 100% right either.

The winning team is often at 90-95% of the perfect setup, when everyone else is 85-90%. So, there is room for improvement in your race cars setup and beat those dominant teams that are not all that they can be.

Never think that you cannot improve what you are doing. Even if your car is perfect with the current level of knowledge, there may be something out there that you can learn and that can take your division to a whole new level. Then you will be top dog, at least until everyone figures out what you are doing.

One critical component in doing well in racing and being successful is consistency. What we will strive for in this school is impressing on you the fact that you have to keep everything under control and in optimum condition. We can never let any of the settings or routines get out of hand. Many winning teams run the very same setup every race of the year, even at different, but similar race tracks.

We test within limits and we always go back to what we know works. This is called the baseline and this school takes you to the baseline. And that baseline is often what wins races and continues to win races. We can step out of that baseline to try and find more performance, but 9 times out of 10 we won't improve and that is why we go back to the baseline.

We often think of the formula one cars as being the ultimate race cars and perfect in every way. If that were so, then there wouldn't be a two second gap between the fastest and slowest cars. There must be a difference in designs that causes that huge gap in speeds. We like to think it is lack of mechanical grip that makes not only the slow turns faster, but the entire lap.

What We Should Have Already Learned – RCT Level Two taught us how to prepare the car for this Level Three Course. If you have not taken Level Two, then you might miss something important that will interfere with what we are about to do in Level Three. It happens all of the time, and you are no exception. Here is what we said about Level Two in that courses introduction.

Race cars don't just come ready to race. You might be buying a brand-new race car, or starting to work with a used race car you just acquired, or maybe you are building your own race car. In any event, you must go through the car and setup each system of that car properly and in a specific order.

No one gets away with not doing any of this. You cannot be successful in racing if you don't work with these setup routines. And even if the race car manufacturer or previous owner says it is good to go, you cannot trust anyone but yourselves. We must verify each and every setup parameter of the race car.

If you are racing now, or have watched others race and have asked how they do it to make their cars so successful, it isn't magic and no one ever falls on a winning setup by accident. No matter how easy it all looks, a lot of effort and smart thinking goes into every winning race car.

In the Level Two Course, we performed simple, but necessary functions like determining ride heights. We set the weight distribution, and establish proper caster and camber for your application. We went through all of the various alignment settings and carefully explained why we are concerned with each phase and how those will make the car perform.

We established what spring rates to use and which sway bar will match those spring rates. We matched our shock rates to the spring and sway bar rates too. We chose tires and tire sizes and learned how to select tires and prepare them for competition. Then we put this package to the test and fine tuned it at the race track.

When you got done with the Level Two Course, you were able to take a modern day race car and set it up. When I first started into racing, I could find no good information to tell me how to setup a race car. Even today, information comes at us in parts and pieces and it is very difficult to put it all together into a winning combination. The Level Two school was designed to provide all of the necessary information to guide you through the setup of your race car.

The Ultimate Goals - Our ultimate goals are as explained in the RCT Level One Introduction. For those of you who forgot, or those who chose to begin your schooling with RCT Level Two or Three, we urge you to go back and take the Level One and Level Two courses. In our introduction to Level One, we explained what our overall goals are for designing a high performance, winning race car. It bears repeating, so we will again explain.

The race engineer's goals are very simple in concept, but more complicated to carry out. On any race course, be it with circle track or road racing, on a dirt or asphalt surface, there are key areas of performance where your car needs to be made better than any of your competition and those are the following;

Maximizing Grip - When non-chassis elements like powerplants are more or less equal, Grip is what will help you win races and championships. Motors accelerate you, brakes slow you, but Grip makes you faster through the slower turn portions of the track and relatively small gains in lateral Grip will produce huge gains in speed and performance.

What are the Key elements of Grip? - The tire contact patch is where Grip is produced. Causing the tire contact patch to: 1) be larger, and 2) causing the most vertical loading on that tire contact patch are the two key ingredients for maximizing Grip.

Maximizing the amount of traction that is available from the four tires on a race car, any race car, will make you as fast as you can be, all other things being equal. Everything we present in RCT L3 will ultimately lead to optimization of the race cars Grip and the use of that Grip to go faster.

What are the basic parts that make up Grip?

- Loading On The Tire – The more load we can put on a tire, the more Grip that tire will have, period. But, the gain in grip is not linear as we will explain later on. Load can come from the weight of the car, mechanical downforce from banking, and aerodynamic downforce, all of which will be explained in detail later on.

- Contact Patch Area – The greater the size of the contact patch, the more Grip. If we can find ways to make the tire contact patch larger, then that tire will produce more Grip.

- Tire Compound – The physical and chemical makeup of the tire can provide more Grip. The softer the material, the more Grip we will have within limits. We have rules that govern the softness of the tires, but we need to stay very close to those limits.

- Load Distribution – A pair of tires on the same end of the car, or same axle, will produce more Grip when they are more equally loaded. The most Grip from a pair of opposing tires will come when they are equally loaded. There is a variation to this concept for dirt racing that will be addressed later on.

- Angle of Attack – What is called Angle of Attack, or Slip Angle, is when a tire is pointed slightly to the inside of the arc it is following through a turn. If a tire were to follow the exact tangent line around a curve or arc of the turn, it would not generate any side force to counter the centrifugal force.

So, in consideration of the other items that make up Grip, it is fair to say that none of those would be useful if it weren't for the creation of Angle of Attack. No matter what amount of Loading or what the size of the Contact Patch area is, or how soft the tire compound is, or how equal the load distribution between opposing tires, the car would not stay on the course without the tires developing an Angle of Attack.

So, there you have it. Those five things represent the parts that help make the Grip we seek to make us faster so we can win races. As we go through each part of the race car in the other Lessons, we will explaind how to optimize those parts to enhance our Grip and make us faster through the slow speed turns. And we will understand how that will in-turn make our whole lap faster at every point around the course.

Why Does More Grip Make Us Faster?

When a race car turns, a lateral force called Centrifugal Force tries to push the car to the outside of the turn. The tire contact patches resist this force. The speed we can drive through the turn is limited by the amount of Grip we have in our tires. The more Grip, the faster we can drive through the turns.

The one often overlooked benefit of achieving faster turn speeds is this: the faster you exit a slow speed turn, the higher the speed at which you will start accelerating down the straight part of the track in-between the turns. So, speed gained in the slow speed turns will be carried down the straights too. It's not just that we gain speed in the slow speed turns with more Grip, we gain everywhere around a race track.

To give you an example, on a typical or average length Formula One track, one tenth of a second is about 15 feet on the race track. Some teams are a full second slower per lap than the fastest teams. That is 150 feet per lap that the faster cars move ahead of the slower cars each lap. In a fifty lap race, that equals 1.4 miles, or 2.3 kilometers.

We already know that some teams get lapped in a F1 race, and the average length of those tracks is around 3 miles. So, those lapped cars are on average 2.0 seconds per lap slower than the winning team. And they all run with the same choice of tires. That is hard to fathom.

Is the gain realized by the winning F1 teams all Grip? No, as we have read, some engine packages are down on horsepower and that is a factor in the difference in lap times. But what about the teams who have the same engine package as the winners? Why are they so slow? It could be that they lack mechanical grip in the slower portions of the race track.

Building mechanical grip is one of the most important areas of race car engineering, even over and above aerodynamic grip. The reason is this; To gain maximum aero grip, you must have high speeds and the most gains from mechanical grip happen in the slower turns where the speeds and the aero downforce is lower. So, the only thing we can point to as the reason for the gain in performance for the winning teams is Mechanical Grip.

The Concept Of Mechanical Balance - If your car has more overall Grip than other cars early in the race, that Grip might not stay superior and you might end up being a slower car through the turns later in the race. This is due to the Balance Factor. There are two definitions of balance. They are not the same and they cannot be confused.

One is handling balance which is defined as a car that is neither pushing (understeering) or loose (oversteering). Pushing is when the front set of tires has less Grip than the rear set of tires. Loose is the opposite.

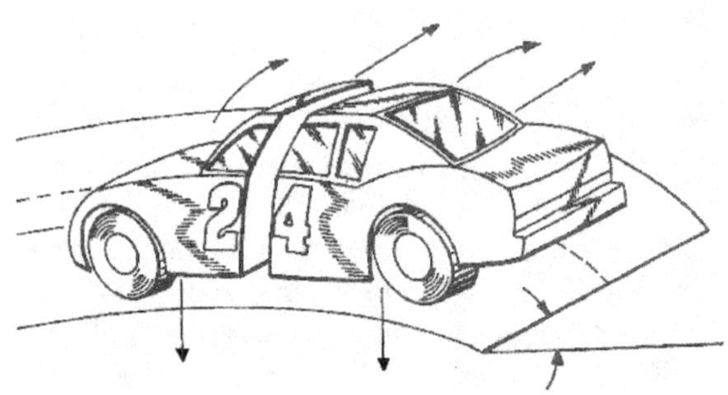

The Mechanical Balance concept is simple. We want both ends of the car to work in sync so that one does not try to cause the other to do what it does not want to do. When the two ends of the car are working in unison, we can achieve the goal of mechanical balance.

A team can attain handling balance fairly easily, but just the fact that the car is neutral in handling doesn't translate to speed or consistency. A car can have less overall Grip than other cars and still be neutral in handling. Although a neutral handling car is a goal, it is not the first and primary goal and not what we are discussing here.

The other balance is Mechanical Balance. Simply put, this is when the two suspension systems, front and rear, are working in sync and where the load transfer is predictable and maximized. This is what teams need to search for. The end result of all we do with chassis engineering must be towards the goal of Mechanical Balance.

Circle track race cars are complicated, but much easier to setup that a road racing car. This is because the setup in a circle track car can be asymmetrical, meaning the spring rates, cambers, etc. from side to side can be different. The circle track car is only turning in one direction and so the setup need only be correct for that direction of turning.

The Search and Desire for Mechanical Balance should be the foremost goal for every race engineer. A lot of good things come with achieving MB. The tires do not work as hard and will last longer. The car is much easier to drive. The car maintains higher turn speeds as the tires wear later in the race, or later in each stint for cars that are allowed to pit and take on new tires.

When you have finished this Level Three Course, you will have all of the tools necessary to complete the setup of your race car. All that is then needed is fine tuning, which we will also cover. If you do it right, your car will have the Mechanical Balance needed to run fast for a long period of time. Your tires will wear more evenly and at a slower rate than if you were not balanced. And you will have the speed at the end of the race that will carry you to victory. So, let's get started.

Race Car Technology – Level Three

Lesson Two – Concepts Of Balance – In Depth

The concept of balance in a dynamic sense must be understood before we can get into any of the other areas of discussion in RCT Level Three. This theme represents the basis of everything we do from this point on. The concept has probably been around for a long time, but in the mid-1990's, a method was invented to improve the way we measure the balance in a race car's suspensions.

Previous attempts to create a measure of balance involved measuring the roll stiffness in each suspension system, front and rear. What we will discuss in this and future Lessons is not that. It is similar, but more refined and takes into account effects that act in addition to the roll stiffness and what has been referred to as roll couple.

The Basic Premise – The basic premise is this. The two suspension systems in a race car, any race car, will have spring rates, spring split and roll / moment center heights that will determine how that suspension system reacts to the lateral forces we encounter during cornering.

The "React" we are talking about in this example is made up of two things:

1) The chassis will, in most cases, dive or compress, especially if the banking angle of the turn part of the track is significant. That is, the center of the chassis will be lower through the turns than is was at normal ride height. If both the front and rear suspension systems dive the same amount, or differently, there is no negative influences associated with the dynamics of the setup.

2) What each suspension system also does as a reaction to the lateral forces that exist when going through the turns, is to roll. Every suspension system will want to roll to its own individual angle that can be calculated. And each pair of suspension systems would do just that except for the fact that we have a strong and resistant chassis connecting the two suspension systems.

IF the two suspension systems want to roll to the same angle, then the overall setup in the race car is what we call balanced. If the two roll angles each suspension system want to roll to are different, then we call that an un-balanced system. And that is not to say that we might, in certain situations, want to purposely cause unequal roll angles for reasons we will get into later on. For now, a balanced setup is one that has equal roll angles.

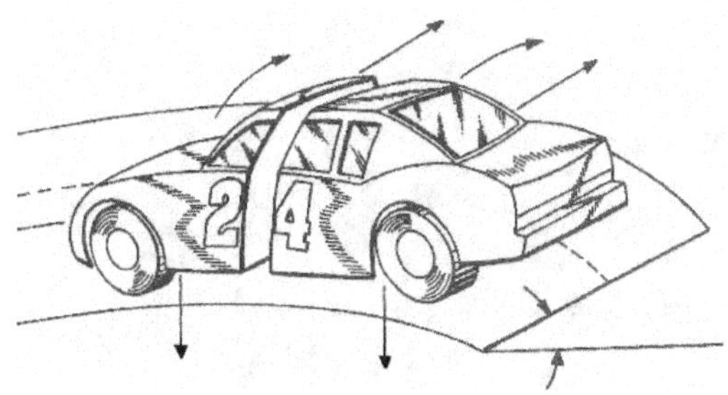

The concept of Roll Angles is how we define balance. Each of the suspension systems at each end of the race car will seek their own roll angle based on the spring rates, moment center heights, etc. The Roll Angle concept takes into account all of the influences that cause the car to want to roll to its desired angle. Our goal is to match these two roll angles.

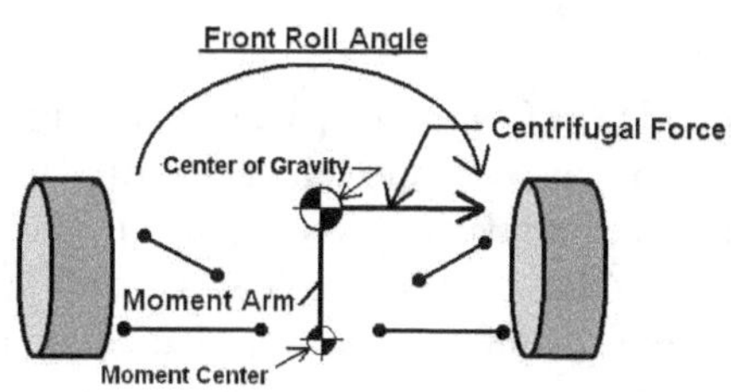

The roll angle in a AA-arm suspension is a product of the length of the moment arm, the amount of centrifugal force and the spring rates translated out to the wheels along with the track width. We want to match this roll angle to the roll angle at the other end of the car. When we do that, the distribution of the weight that transfers becomes very predictable and the handling becomes very consistent.

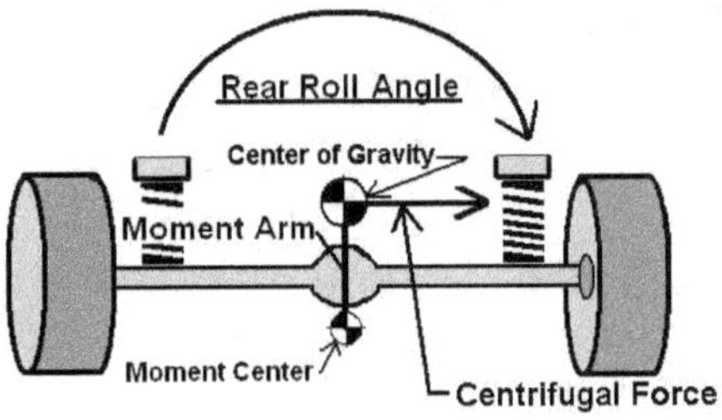

In a solid axle suspension system, we also see a roll angle that we need to match to the other suspension system. This angle is a product of the length of the moment arm, the amount of the lateral force and the spring base width, which acts much like the track width at the front.

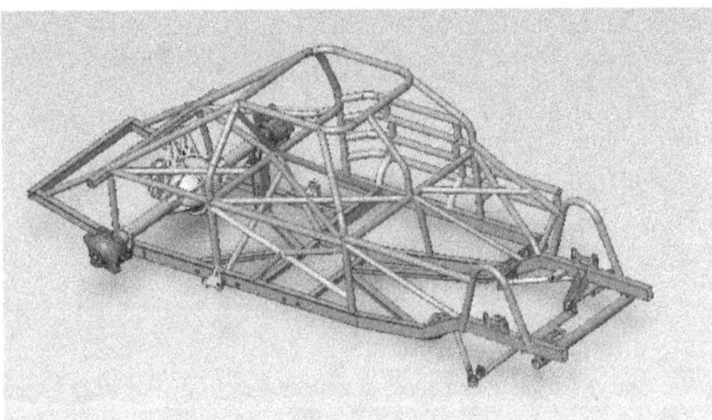

The reason why we need to match the desired roll angles at the front and rear is because each of these suspension systems cannot roll to their own angle when the other suspension system is wanting to roll to a different angle. We have a stiff chassis connecting the two systems that prevents each one from being independent. That is why we need to match the desires of each.

Why Does It All Come Down To Roll Angle? – The reason why we reduce all of the influence of lateral force and banking angle down to the roll angle is because the roll angle is the end result of all of the influences. Roll stiffness does not take into account some very significant influences. And that is why we cannot use that methodology to attempt to measure the balance in race car suspension systems.

The inventor of that method was trying to find a way to give the race car what it wanted for setup. The overall setup involving the front and rear suspension systems was considered and it was concluded that if we can cause the front and rear suspension systems to work together and not fight each other, then that is what the car wants. And this concept proved to be true.

In subsequent testing, it was found that the car reacted very well to the balancing of the roll angles. The first thing good that happened was that the weight transfer became predictable. That is, we could now calculate the weight transfer at each end of the car to know how to set the static weights so that at mid-turn, there would be ideal loading on the four tires. This would produce the most traction possible.

When real race cars were analyzed and the setups configured so that the roll angles were equal, in a stock car racing on a circle track, or on a road racing application, the car became faster, the speed stayed with the car longer, and the tire wear was much better.

So, now that we have a better understanding of what balance is, let's examine what influences the roll angles and in what way.

What Influences The Roll Angles? – The following items influence the roll angles in a race car.

1) The *Spring Stiffness* in the suspension system. The stiffer the overall spring rates, the less roll angle we have. This is common sense, because a race car has very little roll with its very stiffly sprung suspension system while a family sedan wants to roll quite a bit. One has a very stiff suspension and the other a very soft suspension.

2) *Spring Split* – Because of the way the forces work in a race car, using different spring rates on each side of a suspension system will influence the roll angles. A softer spring on the outside will increase the roll angle.

3) *Spring Base* – The width of the spring base does influence the roll angle. The wider the base, the less roll angle. For a straight axle suspension, the spring base is the width between the top of the springs, or coil-over top mounts. For a AA-arm suspension, the spring base is the track width, or distance between the centers of the tires.

4) *Magnitude Of The Lateral Force* – The greater the lateral force, or what we call the G-force, the more roll angle we will have. This makes sense too because as we corner slowly, the measured roll angle is less than when we go faster through the turns. This is why semi-tractor trailers tip over when going too fast on the off-ramp. The roll angle exceeds the suspension travel.

5) *The Location Of The Center Of Gravity* – What acts to roll the car is the overturning moment, something we'll get into in the next Lesson. The CG is at the top of the overturning moment arm and the higher the CG is, the longer the moment arm is which causes more roll angle. The location laterally also has an influence on the roll angle.

The farther left the CG is located, the less the roll angle will be.

6) *The Moment Center Height* – The moment center height is the bottom of the overturning moment arm. As such, the higher the MC, the shorter the moment arm is and the less the chassis will want to roll.

7) *Sway Bar / Anti-roll Bar* – The sway bar, also called an anti-roll bar influences the roll angle. It can be changed to influence the roll angle for cars equipped with those devices. Either the diameter of the bar can be changed, or the arm length can be changed to influence the roll angle. The larger the bar diameter, the stiffer the bar and the less roll angle. The longer the arm, the softer the rate of the bar and the more roll angle the car will have. The reverse is true for each of these changes.

8) *The Track Banking Angle* – The banking angle of the turns from horizontal, does influence the roll angles. Imagine the car racing around a track where the banking is 90 degrees from horizontal. The car would only want to compress into the track and not roll at all. There would be no overturning moment because the lateral forces would be acting straight down from the cars perspective and not to the side. The track banking angles we usually see have a similar effect and act to reduce the roll angle. We'll get a better understanding of that in the next Lesson.

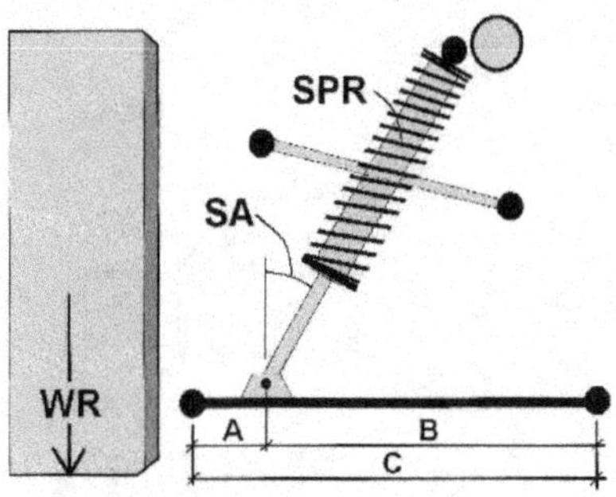

The installed spring rate in a AA-arm suspension is translated out to the wheel through a motion ratio. Here we see a coil-over installation. The track width (distance between the centers of the two tires) helps determine the roll angle. The wider the width, the less the roll angle. Then the wheel rate becomes the spring rate for a AA-arm suspension for the purpose of determining the roll angle.

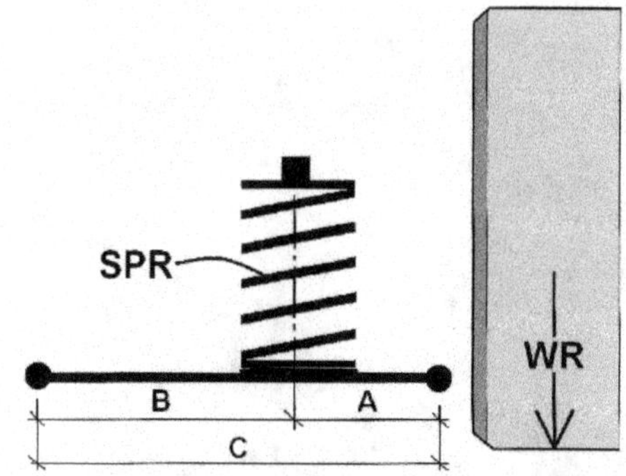

The same motion ratio concept is true of this big spring, or stock spring, design. There is a motion ratio that translates the installed spring rate out to the wheel to become a wheel rate we use in determining the roll angle.

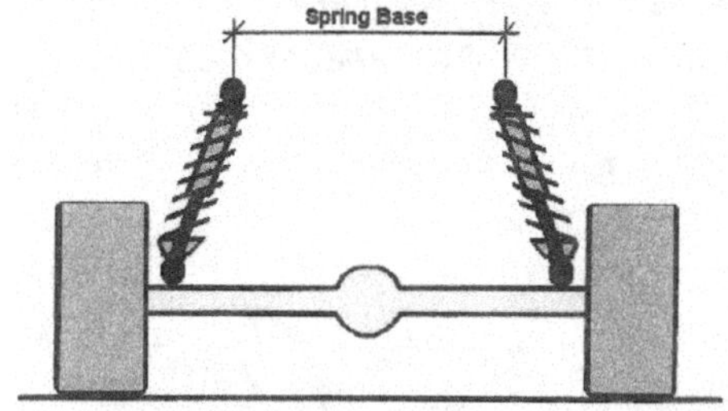

In a solid axle suspension, the spring base is the distance between the top coil-over mounts for this coil-over installation. The chassis rests on the top of the springs and this is all the suspension knows for spring rate and spring base.

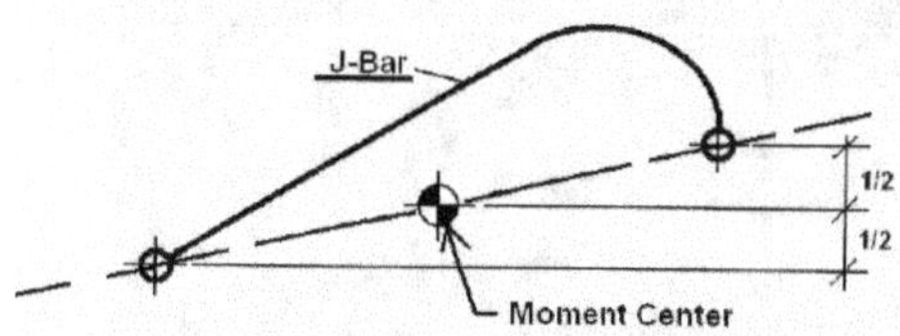

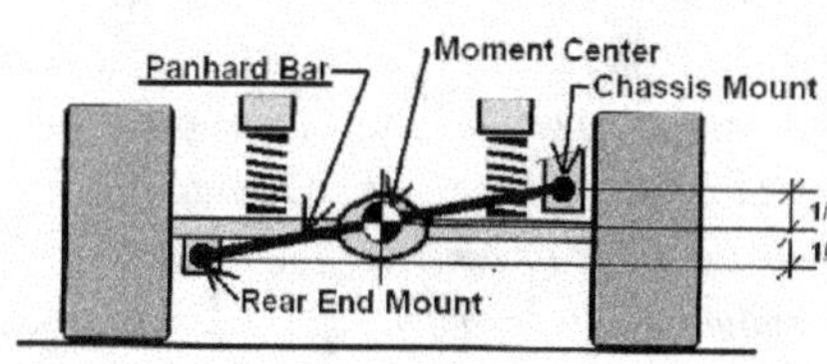

In the rear of a stock car, or any solid axle suspension, the moment center is also the bottom of the moment arm. When we move the panhard bar, or other device used for lateral location of the rear end, up or down, we are shortening or lengthening the moment arm. This has an effect on the amount of the roll angle. A longer moment arm yields a larger roll angle.

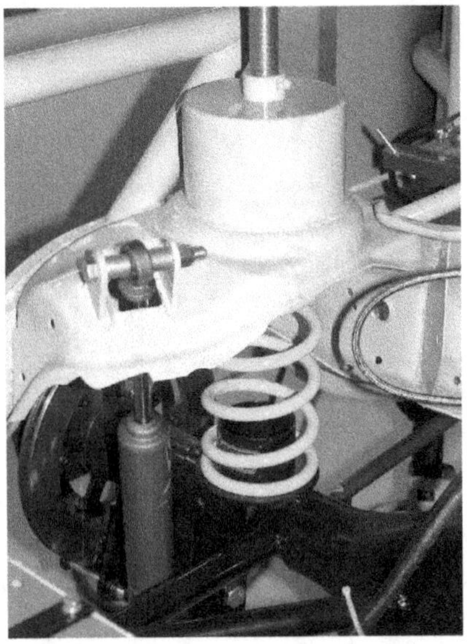

In this stock rear suspension, the springs are mounted straight up, unlike the coil-over springs on a typical asphalt or dirt late model. The distance between the centers of the two springs is the spring base. The wider the spring base, the smaller the roll angle.

Because spring split, or having a spring on one side of the car stiffer or softer than the spring on the other side, is so influential, we need to know the exact spring rate for every spring we install in our race cars. If we install a spring rubber, or bump stop, we then need to rate that spring in the range of motion it will be moving within on the race track so we can know how to arrange the spring rates.

What Can We Change To Make The Car More Balanced? – Out of the above influences, there are some that we can easily change to cause the roll angles to change. These are:

1) The *Spring Rates*,

2) The *Spring Split*.

3) The *Height of the Moment Centers* for both AA-arm and solid axle suspensions,

4) The *Spring Base* – In some designs, especially for dirt cars, the rear spring base can be easily adjusted by moving the tops of the coil-overs in or out to change the spring base width.

5) *Sway Bar* – Teams can change the sway bar settings, or size of the actual bar, to increase or decrease the roll angle.

All of these are easily changed and those are the changes we mostly use to setup our race cars.

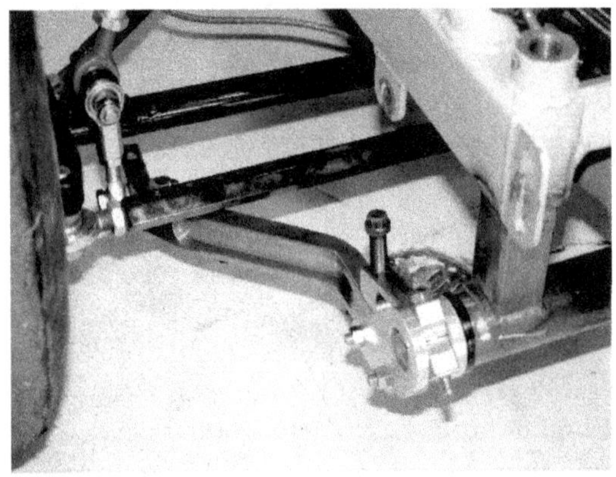

The sway bar influences the roll angle. There are many different types and designs of anti-roll systems. This one is used on an asphalt late model car and is adjustable for pre-load. The sway bar becomes a spring much like the ride spring, when it is made to twist by the dive and roll of the chassis. A larger, or stiffer, bar will result in a smaller roll angle.

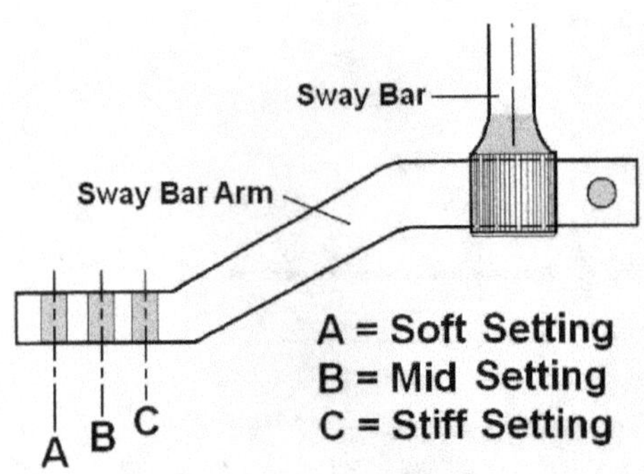

The length of the sway bar arm helps determine the usable spring rate of the bar. In the A position, the arm is the longest and the rate is the softest. In the C position, it is opposite, it becomes the stiffest setting.

In this example, the sway bar system consists of a bar and a set of two blades that act as the sway bar arms. These blades can be rotated so the when flat to the direction of pull they will bend and soften the overall spring rate of the sway bar system. When turned to be long ways to the direction of pull, they do not bend and create the stiffest setting for the sway bar system.

How To Use The Items To Change The Roll Angles – Here are some ways, in general, that we can change the roll angles in the suspension systems. It is good to know, at this point in the RCT Level Three, what can be done to change the roll angles and with those, the balance of the setup.

Spring Rates – We can increase the overall spring rates to reduce the roll angles. The stiffer the spring rates, the less roll angle we will have. And conversely, the softer the spring rates, the more roll angle we will see. So, if we need to reduce the rear roll angle to match the front, we can stiffen the rear spring rates as an example.

Spring Split – We can also increase or decrease the roll angle by changing the difference in spring rates on each side of the car. This is usually only done for circle track race cars. For example, we could soften the left rear spring rate and/or stiffen the right rear spring rate and that would reduce the rear roll angle. This works the same way at the front.

Conversely, if we want to increase the rear roll angle, we can increase the left rear spring rate and/or decrease the right rear spring rate to increase the rear roll angle. This works the same way at the front.

Spring Base – In certain race cars, the top spring mounts can be moved closer or further away from the centerline, mostly on straight axle suspensions, to change the roll angle. Moving them in towards the centerline increases the roll angle by shortening the spring base width. Moving them out away from the centerline reduces the roll angle.

Moment Center Height – We can change the moment center height at each end of the car to change the roll angles. The changes to the AA-arm suspensions will be more difficult, but not impossible. Changes to a panhard bar / straight axle system will be much easier in most cases.

Sway Bar – To tune the balance of the car, we can make adjustments to the sway bar. At the track, we will usually change the actual bar in a stock car running either circle tracks or road racing circuits. Less common changes include changing the length of the arm attached to the sway bar. The shorter the arm, the stiffer the bar rate will be and the less roll angle we will see.

For road racing prototypes and formula applications, there is usually a system that includes a sway bar and arms that are bendable blades with different thicknesses. The designs have the ability to rotate the blade so that they bend more or less easily, which stiffens or softens the bar. If the balance, and roll angles, are way different, then a sway bar diameter change is required.

Summary - This Lesson is relatively short, but very important. The very concept of balance and suspension roll angles is at the very heart of everything we do with the springs, moment centers, etc. If you can grasp this concept and use it to plan out your setup, then you will be way ahead as a race car crew chief, designer and engineer. From here it is a natural progression to talk about the Forces on a race car in the next Lesson Three.

Exam - In The Context Of This Lesson:

The Most Efficient Method Of Measuring Balance In A Race Car Is?

1) By determining the roll stiffness

2) By determining the roll angles

3) When we can measure the shock travels

4) Related to spring stiffness

Roll Angles Are Determined Using Which?

1) Spring rates

2) Moment Center heights

3) Spring base

4) Spring split

5) All of the above

The Center Of Gravity Is The?

1) Top of the moment arm

2) The bottom of the moment arm

3) Always at the center of the car

4) Different for the front and rear of the car

The Front And Rear Suspensions Cannot Roll Freely Because?

1) The spring rates are different

2) The moment arm heights are different

3) They are connected by a stiff chassis

4) The forces at the front are different from the rear

The Spring Base For A AA-arm Suspension Is The Track Width?

1) True

2) False

The Spring Base For A Solid Axle Suspension Is The Width Of The Top Of The Springs?

1) True

2) False

The Sway Bar/Anti-roll Bar Does What?

1) Assists the roll angle

2) Resists the roll angle

3) Acts like a ride spring in roll

4) 2 and 3

A Longer Sway Bar Arm Does What?

1) Makes the sway bar stiffer

2) Makes the sway bar softer

3) Increases the spring stiffness in roll

4) Increases roll stiffness when a smaller bar is used

Race Car Technology – Level Three

Lesson Three – The Forces Acting On The Race Car

In this lesson we will learn about and study the forces that act on a race car. We won't make this overly complicated because we want you to both grasp the concepts and then be able to think through how they affect the race car. This will lay the foundation for much of what follows in the Level Three course.

To be complete when talking about forces on a race car, we have induced forces from two different directions. There is the Lateral Force, which acts on the car when we are cornering and the car is changing direction. The other force is Longitudinal Force that acts on the car when the car is accelerating and when it is decelerating.

We want to now concentrate on only the Lateral Force and we will take up Longitudinal Force later on. Trust me, Longitudinal force is important, but it applies to other areas of setup than what we are getting into within the next few Lessons.

There are some very complicated and sophisticated dynamic and kinetic measurement rigs available to the race teams. These, like this example at Morse Measurements in North Carolina, will try to duplicate the forces we record acting on our race car when we are cornering, braking and accelerating.

What Are The Forces We Need To Know About? – The obvious force we can feel when the car corners is the Lateral Force. It tries to roll the car and it tries to push the driver to the outside of the turn. This force is rated at G-forces. The G-force is the sprung weight of the car in multiples directly related to turn speed. The faster we go through the turns, the greater the G-force. The number, 2.0 G's represents the sprung weight times two.

There is a formula for calculating the G-force. It calculates the total lateral force and then divides that by the sprung weight. It is:

G-force = [(Sprung Weight x Speed in MPH^2)/(14.97 x Turn Radius in feet)] / Sprung Weight.

So, if we had a race car that weighs 2100 pounds in sprung weight, and was going 80MPH through a turn having a radius of 200 feet, the G-force would be the total lateral force (the part in the [ ]) divided by the sprung weight, or 2.14 G's.

The formula cars show us a lot about lateral forces and how cornering produces this force. Many F1 cars can reach or exceed 4 G's in the turns with the aid of Aero Downforce. This represents four times the sprung weight of the race car when we are using lateral G's for the calculation of forces and how they affect the roll and weight distribution of the car.

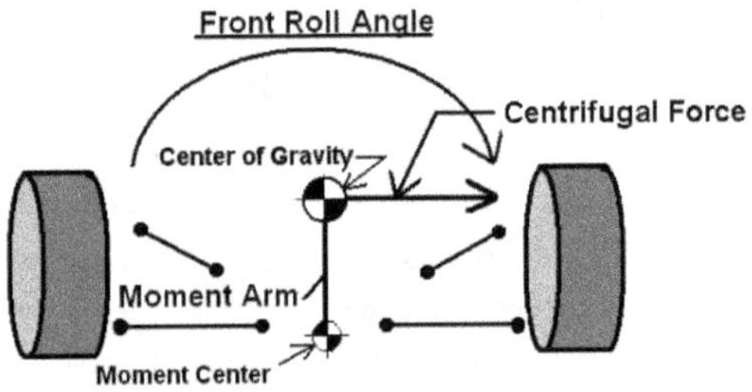

A very simplistic way to look at cornering forces is to just consider the lateral force, or what we call the Centrifugal force. This is what is trying to force us off the turns and what the tires resist. We cannot, in the study of the forces on a race car, just think about only the lateral force. There is more to the equation, so to speak.

There is another force acting on the race car when it is cornering, and everywhere else it is situated. That force is Gravity. Gravity acts on the race car when it is at rest as well as when it is cornering. It never goes away. When we analyze the forces acting on the race car, we have to consider both the lateral force and the gravitational force.

In the study of physics, there is a term called Resultant Force Vector. This Resultant vector is a magnitude and direction of a force that represents a combination of two forces. That is why it is called Resultant, it is a result of the existence of two forces. In our race car we do have two forces, lateral force and gravitational force and they both act at the center of gravity of the sprung mass. So, we too have a Resultant Force Vector.

The concept and existence of a resultant force vector helps us understand a lot of what goes on with a race car during cornering. And we will get into all of that. But first, I don't want to lose you on this concept and how it works. It does sound complicated from a distance, but once we get closer to it, everything will make perfect sense and be easy to understand, I promise.

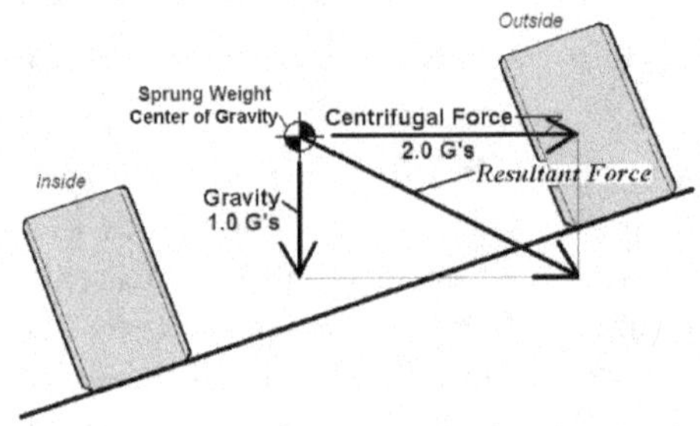

This is a more accurate and complete way to look at the forces acting on our race car and the suspension. There are two forces that act on the sprung mass of the car, and the both act at the sprung mass center of gravity. The first and ever-present force is Gravity. It is always there pulling the car towards the center of the earth, even when we are racing and cornering. The other is the Centrifugal force that is pulling on the car parallel to the earths surface. Its magnitude is determined by the weight of the sprung mass of the car, the speed, and the turn radius. Those create a force and that force divided by the sprung weight gives us the G-force number.

What Does A Force Vector Look Like? – When we have a force vector resulting from two forces, it will look like the sketch in RCT-C3-L3-04. The length of the two force lines are used to form the resultant force length. The length of the gravity force line is then multiplied times the G-force number, such as 2.0, to determine the length of the G-force line.

From the ends of these two lines, we draw lines that are parallel to the opposite line to form a rectangle, or in some cases with force vectors, a parallelogram, meaning all sides are parallel to the opposite side. In our case, with a race car, the object is a rectangle because the gravity line is always at ninety degrees to the lateral force for almost every turn condition at almost every race track. Or, gravity is a force pulling towards the center of the earth and the lateral force is pulling parallel to the earths surface.

How Does The Force Vector Concept Help Us Understand Our Race Car? – Many people who try to understand vehicle dynamics forget about gravity. They only concentrate on the lateral force. The direction and the magnitude of the force acting on the race car matters. If we think of the dynamics in terms of a force going in a much different direction, any conclusion we come to has to be wrong.

Here is an example of why it matters. Suppose we have a typical race car going through a flat turn and gravity was not acting on that car. Only the lateral force is pulling on the car. The car would roll such that weight would come off the inside (inside of the turn) springs and be put onto the outside springs.

In this scenario, the inside of the car would raise up and the outside of the car would compress. But, our observation of most race cars shows us that this is not what happens. For most race cars, the outside of the car compresses, but the inside does not raise up and in some cases with high banked tracks, will compress too.

If both sides are compressing, then some force must be pulling down on the car. Note in illustration RCT-C3-L3-YY that the resultant force vector is pointing down and to the outside and meets the ground at a point between the two tires. Won't this direction of the force, and the fact that it is directed between the tires mean it will be doing two things, one rolling the chassis and the other, pulling down on the chassis.

The higher the banking angle, the closer to the centerline of the chassis the resultant force will move. That is why on very high banked tracks like at Daytona, the cars that ran before coil bind and bump setups rolled very little and compressed a lot. For most medium banked race tracks, the outside compresses and the inside of the car stays relatively at the same height as it is at normal ride height.

On a really high banked track like Bristol Motor Speedway, the direction of the Resultant Force is more towards the center of the car and we then have less roll angle and more mechanical downforce. The G-force is much higher than we see at flatter tracks. This force pulls down on the car more than it tries to roll it over. The setups for this type of track are much different than those for flatter tracks. The springs must have more rate, much more, and the balance is different, all due to the high banking angle.

The Effect Of Track Banking Angle – The banking angle of the turns has three effects on the performance of the car. First 1) the high angle changes the location of where the resultant force intersects the ground keeping more load on the inside tires creating more grip, 2) as a result of the vector angle, there is more mechanical loading on all of the tires causing the car to be able to go faster, which is why the G-force number is so much higher, and 3) the resultant force is higher due to the increased speed creating even more tire loading through the turns on high banked tracks. A faster speed generates the greater G-force number. Note that in the formula for calculating G-force, speed is a determining factor in the calculation.

Since the force is directed more downward in relation to the track surface, and since it is greater, there is more loading on the tires than we would see on flatter tracks. This provided more grip helping the car to go faster. On tracks like this, the turn speeds can increase by almost double over a flat track.

Summary For Lesson Three – Please go back and read this over again until you can fully understand this concept. The way these forces act on the race car chassis determines everything we do to enhance performance. As we race at different tracks with different turn radii and different banking angles, we need to understand how these forces will change in magnitude and direction. We must then make changes to our setup in order to meet our performance goals.

Exam - In The Context Of This Lesson:

The Two Forces Caused By The Turning, Braking and Acceleration Are?

1) Lateral and Longitudinal forces

2) The braking force

3) The accelerating force

4) The Earths rotational force

The Two Forces That Act On The Race Car In The Turns Used For Handling Are?

1) Lateral and Longitudinal forces

2) The braking force

3) The accelerating force

4) Centrifugal and Gravity

The Resultant Force Direction Is Determined By?

1) The magnitude of the Lateral force

2) The weight of the sprung mass

3) The front and rear moment centers

4) 1 and 2

Which Of The Following Are Used To Determine The G'force?

1) Speed of the car

2) Sprung weight of the car

3) The radius of the turns

4) All of the above

High Banked Tracks Cause What?

1) More mechanical downforce

2) Higher speeds

3) Less roll angles

4) All of the above

Race Car Technology - Level Three

Lesson Four – Setup Platform Analysis & Springs

This Lesson is about the setup Platform, as many call it. We can also call it the synchronization of the Roll Angles (front and rear), something we have described before in Lesson Two. This is the Concept of Balance. Now we will work with the roll angles and describe how different spring rates and other settings affect the roll angles to create a perfect setup platform, and see how much of each change affects the setup balance.

The concept of roll angles is useful so we can develop a balanced setup to where the loading on the four tires is ideal and predictable. In some racing circles, this is called the platform, and this would be a good description. In road racing and high-performance formula racing, it is also called the platform and that term includes the aero platform, or attitude that produces the best aero properties.

In this Lesson we will stay with the roll angles and platform as it relates to the setup of the race car. This basic concept is not a new one, but it has been refined and perfected over the past twenty years or so. And, it has been tested and proven to be an accurate and helpful way to achieve a dynamic balance for any race car.

Every race car needs to have a balanced setup. This car has a balance, but also some roll angle to it. There is no advantage to eliminating chassis roll altogether. We can easily configure the body and front valance so that the air is sealed off from coming under the front of the car, even while it is rolling a little. We still need some roll angle and compliance in our race cars. Even when observing Formula One cars, we see that they roll going through the corners. Here we will work with a race car and see how we can change the roll angles.

A Short History Of Chassis Research - In years past starting just after World War II, automotive engineers sought ways to measure the performance and balance of an automobile and later on in the late 1980's and early 1990's, the interest shifted to race cars. A method was developed that analyzed the roll stiffness and what was called the roll couple for a chassis and its suspension systems.

Roll stiffness related to one suspension system and roll couple was the analysis of the stiffness relationship between the two suspension systems in a car. This was a good approach, but for race cars, it was deficient for several reasons we won't go into now. Just know, for any race car to be both fast and efficient and for that performance to stay with the car longer, we need to balance the setup.

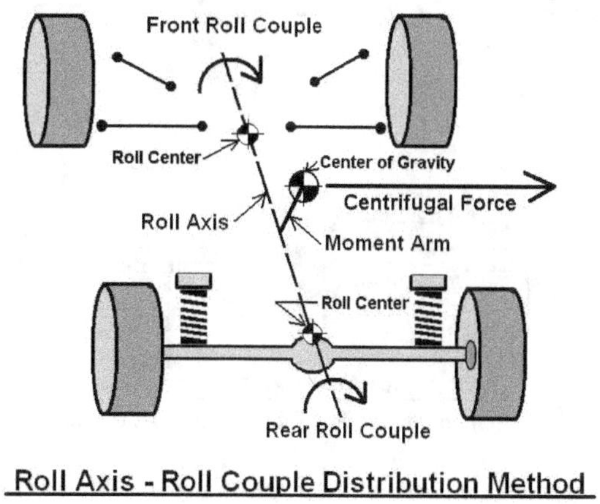

The roll axis and roll couple method of evaluating the "roll stiffness" of a chassis was a good start to understanding chassis dynamics. This line of thinking was limited and probably worked alright for a production car, but not so well for many of our current day race cars. In this day and age, we know so much more about chassis dynamics and how to apply methods that will truly tell us if our chassis setup is balanced.

How Is The Roll Angle Produced? – The actual roll angle in a suspension system is produced when the car is cornering and a lateral force is applied to the Center of Gravity of the sprung mass. The sprung mass is the part of the race car that is not a part of the wheels, tires, solid axle, etc. and that is supported by the suspension.

This sprung mass wants to keep traveling in the same direction the car was going down the straights before it began to travel into and through the turns. Once the car begins to turn, a force pulls on the car towards the outside of the turn away from the radius point. This is called the Centrifugal force, or what we call CF.

The G-force is the CF divided by the sprung mass weight, or gravitational force on the sprung mass. If the CF were twice the gravitational force on the sprung mass, then the G-force would be 2.0. The magnitude of the G-force helps determine the amount of the roll angle.

As we described in Lesson Three, it is not only the G-force that creates the roll angle, it is the combination of the G-force and gravity which forms the Resultant force and its direction of "pull" on the Center of Gravity.

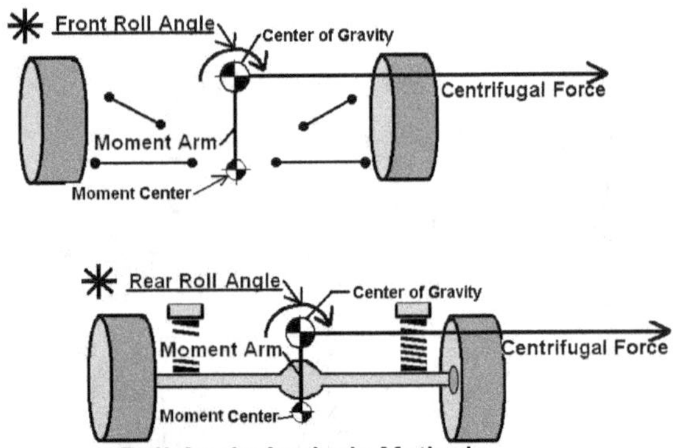

Roll Angle Analysis Method

✱ *Matching the two Roll Angles Balances the car*

The more complete method for determining balance with a setup for a race car is by using what is called the Roll Angle method. This was developed in the mid-1990's and applied to a computer program. The critical dimensions and weights are entered and the program calculates the desired roll angles for the front and rear suspensions. This program was designed for a circle track race car with a AA-arm front suspension and a solid axle rear suspension. We will try some setups to see how we can balance the two roll angles.

What Determines The Degree of The Roll Angle? - So, we have a lateral force which pulls on the CG which produces a roll angle in a suspension system. We can then measure the angle if we know certain things about the suspension system. Here are the things we need to know:

1) The magnitude of the Resultant force (G-force and Gravity combined)

2) The height and width of the Center of Gravity

3) The height of the moment centers

4) The spring base width front and rear

5) The stiffness of the four springs

6) The spring split if one spring is a different rate than the opposing spring

The formulas, or algorithms as they are often called, are very complicated, but we have a tool that can measure the roll angles. We will use that tool to demonstrate how we measure and adjust the roll angles. But what do we use these roll angles for?

How Do We Use Roll Angles? – The process of using roll angles to measure the balance of a setup has been in existence for some twenty years. This process has been used by thousands of race teams, but most race teams still do not know how to use this method, and that is part of the reason you are taking this course, to learn new things.

We calculate the roll angles at each end of the race car and then compare those angles. In most asphalt circle track racing, and road racing with prototypes and formula cars, we match the angles precisely. That is to say, we match them to within a tenth of a degree.

For example, to help you understand what that tenth of a degree means, it is the movement of one wheel by one-eighth of an inch in a system with a 66 inch track width. That is a small amount indeed, but those limits are what we need in order for the system to perform.

In other forms of racing including dirt racing, the approach is very different. Because the relationship of the front and rear roll angles determines the load distribution, we might want to cause the roll angles to be different so that our loads will be biased towards grip for one end or the other.

The end we usually want to have more grip in a dirt car is the rear when the track is very slick. This gives the car a greater acceleration, but at the expense of lateral grip, of which there is very little to begin with. If the gain in acceleration outweighs the loss in lateral grip, then the car goes faster around the track.

For dirt cars, the teams will vary the difference in roll angles depending on the track conditions. Naturally, if the track has a lot of grip, then the front and rear roll angles will be made to be much closer to equal, much like on asphalt with its high grip level. Then as the track changes to become more slick, which often happens, the team will make changes to the setup to cause the rear roll angle to be more than the front roll angle giving the car more rear grip.

How Much Roll Angle Do We Need? – There is generally no perfect roll angle, but there are considerations that differ for different types of racing. Some race cars need to be more compliant than others, meaning that the suspension systems need to move more. So, for those cars, we need for the suspension springs to be softer, like with dirt cars.

For most of today's race cars that race on asphalt, the roll angles are minimal. The goal should not be to eliminate the roll angle, but to limit it for various reasons. Those reasons include the following:

1) To provide a more level and low attitude for stock cars to improve aero properties

2) To reduce camber change in a AA-arm suspension

3) To provide very little vertical movement for a more stable aero platform in both stock cars and prototypes and formula cars

4) To reduce camber change for race cars with wide tires such as prototypes and formula cars

5) In some cases, to increase the heat in the tires for shorter runs

How Do We Make Changes To The Roll Angles? – Because we need to match the roll angles, or miss-match them as the case may be, we need to know how to change the roll angles once we know what they are for our particular race car. We do that by making changes to the following:

The Spring Rate – We can change the overall spring rate, or stiffness in a suspension system. Making the springs stiffer will reduce the roll angle. Making the springs softer will increase the roll angle.

Spring Split - Causing a spring split, or difference in spring rates on each side of a suspension system, will cause a difference in roll angle. For a AA-arm suspension, softening the inside spring rate, or wheel rate for this system, will decrease the roll angle. Softening the outside wheel rate will increase the roll angle. We can stiffen and soften this system by changing the motion ratio, spring angle, or actual spring rate of the installed spring or bump device.

For a solid axle system, the same applies with a difference. That is, the spring base is the distance between the top of the springs, not the wheel rate. For the purpose of roll angle determination, wheel rate has no value. The chassis only knows that it sits on the two springs.

So, for this solid axle system, we can not only change the spring rates and spring split to make roll angle changes, much like the AA-arm suspension systems, but we can also make changes to the spring base width. This is more common in dirt racing where, as we stated above, changes to the roll angles are needed to compensate for changing track conditions.

Moment Center Heights – We can change the roll angles by altering the moment center height. In a AA-arm suspension system, it is harder to do because we would need to make changes to the control arm angles, and that is not a quick and easy change. It can and is done, but not usually at the race track.

For solid axle suspension systems, the changes can be made much easier. If the solid axle is held in place laterally by a panhard, or J-bar, most race cars with this system have slotted mounts on each side so that the bar can be easily raised or lowered by some amount.

If we raise the bar, we shorten the moment arm and the suspension becomes harder to roll. Therefore, we see less roll angle when we raise the bar.

If we lower the bar, we lengthen the moment arm and the suspension becomes easier to roll. Therefore, we see more roll angle when we lower the bar.

How Do We Balance The Setup Using Roll Angles? – Now that we know what influences the roll angles, we need to go to the car and make changes to balance the car. Later on we will tell you how to read your roll angles directly and/or read the tire temperatures to know how to balance the roll angles. We will be making changes to actual race car roll angles to show you how to do it yourself.

For Circle Track Race Cars - In a circle track car using a AA-arm front suspension and a solid axle rear suspension, the front is not easily adjustable and we usually set the front spring rates and bumps based on knowing what type of race track we will be racing on, such as high or low banking. So, the front is not considered an easily adjustable part of the car for at-the-track adjustments in most cases.

As for the rear, the solid axle is highly adjustable in several ways. We can change one or both spring rates, the moment center height, and the spring base if the car was constructed to do that. Let's say we had too much roll in the rear and we need to make changes to reduce it. As we stated above in our explanation on how to make roll angle changes, we could increase both spring rates equally, decrease the LR spring rate, and/or increase the RR spring rate to create a larger spring split. Also, we could raise the panhard or J-bar, or widen the spring base. All of those would decrease the roll angle in the rear.

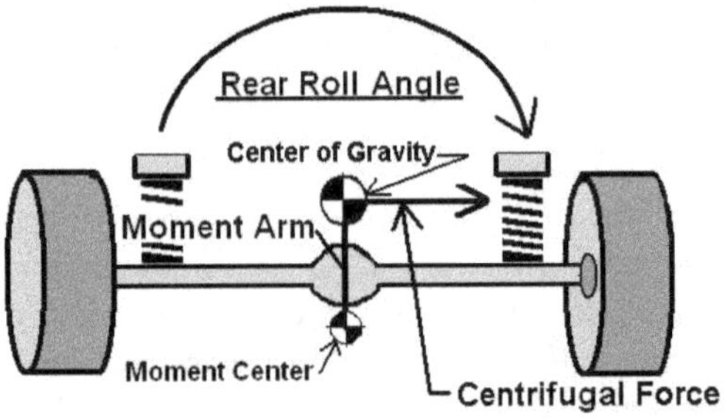

In a typical circle track race car, it is easier to adjust the roll angles to be the same by working with the rear roll angle only and matching it to the front roll angle. The roll angle is much harder to change in a AA-arm suspension. Here we will work with spring rates, spring split and rear moment center heights, i.e. the panhard/J-bar heights.

Which Of The Changes Make The Most Difference? – Now we will tell you how much each change makes for a typical 2800 pound asphalt late model race car with a 15.0 inch high center of gravity, a 1.0 inch high front moment center, and running on a race track that has 12 degrees of banking and where the G-forces are 2.0 G's.

The front spring rates are 1650ppi (pounds per inch) left and right (using a 150ppi ride spring and a 1500ppi bump spring, or bump stop). The LR spring is a 175ppi and the RR spring is a 225ppi for a 50ppi spring split. The panhard bar heights off the ground are 9.0" left and 10.0" right and we are using a 7/8" hollow sway bar. The roll angles are 1.29 degrees front and 2.49 degrees rear. So, in order to match the roll angles, we will need to reduce the rear roll angle. Each of these changes will tell you approximately how much roll angle difference each of the changes makes.

Here is our starting setup. We can see the four spring rates, the sway bar diameter, the panhard bar heights, the Lateral G-forces and the track banking angle. We want to match the front and rear roll angles by making changes to the numbers. Let's see what happens.

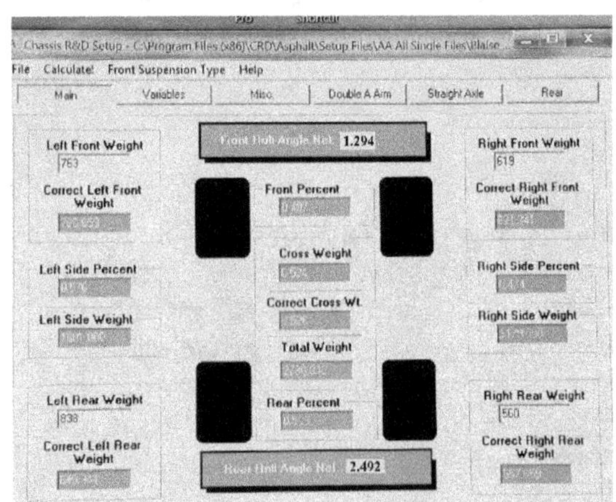

When we calculate the roll angles for our base setup, we get a front roll angle of 1.3 rounded off and a rear roll angle of 2.5 rounded off. We need to reduce the rear roll angle in order to match the front roll angle. Let's try some things.

Spring Stiffness – If we first stiffen both rear springs by 25ppi, for a LR spring rate of 200 and a RR spring rate of 250, the rear roll angle changes to 2.32. This is in the right direction, but too little of a change.

If we continue to raise the rear spring rates and not change the spring split, we would need to install a 475 LR spring and a 525 RR spring to get the rear roll angle down to 1.27 degrees. This is not a viable solution because the rear spring rates end up being way too stiff.

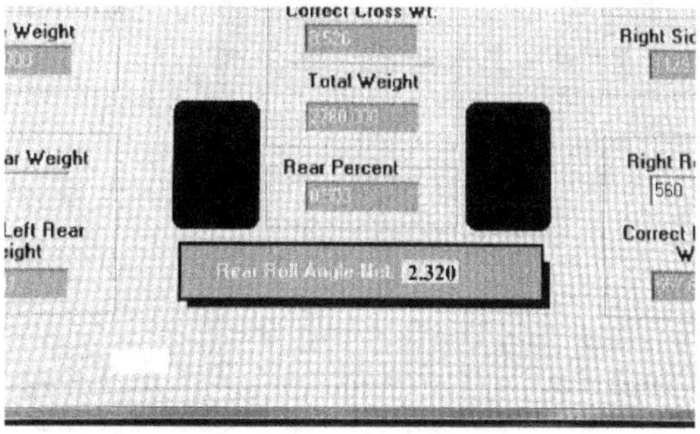

If we stiffen both the rear springs by 25ppi (pounds per inch), we get a rear roll angle of 2.3. This is less than we started out with, but not near enough of a change. Let's keep going.

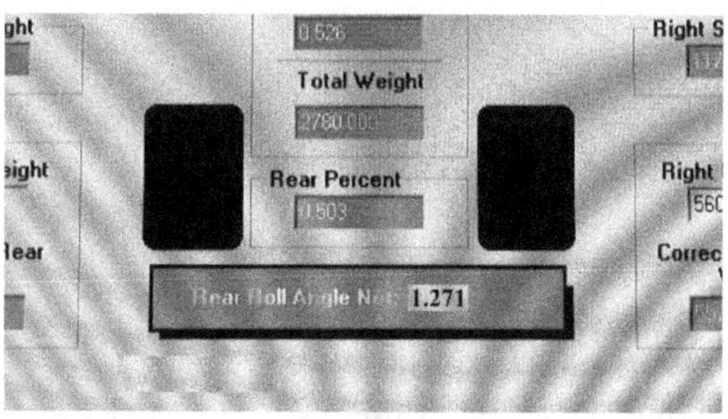

We kept increasing the rear spring rates while keeping the same 50 pound spring split and we had to go all of the way up to a LR rate of 475ppi and a RR rate of 525ppi in order to get to a rear roll angle of 1.27. This is not feasible and we need to find another way to reduce the rear roll.

Spring Split – Going back to our original spring rates, if we now reduce just the LR spring rate by 25ppi, the rear roll angle goes down to 1.92. This was a more significant change, but not enough either.

Going back to our original spring rates, if we then increase the RR spring rate by 25ppi, our roll angle goes to 1.90, about the same as when we reduced the LR spring rate.

If we increase the spring split and then go to 150 LR and 250 RR spring rates, our rear roll angle is now 1.32, or very close to the front and good enough to go racing. But we may not want that much rear spring split.

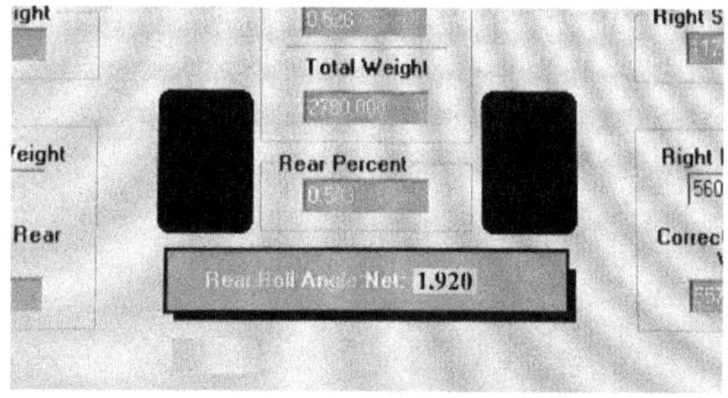

If we go back to the original setup and reduce the LR spring rate by 25ppi to a 150ppi spring, we then have reduced the rear roll angle to 1.9, a more significant change, but not nearly what we need.

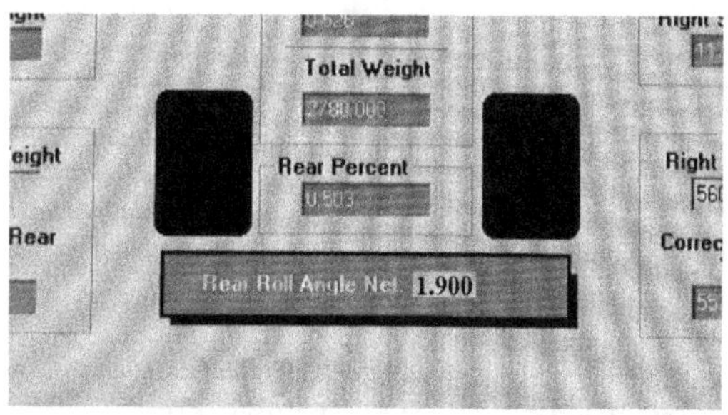

If we go to the RR spring and increase it by the same 25ppi, leaving the other setting the same as the original setup, we get a rear roll angle of 1.90, similar to the change we made to the LR spring rate, but again, not enough.

Panhard Bar – Going again back to our original setup, if we raised the panhard bar by ½" on both sides, we will have raised the rear moment center by that same ½". The rear roll angle is now 2.07, again not enough, but better.

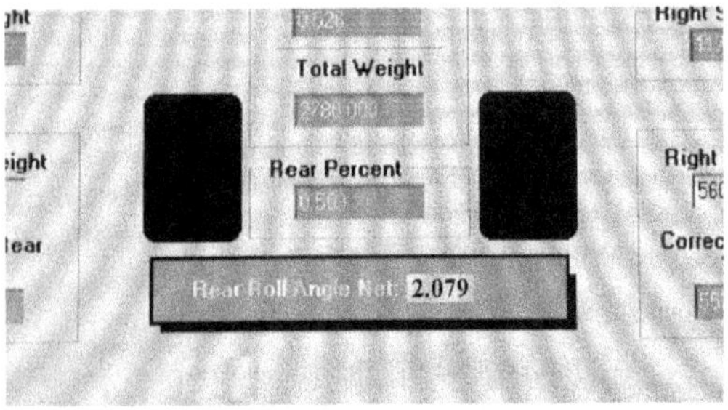

Again, going back to the original setup, if we raise the panhard bar by ½ inch on both sides keeping the same split, we then have a rear roll angle of 2.07, better, but not what we need. Maybe a combination of increased spring split and raising the panhard bar would get us to where we need to go.

Refining The Setup – To get the rear roll angles to match, we can use one or more of these three different methods. We can use spring stiffness, spring split, raising the panhard bar, or some combination of those. Let's see how that works.

If we increase the spring split from 50ppi to 75ppi by installing a LR spring rate of 150ppi and a RR spring rate of 225ppi, and if we also raise the panhard bar by ¾" on each side, our rear roll angle now becomes 1.24 degrees, or just 0.06 degrees different than the front roll angle. We now have a balanced setup.

For a race car with AA-arm front and rear suspensions, the teams work with both spring stiffness and/or sway bar rates to balance the roll angles. In most cases, they want to maintain a certain ride stiffness to help maintain the aero platform and so, changing the sway bar rates offers the best way to balance the roll angles.

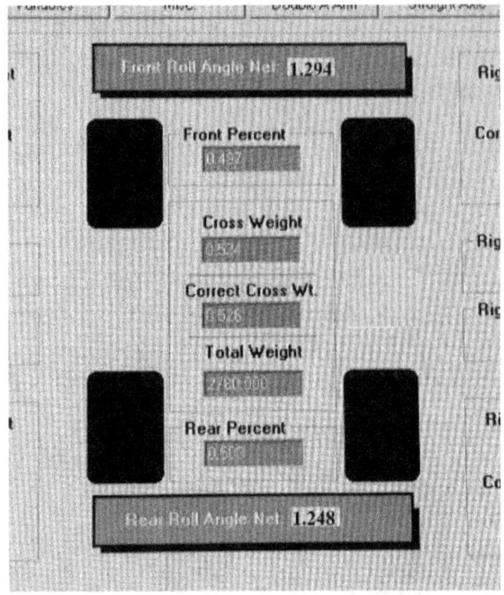

From a starting point being the original setup, we now increase the spring split by reducing the LR spring rate by 25ppi for spring rates of 150 LR and 225 RR. We then raise the panhard bar by 0.75" to 9.75" left and 10.75" right keeping the one inch bar split. Then our roll angles are calculated to be the 1.29 front and 1.248 rear. This is close enough to go racing with. This car now has a balanced setup and it was not that hard to change it to get there.

Summary Of Lesson Four – You have been introduced, or re-introduced if you have come across this method before, to one of the most significant tools we have available to us for setup design. By now you should be able to see where we can predict the balance condition of a race car by observing the roll angles.

This method is only as useful as the accuracy of the information entered into the software. There are other ways we can read the balance of the car by reading tire temperatures and we will get into that later on.

Next up we will discuss Weight Transfer and Tire Loading, and how the roll angels can influence those two. We are closing in on a full understanding of how to perfect the race car setup.

Exam - In The Context Of This Lesson:

The Platform and Matched Roll Angles Are What?

1) Different in concept

2) The same basic concept

3) Used to perfect shock values

4) Both derived from early development in chassis dynamics

Roll Stiffness And Roll Angles Are Affected By What?

1) Spring rates

2) Moment center heights

3) Spring base

4) All of the above

We Use The Roll Angle Calculations To Do What?

1) Work for zero roll angle

2) Create a balanced setup

3) Improve the ride on bumpy tracks

4) Determine what shocks to use

Which Change Produces The Greatest Difference In Solid Axle Roll Angles?

1) Raising overall spring rates

2) Raising the panhard bar

3) Increasing the spring split

4) Changing the spring base

How Close Do We Need To Make The Roll Angles For Asphalt Racing?

1) Within 1.0 degree

2) Within 0.01 degree

3) Within 0.10 degree

4) It varies with the track conditions

How Close Do We Need To Make The Roll Angles For Dirt Racing?

1) Within 1.0 degree

2) Within 0.01 degree

3) Within 0.10 degree

4) It varies with the track conditions

Race Car Technology – Level Three

Lesson Five – Weight Transfer & Static Loading

Weight Transfer is a process whereby the loading on the four tires of a vehicle, in this case a race car, changes when there is acceleration. Acceleration isn't only a change in forward speed as we commonly know it as, but also when forces are acting on the race cars Center of Gravity in the turns. When we have a change in direction such as going through the turns, that change generates a lateral acceleration as well. We call any acceleration a G-force.

In this Lesson, we will be discussing both lateral and longitudinal acceleration. Both create weight transfer and the re-distribution of loading on the four tires. While braking and accelerating happen mostly on the straighter portions of the race track, the acceleration from cornering is all about a force that is pulling out in a direction away from the radius of the turn. This transfers weight from the inside two tires to the outside two tires.

When we are studying how to make a car handle more efficiently, we are interested in how the tires are loaded. The load distribution on the four tires is what determines how our car is going to handle and how much speed it will be able to carry through the turns. To better understand dynamic loading, we need to first understand how weight transfers and what affects that transfer.

In this example, the car on the inside is keeping all four wheels on the ground, so the inside front tire still has load on it despite the weight transfer. The outside car is lifting the left front tire off the ground, so all of the front weight has been transferred to the right front tire. This is 100% weight transfer at the front. This is basically a three-wheel car and not the ideal situation.

Weight Transfer during cornering takes weight off the left, or inside, tires and puts that weight onto the right, or outside, tires. The amount of weight can easily be calculated, but where the loads end up on the four tires can be very unpredictable. With a balanced setup where both roll angles are matched, the distribution of weight is predictable and can be determined.

Race cars with a low center of gravity, like these modifieds at Boman Gray Stadium, transfer less weight and therefore retain more inside weight on the tires through the turns. This provides more grip due to more equally loaded tires.

Formula and prototype race cars are very stiffly sprung and therefore the balance of the setup is more critical as it applies to weight transfer. The slightest miss-match in the roll angles causes a greater change in weight distribution.

What Affects Lateral Weight Transfer? – Weight transfer happens in all vehicles that are turning. Even vehicles that don't have a suspension will transfer weight. A good example of that is when we corner too fast in a car with a high center of gravity, it tips up onto two wheels. That represents 100% weight transfer, and all of the vehicles sprung weight must be supported by the outside tires because they are the only ones on the ground.

The following are items that affect the amount of weight transfer:

1) *Weight of the Sprung Mass* of the race car. The greater the weight, the more that is transferred.

2) *Height of the Center of Gravity.* The higher the CG, the more weight that is transferred.

3) *Track Width* – dimension from center to center of the tires. The wider the track, the less weight transfer.

4) *Magnitude of the Lateral Force* measured as speed over change in direction. The more force, or speed and the shorter the radius, the more weight transfer.

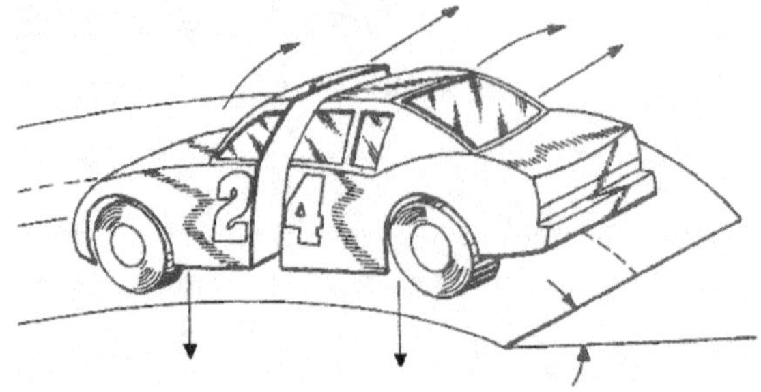

We calculate weight transfer by analyzing the front and rear of the car separately. There is a weight transfer happening at each end. For a car with AA-arm front and rear suspensions, the calculations are fairly simple. For a car with one or more solid axle suspensions, the calculation is different. For the solid axle, there are two calculations, one for the sprung weight and one for the un-sprung weight. These are then combined to yield a total weight transfer at that end of the car.

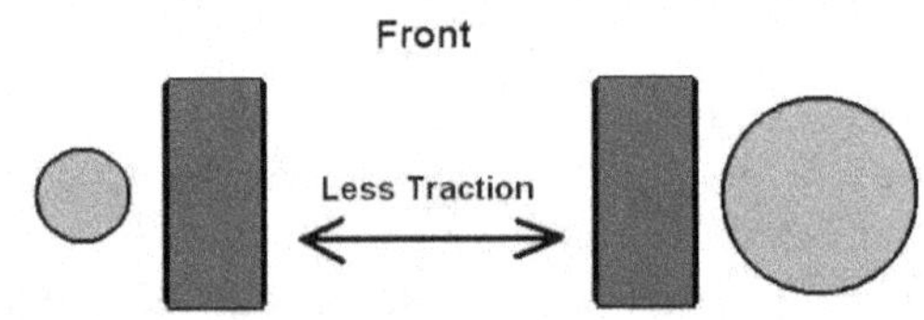

The result of an unbalanced setup where excess load transfers to the Right Front tire causing the rear tires to be more equally loaded. This would be a very tight handling car.

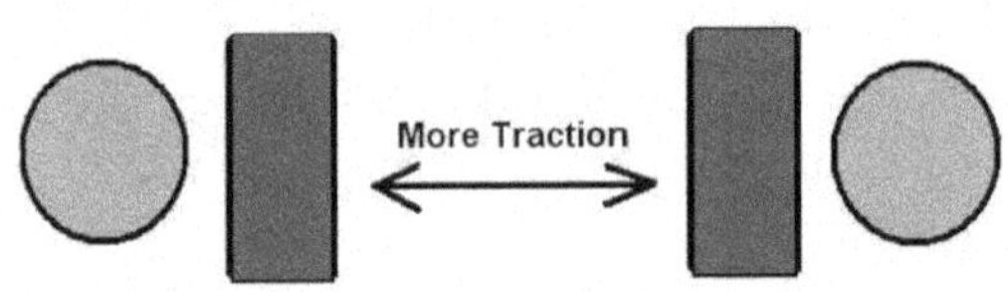

If the roll angles are different with the rear rolling more than the front, then we have excess weight transfer at the front causing the right front tire to do more work and the left front tire to do less work resulting in more un-even loading on the front tires and therefore less grip. The rear tires are more equally loaded and therefore gain grip. Loss of grip in the front combined with the gain in grip at the rear makes this a tight setup, or one that produces under-steer.

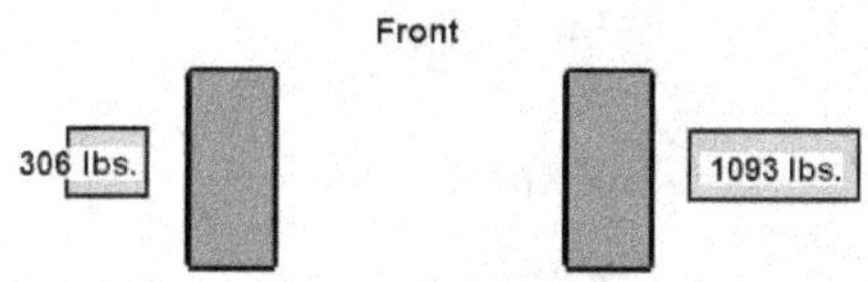

Ideal Dynamic Weight Distribution Has Equal Weight On Both Inside Tires and Equal Weight On Both Outside Tires

The ideal situation is when we balance the setup by matching the roll angles front and rear. Then the weight transfer becomes predictable and if we can arrange the static weights correctly, we can end up with equal weight on the inside tires and equal weight on the outside tires. The level of grip at the front and rear are now the same, if the car has the same size tires on the front and rear of the car.

How Do We Calculate Weight Transfer? - The formula for weight transfer is:

WT = (Sprung Weight x CG x G-force) / Track Width

In this exercise, we are going to calculate weight transfer for a car turning in one direction. We will then back out what we need for static weight distribution, or as we call it in circle track racing, cross weight, in order to end up with the ideal loading on the tires at mid-turn. This is a big deal. You can now know what weight distribution you need to start with in order to have the ideal weight distribution at mid-turn. This will only work if the roll angles are matched up and the setup is balanced.

For this example, we have a race car that weighs a total of 2,800 pounds. We are going to divide that into percentages of front to rear. Since this car is a 50-50% car front to rear, we will have 1,400 pounds resting on the front tires and 1,400 pounds resting on the rear tires. We also have 54% left side weight, so the left side weighs 1,512 pounds (2800 x .54 = 1,512) and the right side the remainder, or 1288 lbs. We'll use those numbers later. Here are all of the numbers we will use:

Vehicle Sprung CG = 15.0"
Front Weight = 1,400
Front Un-sprung Weight = 150
Front Track Width = 66.0
Rear Weight = 1,400
Rear Un-sprung Weight = 300
Rear Track Width = 65.0
Rear Un-sprung CG = 13.5 (this is the height of the axle)

Since we subtract the un-sprung weight from the total front and rear weights to get sprung weights:

Front Sprung Weight = 1,400 – 150 = 1,250
Rear Sprung Weight = 1,400 – 300 = 1,100

We now have to calculate the weight transfer for each end for the sprung weight, and at the rear for the un-sprung weight of the solid axle assembly. The front, or for any AA-arm suspension, un-sprung weight does not transfer. It stays right where it is. Using the formula:

For the Front Sprung Wt /

$$WT = \frac{1250 \times 15.0 \times 1.5}{66.0} = \underline{426.13 \text{ lbs.}}$$

Remember this number.

For the Rear Sprung Wt /

$$WT = \frac{1100 \times 15.0 \times 1.5}{65.0} = 380.76 \text{ lbs.}$$

For the Solid Axle /

$$WT = \frac{300 \times 13.5 \times 1.5}{65.0} = 93.46 \text{ lbs.}$$

The total rear WT = 380.76 + 93.46 = <u>474.22 lbs.</u> Remember this number.

Adding the front and rear weight transfers, we get a total weight transfer for the race car of 900.35 pounds. Let's just round that off to 900 pounds. In our example, the left side weighs 1512. If we transfer 900 lbs. at the front, there will be 612 remaining on the left side. The right side weighs 1,288 lbs., so if we add the 900 lbs. of transferred weight to that, we get 2188 lbs. ending up on the right side tires at mid-turn.

Ideal Weights mean that equal weights will end up on the two right side tires and equal weight will end up on the two left side tires. By that rational, we divide the right and left side dynamic weights by two (2) to end up with:

Left Side Dynamic Corner Weights = 306 (612 / 2)
Right Side Dynamic Corner Weights = 1,094 (2188 / 2)

Those weights are what the corners will have on them if we start out with the correct corner weights and weight distribution. Calculating that is easy. If we know the final weights we want, and we do, then we just subtract the amount of weight transfer at each end to know the static corner weights we need to start out with. Let's do it.

RF = 1,094 – 426 = 668
LF = 1,400 – 668 = 732 (this is the remainder once we subtract the RF from the total front weight)
RR = 1,094 – 474 = 620
LR = 1,400 – 620 = 780 (this is the remainder once we subtract the RR from the total rear weight)

Now we have the corner weights we need to start out with at static ride height in order to have the perfect loading on the four tires at mid-turn after the weight has transferred. If we add the RF and LR weights, we get 1,448 lbs. If we divide that by the total weight of 2,800 lbs. we get 51.7% cross weight. If we dial in 51.7% cross weight reading on our scales, we will have the perfect load distribution on our tires starting out to end up with the perfect load distribution at mid-turn, ONLY IF, we have a balanced setup with the roll angles being matched up.

Coincidentally, the program we used to calculate the roll angles in the last Lesson does this calculation and gives the user the correct cross weight that will produce the ideal dynamic loading. But now you can do this yourself. All you have to do is enter your race cars numbers and do the calculations.

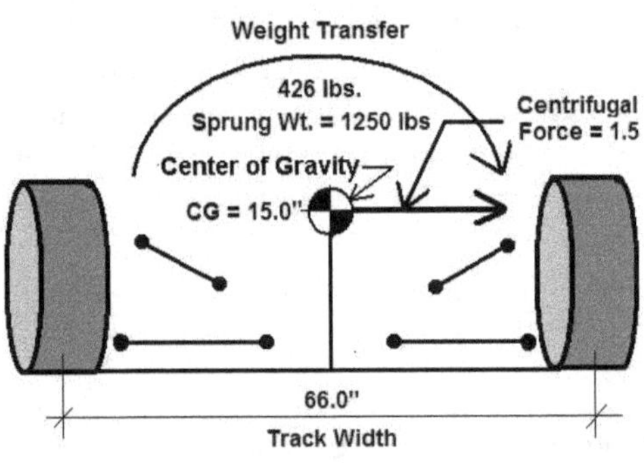

We can calculate the weight transfer at the front, or any, AA-arm suspension. We need to know the Sprung Weight, the Center of Gravity height, the Track Width, and the lateral G-force. Then we can calculate, using the formula shown, the weight transfer, and then the correct static tire loading that will result in the ideal tire loading at mid-turn.

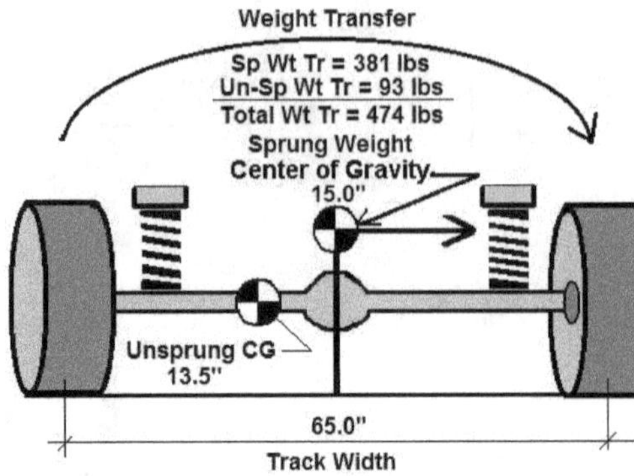

Calculating the weight transfer at the rear of a stock car, or any solid axle suspension, is different than what we do for a AA-arm suspension. We need to do two weight transfer calculations. One is for the sprung weight transfer and one is for the un-sprung weight transfer of the solid axle assembly. These two are then added together to come up with a total weight transfer at this end of the car.

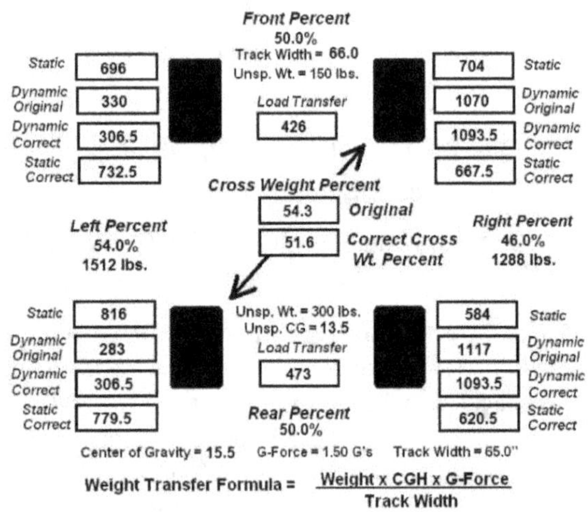

Here are all of the numbers for our Sample weight transfer calculation. We have demonstrated that we can find the correct static weight distribution that will yield the correct dynamic weight distribution after the weight has transferred. For this to work, the setup must be balanced.

How Do Roll Angles Affect Weight Transfer? – The calculations we made to determine the static tire loading that would result in the ideal dynamic loading of the four tires through the turns is only possible if the roll angles are equal. If the roll angles were different, then the weights would end up a different distribution and what that would be depends on how much the angles were different. It would be impossible to calculate what static weights would be needed in order to end up with the ideal dynamic weights if the setup were not balanced.

Plus, an unbalanced race car with un-equal roll angles is very unpredictable. If the banking or G-forces change slightly, the weight distribution changes too because the roll angles will change. The dynamic load distribution would be very unpredictable and so would the handling of the car.

Longitudinal Weight Transfer – Longitudinal weight transfer moves weight from the rear of the car to the front when we are braking. It then moves weight from the front to the rear when we are accelerating. These actions will affect the loading and the grip of each end of the car.

The front will gain a lot of grip from braking and this helps the car turn into the corner. As soon as the brakes are released, the loads equalize again and we are ready for mid-turn lateral weight transfer.

The rear gains grip from accelerating in much the same way it gains front grip from braking. The G-forces are much higher as a result of braking than what is generated from acceleration. How do we know that? It's easy.

The car takes a full straight to accelerate to the speed where the car will begin braking. Then it is a short distance, compared to the length of the straight, to where it again returns to turn speed while braking and decelerating. The short time to go from full speed to turn speed verses the longer amount of time it takes to accelerate to full speed means the G-forces generated have to be much more for braking. Acceleration force is determined by the amount of speed gained or lost over the amount of time. Less time equals more force.

The acceleration that adds loading to the rear tires while the car is accelerating helps those tires gain the grip that is needed to counter the work load the tires are asked to do. They must resist the lateral acceleration caused by the car still turning plus the torque of the motor pushing the car faster. This added grip is used to keep the tires from slipping and losing traction.

Since we are neither braking or accelerating at mid-turn, the Longitudinal weight transfer is of little concern for setting up for the mid-turn balance. But it does affect the corner entry and corner exit.

Just to be complete, there is a portion of the entry segment and the exit segment of the turns where there is a combination of longitudinal and lateral acceleration. On entry, this loads the right front corner a greater amount and on exit, it loads the right rear corner a greater amount.

On braking into the corner, there is longitudinal weight transfer and we can see the front of this car has moved down, probably onto the bump stops or bump springs. Since the longitudinal G-force is greater when braking harder, there is more weight transfer from the rear to the front. This can upset the balance of the car more so than easier braking. Driver instructors will caution against late, hard braking on entry for that very reason.

When the car is accelerating off the corners, weight transfers from the front to the rear. This car shows this effect as we can see how high the front valance is off the ground. When the car was at mid-turn, that same valance was on, or very near, the ground. The weight transfer has taken weight off the front springs and put that weight onto the rear springs. Even if this car had shocks with more rebound to where it would not rise up in the front, the weight transfer would still occur.

So, let's calculate the weight transfer from braking and accelerating. We use the same basic formula for weight transfer that we used for the lateral weight transfer. We just adjust the numbers somewhat.

Here are the numbers we will use for Braking:

Total Vehicle CG = 15.0
Longitudinal G-force = 2.5
Total Vehicle Weight = 2,800
Front Un-sprung Weight = 150
Wheel Base = 110.0
Rear Un-sprung Weight = 300
Race Car Sprung Weight = 2350

A Note About Longitudinal Weight Transfer: In longitudinal weight transfer, the un-sprung weight of the front and rear do not transfer. They stay where they are.

Using this formula:

WT = (Sprung Weight x CG x G-force) / Wheel Base

Because race cars can brake with significant force, especially the formula cars, our G-force number will probably be higher than our lateral force number. We will be using 2.5 G's of braking G-force. Let's Calculate the weight transfer on braking.

$$WT = \frac{2350 \times 15.0 \times 2.5}{110} = 801 \text{ lbs.}$$

Because race cars accelerate at a much lower G-force than when braking, our G-force number we use for acceleration will be lower than the number for braking. We will be using 0.75 G's. Let's Calculate the weight transfer when accelerating.

Here are the numbers we will use for Accelerating:

Total Vehicle CG = 15.0
Longitudinal G-force = 0.75
Total Vehicle Weight = 2,800
Front Un-sprung Weight = 150
Wheel Base = 110.0
Rear Un-sprung Weight = 300
Race Car Sprung Weight = 2350

$$WT = \frac{2350 \times 15.0 \times 0.75}{110} = 240 \text{ lbs.}$$

In these examples, we see how much weight transfers due to longitudinal forces for both braking and accelerating. The braking weight transfer is a lot. The harder we brake, the more G-force we generate and the greater the weight transfer. That is exactly why we cannot brake super hard into the corners. If we do, the weight balance shifts to where we won't be able to control the car.

For acceleration, we can see what we have to work with for distribution on the rear tires. If we transfer 240 pounds when accelerating, then maybe we can move more of that transferred weight to the inside tire to gain more grip.

Summary For Lesson Five – This discussion and the calculations will be most useful for the next Lesson in RCT Level Three. Now that we understand how weight transfers and how we can determine the desired tire loading at mid-turn, we can also calculate the spring force we would need to produce that tire loading. On to Lesson Six.

Exam - In The Context Of This Lesson:

Weight Transfer Is Caused By What?

1) Acceleration of the race car in some direction
2) An un-balanced setup
3) Soft ride springs
4) Gravity acting on the sprung weight mass

Weight Transfer Becomes Predictable When?

1) The spring rates are the same front and rear
2) The setup is balanced
3) The roll angles are matched
4) 2 and 3

What Affects Lateral Weight Transfer?

1) The weight of the sprung mass
2) The lateral G-force
3) The height of the Center of Gravity
4) The track width
5) All of the above

A Setup Is Balanced At Mid-turn When?

1) There is equal weight on all four tires
2) There is more weight on the inside tires
3) More weight on the outside tires
4) Equal weight on the outside pair of tires

Braking Causes Weight Transfer Due To?

1) Lateral acceleration
2) Centrifugal force
3) Longitudinal acceleration
4) A higher center of gravity

Accelerating Off The Corner Does What?

1) Un-loads the front tires
2) Adds load to the rear tires
3) Provides more rear grip
4) All of the above

Race Car Technology – Level Three

Lesson Six – Weight Distribution Front To Rear

In our last Lesson, we learned about weight transfer and how to calculate it. We used, for an example, a car with 50/50 front to rear percent of weight distribution. Now let's see what happens when we change the front to rear weight.

The static weight distribution is one of the most important settings on a race car. It can determine how the other parts of the setup will work. Ultimately, we need the perfect weight distribution on the four tires at mid-turn. As the rear percent goes up, so does the cross weight for a circle track race car. We'll show you how that works.

Importance Of Front To Rear Percent – There are many teams who stand by and maintain a certain cross weight percent. Even if that percent worked with the old car and/or old setup, changes to the car might produce a change to the front to rear percent number. And chances are very good that the weight distribution being used is not correct.

If you install a lighter wheel assembly or a lighter rear end unit, you will have changed the front to rear percent and along with that, the relationship of the un-sprung weight to the sprung weight. Going from a General Motors engine to a Ford in circle track late model racing can change the front to rear percent because the Ford motor is usually heavier.

If we change the front to rear percent from 50/50 to 48/52, that being 52% rear weight, we will demonstrate how by doing the calculations, the cross weight must be increased so that we will still have equal loading on the side pairs of tires.

If we go to 52% rear weight distribution, then we have 2% less weight on the front tires and 2% more weight on the rear tires. Two percent of a car weighing 2800 pounds is 56 pounds. That doesn't seem like a lot of weight, but when we do the calculations, you'll see that the cross weight percent must go up considerably to compensate for that change.

The static weight distribution helps us achieve the proper tire loads when we negociate the turns. Every race car needs a given weight distribution that can be calculated easily.

Doing A Sample Calculation With A Higher Rear Percent - In this Lesson, we will be doing the calculations for weight transfer again using a different rear percent than we did in the last Lesson. And, we will show you approximately how the cross weight percent must go up as the rear percent goes up. Conversely, if the rear percent goes down, then the cross weight must also go down.

Since we are using the same total weight, the same un-sprung weights, same Center of Gravity height and the same side percentages, etc., our weight transfer numbers will be the same. What will change will be the front to rear weight distribution and the static loading on the tires which will produce a higher static cross weight.

Here are my calculations that are shown on a data sheet I made up.

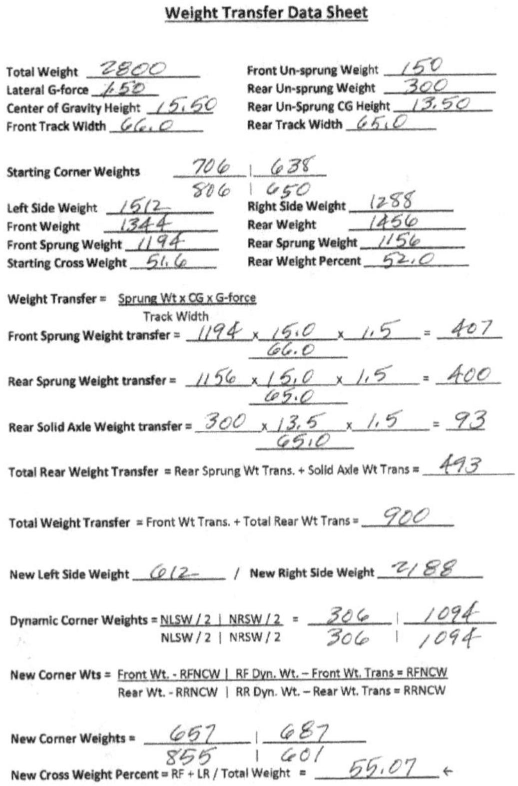

Weight Transfer Data Sheet

Total Weight 2800 | Front Un-sprung Weight 150
Lateral G-force 1.50 | Rear Un-sprung Weight 300
Center of Gravity Height 15.50 | Rear Un-Sprung CG Height 13.50
Front Track Width 66.0 | Rear Track Width 65.0

Starting Corner Weights 706 | 638 / 806 | 650

Left Side Weight 1512 | Right Side Weight 1288
Front Weight 1344 | Rear Weight 1456
Front Sprung Weight 1194 | Rear Sprung Weight 1156
Starting Cross Weight 51.6 | Rear Weight Percent 52.0

Weight Transfer = Sprung Wt x CG x G-force / Track Width

Front Sprung Weight transfer = 1194 x 15.0 / 66.0 x 1.5 = 407

Rear Sprung Weight transfer = 1156 x 15.0 / 65.0 x 1.5 = 400

Rear Solid Axle Weight transfer = 300 x 13.5 / 65.0 x 1.5 = 93

Total Rear Weight Transfer = Rear Sprung Wt Trans. + Solid Axle Wt Trans = 493

Total Weight Transfer = Front Wt Trans. + Total Rear Wt Trans = 900

New Left Side Weight 612 / New Right Side Weight 2188

Dynamic Corner Weights = NLSW / 2 | NRSW / 2 / NLSW / 2 | NRSW / 2 = 306 | 1094 / 306 | 1094

New Corner Wts = Front Wt. - RFNCW | RF Dyn. Wt. – Front Wt. Trans = RFNCW / Rear Wt. - RRNCW | RR Dyn. Wt. – Rear Wt. Trans = RRNCW

New Corner Weights = 657 | 687 / 855 | 601

New Cross Weight Percent = RF + LR / Total Weight = 55.07 ←

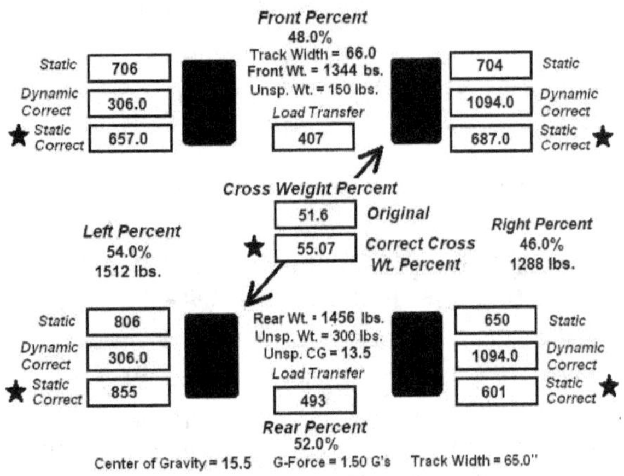

As this data sheet shows, all of our original numbers stayed the same as in Lesson Five, except for the corner weights. This car has 52% rear weight and so the static corner weights are different. What we end up with is a correct cross weight of 55.07 percent. That cross weight, or weight distribution, will give us the ideal tire loads at mid-turn once the weight transfer has taken place.

Some asphalt, and most dirt cars, run a higher rear percent of weight than 50/50. This modified runs anywhere from 50.5 to 51.0 percent. At 50.5 percent, with its numbers, it runs 53.2 percent of cross weight. It won 14 races in a row at five different race tracks in Florida including New Smyrna Speedway 2016.

On flatter tracks, there is more need for bite off the corners and using more left rear loading on the tires helps achieve this. If the car has a higher rear percent, up to a point, then the cross weight will need to be higher and that puts more left rear weight onto the tire. At mid-turn, the loading is normal, but coming off the turns when the car is running more straight ahead and turning less, the left rear load increases the bite for better acceleration.

Road Racing Applications – In road racing, the cars turn both ways. Since we are only turning one way when we are circle track racing, we can choose any static weight distribution we want. For road racing, the cars turn both ways. The cross weight must be the same for both directions or else the car won't handle the same for right and left hand turns. So, these cars must run a 50/50 diagonal weight distribution.

If the cross weight is dependent on the front to rear percent, and we know it is, then as we setup the car for road racing, we then need to change the only thing we can change, and that is the front to rear percent in order to end up with a 50/50 diagonal weight distribution.

For road racing, the diagonal weight distribution must be 50/50. These cars turn both ways and if the distribution were not that way, then the car would handle much differently between right and left hand turns. If the front to rear percent of weight distribution were not correct, then 50% cross might not work for having correct weight distribution through the turns.

For road racing, in order to end up with the ideal dynamic weight distribution, we need to adjust the front to rear percent to a number that needs 50% static cross weight, or 50 percent both ways. That would help the car to handle in a neutral balance no matter which way it is turning.

Let's assume we are taking this same sample circle track race car road racing. First off, we will need to move weight side to side to make the left and right side percent numbers the same, or 50%. We also need to reduce the rear percent. Since the cross weight percent is at 51.6 for our Lesson Five car that had a 50/50% front to rear distribution, in order to get to where it needs the 50% cross, we would have to move weight and go down on our rear percent to arrive at the lower 50/50 weight distribution that would be ideal for road racing.

When we went from 50% rear to 52% rear, we had to increase the cross weight from 51.6 up to 55.07 percent. Going down from 51.6% cross to 50% cross is about half the gain when we needed when we went from 50% to 52% rear percent. So, let's try using 49% rear to see how it goes.

Our calculation is for a left or right hand turn since our side percent numbers are the same. When we go through these same calculations, we end up with very nearly 50% static cross weight in order to have the ideal tire loading through the turns.

Weight Transfer Data Sheet

Total Weight 2800 — Front Un-sprung Weight 150
Lateral G-force 1.5 — Rear Un-sprung Weight 300
Center of Gravity Height 15.5 — Rear Un-Sprung CG Height 13.5
Front Track Width 66.0 — Rear Track Width 65.0

Starting Corner Weights 714 | 714
686 | 686

Left Side Weight 1400 — Right Side Weight 1400
Front Weight 1428 — Rear Weight 1372
Front Sprung Weight 1278 — Rear Sprung Weight 1072
Starting Cross Weight 50.0 — Rear Weight Percent 49.0

Weight Transfer = Sprung Wt x CG x G-force / Track Width

Front Sprung Weight transfer = 1278 x 15.5 / 66.0 x 1.50 = 450.

Rear Sprung Weight transfer = 1072 x 15.5 / 65.0 x 1.5 = 383

Rear Solid Axle Weight transfer = 300 x 13.5 / 65.0 x 1.5 = 93

Total Rear Weight Transfer = Rear Sprung Wt Trans. + Solid Axle Wt Trans = 476

Total Weight Transfer = Front Wt Trans. + Total Rear Wt Trans = 926

New Left Side Weight 474 / New Right Side Weight 2326

Dynamic Corner Weights = NLSW / 2 | NRSW / 2 over NLSW / 2 | NRSW / 2 = 237.0 | 1163.0 over 237.0 | 1163.0

New Corner Wts = Front Wt. - RFNCW | RF Dyn. Wt. – Front Wt. Trans = RFNCW
Rear Wt. - RRNCW | RR Dyn. Wt. – Rear Wt. Trans = RRNCW

New Corner Weights = 715 | 713
685 | 687

New Cross Weight Percent = RF + LR / Total Weight = 49.93 ←

These are the calculations for a road racing car with the weights changed to equal side percentages and different rear percent. We chose 49% rear weight and used the same total weights, same un-sprung and CG numbers, etc. We ended up with 49.93 percent of cross, or very near the desired 50/50 weight distribution we would need on a road course.

Using Different Cross Weight Numbers – Suppose for example we have a circle track race car that needs 51.6 percent cross weight in order for the dynamic tire loadings to be correct. What if we increase the cross weight percent, then what happens? Let's take the static cross weight up to 56.0%.

When we increase the static cross weight percent, we are increasing the static RF and LR weight. We are also decreasing the LF and the RR weight. Since our weight transfer at each end stays the same, then we end up with more RF dynamic weight and less LF dynamic weight.

At the rear, we end up with less RR dynamic weight and more LR dynamic weight. In this situation, the front becomes less equally loaded and the rear becomes more equally loaded. For our 51.6% cross example, we saw a difference in loading between the left and right side tires of 788 pounds at each end, or the same amount.

In our 56.0% cross weight example, we ended up with a front difference of 912 and at the rear we had a difference of only 664. That means that the rear tires are more equally loaded by 248 pounds and will therefore have more grip. Since more equally loaded tires have more grip, then this car will be tight, or understeer, going through the turns.

What Happens Next? – Typically, if the team does not consider a cross weight change as a fix to the tight setup problem, then they might do other things to make the car more neutral. They might install spring rates that would otherwise make the car loose, but now bring it to a more neutral handling balance.

What if the original springs were the correct springs that would make the roll angles the same and balance the setup? When the team changes the spring rates to make the car more neutral, the setup is no longer balanced. It might be neutral in handling, but not dynamically balanced.

This syndrome is all too common in todays racing. That is why learning about weight transfer and mid-turn tire loading is so important. Setting up a race car to be as fast as it can be and to win does not means just making it neutral in handling. There is so much more to it. And providing the proper load distribution at mid-turn is the way to high performance.

Here Is How It Works – OK, a team puts too much cross weight in the setup. They get to the track and then realize that the car is tight, or won't turn very well. They can do several things to make the car more neutral, other than the obvious which is lowering the cross weight.

First they might raise the panhard bar, by far the quickest and most effective way to make a car less tight. In extreme cases they might raise the RR spring rate, or lower the LR spring rate to increase the spring rate split and loosen the car. Some crew chiefs have been known to roll back the RR wheel by lengthening the RR trailing arm on a three link suspension. This is rear steer and can make the car very loose off the corners.

All of those changes will make the car more neutral, but will take the setup farther away from balanced. The panhard bar/spring changes only help the car in the very middle of the turns On entry and exit, the car will still be tight, but if the driver likes the feel through the middle, he might not complain much, but the car will be slower.

If the grip in the track surface changes, or the driver is forced to run a different line, then the un-balanced setup will show its ugly head and not feel the same. And, the consistency will not be there like it would with a balanced setup. The longer the race is, the more the tires will give up grip and the slower the car will be towards the end.

There is no escaping the problems that incorrect static weight distribution can cause. In the very beginning of this Race Car Technology school, it was explained that performance is about proper loading on the four tires along with having the largest contact patch among other things. The proper loading is right up there at the top of the list.

Summary – Don't be fooled by the brevity of this Lesson. This information is at the very root of making a race car fast and consistent. Along with the roll angle balance, we cannot realize the potential of dynamic balance if the weights are wrong. Now, with the information presented in this and the last Lesson, you can calculate and know exactly what weights to run in your race car.

Next up we will learn how to set our spring loading to the proper force value for static ride height, plus how to read the spring loading so that we can obtain proper tire loading. The possibilities are many and the process is exciting and fun.

Exam - In The Context Of This Lesson:

The Cross Weight That Provides Ideal Mid-turn Distribution Is Dependent On?

1) The total weight of the car

2) The left side percentage of total weight

3) The front to rear percent

4) The center of gravity height

In Our Examples, We Changed The ?

1) The total weight of the car

2) The left side percentage of total weight

3) The front to rear percent

4) The center of gravity height

If We Increase The Front Percent, We Also Must?

1) Reduce the left side percent

2) Reduce the rear percent

3) Reduce the cross weight percent

4) Increase the cross weight percent

When We Increased The Rear Percent, We Also Had To?

1) Reduce the left side percent

2) Reduce the front percent

3) Reduce the cross weight percent

4) Increase the cross weight percent

Road Racing Cars Must Have?

1) Equal side percentages

2) Equal diagonal percentages

3) A front to rear percent that makes 50/50 cross weight

4) All of the above

Race Car Technology – Level Three

Lesson Seven – Dynamic Loading – Force Analysis

Now that we have learned all about weight transfer, we can move into using that information to actually setup the race car. We will now study the art of setup concerning loads and forces. This is where we physically measure the loading on the four tires while the car is on the track.

This presentation represents advanced analysis of what is going on while the car is going through the turns. We are going from predicting and calculating the forces to measuring the actual loads and forces. This is an important step up in our understanding of how to setup a modern day race car.

The loads on the tires at mid-turn can be calculated as well as measured. We will now show you how to use shock and spring measurements to know the spring loads that translate to the wheel loads, and wheel loads to spring loads. This is known as dynamic loading of the four tires. This car uses ride springs as well as bump devices to generate the force needed to support the car at mid-turn.

In some classes, like this asphalt modified, there are no bumps, so the car must use the spring force alone to support the car. For any race car, once we know the wheel force, or tire loading in the turns, we can measure the spring force being generated so that we can use that to produce the ideal wheel load.

In the recent past, the Cup cars were not allowed to use bumps, so they ran the cars into coil bind where the springs compressed all of the way and then the tires became the suspension springs. Tire spring rates of 3,000 to 3,500 ppi were not uncommon. In today's Cup racing, the teams use bump springs surprisingly close to those same rates as the old tire spring rates.

As you travel through this Lesson, it may seem complicated, but in reality, it is more simple and easy to understand than you might think. The calculations we are going through are not necessary for you to use this information on your race car. It is presented to help you understand the importance and need for this approach and to help racers balance the setup.

The methodology presented here has been used with great success for around the past ten years. The process was commercialized about six years ago by Gale Force Suspension. Now, many top teams are using these routines to measure and adjust the setups of their race cars. This lesson will explain not only how it works, but how to properly apply it.

So, let's get started. When a tire carries a load, it must be supported by a spring device in a car with a suspension. Examples of spring devices include coil springs, torsion springs, leaf springs, bump springs, and bump stops. Those are what we see primarily used in race cars in various combinations.

Wheel Rate To Spring Rate – In previous lessons, we have talked about how we can translate the spring rate to a wheel rate. In a similar way, we can translate a wheel force, or tire load, to a spring force. The weight the scales read when we are scaling the car in the race shop must be supported by springs. And those springs must produce a force sufficient to support that scale weight. The key word here is FORCE.

The definition of Force as it applies to this Lesson is this: A spring at rest exerts no force. If we put a weighted object on top of the spring, the weight of that object compresses the spring and then the spring has force. It is holding up that weighted object. If we remove the object quickly from the spring, it would recoil suddenly and return to its original length and again have no force. The force generated is the weight of the object that caused it to compress to a length less than its original length.

So, for a spring under a load, the force is the compressed distance that the spring is verses what it was at rest times the rate of the spring in pounds per inch of displacement. A 200 ppi spring compressed one inch has a force of 200 pounds.

We used the "motion ratio squared" method to translate the spring rate to a wheel rate, and now we can then reverse that process to find the spring force that supports the wheel weight. If, for example, the scale under the right front tire reads 600 pounds, that force of 600 pounds is then divided by the motion ratio squared and that equals the spring force needed to support that 600 pounds.

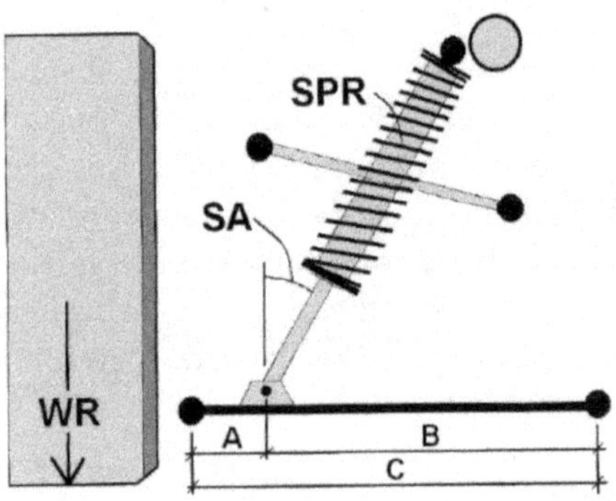

The wheel rate, as we have previously learned about in past Lessons, can be translated to a spring rate. The wheel loads can also be translated in a similar process to a spring force. We use the very same motion ratio. Here is a drawing depicting the measurements we use to determine a coil-over springs motion ratio. We can also just measure the wheel travel verses the spring travel to come up with the motion ratio. When we are translating the forces, we use the motion ratio squared.

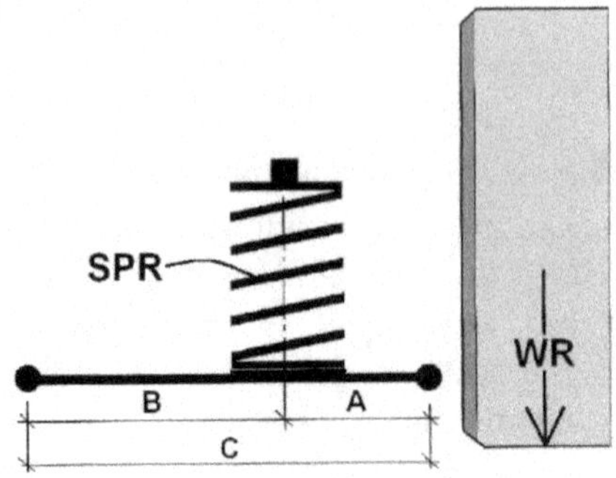

The process to find the motion ratio for a big spring car is the same as we use for a coil-over car. The spring moves at a decimal ratio to the movement of the wheel. To translate spring rates or forces, we use the motion ratio squared.

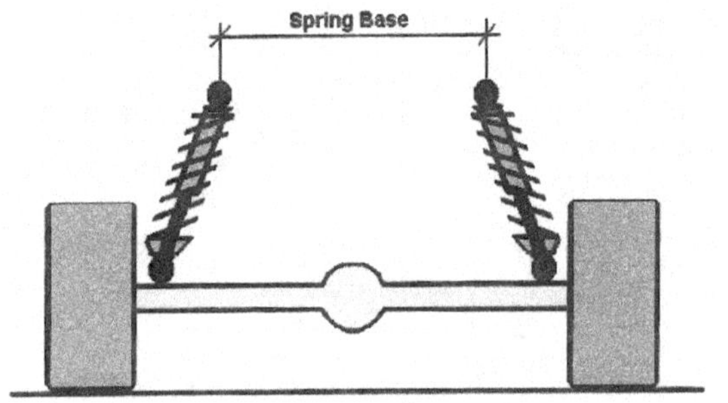

The rear has no wheel rate for the purpose of dynamic evaluation. It does have a wheel force, but that is not easily translated to a spring force. If we know the four dynamic wheel loads that will balance the setup, we only need to make sure one of the wheels, or tires, have the correct loading at mid-turn. That way, the other three loads must be correct too. If one were not correct, then the other three would not be correct either. The total loading of the four tires combined is a finite number and the distribution is dependent on the loads on the four tires.

For a typical circle track late model, the motion ratio of the RF suspension might be something like 0.8152. Remember that motion ratio is simply the distance the spring moves verses how far the wheel moves. In this case, if the wheel moved 1.0", then the spring moved 0.8152". It's as simple as that and any team member can measure the motion ratio.

If we square that number using a calculator (0.8152 x 0.8152), we get 0.6645. Our scale weight of 600 for the RF tire divided by 0.6645 equals 902.88 pounds of spring force just to hold up the right front corner of the car. And, if that spring were rated at 400 ppi (pounds per inch), then it would be compressed a distance of 902.88 / 400 = 2.257 inches from its free height when out of the car.

If we put that coil over spring and shock on a force measuring machine, and then compressed the shock to the length it was while on the car on the scales, the force should read 902.88 pounds. Compressing the spring to its installed, or measured length, gives us the force it is producing at that length. Remember this process because this is how we can very easily read our dynamic force while on the race track.

For a conventional setup like this one, the shock travel translates into a force that can be translated to a tire load at mid-turn. The only advantage the bump setups have is in lowering the car for a lower CG and better aero properties, and minimal camber change. In a test comparing conventional setups to bump setups, there was only a tenth of a second per lap difference between the two, but a tenth can be the difference between winning and losing.

Reading Shock Travel To Get Spring Force - Suppose we can read the shock/spring travel of the RF corner while the car was on the race track going through the turns. We can do that by using a couple of different methods. We can use a simple shock travel indicator, or we can use data acquisition, both of which measure shock travel. The maximum travel that we see that is due to dynamic forces happens when we are at mid-turn at maximum speed. There could be an influence on this travel from entry braking and that might affect the numbers we see, but we will address how to take care of that later on in this Lesson.

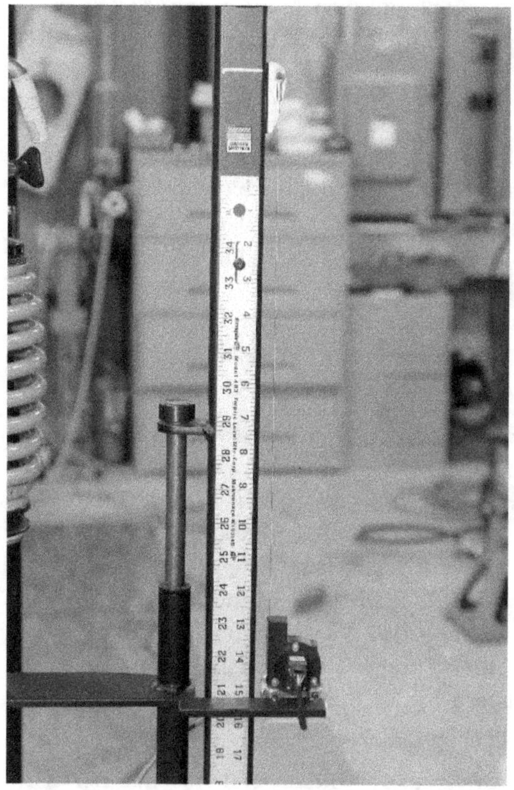

To find the force that the coil-over spring and/or bump produce at mid-turn, we only need to measure the length of the shock while on the car at mid-turn and then compress that shock and spring to the same length in a force measurement machine. On the Gale Force machine, the shock length is shown on the graduated scale in inches on the rig. Once we compress that coil-over to that same length as was recorded on the track, we will then be able to read the force in pounds.

When the car has run some laps on the track at speed, we bring it in and measure the additional shock travel from when it was on pit road at rest and we get 1.859 inches. Since our spring rate was 400 ppi and our spring travel at ride height was 2.257", then if we add those two travels together, we get a total travel of the spring of 4.116". That travel amount multiplied times the spring rate of 400 ppi equals a force of 1,646.4 pounds.

If the spring/shock were 18.5" at free height, then the length at mid-turn would be 18.5 – 4.116 = 14.384". If we put the shock/spring in the force machine and move it to the length, then we could read the force directly. And it would read 1,646.4 pounds, just like we calculated.

That is the spring force we need to hold the RF corner up at mid-turn. Now, we want to know, what is the tire loading? If we multiply that total force number by the motion ratio squared, we get (1, 646.4 x 0.6645) 1,094 pounds. The RF tire supports a load of 1,094 pounds at mid-turn.

To further explain this, if we could put a scale under the tire at mid-turn, that is what the scale would read. Going back to our weight transfer lessons, we can know the RF ideal loading by doing calculations. In Lesson Five, our calculations yielded an ideal RF tire loading of 1,094 pounds for the example car.

Since we match our intended RF tire loading as evidenced by the spring force we are reading, then we have proven that this car is setup to be balanced and that means that the roll angles must be matched. We have now demonstrated how we can measure the balance of the setup by knowing the spring force. We simply read the shock travels and translate that force to a wheel load at mid-turn and then compare that load to what we know is correct for this car.

The tire loads are interdependent. That means, if one tire load not ideal, then all of the tire loads will be wrong. That also means, we only have to measure one tire load to know the entire picture, or layout, of the other tire loads. If the RF carries 100 pounds more load than necessary for a balanced setup, then the LR tire will also carry 100 pounds more, and the LF and RR tires will each carry 100 pounds less than necessary.

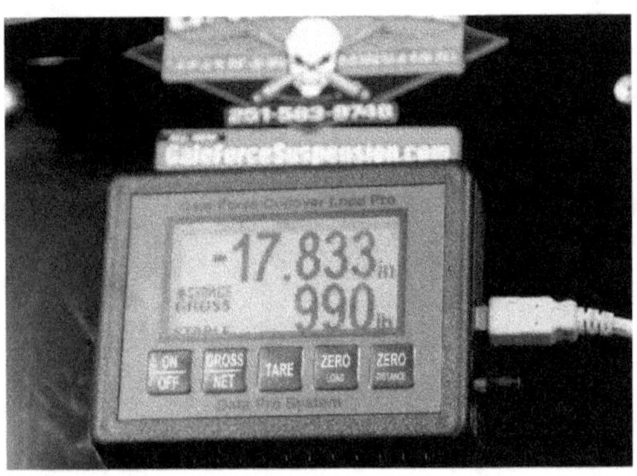

This readout from the Gale Force machine shows us that at a length of 17.833 inches, the force of the spring, and maybe the bump too, measure 990 pounds. If this was our example car, then the wheel load would be 657.86 pounds, or our motion ration squared times the spring load (0.6645 x 990 = 657.86 lbs.).

Modern Day Soft Spring Setups With Bumps – The example car we looked at above was what we call a conventional setup. It uses only ride springs to support the car. Many modern day race cars use what we call bumps in addition to the ride spring. Bumps can be a coil spring with a consistent and linear spring rate, or some form of bump stop, or compressible material with a variable spring rate.

When running on bumps, we can also see when the shock body contacts the bumps and then read how the force increases as the travel further compresses the bump stop. For bump stops, the rate of increase in force, or load, is not linear. It rate of increase rises as the bump compresses. On a graph it produces a curved line. With a bump spring, the line showing the increase in load is a straight line.

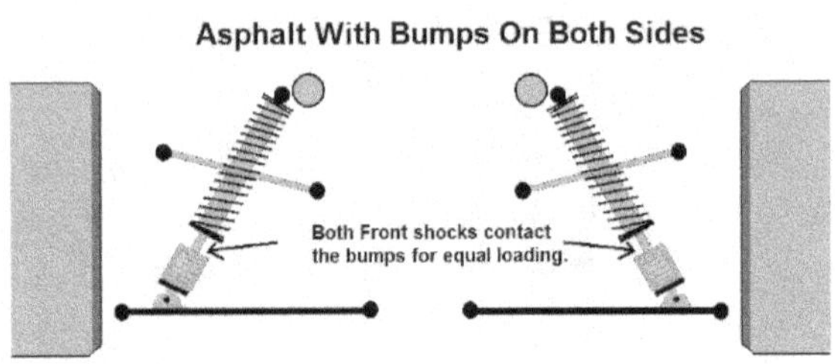

The bumps shown here in red support additional loads that if supported by the softer ride springs would cause the chassis to travel more than it has room to travel. Without the bumps, the car would be hitting the race track. We do the calculations to show you how that works.

This shows a typical bump stop. The initial force needed to compress this particular bump is not a lot. As the bumps is compressed, it resists the compression more and more and therefore the spring rate increases. At some point, it becomes "coil-bound" so to speak and cannot compress any more.

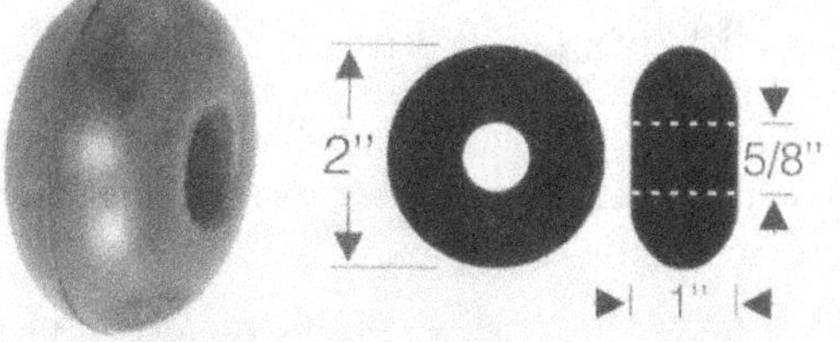

This much stiffer bump device is used on smoother tracks to reduce overall shock travel. These can be stacked in any numbers to change the progression of force and regulate the travel.

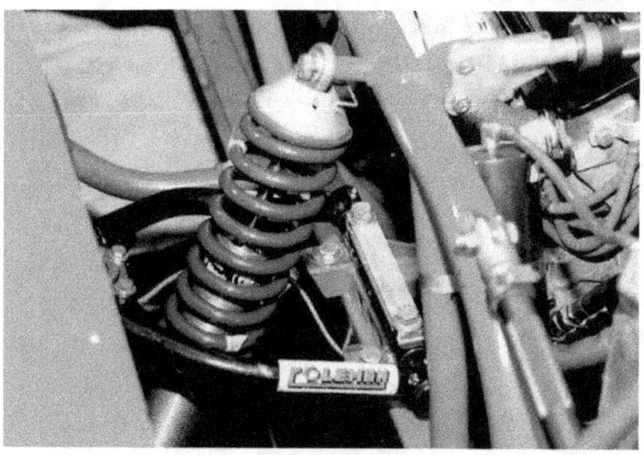

When soft springs are used as ride springs, they must be pre-loaded in order to provide the necessary force to support the car at static ride height. Once on the track, we then use bumps to stop the movement of the chassis to prevent the chassis from bottoming out on the race track.

The bump devices actual spring rate, when activated, are added to the ride spring rate. The bumps are placed on the shock shaft and when the shock moves in compression, the shock body contacts the bump device and the spring rate of the suspension goes from that of the ride spring to a rate equal to the bump spring rate plus the ride spring rate.

If for example, we are using a ride spring with a rate of 150 ppi and a bump spring rate of 1500 ppi, then the spring rate of that suspension when the shock has traveled onto the bump is 1650 ppi. Once the shock comes off the bump, the rate then goes back to 150 ppi.

A note about the bump stops. The nature of bump stops is this. The spring rate of a bump stop increases as it is compressed, and if compressed enough, will go up to a very high rate or even solid. The same is true of the bump spring when it goes into coil bind, or at the maximum travel. But the bump spring rate stays consistent all of the way up until it goes into coil bind. We call this linear progression.

Just as with the conventional setup, we can record the shock travel and then duplicate that in a force machine to read the spring force at mid-turn. In a bump setup, the force is a combination of the ride spring force and the bump force.

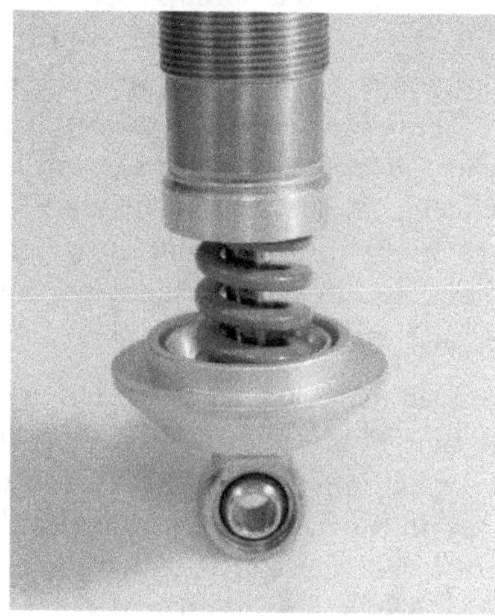

The bump spring is used to control the amount of shock travel as does the bump stop. It will compress and provide the force needed to support the chassis at mid-turn. Its force is added to the ride spring force that comes from both the shock travel and the pre-load forces.

Reading Spring Force Conventional Verses Bump – We might think there is a difference in the force readings between the conventional setups and the bumps setups just because the bump combined spring rate is so high. That is not true. The tire loads that are produced by a balanced setup are the same whether the car is running ride springs only, or ride springs plus bump springs.

The only advantages that the bumps setups have over the conventional, ride spring only, setups is that the bump setup produces a lower attitude that creates a lower center of gravity, possibly more aero downforce, and minimal camber change due to limited vertical motion. All of that was covered in earlier Lessons in Race Car Technology.

As to aero downforce, if the front of the car produced an additional 200 pounds of downforce due to the lower attitude of bump setups, then that 200 pounds would be equally distributed between the two front tires and our RF tire load would go up by 100 pounds. This gets into the finer points of reading tire loads to help us to produce a balanced setup, but for now, let's stick to the basics.

Bump Setups Using Soft Ride Springs – If we do the calculations for spring travel to produce the force necessary to carry the mid-turn loads, then with the softer ride springs used on bump setups the travel would exceed the available suspension travel. So, the teams running those setups must pre-load the softer ride spring. Pre-load means to compress the spring while it is on the shock with the shock fully extended. This puts force on the spring without any movement of the shock. Its rate remains the same, but a lot of the force it will need is already there.

If we pre-load that 150 ppi ride spring on the shock by four inches, it would have a pre-load force of 4 x 150 = 600 pounds. If we subtract that 600 pounds of pre-load from the 902.88 pounds of force we need to hold up the car at normal ride height, we end up with a shock/spring travel of 2.0192" at normal ride height. This is what teams actually do when working with relatively soft front springs.

So, we have a mid-turn spring force, from previous calculations, of 1646.4 and a force to hold the car up at ride height of 902.88. The additional force needed to support the car at mid-turn is 743.52, or the difference between the static force and the dynamic force.

If our ride spring is 150 ppi, then 743.52 / 150 = 4.96". Taking that travel out to wheel travel means that the wheel would move 6.08 inches. There isn't that much travel left in the suspension if we set the ride height at 4.0". The car would hit the track big time. We need another spring to work with that force so that we don't make contact with the track. We need a bump spring or bump stop. In this example, we'll use a bump spring.

So, the wheel would move 6.08" with just the ride spring, and we have maybe 3.5" of travel available to stay off the track by 1/2". Doing the math, 6.08 – 3.5 = 2.58". 2.58 x 150 = 387 pounds of force we need to add to the suspension to hold it up off the track.

If we install a 1,500 ppi bump spring onto our shock, then it would travel 387 / 1,500 = 0.258 inches, or about a quarter inch, to create the force needed to keep the car the ½ inch off the race track. But if it travels ¼" under that load, we must subtract ¼" from ½". Now our car is ¼" off the race track. And that is what we would normally do. We would install a bump device that would create a much stiffer spring rate to hold the car off the race track.

That bumps spring, or bump stop, doesn't have to be a 1,500 ppi spring rate. It just has to develop a force necessary to hold the car off the track. If it were a 3,000 ppi spring, then the travel would be half the 0.258 inches, or 0.129, or about an eighth of an inch.

The Influence Of Braking On Shock Travel – We would be leaving something very important out of this discussion if we didn't cover what could influence the RF shock travel, or the shock travel at any corner of the car. For dynamic load analysis, we need to make sure that the shock travel readings we are seeing are truly created from the dynamic forces the car experiences at mid-turn.

For example, if we have very little anti-dive in the RF suspension, and/or pro-dive in the LF suspension, then on braking into the corner at most short tracks, the loads the RF and LF suspensions will show might be wrong.

The loading in the RF and LF under braking is usually more than we would see at mid-turn. So, the reading we take from shock travel might be wrong. If the shock travel was 0.25" more from braking than what it is at mid-turn, then with a combined spring rate of 1650 ppi at the RF, we would be recording a load that is 412.5 pounds more than is actually what the mid-turn dynamics produces. In other words, the shock travel method is useless if we allow the shock/spring to move that far under braking.

There is a solution. Back in the day, teams used Anti-dive to help reduce the nose dive while under braking to help reduce the camber change that motion produced. In recent years, teams have reduced the use of Anti-dive and even introduced Pro-dive into the LF suspension. Both of those changes have allowed erroneous readings to occur.

It is now time to reintroduce Anti-dive on both front suspension systems. We won't be reducing dive because we'll already be on the bumps, but we won't be over-loading the bumps from heavy braking and then the readings will truly represent the mid-turn shock/spring travel. Then we will be able to rely on our shock/spring travel amounts knowing they represent only mid-turn forces.

You have just discovered the secret of using bumps to setup a race car. In conjunction with the soft ride springs and stiff bumps, we use shocks that will work with the bump spring rates and not the ride spring rates. They will be much stiffer on rebound than the ride springs and that keeps the front suspension riding on the bumps much longer. We'll get into the design of shocks for various types of setup in a later Lesson.

Sway Bar Loading And Force Influence – The sway bar is a spring and adds a force to the ride and bump spring forces to assist in holding the car up at mid-turn. At ride height, with no pre-load, the sway bar produces no force, but it still has a spring rate. When the car dives and rolls, the sway bar may be forced to twist, and at that point, it produces a force that adds to the ride spring and bump spring forces.

In the previous examples of forces in a conventional and bump type of setup, we purposely left out the influence of the sway bar so that the explanations wouldn't be overly complicated. And the addition of the sway bar force is not really complicated either. It just needs to be included in the overall consideration for force needed to hold the car up at ride height.

As you would expect, the sway bar force is directly related to the spring rate, or stiffness, of the sway bar system, and the amount the chassis is twisting the bar, just like what happens with spring travel. The more travel, or twist, the more force that is produced. Many designs of late model circle track cars will cause the sway bar to twist when the chassis dives without any roll at all.

Gale Force Suspension makes a sway bar force measuring tool that will measure the sway bar force, if you know the suspension travel at mid-turn. With this tool, you can re-create that movement and then read the sway bar force at the shock/spring position. This provides a force number we can include in the overall force that is needed to support the car at mid-turn. Once we know the sway bar force, we then subtract that force from the force we need to hold the car up at mid-turn. The remainder force is what the ride and bump spring must produce.

Here we see the Gale Force sway bar force measurement tool. When we replicate the suspension travel to what it is on the race track at mid-turn, we can isolate the sway bar force that acts like a spring. This force is then included with the ride spring and bump forces to make up the required RF corner force needed to support and balance the car at mid-turn. The tool is mounted where the shock/spring is mounted, so the sway bar force is measured at the same place as the coil-over spring is measured at. Then it can be directly added to the spring/bump force.

Since the sway bar can easily produce 150 to 300 pounds or more of force, it is a serious consideration when we are trying to develop the total suspension force to support the car at mid-turn. If we don't include the sway bar force into our suspension force calculations, then we would end up with much more spring/bump force than we need.

Using The Sway Bar As A Spring – In modern asphalt circle track late model racing, the use of bump devices includes running a bump on both sides in the front suspension, or only running a single bump, usually on the left side. When running the bumps on both sides, the sway bar acts mostly in roll.

For setups where the bump is only installed on the left side, the size of the sway bar must be increased dramatically in order to provide a sufficient Right Front spring rate to hold up the car going through mid-turn. Whereas with a two bump system that can more efficiently run a much smaller sway bar, with the single bump system, we need much more spring rate from the sway bar and need to install a much larger sway bar. Sizes ranging from 1.50 inch diameter up to 2.00 inch diameter sway bars is not uncommon. We will further discuss the use of a sway bar as a spring in Lesson 14.

Summary – We hope that this has been a big learning experience for those of you who have not been aware of how bumps actually assist us in setting up a race car. We have calculated the ideal loads we want to see on our tires. We translate those loads to a spring load we need to produce the wheel loads at mid-turn. And then we decide to run bumps with soft ride springs, so we then need to preload the ride spring to produce nearly the force needed to keep the car off the track at mid-turn. We then add in bump stops, or bump springs to provide the added force to finish the job.

This whole tire loading/spring force theme represents just one of the three primary elements that serve to create the overall grip in our race car. Next, we will learn all about element number two, the Angle Of Attack.

Exam - In The Context Of This Lesson:

A Spring Has A Force When?

1) Preloaded on a shock

2) It is a loaded sway bar

3) It is compressed

4) All of the above

A Tire Has A Force When?

1) The chassis is at ride height

) The race car is at mid-turn

3) The suspension is on the bumps

4) All of the above

The Ideal Load On The RF Tire Is?

1) When that load equals the RR tire loading

2) When that load equals the LF tire loading

3) When that load equals the LR tire loading

4) When that load equals the spring and bump forces

The RF Tire Load Can Be Translated To?

1) The RR tire load

2) The RF suspension spring load

3) The LF suspension spring load

4) The LR tire load

To Calculate The Spring Force From The Tire Load, We?

1) Use the motion ratio

2) Use the motion ratio squared

3) Use the spring/shock travel

To Calculate The Tire Load From The Spring Force, We?

1) Use the motion ratio squared

2) Use the spring/shock travel

3) Add the spring force, bump force and sway bar force

4) All of the above

Bumps Are Primarily Used To?

1) Provide more force

2) Increase the spring rate of the suspension

3) Keep the suspension from bottoming out

4) All of the above

The Following Can Interfere With The Force Readings

1) Excess camber

2) Heavy braking

3) Acceleration off the corners

4) Running softer springs

Race Car Technology – Level Three

Lesson Eight – Angle Of Attack & How Tires Generate Grip

The following is a basic repeat of the second Lesson in RCT Level One. Some of you might have skipped Levels One and Two, so we needed to make sure you understand this very important concept. It was presented early on in the RCT courses because this concept first of all has never been fully explained before and you are now seeing it again because it fits right in and is a major part of understanding the three main ingredients of how tires develop grip.

One of the first things we talked about in Level Three was loading on the tires. We told you how weight transfers and how it is distributed on the tires going through the turns. We explained how equally loaded tires offer the most grip for a pair of tires on the same "axle". We went on to tell you how a balanced setup will offer the best weight distribution possible and why.

But, here is the clincher. Loading of the tires is not what makes them develop grip to resist lateral forces when we are going through the turns. A car going straight can be easily pushed to the side with relatively little force and it goes there because it has no resistance to that force even though it may have a lot of loading.

An example is the PIT maneuver used by law enforcement to stop a car that is involved in a chase. The police officer goes to the rear quarter panel and moves over against the offending car and pushes it until it spins out. The rear of the car seems to go sideways fairly easily. Its tires have no AOA initially, and so they cannot resist the lateral force that is the police car.

What the tire needs to resist lateral forces is Angle of Attack. Some might call it slip angle or some other terminology, but we will call it by a name that better explains the concept, and that is angle of attack, or AOA for short.

The concept of AOA as we explain it has never been fully examined as far as we know. It is a simple, but effective, description of why a car can drive through a turn at high speeds. It is one of those items of new knowledge that we promised the veteran racers they would get when considering enrolling in the RCT Courses.

One of the key ways, and actually a necessity in, a tire developing grip is from AOA. A tire running parallel to the path of the arc of the curve cannot generate the Grip it needs to maintain its direction along the arc. It would "slip" off the arc, hence the name "slip angle". It is the angle the tire needs to keep from slipping off the curve.

How The Steering Creates AOA - Every race car in the world is a front-end steering car. That is, the driver turns the car using a steering mechanism that is used to change the direction and angle of the front wheels. The rear wheels seem to follow along. But why do the rear wheels stay on course? The are not being steered.

The driver creates an angle of attack for the front tires so they can develop the side force necessary to change direction and also to stay on the track. AOA is needed to counter the lateral force we know as the centrifugal force that is created once the change in direction has taken place. Without this Angle of Attack, the car would not turn no matter how much the tire is loaded, or no matter what the size of the tire contact patch is.

Rear Tire AOA Study - OK, so we have learned how the front end turns and develops AOA, but what about the rear tires? We cannot steer them, setting aside the effects of mechanical rear steer that will be covered later on. How do the rear tires develop the angle of attack needed to stay on the arc of the turn? The answer is: they already have their angle of attack, and here's why.

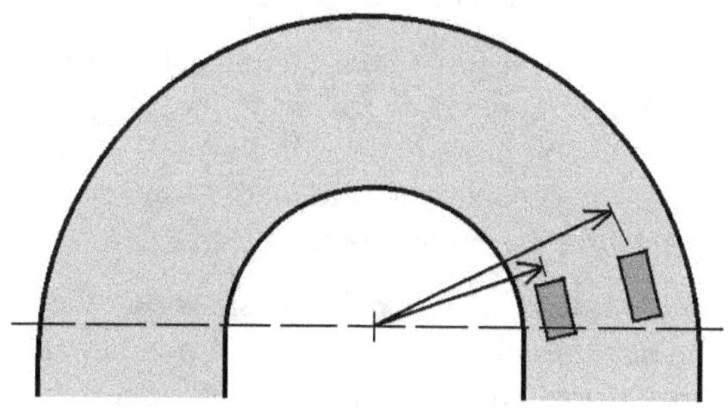

In this illustration, we show how the inside and outside tires have different radii and must follow different arcs when driving through the turn. A line perpendicular to the line extending to the radius point is called a Tangent. To develop the Grip needed to counteract the Centrifugal force, all of the tires must be pointed to the left of the Tangent line.

In the last figure, you will see how each tire follows its own arc. The arc radius of the inside (inside of the turn) tires is less than that of the outside (outside of the turn) tires. In the next figure, each tire is placed over an exaggeration of the arc lines with a radius equal to the distance from each set of tires to the common radius point of the arcs.

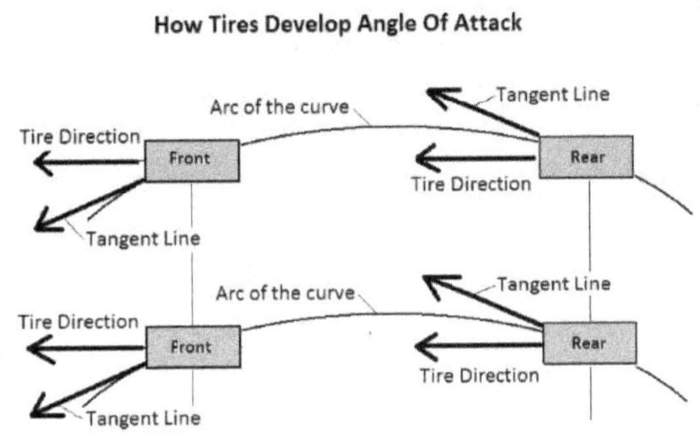

This sketch shows how when the four tires are pointed straight ahead, they form angles to the tangent of the curve that are much different. The front wheels need to be turned left to reach the direction of the tangent line, and more to turn the car. The rear wheels are already pointed left of the tangent line and therefore already have some degree of AOA.

The front and rear tires are pointed in different directions in relation to the Tangent of the arc they are following. In the front, the tires are pointed away from the Tangent and in the wrong direction needed to turn the car.

We must steer those tires left in this left turn in order to not only track parallel to the Tangent line, but a bit more to develop an Angle of Attack needed to generate the amount of Grip necessary to keep the car on the course.

At the rear, we have a different situation. With the rear tires positioned straight ahead, they are already pointed left of the Tangent line and have some Angle of Attack.

If the AOA angle in the rear is less than what is needed, the rear of the car will slip out until the tire angle is enough to provide the sufficient AOA needed to stay on the course. If the angle is more than is needed, then the driver will need to turn the steering wheel more to compensate and to make the car turn.

The front tires react the opposite if they slip. The angle of attack decreases and the steering must be increased in order for the car to stay on the course. We will tell you more about steering angles and front AOA later on. There are limits to how much angle the front tires can be steered to.

When you come to understand this concept of how the car is able to make it around the turns, it will answer a lot of questions people might ask about vehicle dynamics. I have often wondered, given that we have a lot of setup tools to make the front ends turn, why and how does the rear turn with the front? Now we know.

Here we can actually see the lines of the arc of the turn. The front wheels are steered to create the Angle of Attack needed to develop the force to counter the G-lateral forces. The rear tires are already pointed to the inside of the turn and have their own Angle of Attack. This is why both the front and rear tires develop the grip needed to turn the corner.

Angle Of Attack, How Much Is Too Much?
There is a limit to how much AOA a tire will tolerate. We know from tire testing that a tire will gain lateral loading when the AOA is increased, up to a point. At some predetermined angle, and it is different for different designs of tires, the lateral force the tires provide goes away. They will stop generating lateral force and then slip.

The design goal for creating maximum lateral acceleration, or grip, is to find the angles that are just less than the angle where the tire loses lateral grip. If we can make the front and rear tires work at that maximum AOA, then the car will be the fastest it can be through the turns.

And, know this, every race car, no matter the design or classification, is subject to the concept explained above. Farther along in this Course, we will get into various design effects, goals and geometry for the rear suspension where we will explain how you can change and fine tune the angle of attack for the rear tires. We can introduce rear steer to gain AOA in the rear in measured amounts.

Now we have covered two of the three elements of how tires develop grip. We have covered loading and now Angle of Attack. Next, we'll discuss tire contact patch optimization. It is the third and final element of grip. Like the two elements we have already covered, without the contact patch, all of the loading and all of the AOA in the world will not maximize the cars mid-turn performance. We must have all three of these working to maximum capacity.

Exam - In The Context Of This Lesson:

Which Element of Grip Is Essential Above All Others?

1) Tire Loading
2) Tire Compound
3) Angle of Attack
4) Contact Patch Area

How Do We Make The Front Tires Develop Angle of Attack?

1) Add More Camber
2) Add Tire Pressure
3) Drive Slower
4) Turn the Steering Wheel

How Do We Make The Rear Tires Develop Angle of Attack?

1) Add More Camber
2) Add Tire Pressure
3) Drive Slower
4) We don't need to, they already have Angle of Attack

How Do We Adjust The Lateral Grip For The Front Wheels?

1) Lengthen the Tie Rods
2) Change The Speed
3) Change The Tire Cambers
4) Turn the Steering Wheel Either Way

How Do We Adjust The Angle of Attack For The Rear Wheels?

1) Change The Tire Stagger
2) Change The Tire Pressure
3) Adjust The Speed
4) Through rear steer geometry

Race Car Technology – Level Three

Lesson Nine – Advanced AA-arm Geometry & Contact Patch Optimization

The third and final of the elements that make up overall grip in a race car is contact patch optimization. If we have designed the setup so that the forces on the tires are optimum, and the angle of attack of the tires is sufficient to create a resistance to the lateral forces going through the turns, then we now need a large contact patch.

It is a fact that a tire with a fixed loading and angle of attack will generate more side force to resist the lateral forces during cornering when the contact patch is as large as possible. And, the optimization of the contact patch size comes from tire camber and tire pressure.

Our goal is to find the correct camber, and camber change that will go along with the correct tire pressures to create a larger contact patch during cornering. And all of that is dependent on the tire we are using.

In circle track racing for instance, it is generally known and accepted that Hoosier tires are constructed with a softer sidewall than a Goodyear tire. A team cannot run the same cambers for one tire as the other because of how the sidewall reacts to the camber. Other brands of tires have their own characteristics for sidewall stiffness.

The camber design for race tires has evolved in order to maximize the contact patch area. We will show you how to create ideal camber change and contact patch size for a modern day race car.

Camber Change – As the race car goes into and through the turns, the chassis will dive (or raise up) and roll. These two motions combined are what can cause camber change. If the chassis did not change its height along the centerline, then the roll of the chassis would change the camber by roughly the same angle as the roll angle.

As the chassis dives, and most race car chassis run lower through the turns, there is a "dive" camber change for the outside wheel that is opposite of the roll camber change. In many race cars, we can design the control arm angles so that there is very little, or no, camber change for the outside tire through dive and roll.

The inside tire and wheel will lose camber in most race cars, and with road racing cars, the static camber is opposite of what is needed for the direction we are turning. So, it is beneficial to run as little camber as possible in road racing cars so that the contact patch is somewhat useful for the inside tire.

Tires Don't Like Camber Change - In most cases, race tires do not like camber change. When entering a turn, the tire takes a set, or rolls over and begins to develop Grip. If the cambers do not remain stable, the tire will take more time to adjust, and frankly, we don't have a lot of time for that.

The result is loss of Grip and loss of speed as the tires struggles to find Grip. So, in modern race car design, we need to plan out our AA-arm geometry so that there will be minimum camber change. How do we do that? There are several ways to accomplish the goal of minimum camber change for both the inside and outside tires.

First, we can arrange the control arms with angles so that with the dive and roll we expect to result from going into and through the turns, will result in minimum camber change. This has been done with great success in what we call conventional setups with both stock cars and some formula cars.

Another way to achieve minimum camber change is to reduce the amount of dive and roll. The setups in Formula One cars are very stiff and the chassis does not move vertically or roll very much. This reduces the amount of camber change. The reverse angle of the upper control arms in a F1 car (meaning that the inside chassis mount is higher than the ball joint) may be beneficial and cause a small amount of beneficial camber change.

The third way we can minimize camber change is through the use of what we call bump stops. In modern race car design, some chassis are designed to travel a predetermined amount and then the shocks contact a bump material and basically stops in both dive and roll motions.

These bumps can be made of hard rubber, polypropylene or actual steel springs. In each case, when the shock body contacts the bump device, the vertical chassis motion is reduced to very small movements and the tire cambers change very little while the suspension is on the bumps. The reduction in camber change is one of the primary reasons why the bump setups are beneficial and help the car to turn.

Bump Setup Camber Change – With the advent of the bump setups in circle track and other forms of racing, camber change of any significant amount while on the bumps is a thing of the past. Even so, there is still a lot of camber change going on from when the car is on the grid to when it is going through the turns. That camber change can cause other setup changes we might not easily see.

The tires provide better grip when the cambers are not changing and will stay very close to ideal due to very little vertical chassis movement. But what about the transition onto the bumps? What happens during that process? Few teams consider what happens then. The associated movement can cause other problems.

As we have discussed above for a more conventional car, when we change the cambers, we are also changing the corner heights and along with that, the distribution of loading on the four tires. If the chassis travels some three or three and a half inches down onto the bumps, where does the wedge, or cross weight, go to then? It has to change from where it was statically based on the fact that the tires change their height relative to the chassis.

Weight change at one corner of the car due to the jacking effect of camber change (and that is essentially what it is, the tire jacks up, or down depending on the change) also changes the loading on the four tires.

We already know that the extreme travel associated with bump setups, from ride height to on-track height will also most times pre-load the sway bar. This adds cross weight to the setup. This weight distribution change must be taken into account with the initial setup when we set the cross weight and sway bar preload.

Some Examples Of Camber Change – It is time to run some experiments setting the car up with various cambers and control arm angles. In our first example, we will setup the front end geometry with low upper control arm angles, meaning they are close to level.

This car is a road racing car with the arm angles being equal on both sides for the upper and lower control arms. This car will lose negative camber at the outside tire (car is turning left in this example) and gain negative camber with the inside tire.

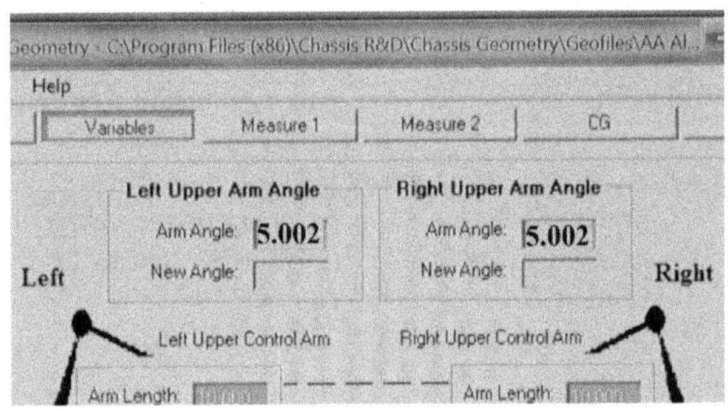

This road racing car has equal upper control arm angles and they are only 5.0 degrees off of horizontal. We'll see how the cambers change when the car turns left and dives and rolls.

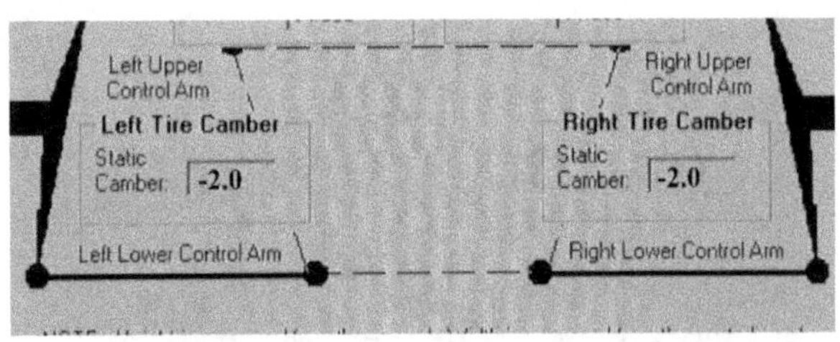

The cambers for the front wheels are equal too because this car must turn both ways. We have a negative (-) 2.0 degrees for the static cambers on both sides.

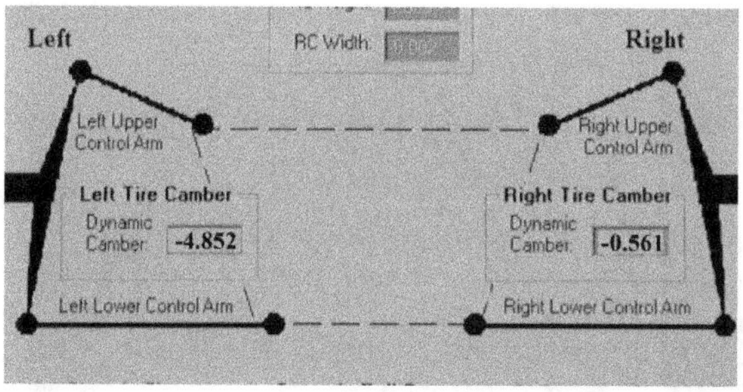

When we calculate the dynamic cambers by entering 1.0 inches of dive and 3.0 degrees of roll, we end up with a left tire camber of -4.85 degrees and a right tire camber of -0.56 degrees. We lost about 1.5 degrees of camber at the right front, or outside tire, and gained 2.8 degrees of negative camber at the inside, or left front tire.

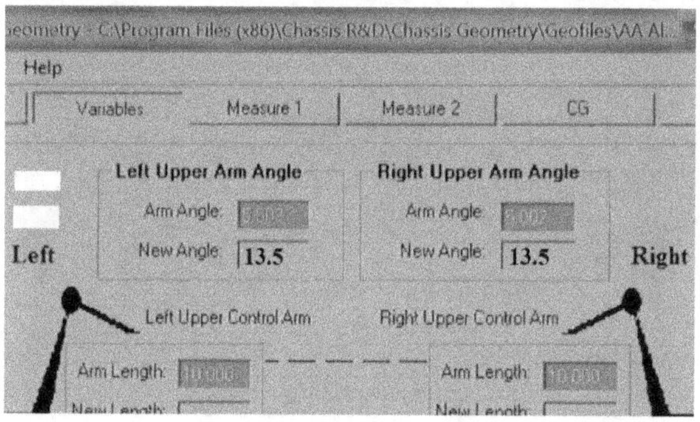

Here we added angle to the upper control arms and now have 13.5 degrees for both upper control arm angles. Look at the "New Angle" box. This should improve, or lessen the outside tire camber change.

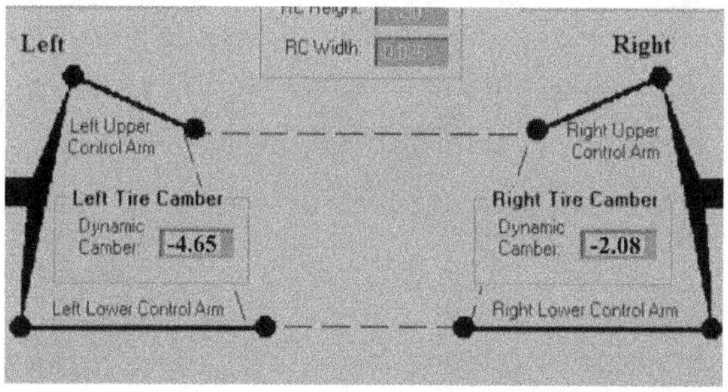

The new dynamic cambers show a -2.08 degrees for the outside tire, or basically what we started with, and a -4.65 degrees of inside tire camber. Our outside tire camber did not change at all and the inside tire camber stayed basically the same as before. The zero change in the outside tire on this car will improve how it turns.

Now let's look at a typical circle track race car running a conventional setup. This will simulate a Tour Modified because this is what we see a lot with those cars. It could still be a late model car, and the results speak to the importance of controlling the camber change.

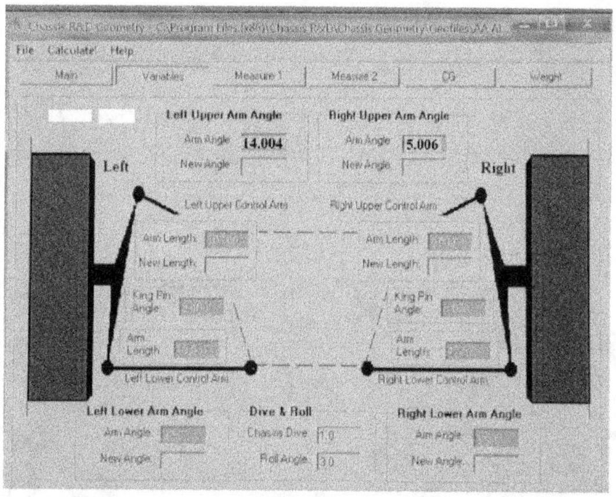

For this example, we use a circle track car that could be a Tour Modified or a late model running a conventional setup. The upper control arm angles are 14.0 degrees for the left upper and 5.0 degrees for the right upper. We also used a common set of dive and roll numbers that relate to more conventional setups of 1.0 inches of dive and 3.0 degrees of roll.

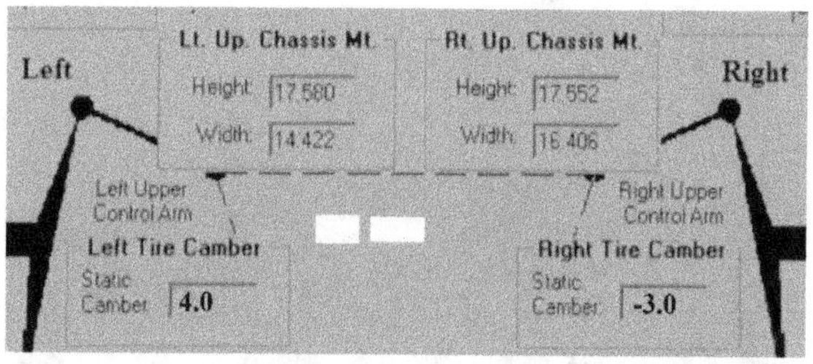

The static cambers for this car are a positive 4.0 degrees on the left front and negative (-) 3.0 degrees on the right front. We'll calculate this and see how our cambers come out.

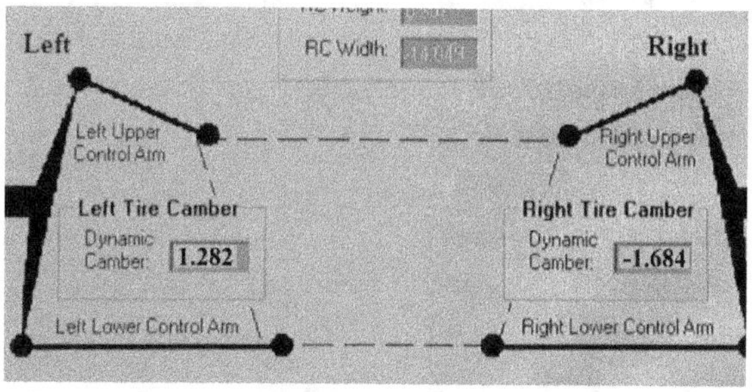

When we calculate the cambers with those upper control arm angles, we get these numbers. The RF lost about 1.4 degrees of negative camber. The LF ended up with 1.3 degrees of positive camber, a number that is not bad at all. But the loss at the RF will be detrimental to how the car turns. Let's see if we can fix that with new upper control arm angles.

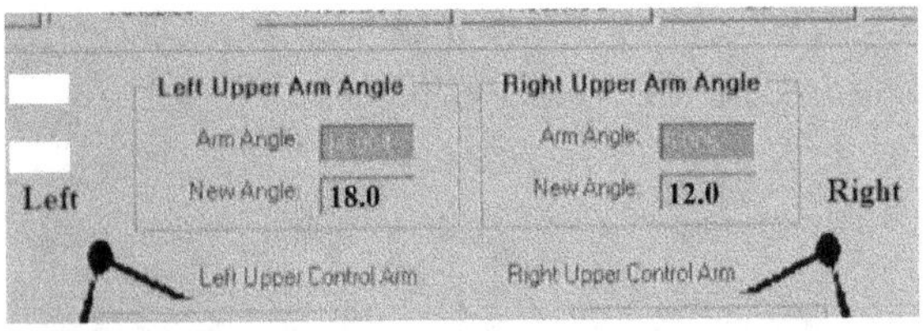

We enter new upper control arm angles and they are 18.0 for the left upper and 12.0 for the right upper. These should improve the camber change, especially at the RF.

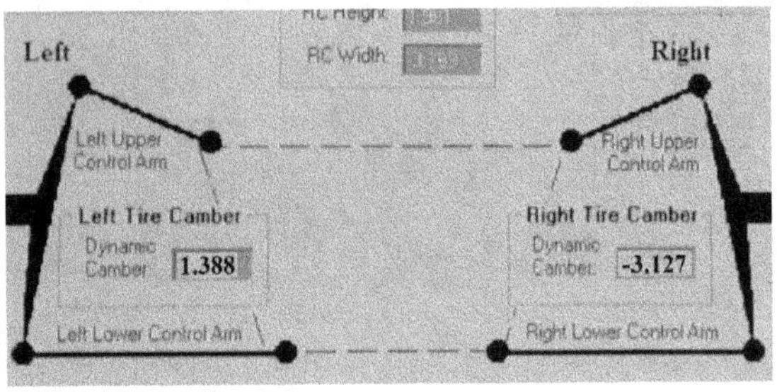

The new cambers show a negative (-) 3.1 degrees at the right front, nearly the same as the static camber, and 1.39 at the left front, nearly the same as before. The fact that the right front camber, or outside in this example, never changed through dive and roll means that this tire will take a set and then provide grip all through the entry and middle of the turn. This provides a much better contact patch to give the car more front grip.

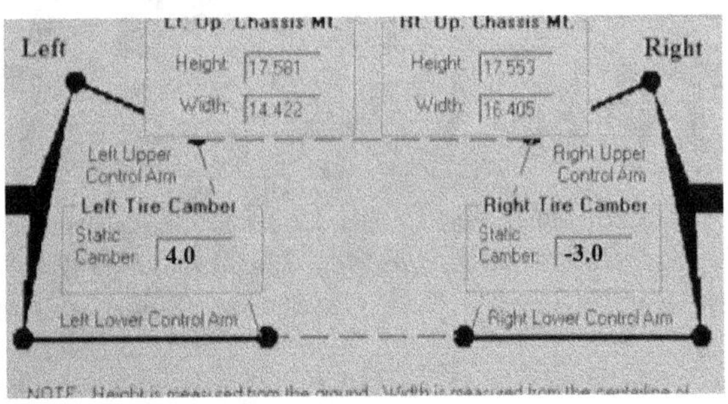

Now we need to look at a typical bump setup. We'll start out with the same front cambers in this example. We will change the dive and roll numbers to simulate the travel onto the bumps and the low roll angle the bumps provide.

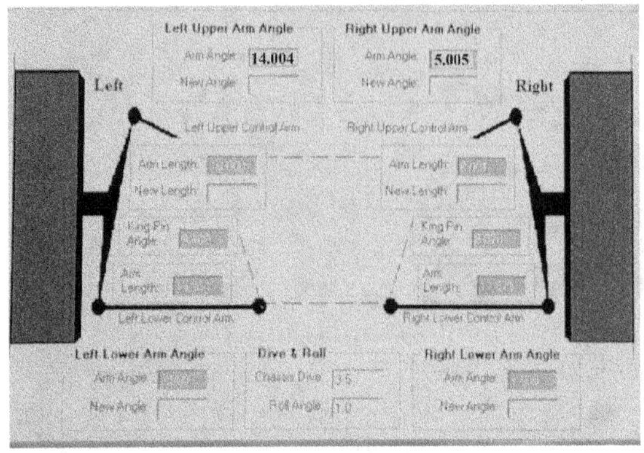

Because the front travels so much with the Bump setups, most teams will put low upper control arm angles into the front end. Here we have 14.0 degrees at the Left upper arm and 5.0 degrees at the Right upper arm. Our dive is 3.5 inches and the roll angle is 1.0 degrees.

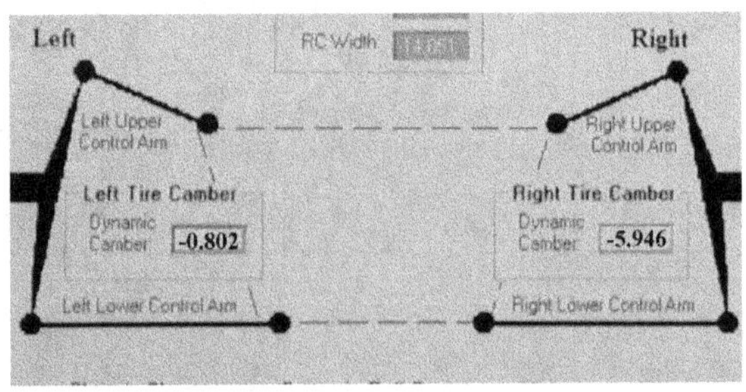

When we calculate the camber change using the upper control arm angles and the dive and roll that simulates the bump setups, we get a RF camber of negative (-) 5.9 degrees and a LF camber of -0.80 degrees. We obviously need to take static negative camber out of the RF wheel and put more positive camber into the LF wheel. But maybe there is a better way to set this up using much different upper control arm angles? Let's take a look.

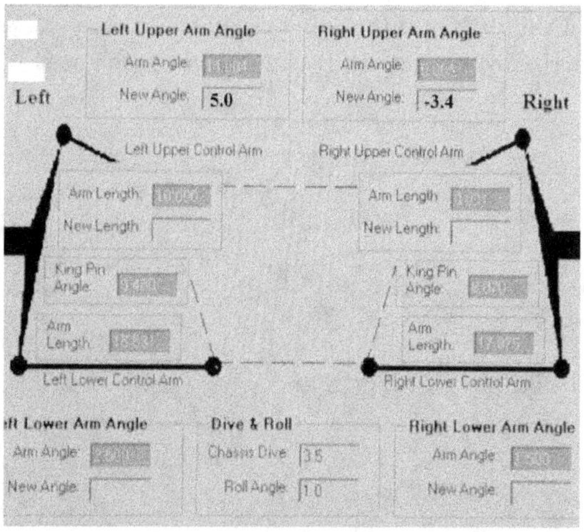

Let's experiment with upper control arm angles and go unconventional. We put 5.0 degrees in the left upper arm and a negative (-) 3.4 degrees in the right upper arm. This means that the right arm chassis mount is higher than the ball joint, which is unconventional. But this car will travel down onto the bumps right away once it is on the track and will really never leave that attitude. We also use the same dive and roll numbers, 3.5 inches of dive and 1.0 degrees of roll.

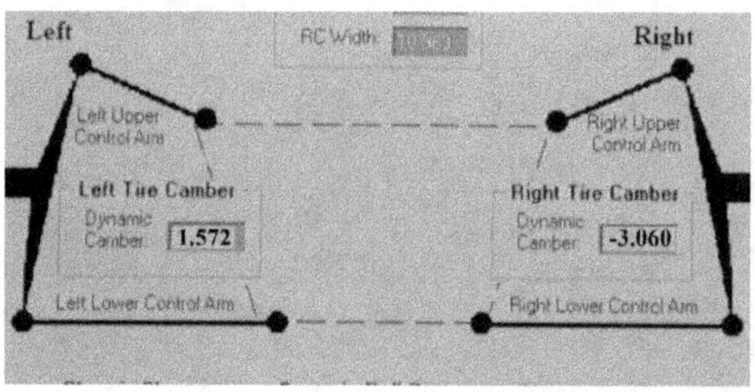

Once we calculate the cambers, we end up with a positive 1.5 degrees at the LF wheel and a negative (-) 3.0 degrees at the RF wheel. These numbers are very close to what we need. So, instead of using a very high positive LF camber and a very low RF camber, we just need to set the car up with more normal static cambers to end up where we need the front cambers to be for the best tire contact patch.

These examples are intended to show how camber change works in the real world. We're not sure anyone has installed exactly the same upper control arm angles as the example, but with what we know about what the tires like, and what happens with the weight distribution when the cambers do change, we might consider the benefits of the examples.

We introduced the concept of weight distribution changes when the camber change, so let's take a look at how that happens. It concerns the height of the spindle related to what camber the tire and wheel are set at.

Weight Change From Camber Change - When we change the static front cambers, or when the cambers change with the car on the race track, we also might be changing the height of the chassis at that corner and we need to bring the weight distribution back to where it was before the camber change. The illustrations show how this happens.

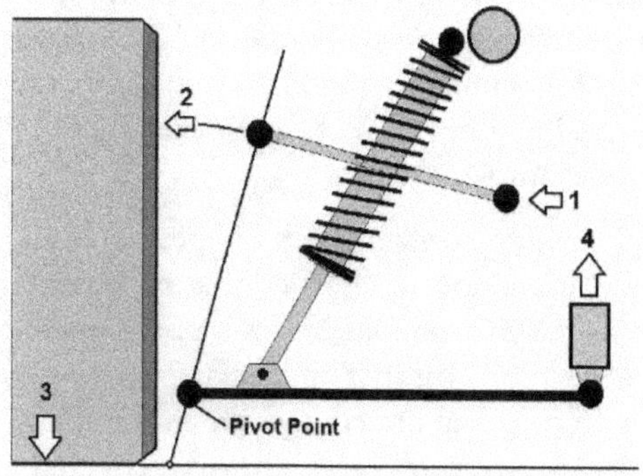

On the left side, we can see here how the intersection of a line through the centers of the ball joints, upper and lower, intersects with the ground inside the tire contact patch. Not all suspensions are setup exactly this way, this is just a demonstration of approximately how this works. If we add positive camber by moving 1, 2 moves in the same direction. Then as 2 rotates around the Pivot Point, 3 moves down lifting 4 the chassis up. This adds more of the total weight of the race car to this tire. We would need to adjust the spring length to bring the chassis back to normal ride height. On the track, we cannot do this, so there is load change happening when the cambers change on the race track.

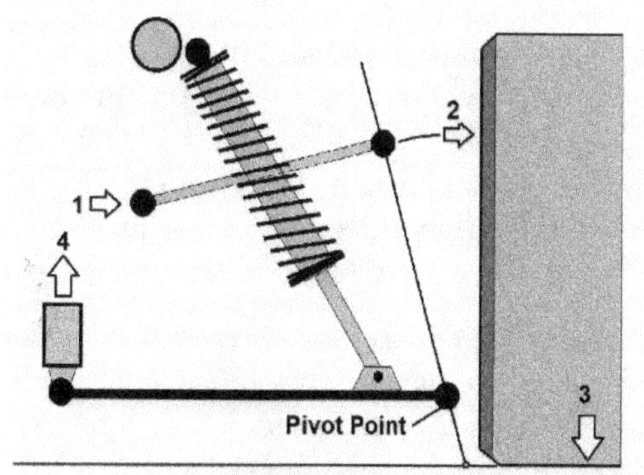

On the right side, what we are seeing is that as we reduce negative camber, or move the wheel towards positive camber, we are also raising the chassis and adding cross weight percent if we don't adjust the spring length to compensate for the chassis height change. In a suspension where the camber changes through dive and roll, no only does the tire not have a consistent contact patch, it also starts carrying a different load.

The concept of weight distribution caused by camber change means that we need to strive for minimal camber change for not only the purpose of contact patch optimization, but also to reduce the change in weight distribution on the four tires.

In the past when I thought about all of the cars I designed and re-designed, the best and most successful cars were the ones where the camber change for both tires was minimal. In todays racing for both conventional setups and bump setups, the same is true as evidenced by the above examples.

Modern Day Contact Patch Shapes - In the above discussion, we learned that certain control arm angles in a double A-arm suspension will create minimal camber change and help us end up with the ideal tire contact patch for a particular race car. We also have discovered through testing and evaluation that a larger contact patch provides more traction in a tire, much like increasing the loading, but not needing to.

Early on we talked about what the tire needs for camber. We are talking about a AA-arm suspension and for stock cars, that is the front tires obviously because there is little we can do to adjust the rear tire cambers for a straight axle suspension. Each brand of tire has different sidewall stiffness and tire construction for the tread support and so need different cambers for each type of tire so that we can end up with the largest contact patch possible.

Like we also pointed out, you may be able to extract a larger contact patch from one tire brand than any other tire brand that has the same tread width and sidewall height. Tire temperatures can point us in the right direction, but we are finding that the driver may be the best litmus test.

As for tire temperatures, a softer sidewall tire can be cambered so that the tire surface temperatures on the side closest to the inside of the turn are 20-25 degrees hotter than the surface temperatures on the outside away from the turn. This difference is achieved with higher camber settings which tend to cause the inside of the tire to flatten out.

Lower tire pressures help us accomplish the goal of producing a larger tire contact patch, but we are not advocating low tire pressures that compromise safety. We do know the trend in racing is to run as low a tire pressure as we can get away with and this is an indicator of the importance of increasing the contact patch area.

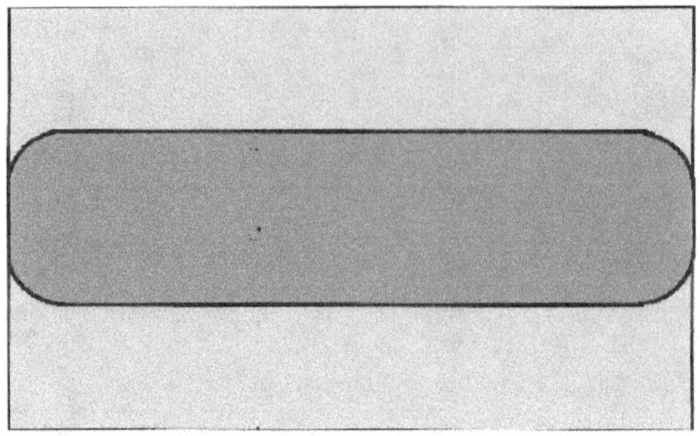

In more conventional times, and when using ultra-wide tires, this might be the contact patch that results from having equal tire temperatures across the face of the tread. If we attempted to put too much camber into this wide tire, then we would lose contact loading and area at the outside (meaning outside of the turns) edge of the tire. For more conventional sized tires, this does not represent the ideal camber we need for maximum contact patch area.

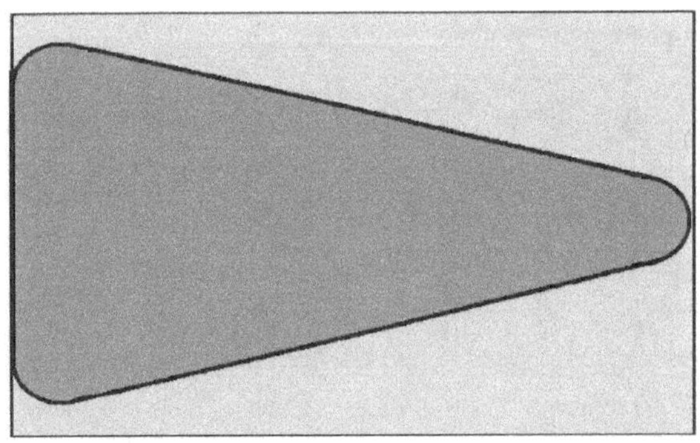

If we add camber to the outside tire and reduce the pressure in that tire, our contact patch might well look like this. The inside (inside of the turns) part of the contact patch is much wider while the outside part is very close to what we showed in the previous sketch. The area of this contact patch is much larger than if we set a camber angle that resulted in even tire temperatures across the face of the tire. For this example, we usually see 20-25 degrees hotter at the inside edge of the tire than we see at the outside edge.

The above example of contact patch resembling a triangle could be applied to this formula car, only in reverse. The outside tires are going into positive camber in this high speed turn. It is possible that the triangle contact patch shape is reversed from what we need for circle track cars. The end result is the same, a much larger contact patch. In this example, this team can set the inside tire camber much more straight up increasing its contact patch size too.

Summation – When you purchase a race car, it is yours. What you do with it is entirely up to you and your team. If you think you need to make changes to the control arm angles based on what we have learned here, then by all means do it. I have personally cut up many hundreds of race cars so that the geometry, mostly for camber change, is better and to help the tires gain contact patch area and with that, grip.

The front of a stock car is hard to tune for dynamic balance, but easy to tune for contact patch optimization. Then when we get the front end as good as it can be, we can easily tune the rear suspension to match the front for grip.

A race car where the front is deficient has to be de-tuned at the rear so that the car will handle neutral. This car will be slower due to lack of overall grip compared to a car that has been optimized.

Exam - In The Context Of This Lesson:

We Need Camber Because?

1) The Tire Distorts When Lateral Force Is Applied

2) The Suspension Points Move In Dive And Roll

3) Camber Helps Generate More Lateral Force

4) All Of The Above

Camber Change Occurs Because?

1) The Wheel Moves Up and Down

2) The Car Rolls

3) The Tire Distorts From Lateral Force

4) A Combination of 1 and 2.

What Changes To The Suspension Will Affect The Amount of Camber Change?

1) Changing the Spring Rates

2) Installing different length control arms

3) Changing the angles of the control arms

4) All of the above

How Much Camber Changes Do We Need?

1) Enough to get what the tire needs

2) An amount equal to the tire roll over

3) As much as possible

4) As little as possible

What Are Common Ways To Minimize Camber Change?

1) By running very stiff setups

2) Through special design of control arm angles

3) With the use of bump devices that limit suspension movement

4) All of the above

The Ultimate Cambers We Need Are Determined By?

1) The Amount of Lateral Load

2) The Degree of Camber Change In Our Suspension

3) The Stiffness of the Tire Sidewall

4) All Of The Above

Camber Change Affects The Tire Loading?

1) True

2) False

Even Tire Temperatures Across The Tire Is Always Optimal?

1) True

2) False

Race Car Technology – Level Three

Lesson Ten – Jacking Force Concepts and Uses

Jacking Force (JF) is a concept that was introduced about ten years ago in an attempt to more accurately define basic vehicle dynamics for a AA-arm suspension. It was thought by some to be the primary force that generates and regulates the chassis dynamics in those suspensions. It is not. It is, in fact, in most cases with modern race car geometry, a system that produces an Anti-roll effect, much like a sway bar.

In this Lesson, we will explain how JF works, how it is calculated, how we might use it to our advantage, and how it might be miss-used. It is extremely important to understand this subject mainly because there is so much dis-information being taught that is floating around the internet. Further, the concept of JF is sometimes made more complicated than it really is. And some of the information is just plain wrong.

Modern day race cars running on circle tracks try to utilize Jacking Forces to create an Anti-roll effect very similar to what a sway bar does. This helps keep the front of the car more level to the racing surface and helps improve the aero effect.

History Of Jacking Force – The effect we call Jacking Force has always been around, but, in production automobiles, the control arm angles are small, so the industry paid little attention to it historically. As we will see in this Lesson, when the control arm angles are small, there is very little JF present in the system.

In the past as far as race cars are concerned, the control arm angles were more than production automobiles, but not sufficient to cause much JF effect. More recently, we see designers and race teams using much greater upper control arm angles with the desire to create higher Jacking Forces.

There are various design reasons why the higher JF is desirable, but know that we only really became concerned with JF over the past decade or so. I myself have a history of discounting JF as a significant effect mainly because the cars I was working with had very little JF incorporated in the design of the geometry. Then I did some experimenting and began working with a unique scale model that could test the JF concept for AA-arm suspensions. Now I can tell you with confidence how JF is produced and how to calculate it.

How This Lesson Came About - In creating this Lesson, we first conducted a series of tests using a unique scale model of a double A-arm suspension whereby we could apply forces to the contact patches and the Un-sprung mass, change the control arm angles, and then measure the roll stiffness and ultimately the chassis roll angle.

The roll angle of the chassis demonstrates the roll stiffness related to sum of all of the effects present in the system and differences in roll angle between the different arrangements of control arm angles tells us how much influence JF has on the dynamics of the suspension.

As a direct result of this testing, we were able to develop roll angle calculations combining the overturning moment method and the JF methods. These calculations resulted in almost exact replication of the actual roll angles we recorded on the scale model.

What Produces Chassis Roll? – The JF does, or can, have an effect on the amount of chassis roll, but what causes chassis roll in the first place? It is what we call the Overturning Moment that is the primary cause of chassis roll. That is, when we enter and negotiate a turn, a lateral G-force is created which tries to pull the car to the outside of the turn. The tires resist this force.

This lateral force causes some weight to transfer from the inside tires onto the outside tires. The amount of weight that transfers is dependent on how fast we go through the turn, i.e. the amount of G-force, the length of the Moment Arm and how wide the track width is at that end of the car, track width being the center to center distance between the two contact patches of the tires. The formula for calculating weight transfer is:

$$\text{WtTr} = \frac{\text{MA x Sprung Wt x G-force}}{\text{Track Width}}$$

Real numbers might look like this for a modern super late model asphalt car:

$$\text{WtTr} = \frac{15.0 \times 1280 \times 1.50}{66.0} = 436 \text{ lbs.}$$

If the chassis is sprung, the chassis will roll as this weight transfer takes place. As the weight transfers, the left front corner of the car looses 436 lbs. and the right front corner gains 436 lbs. The suspension springs for each side must adjust to the new weights, and so the left side rebounds and the right side compresses. If our wheel rates were say 250 lb./in. (340+/- ride spring rates), then each side would move 436 / 250 = 1.74 inches. If we take the ArcTan of [(1.74 x 2) / 66.0], we get an actual roll angle of 3.02 degrees for this car. For a bump setup using 2,000 lb./in. bump springs, the wheel rate would be about 1,520 lb./in. and the roll angle would then be much less, or about 0.50 degrees.

The above is what the chassis would roll to if we don't include the sway bars Anti-roll properties, or the JF properties. Each of those, the sway bar and JF, produce Anti-roll effects that reduce the chassis roll. If we are trying to balance the chassis' front and rear suspension systems, then we need to know something about how much reduction in roll angle each of these Anti-roll effects has. If not, we may think we have a balanced setup when in fact we do not.

A chassis running 52% cross weight and 56% left side percent has a LF static weight of 756 pounds and a dynamic LF weight after weight transfer of 320 pounds. The RF has a static weight of 644 lb. and a dynamic RF weight after weight transfer of 1,080 lbs.

These weights are used to calculate the lateral forces at the contact patch and at the chassis Center of Gravity. We simply multiply the corner weights times the lateral G-force. If the lateral force changes, so does the amount of weight transfer (wheel loads) and so do the Jacking Force numbers. Keep this in mind as you travel from track to track. When your lateral G-forces change, so do the JF amounts. And, If the chassis roll angle changes, so to do the dynamic control arm angles and therefore the JF numbers.

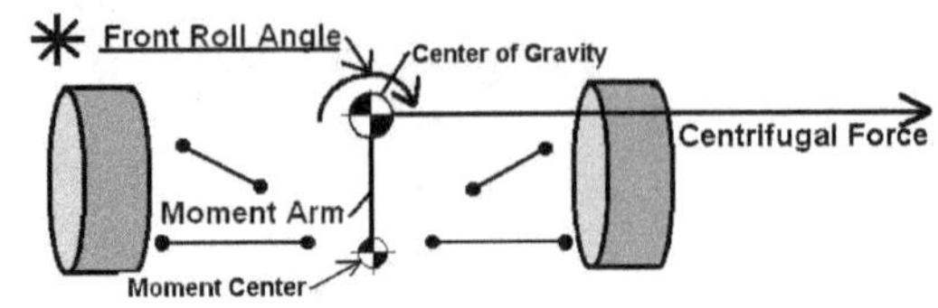

In this illustration, we see where the Centrifugal force acts on the Center of Gravity of the sprung mass and creates chassis roll. This is referred to as the Overturning Moment that causes chassis roll. In a race car, this roll is somewhat restricted by sway bars and possibly Jacking Forces.

Two Types Of Jacking Force – There are two distinctly different types of JF. These two types combine into a single force that can enhance or restrict chassis roll. The first is what we will call Contact Patch Initiated JF because the initiating force comes from the contact patch resisting the lateral forces created when cornering. The second type we will call the Un-Sprung Mass Initiated JF because it is initiated from the weight, or mass, of the un-sprung components in the suspension.

Contact Patch Initiated JF - The first type is a product of the forces acting on the chassis and the tire contact patches of the outside (right in a left turn) and the inside (left side in a left turn) tires that produce a lateral force on the control arms. If the control arms are mounted at some angle from horizontal, then there will be as a result a vertical force acting on the chassis. These forces will add to, or subtract from the chassis roll.

In Illustration 1, we see that as the lateral force is pulling to the right, the contact patch is resisting that force in the opposite direction. This tries to rotate the spindle around the lower ball joint. This motion puts a force into the upper ball joint in the opposite direction of the contact patch force. This force is proportional to the distance from the lower ball joint (the point of rotation) to the ground verses the distance from the lower ball joint to the upper ball joint.

Then, if the upper control arm is mounted at an angle from horizontal, then vertical forces are created. If the angle is down to the centerline, or the chassis mount is lower than the ball joint, then the vertical force would be down onto the chassis.

If the chassis were rolling, then this force is counter to the rolling action and the roll angle is reduced. If the angle of the control arm were reversed with the chassis mount higher than the ball joint, then the roll angle would be increased.

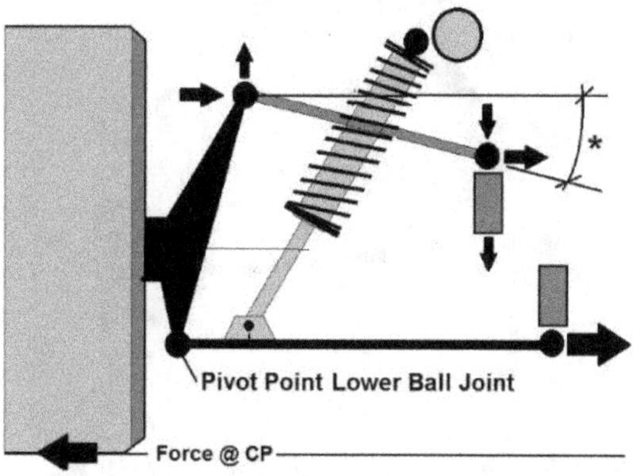

Illustration 1

Here we see the left side of a AA-arm suspension in a left turn from the drivers view. The lateral force at the Contact Patch acts to try to pivot the spindle in a clockwise direction around the lower ball joint, putting a force upon the upper ball joint. This inward lateral force pushes on the control arm and creates a vertical force being up on the BJ and down on the chassis mount. This creates an Anti-roll effect.

In the right side suspension (see Illustration 2), the lateral force is again to the right through the lower ball joint and resisted by the contact patch with that force going to the left. Again, with these control arm angles, we see an upward force at each chassis mount that together resists chassis roll.

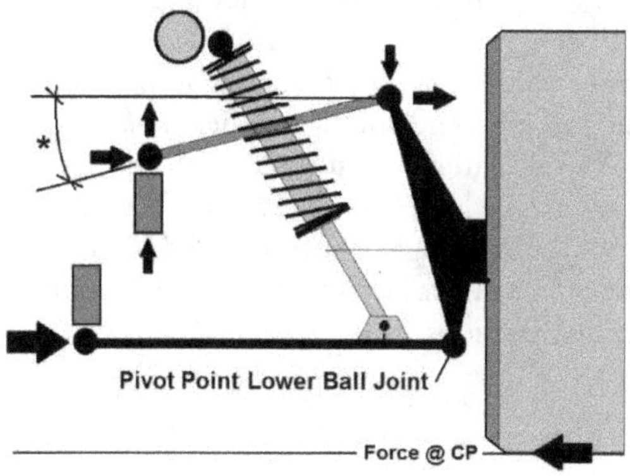

Illustration 2

This is the right side of a AA-arm suspension in a left turn from the drivers view. The lateral force at the Contact Patch acts to try to pivot the spindle around the lower ball joint in a clockwise direction, putting a force upon the upper ball joint. This outward lateral force pulls on the control arm and creates a vertical force being down on the BJ and up on the chassis mount. This creates an Anti-roll effect.

It is a similar action with the lower control arm (see Illustration 3). In this case, a lateral force equal to the force at the left side contact patch acts through the lower ball joint. If the lower control arm is mounted at an angle, then a vertical force is created that, like with the upper control arm, either increases, or reduces chassis roll. In this case, the force is upwards and adds to chassis roll.

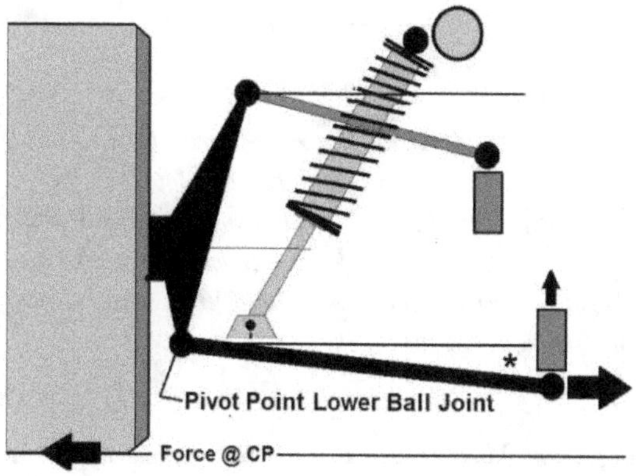

Illustration 3

Here we see the left side of a AA-arm suspension in a left turn from the drivers view. The lateral force at the lower chassis mount acts to try to pull on the lower control arm and is resisted by the lower ball joint. This then puts an upward vertical force upon the lower chassis mount. This creates a Pro-roll effect.

At the right side showing the lower control arm angle (see Illustration 4), the force on the lower ball joint is to the right and the resulting vertical force on the chassis is down, or adding to chassis roll as was the case with the left lower control arm.

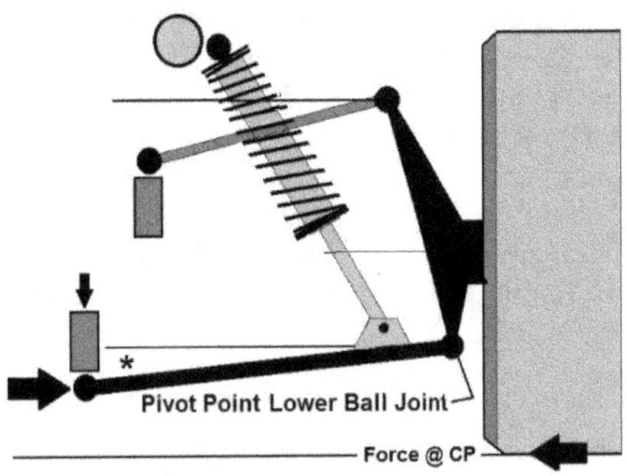

Illustration 4

Here we see the right side of a AA-arm suspension in a left turn from the drivers view. The lateral force at the lower chassis mount acts to try to push on the lower control arm and is resisted by the lower ball joint. This then puts a downward vertical force upon the lower chassis mount. This creates a Pro-roll effect.

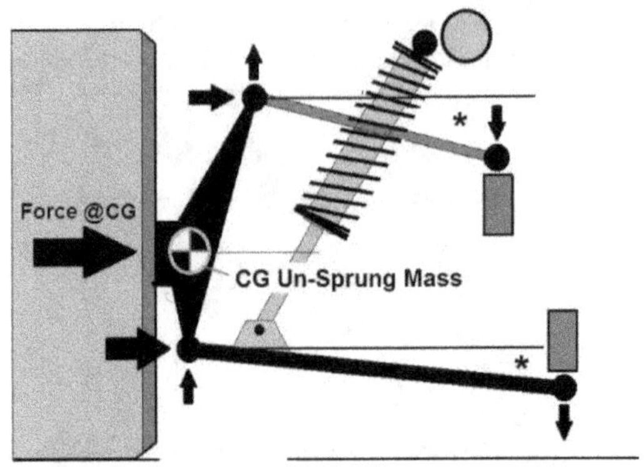

Illustration 5

In the case of the Un-sprung Initiated JF, in this view of a AA-arm suspension in a left turn from the drivers view, we concentrate on different forces. The lateral force at the Center of Gravity of the un-spring mass acts to push on the control arms. In this example, this then puts downward vertical forces upon the upper and lower chassis mounts. The net result creates an Anti-roll effect.

Un-Sprung Mass Initiated JF - The other somewhat less significant JF is produced by the lateral G-force acting on the un-sprung components of the AA-arm suspension. The mass, or weight, of the wheels, tires, spindle, brakes, etc. are in themselves producing a lateral force as a result of the G-force going through the turns and therefore put a force on the control arms separate and apart from the forces applied as a result of the contact patch forces.

In Illustration 5, we can see where the lateral force created by the G-force times the un-sprung mass weight is then creating lateral forces onto the ball joints. The force for both the upper and lower ball joints are in the same direction, similar to the contact patch JF scenario except that the center of the originating force is concentrated at the Un-sprung Center of Gravity and not the contact patch or the lower chassis mount.

These forces are basically pushing on the control arms and as such are trying to push up on the ball joint end and down on the chassis mount end. We calculate the actual vertical force by multiplying the lateral force times the Tangent value of the arm angle. The calculated force is the same at the ball joint as it is at the chassis, only in opposite directions.

Then, with this layout of forces, the amount of the lateral force for each ball joint is divided between the two ball joints and are proportional to the distance each is to the Un-sprung CG. More of the proportion of lateral Un-sprung initiated force will be upon the lower ball joint and less on the upper ball joint because the upper BJ is farther away from the center of force, or Un-sprung CG.

In Illustration 6, we see the right side suspension and how the Un-sprung forces act on the control arms. These forces are basically pulling on the control arms and as such are trying to push down on the ball joint end and up on the chassis mount end. We calculate the actual vertical force by multiplying the lateral force times the Tangent value of the arm angle. The calculated force is the same at the ball joint as it is at the chassis, only in opposite directions.

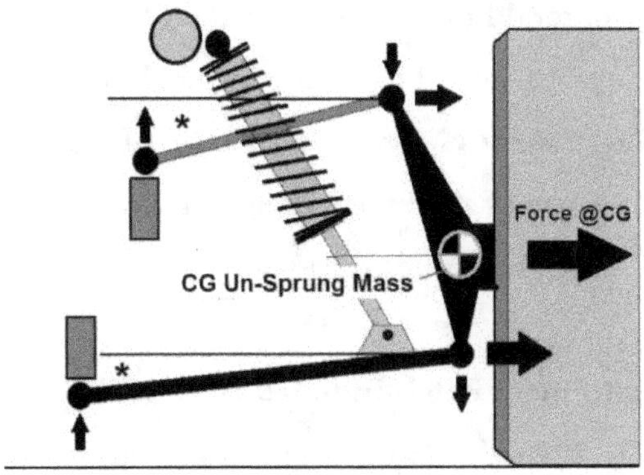

Illustration 6

In this view of a AA-arm suspension in a left turn from the drivers view, we concentrate on the right side Un-sprung Initiated JF forces. The lateral force at the Center of Gravity of the un-spring mass acts to pull on the control arms. In this example, this then puts an upward vertical forces upon the upper and lower chassis mounts. The net result creates an Anti-roll effect.

Combining Jacking Forces - To arrive at a net jacking force for each wheel, we add all of the JF forces, and translate them out to the wheel as a positive or negative wheel load that is then added or subtracted from the normal sprung mass weight transfer. The number representing weight transfer in our example being 436 would be smaller for the left side suspension and also smaller for the right side suspension.

Based on the new weight transfer numbers that now include the JF weights, our new roll angle will be less because "less weight has transferred". In reality, the same amount of weight transfers, but the weight numbers that we use to calculate roll angle have in fact changed. The JF has restricted chassis roll, but not the actual weight transfer.

This sounds complicated, and the actual computations are a little involved. We won't be going into how to exactly calculate the JF numbers, but we have explained here how ending up with certain control arm angles can have an effect on the amount of JF our chassis will have.

Chassis Roll Changes Arm Angles – We cannot use the static geometry for the suspension to analyze the Jacking Forces because as the chassis rolls, in reaction to the overturning moment dynamics, the control arm angles will change. It is important to know that as these arm angles change, so do the Jacking Force, they have to.

So, there is some back and forth calculating that needs to be done in order to arrive at accurate numbers for JF. You may have seen illustrations showing force lines acting on static control arm angles. You now know that any analysis of Jacking Force must be applied to dynamic control arm angles, and that is control arm angles as they end up at after the chassis rolls.

Jacking Force Is Dependent – One very important thing to remember is that JF is entirely dependent on how much lateral force there is on the contact patch. That is dependent on the lateral G-forces, track banking angle, tire loading, and contact patch size and weight distribution.

The Jacking Forces will change as the above described conditions change, so JF is very dynamics and moving all of the time. We can only approximate the JF for a particular setup, geometry and track layout.

The more consistent we can make the setup and the more balanced we have between the two suspension systems, the more reliable our estimates of JF magnitudes. And that is our overall primary goal with chassis setup. Just know that JF calculations is a moving target. If we know where the target is moving to, we can more easily hit the target.

Modern Applications Of Jacking Force – It seems that many teams today desire high JF numbers in the Anti-roll direction. This reduces the chassis roll and that might be a desired effect. Since the dynamic lower arm angles for most dedicated production race cars end up at a high value, a lot of Pro-roll is created and opposes the desired Anti-roll effect of JF.

What we see now is a trend to reduce the dynamic lower control arm angles so that there is less Pro-roll effect in the suspension that might counter the JF Anti-roll teams desire. The decision to use JF or not is a personal choice for race teams. We are here to present knowhow and not to influence a race teams thinking about which way to go. Hopefully our students will now have more information that will help them make critical decisions about Jacking Force.

Exam - In The Context Of This Lesson:

Jacking Force Does What?

1) Causes chassis roll

2) Provides Anti-roll properties

3) Provides Pro-roll properties

4) Adds weight to the inside front tire

5) 2 and 3

The Amount of Weight Transfer Does Not Depend On?

1) The Jacking Force amounts

2) The Moment Arm length

3) The lateral G-force

4) The sprung weight

5) The track width

Jacking Force Has Been Around For How Long?

1) Forever

2) Just the past decade

3) Since race cars became popular

4) Never before

Jacking Force Is An Affect As Long As?

1) There is lateral force

2) The control arm angles are not level

3) The car has a sprung suspension

4) All of the above

The Lateral Force At The Contact Patch Is?

1) Equal to the G-force times the tire loading

2) Equal to the G-force at the Center of Gravity

3) The sum of all of the Jacking Forces

4) G-force times the Gravity force

The Contact Patch Initiated JF Uses Which?

1) The lateral force on the un-sprung mass

2) The lateral force on the sprung mass

3) The force at the contact patch

4) 2 and 3

The Un-sprung Mass Initiated JF Uses Which?

1) The lateral force on the un-sprung mass

2) The lateral force on the sprung mass

3) The force at the contact patch

4) 2 and 3

To Determine Accurate JF Numbers, We Don't...

1) Use the static control arm angles

2) Use the dynamic control arm angles

3) Use dynamic wheel loads

4) Combine all of the JF affects

In Modern Racing, Teams Desire The Following...

1) Anti-roll effects of JF

2) Pro-roll effects of JF

3) More chassis roll

4) Less upper control arm angles

Race Car Technology – Level Three

Lesson Eleven – Rear Steer Concepts & Uses

We have talked about front geometry (Lesson 9) and we have talked about Angle of Attack (Lesson 8). Now we need to talk about Rear Steer and rear geometry. There are two primary purposes for rear steer in an asphalt race car and an additional purpose for dirt cars.

The first primary purpose for Rear Steer is that

1) it helps develop and tune the AOA, or angle of attack, for the rear suspension mostly in a solid axle rear suspension, but also in a AA-arm rear suspension to some extent. The second purpose is

2) it can help develop additional AOA for drive off the corners independent of, and in addition to, the mid-turn rear steer. The third purpose, and this is mostly for dirt cars, is

3) to develop a rotation of the car body in relation to the direction the car is traveling to produce flat plate aero side force. This is somewhat useful for some asphalt cars too to a much lesser degree.

This Florida modified car used rear steer under power to enhance the grip off the corners. These cars are limited to an 8 inch tire and need all of the added grip they can get for bite off the corners. The mid-turn rear steer was near zero.

The dirt late model cars usually have four-link rear suspensions that are setup to steer quite a bit to the right. This pushes the left side of the rear end up towards the driver and steers the rear of the car out to the side by a foot or more. The benefit is in the aero package. The flat right side of the car creates flat plate aero sideforce helping to push the car towards the inside of the turns. The angle of the rear end compared to the alignment of the front wheels is not that much different. But the alignment of the contact patches side to side is miss-aligned by quite a bit.

Why Don't Cars Have Rear Steer Instead Of Front Steer? – Steering from the rear is very difficult because the effect is so pronounced. A very small amount of rear steer goes a long way. That is why we don't see any production or race cars that steer from the rear, and that makes sense.

If, as we learned in Lesson 8 dealing with Angle of Attack, we only steered with the rear wheels, then how would the front tires develop their AOA? We know they need to be steered quite a bit to attain an AOA, so the rear would also need to be steered quite a bit just to give the front tires their needed AOA.

Once we had steered the rear wheels a sufficient amount to give the front tires their AOA, the rear would be steered way too much to the outside of the turns to ever develop the AOA they need. It just doesn't work. So, we have front steered cars.

What Is Rear Steer? - Rear Steer in a race car is an effect that is mostly caused by suspension movement. We don't have a steering mechanism the driver can use to steer the rear. Under the right conditions in measured amounts, RS can be beneficial and enhance performance. Under the wrong conditions, it can ruin your handling. We need to have a solid understanding of what produces RS and what effect RS has on the handling in our cars in order to have the best performance. And we need to know how much to use.

How Rear Steer Works - When we have rear suspension movement, as the rear suspension of the car moves, along with the controlling arms that locate the rear end fore and aft, the components for each side can move the wheel either forward or to the rear. We will use, for example, a common three link rear suspension.

If the trailing arms for a common three link are mounted at an angle from a top view where the chassis mounts are closer together, or closer to the centerline of the chassis, than the rear end mounts, then as the rear end moves laterally, there can be rear steer. This would mostly come from the movement of the panhard bar, which moves the rear end laterally as the bars chassis mount moves through an arc with the rear end mount as the radius.

In most circle track race cars turning left, the panhard bar is mounted to the chassis on the right side of the car, or to the outside of the turn. In a left turn, if the chassis mount is higher than the rear end mount, then as the right side of the chassis moves down in dive and roll, the panhard bar will push the rear end to the left until is gets to an angle that is parallel to the racing surface. This steers the rear end and rear tires so that they are pointed to the right of the chassis centerline. This tends to loosen the rear of the car.

As the chassis mount moves farther, the chassis mount is now moving below parallel to the racing surface and will now pull the rear end to the right back towards where it started out. If we know the amount of right side chassis movement vertically, we can simulate the movement of the rear end laterally.

If the panhard bar were mounted level to the racing surface, or with the chassis mount lower than the rear end mount, then the rear end would be pulled to the right with the normal chassis movement in dive and roll. This would then steer the rear end and rear tires in a direction to the left in relation to the centerline of the chassis and that would tighten the car.

What Does Rear Steer Do? - A rear end that is steered to be pointed to the left of centerline will cause the thrust angle and AOA to be left of centerline and make the car tighter on entry, tighter in the middle of the turn, and tighter on exit under acceleration. A rear end that is steered to the right of centerline causes the thrust angle and AOA to be pointed to the right of centerline and will generally make the car looser on entry, looser in the middle, and looser on exit under acceleration.

The asphalt racing surface provides a lot of traction, even on those flat "slick" tracks. Because there is very little slip of the tires on asphalt, the range of useable rear steer is very small. There are six predominant rear suspension systems used in stock car racing and all of them can produce some amount of rear steer. They are:

The Three Link System – This system has two trailing arms mounted near the rear tires and one third link, usually mounted atop the rear differential. The third link controls rear end wrap-up. The trailing arms can be mounted parallel to the centerline of the car or angled with the front mounts closer to centerline. They can also be mounted at different angles from a side view.

RS (rear steer) in this system is caused by chassis movement which can produce several secondary effects. Usually, the right rear corner of the chassis moves more than the left side, and on most flat to medium banked tracks, the left rear moves very little.

On most asphalt 3-link cars, the RR trailing arm mostly controls RS due to body roll. We usually need to position the angle of the trailing arm so that the front mount is higher than the rear mount by roughly one-third the distance that the front mount will move down during cornering. This is a generalization though and we will get more into control arm angles later on in this discussion.

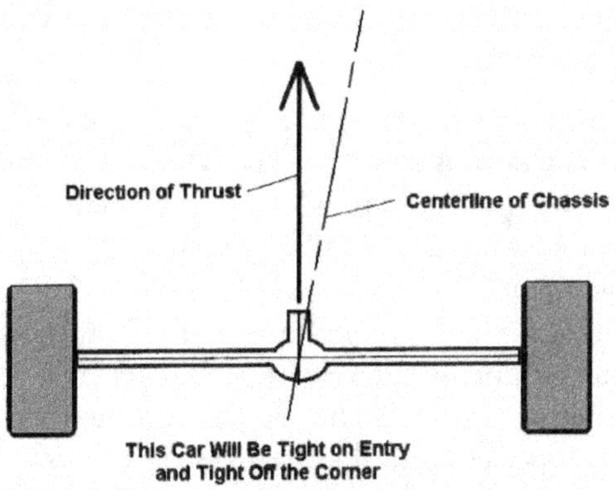

With the rear end pointed to the left of centerline, the steering will cause the rear of the car to want to run left of the front end causing a very tight condition, especially under acceleration. With the rear end pointed to the right of centerline, the car will be freed up going in, through the middle and possibly loose off the corners with the rear end wanting to run around to the right of the front. Rear steer can be tolerated in very small amounts.

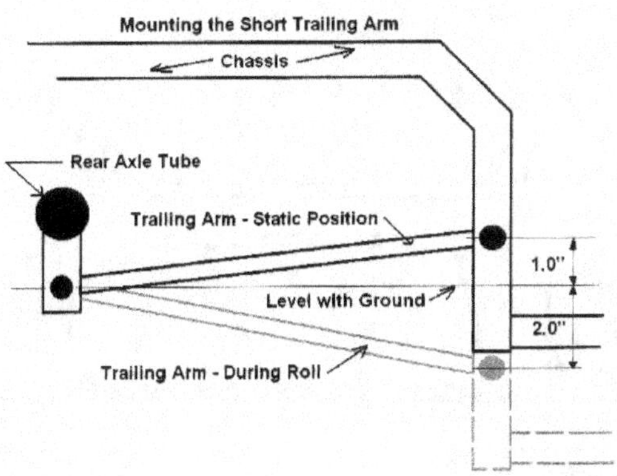

The three-link rear suspension system can produce rear steer in both directions. As the chassis moves down on the right side, the RR wheel will be moved back as the front mount approaches the height of the rear mounting point. As the front mounting point continues to move down, the RR wheel will be pulled forward. Normally, to produce a small amount of rear steer to the left on asphalt, we mount the front pivot point 1/3 of the total travel distance higher than the rear pivot.

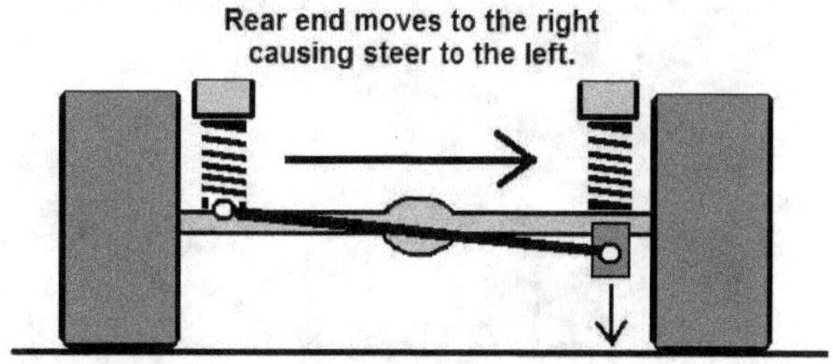

In this illustration, if the panhard bar chassis mount (to the right side) is mounted lower than the rear end mount (left side mount), then when the chassis dives and rolls moving the right chassis mount down, then the rear end will be pulled to the right. If the trailing links are angled from a top view with the forward mounts narrower, then this movement will steer the rear end to the left of centerline. At the same time, if the right link were angled so that the front chassis mount was higher than the rear mount, with the same chassis dive and roll, the right side of the rear end will be pushed back creating rear steer to the right. If we setup the movement to cancel each other out, we will end up with Net Zero rear steer. Tuning the angles of the panhard bar and the right link will allow us to fine tune the rear steer.

The Truck Arm System – The truck arm system has been adapted from the design for a 1964 Chevy pickup truck and is used on many Late Model Stock cars as well as the three premier divisions of Nascar, the Camping World Trucks, and the Xfinity and Monster Energy Cup cars. These systems react to dive and roll in a similar way to the three link suspension we described earlier. These systems can roll either to the left or to the right. The roll of the chassis and the movement of the panhard bar are the two components that influence the amount of steer in these systems.

As far as geometry related to rear steer is concerned, this can be a good system for asphalt cars. The amount of rear steer is determined by two things, 1) steer due to body roll is regulated by the height of the front mounts of the arms which are always mounted lower than the rear point of rotation which is the axle. Rear steer due to the panhard bar vertical movement are regulated by the angle of the bar, similar to the three link system.

The truck arm rear suspension system is still being used as of the date of this publication with Cup cars of NASCAR. It is a very strong system, but one that lacks proper adjustability for geometry and dynamics.

The Standard 4-Bar System – The 4-bar suspension is highly adjustable and can be made to steer in both directions. The rule about never steering the rear end to the right for an asphalt car does not apply to a dirt car. There are times when we definitely want the rear to steer to the right.

To determine the amount and direction of rear steer in this system, we must know something about the vertical movement of the corners of the car. On dirt cars, this movement can be very different than what we see on asphalt cars.

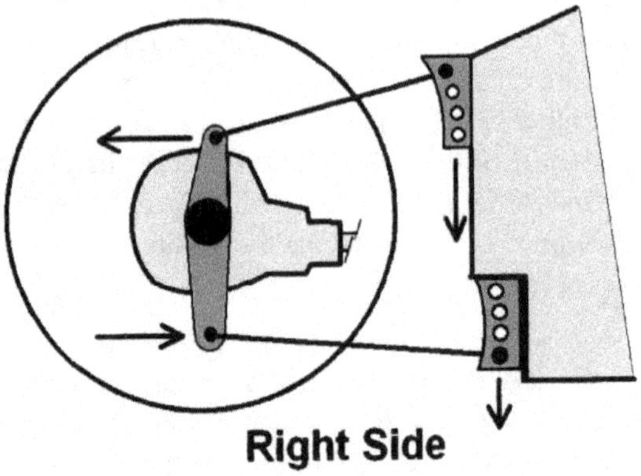

If the bars on a 4-bar car are set in the correct holes, the movement of the top and bottom mounts at the rear end will compensate fore and aft resulting in zero rear steer. This design is best for tight, tacky tracks where we do not need rear steer and we need for the thrust to be pointed straight ahead.

As previously stated, in most cases for dirt cars, a high degree of rear steer to the right causes the body of the car to run at a high angle to the direction of the path of the car. This enhances flat plate aero forces on the flat sides of the car to push the car towards the inside of the turns when going through the turns.

The Metric 4-Link System – The metric four link is a widely used system that comes with some models of stock automobiles and is mandated for use in the stock divisions for team using those cars where it is installed from the factory. It uses four links, two on top and two on the bottom, as the name implies that are not parallel to the centerline of the car. The top links are angled from a top view with the front chassis pivots much wider than the rear pivots. The lower links are angled from a top view with the front chassis pivots mounted much narrower than the rear pivots.

With this system, the rear end stays located laterally by virtue of the opposing angles of the upper and lower links. There may be rear steer inherent in this system depending on the ride heights because the original car was designed for stock ride heights. When we race a stock car, we usually change the ride heights. Under most current rules, there is no adjustment for amounts of rear steer with these systems.

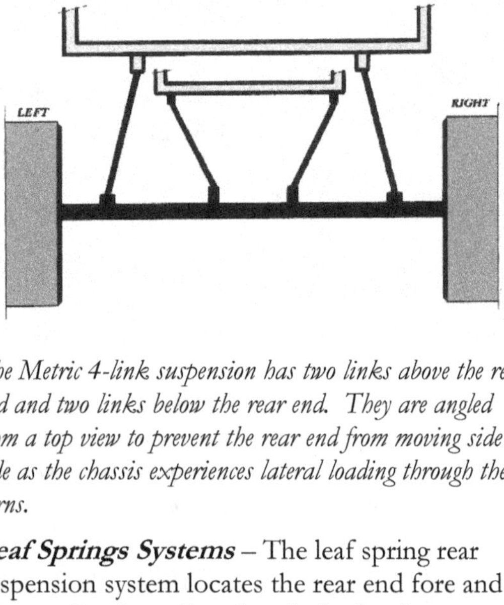

The Metric 4-link suspension has two links above the rear end and two links below the rear end. They are angled from a top view to prevent the rear end from moving side to side as the chassis experiences lateral loading through the turns.

Leaf Springs Systems – The leaf spring rear suspension system locates the rear end fore and aft as well as laterally using the leaf springs. There can be a small amount of rear steer as the chassis rolls and squats, but it is both minimal and mostly fixed as far as adjustability. The advantage of this system under certain conditions is that it keeps the rear end squared up and the thrust under acceleration straight ahead. It is useful if that is what is needed for a particular form of racing on certain race tracks.

The height of the front eye of the leaf spring in relation to the axle connection height determines the amount of rear steer. As the leaf flattens out from chassis travel, the distance between the axle and the chassis mount increases causing rear steer to the right. It pushes the right wheel back. Spacing the leaf different heights from the axle tube does not affect this action, and so the leaf spring system is very hard to adjust for rear steer.

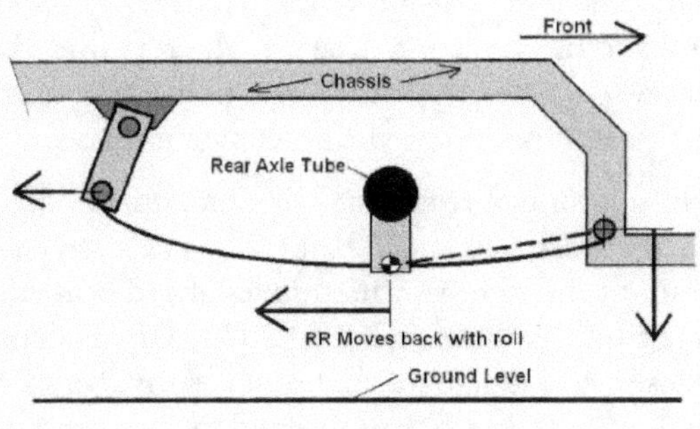

In the leaf spring system, as the front leaf mount moves down, the spring flattens out and moves the axle back. This creates rear steer to the right. Because the rear mount for the leaf spring is on a shackle, its motion will not affect the fore and aft location of the axle. Rear steer is hard to adjust for a leaf spring system.

The Z-link System – The Z-link rear suspension, or swing arm as it is sometimes called, is another system used mostly on dirt cars, but has been used on asphalt cars as well. Compared to the 4-bar cars, it has more limited adjustment for rear steer and historically has worked well on the tighter and more highly banked race tracks because the direction the rear end is pointed is more straight ahead.

Some manufacturers have added multiple mounting points on the front and rear chassis mounts on these systems to make them more adjustable like the four-link systems. This helps make the rear steer characteristics more adjustable for changing conditions, but can never make them steer like the four-link suspensions.

A Z-link suspension system uses a link extending from the under the rear axle tube forward to the chassis and one from the top of the rear axle tube rearward to a mount on the chassis. Most designs use very few mounting holes that would enable the team to adjust for the amount of rear steer in these cars. This one has quite a few. The Z-link was originally designed for reduced rear steer.

Measuring Your Rear Steer – To measure the rear steer in your car, first know the shock travels at the four corners of the car. Then you can recreate the rear attitude at mid-turn. You can use a tape measure for simple measurements or a laser system for more detailed analysis. With the tape, simply measure, at ride height, from the rear wheels towards the front of the car level with the ground to a point. This can be a mark on a tape near the front wheel, or the actual edge of the wheel well. For the laser systems, follow the manufacturer's instructions.

Move the rear wheels in any increment you choose to the position they are when the car is at mid-turn and then re-measure the distances. If the movement for the left and right wheels are in opposite directions, add them to determine the amount of rear steer. If the movements are in the same direction, subtract them. You might be surprised at how much rear steer is going on.

With asphalt circle track cars, we are finding ways to minimize rear steer on entry and through the middle of the turns, and then creating rear steer to the left to enhance bite off the corners through other means.

Creating Zero Rear Steer – Because in some systems the panhard bar causes rear steer as well as the rear link angles also cause rear steer, we can adjust the angles of those two in order to create offsetting steering for a Net Zero rear steer. In my many years in racing, I have never considered this concept until recently. Better late than never.

Suppose we can determine that the panhard bar, with an angle caused by a higher chassis mount than the rear end mount in a three link suspension for example, causes rear steer to the right. In this same system, our right trailing link, by having the front mounted level to, or lower than, the rear mount, causes rear steer to the left. If I adjust those two, the panhard bar angle and the link angle, to create equal but opposite steering angles, I will end up with NZ rear steer. And maybe that is what I want.

Or, maybe I want a combination of the two movements that will lessen the rear steer that might occur from just one of the two movements. I can then fine tune my rear steer if I know how the two movements are working.

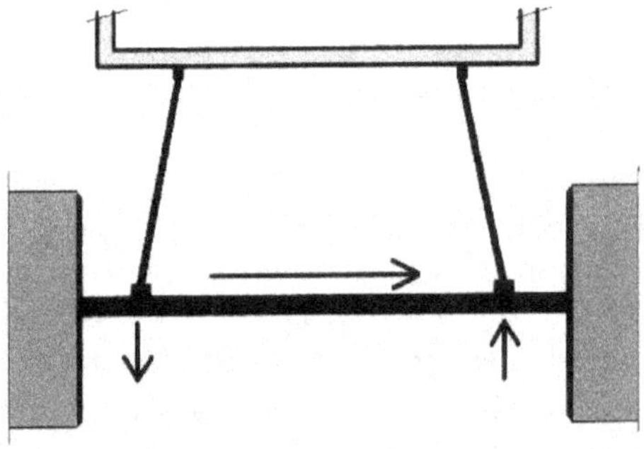

Most three link rear suspensions are designed with the forward mounting points for the outer links to be narrower, or closer to centerline, than the rear mounts attached to the rear axle tubes. With this design, as the rear end moves laterally, the rear end steers. If the rear end moves to the right, the wheels steer to the left.. The left link causes a greater distance from the axle tube to the chassis and the right link causes a lesser distance from the axle tube to the chassis. This motion steers the rear end and wheels to the left. The static panhard bar angle determines which way the rear end moves and which way it will be steered.

Rear Steer Under Power – Because the rear tires are providing all of their grip to keeping the rear of the car on the track through the middle of the turns, when the car begins to accelerate off the corner, the rear tires need more grip to handle the extra work. There are several ways to do that in keeping with the idea of not changing the middle handling characteristics.

Way Number One – The first way to create rear steer under power only is to use a pull bar, or lift arm. This causes the rear end to rotate when power is applied. The rotation is a product of the pinion gear trying to climb the ring gear in the rear end. This put a force into the rear end trying to rotate it clockwise from a left side side-view.

So, the rear end will rotate some amount when using a pull bar or lift arm. If we stagger the height of the trailing arms on a three link rear suspension, then when the rear end rotates, the left axle hub will move rearward more than the right axle hub. This produces rear steer to the left and adding Angle Of Attack to the rear tires. This provides more grip for the rear tires to counter the lateral forces and the accelerating forces.

The amount of rear steer is limited, but we don't need, or can use, much rear steer. The average amount the right rear tire moves ahead of the left rear tire is in the range of 0.050" to 0.075". This doesn't sound like much, but it is sufficient to do the job. Any more than this and the car becomes too tight off the corners.

When using a pull bar third link such as this one, as the car accelerates, the bar is pulled back due to the engine torque trying to rotate the rear end. With this movement, the rear end moves back. To produce rear steer, we stagger the heights of the trailing links, the right side higher than the left side links. With this staggered height, the left side moves back more than the right side creating rear steer to the left. The actual difference in movement between the left and right wheels is in the range of 0.050" to 0.075". This steer adds to the AOA of the rear tires and helps provide more rear grip.

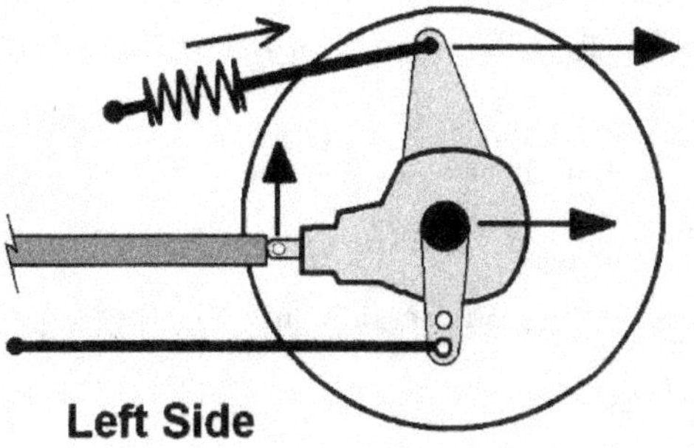

As described above, we mount the left side link in the bottom holes on the bracket coming off the rear end axle tube on the left side.

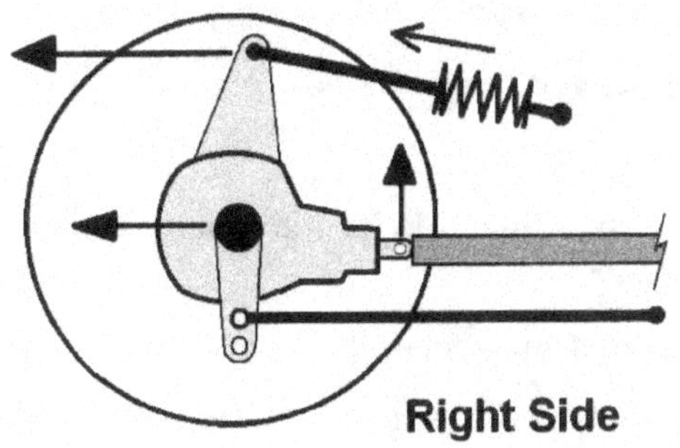

On the right side, we mount the link in the top hole of the bracket coming off the axle tube. The distance that the right side moves rearward comparted to the left side is less. This steers the rear end to the right of the chassis centerline creating rear steer to the left.

Way Number Two – Another way to produce rear steer only under acceleration is to install a push-link or compressible right rear link in a three link suspension. This can also be done with the right side links on a Metric four link suspension.

As the rear tires drive forward under acceleration, a high amount of pushing force is transmitted into the suspension links. If the right side link, or links, are compressible, or able to compress, then the link becomes shorter and the right wheel moves forward and the rear end steers to the left providing more AOA.

The amount of movement of the right wheel forward using this method must be regulated so that it does not become excessive. With the Way Number One method, we see where we only need an average of 1/16th of an inch of movement to do the job. If your compressible link moves ¼ inch or more, obviously that is too much in most cases.

This is called by several names including push rod, spear rod, or just compressible link. As the right rear tire pushes forward during acceleration, the rubber biscuit compresses and the link shortens. This movement steers the rear end left of centerline and creates more AOA for the rear tires. The amount we allow this link to move is critical to how much effect it will have. It can be overdone by far. Movement of 0.250" to 0.375" is not uncommon and that is entirely too much in most cases. The rubber can be pre-compressed so that the movement is reduced.

The beauty of these two methods is that the rear steer only occurs under acceleration and does not interfere with the balance of the setup through the middle of the corner. And, these two can be tuned so that you get the exact amount of rear steer you need for your car at your race track.

Conclusion – Rear steer is an important concept to understand and apply and in reality, one that can either enhance our setups, or ruin an otherwise great balanced setup. It's use and magnitude is specific to the type of racing you will be doing, but to ignore this mechanical effect means you are tossing the dice on how your car will perform on the race track.

Rear steer is one of those chassis setup areas that needs to be addressed, evaluated closely and applied in measured amounts. Through careful analysis and planning, rear steer under power can enhance the overall performance of any race car.

Exam - In The Context Of This Lesson:

The Primary Purpose for Rear Steer Is To?

1) Loosen the car

2) Tighten the car

3) Create additional angle of attack

4) Make it easier for the driver

Virtually All Cars Driver Steer From The Front Because?

1) It is too expensive to design rear steering cars

2) The technology isn't well developed

3) It is closer to the steering wheel

4) The AOA geometry is all wrong when steering from the back

Rear Steer Can Be Caused By?

1) Chassis dive

2) Chassis roll

3) Acceleration forces

4) All of the above

Is Zero Net Rear Steer Possible?

1) Yes

2) No

We Can Regulate The Degree Of Rear Steer By?

1) Changing link angles

2) Changing panhard bar angles

3) The use of compressible links

4) All of the above

Acceptable Levels Of Rear Steer Are Measured As?

1) Ten thousandths of an inch

2) Hundred thousandths of an inch

3) Inches and fractions of inches

4) Degrees and decimals of a degree

The Least Adjustable Rear Suspension For Rear Steer Is?

1) The three link system

2) The four-link system

3) The truck arm system

4) The leaf spring system

Rear Steer Can Be Used To?

1) Provide more AOA

2) Provide less AOA

3) Help create flat plate aero for added side force

4) All of the above

Race Car Technology – Level Three

Lesson Twelve – What Makes Traction? A Review

Here we want to review the three major contributors to making traction. This is a combination of Lessons 2, Lesson 6 and Lesson 8. We want to emphasis the basic methods we use to gain grip, or traction. Once we have these three methods firmly in our grasp and fully understood, we can proceed to working out our setups.

Race car performance is all about gaining speed in the slowest portion of the race track and then carrying that speed all of the way around the race track for a faster overall average speed. That is what makes up a lower lap time.

A race car that has all of the characteristics of what makes traction will carry a lot of speed through the turns and this will make the speed around the entire lap faster. In this Lesson we will review What Makes Traction.

Load Distribution On The Four Tires – In almost every situation for every type of race car, we will never have equally loaded inside and outside tires. This is due to the concept of weight transfer during cornering. Even with a biased left side weight common in circle track racing, we will never be able, under the current rules sections, to have enough left side, or inside, weight to overcome the weight transfer amount. So, we will never have dynamically equally loaded left and right side pairs of tires.

If you remember in Lessons 4 and 6, we discussed how load transfers during cornering and how we needed to end up with equal loading on the outside tires, and equal loading on the inside tires in order to get the most traction out of each pair of tires at each end of the car.

Once we have developed a balanced setup where we can predict and plan for the ideal weight transfer and load distribution on the four tires, we will then have maximum grip from those four tires. But it doesn't stop there. There are two more concepts of making traction that we need to address and perfect.

In short, equally loaded tires provide the most traction from an opposing pair of tires. Equal un-equal loading provides the most traction for a set of four tires on a race car. If we have equal amounts of un-equal loading on the pairs of tires on the two axles, then we have the most overall race car grip possible from a tire loading standpoint.

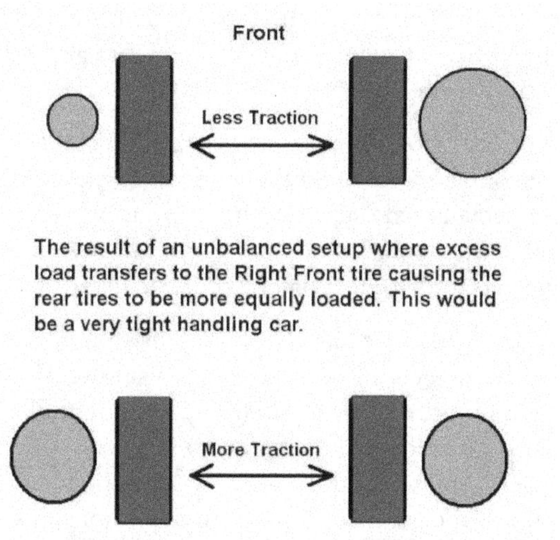

Two tires on the same "axle", or same end of the car, will have the most traction when they are equally loaded. We covered this before explaining that even though two tires have the same total load, when they are un-equally loaded, they have less grip as a pair. On this car, the rear has more equally loaded tires and therefore more grip than the less equally loaded front tires. This car will be tight and not want to turn very well.

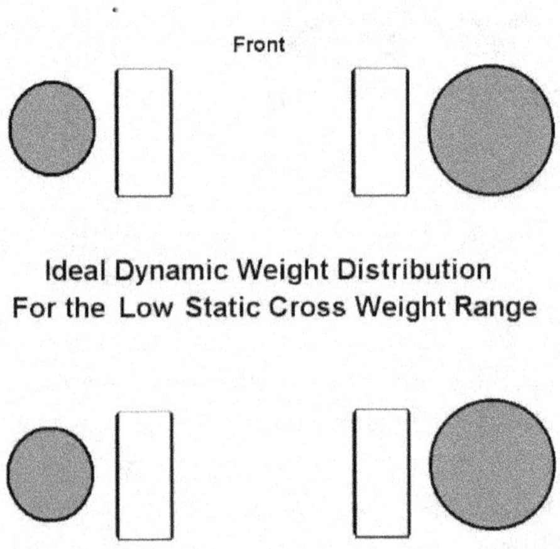

With a few changes to the setup we can create a setup where the tire pairs at each end are what we call equally un-equally loaded and therefore have the same grip level. This car is neutral in dynamic tire loading and neutral in handling.

Contact Patch Size - In the Lesson 20 in RCT Level Two school, we discussed how having the right camber can increase the tire contact patch size. If we have maximized the loading on the four tires, then if we can increase the tire contact patch area, we will then have even more grip from that tire.

The discussion mostly involves maximizing the tire contact patches on the AA-arm suspensions. The straight axle suspension is hard to work with and limited on what you can do to improve the cambers and contact patch for those tires.

So, we work to get a proper load distribution on the four tires, and now we work to get the maximum tire contact patch size to better utilize that perfect loading. But we also need to understand how a tire makes lateral grip to counter the centrifugal forces trying to push us off the track.

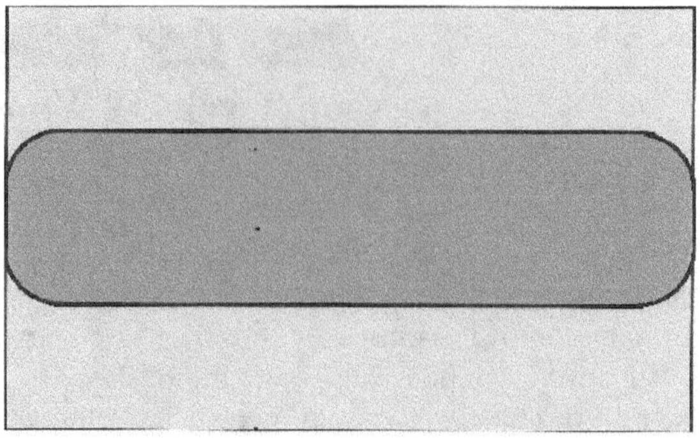

A front tire contact patch can be maximized by adjusting the camber of the tire. We want to adjust the front cambers using tire temperature and/ or wear patterns to tell if the camber can be improved. We then need to maximize the pressure distribution on the contact patch to create the largest contact patch possible. This sketch shows how our contract patch pressure distribution would look if the tire temperatures were even across the face of the tire. This is not ideal for the front tires on a circle track car. It may well be perfect for a race car with wide tires with a low sidewall profile.

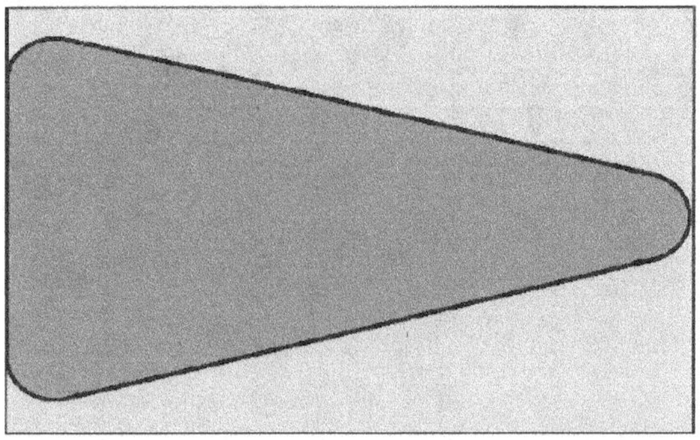

If we introduce more camber to a tire run on most circle track race cars, where the inside edge temperatures are higher by more than 15 degrees from the outside edge, and reduce the pressures, then the contract patch would look somewhat like this. The area of this contact footprint is larger than when we had less camber and uniform tire temperatures across the contact patch. Therefore, it has been proven that this cambered pattern will have more grip for those cars. For wider tires with a low sidewall profile, this may not provide the stated banefits.

Angle Of Attack – The third ingredient for making traction is creating angle of attack, or AOA. When we have the perfect loading on the tires and the maximum contact patch we can possibly achieve, we will still not have traction and lateral grip if we don't have AOA in our tires.

AOA is, as discussed in Lesson 8, how we produce Centripital force, which is the force the tires use to counteract the centrifugal forces pulling the car to the outside of the turns. Something must resist the lateral forces, or G-force caused by cornering. That something is AOA.

Both the front and rear tires must have some degree of AOA in order to negotiate a turn at high speeds. The exact amount of AOA needed depends on the speed, banking angle, tire grip levels, and track surface grip levels. Our speed is really regulated by how much lateral grip our tires can produce. The more grip the tires produce, the faster we can go.

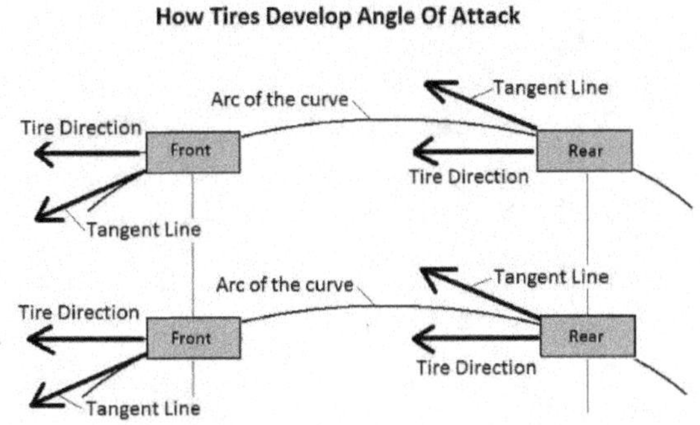

This sketch used in a previous Lesson shows how we need Angle of Attack to generate the resistance to lateral forces commonly called lateral G-forces. A tire cannot resist the G-forces if they are not steered, or pointed, to the left of the arc tangent line. The front tires must be steered to a slight angle past and inside the tangent line. The rear tires already have an AOA angle and we may need to add to, or take away from that angle to provide the tire with the proper AOA for the race track we are racing on.

Make sure you fully understand these concepts of how traction is made. If you need to go back and review the previous lessons that explain in detail these concepts, by all means do that. We will now proceed into the setup part of the Level Three of the ORS where we will plan out and evaluate our setups.

Exam - In The Context Of This Lesson:

The Ideal Dynamic Load Distribution Is?

1) When all four tires are equally loaded

2) When the front and rear pairs of tires are equally un-equally loaded

3) Never possible under the current rules packages

4) Always possible for every setup combination

For Circle Track Race Cars, The Best Contact Patch Is?

1) When the tire pressures are ideal.

2) When the heat on the inside of the tire is more than at the outside

3) When there is even tire temperatures across the face of the tire

4) When there is even wear across the face of the tire

To Provide Proper Angle of Attack, The Front Tires Are?

1) Already at some angle of attack

2) Needing to be steered parallel to the tangent line

3) Needing to be steered just inside of the tangent line

4) Needing to be steered to follow the arc

To Provide Proper Angle of Attack, The Rear Tires Need?

1) To be steered just outside of the tangent line

2) To be steered parallel to the tangent line

3) To be steered just inside of the tangent line

4) To be pointed to follow the arc

Race Car Technology – Level Three

Lesson Thirteen – Force Verses Weight

In the process of setting up and evaluating a setup, we need to understand the relationship of Force verses Weight. Once we have knowledge of how they relate and what each one is, we can then move on in our studies. This is a critical step that is necessary to go through before we move on. It will become very apparent once you begin the setup portions. This is a short presentation, but one where the concept will be referred to over and over again.

Weight is easily measured and read on common racing scale pads. We weigh the car at static ride height on a level surface. This provides the gravitational loading on each tire. We can adjust the distribution of the loading easily on a race car.

Weight Defined – Weight is the measure of the force of the gravitational pull on a body and is read in pounds, or any other denomination of measurements. Everything on Earth is pulled towards the center of the earth by a force called gravity. In a race car, we can measure static weight using scales. That is, when the car is not moving, we place the car onto scales designed to provide a reading of the amount of weight that tire supports.

In a race car, going around a turn, the weights on the four tires change. They change because of 1) weight transfer, and in the case of a banked race track, 2) mechanical downforce. The mechanical downforce is added to the gravitational weight. And, mechanical downforce is the combination of gravity and the lateral force we know as centrifugal force.

If we could measure that combination of centrifugal downforce and gravitational weight that is a result of weight transfer, we would be able to determine the spring force we would need to support that combined weight. But, I'm getting ahead of myself. Let's now define Force.

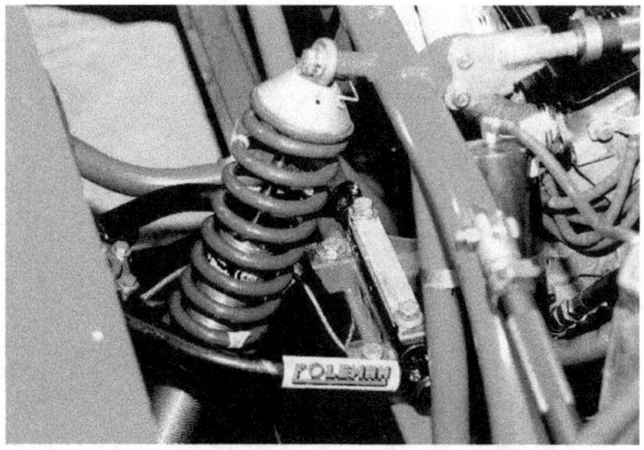

We can easily measure the length of the spring when it is installed in the car at normal ride height. We subtract the installed length from the free height length (while not under load) and multiply that difference times the spring rate in pounds per inch. This gives us the force the spring is exerting on the suspension to support that corner of the car.

Force is force and a solid link will exert the same force as a spring to hold up this corner of the car at normal ride height. If we had a load sensor installed in the solid link, we would read the same force as the spring would provide to hold this corner of the car at ride height.

Force Defined – Force is the energy needed to support weight. If the load on the right front tire on our race car weighs in at 650 pounds on the scale, it would take some amount of spring force to support that weight. For a normally setup asphalt late model, the spring force would be in the neighborhood of 985 pounds. We found that by squaring the motion ratio of wheel travel to spring travel (0.660) and dividing that number into the scale reading.

For a balanced setup, there is a load at each wheel that represents the ideal loading for the four tires at mid-turn at speed. If we can determine the spring travel that provides that loading, we can then determine the spring force being used by the suspension.

In the place of data acquisition, we can record the maximum spring travel that represents the total spring force needed to support this corner of the car. We need to make sure we don't influence this reading by heavy braking into the corners that might present a greater amount of spring travel than what occurs at mid-turn.

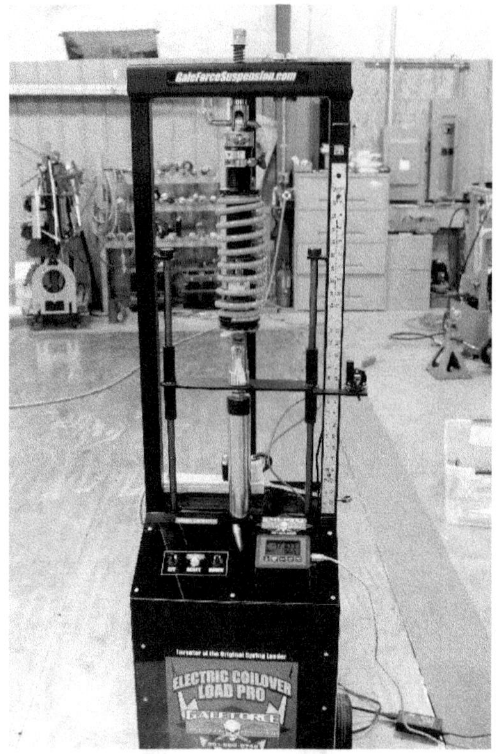

The Gale Force Load Pro measures the loading on the spring at mid-turn shock travel for a coil-over spring. By compressing the spring to the mid-turn length, the loading is then shown and compared to what we determine to be ideal loading.

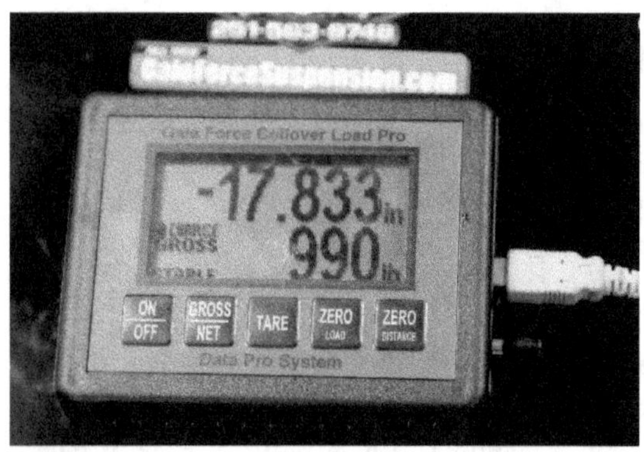

Here is the readout on the Gale Force Load Pro with a car at ride height. This number comes very close to our sample car force calculation of 985. On this machine, the team would continue to travel the spring until it reached the mid-turn length and then read the force directly. There is no other way to measure the dynamic spring force that represents the car on the track.

Calculating and Measuring Force - If we installed a spring with a rate per inch of 250 ppi (pounds per inch), then if we divide 985 by 250, we get a spring travel of about 4 inches. In fact, that is exactly how we can determine the force a spring is providing. If we measured the installed

250 ppi spring at 10 inches in length with no load on it, then if we measured it at a length of 6.0 inches on the car, holding up the RF corner, we would know the spring force was 985 pounds.

If we somehow knew, or could calculate the load in pounds that the RF tire would support at mid-turn, at speed, we could then determine what spring force would be needed to support that dynamic weight. In previous calculations using a sample race car, we came up with a RF dynamic tire load of 1575. The force needed to support that load is calculated the same way as when it was at rest. The ratio is the same.

With the static weight verses dynamic weight ratio of 0.6600 (650 / 985), if we divide the RF corner dynamic load of 1575 by that ratio, we come up with 2,386 pounds of force. The ride spring force, plus the bump force, plus the sway bar spring force would all need to add up to 2,386 pounds to support that weight.

This is the essence of measuring and evaluating a setup. If we can determine the ideal loads the tires would support that would create the ideal dynamic load distribution, we could then compare the ideal spring force to what the car was actually doing and then make adjustments to bring the setup into balance. See Lesson 4 and 5 for more information on that.

The sway bar can provide a spring force that will help support the Right Front corner of the car. The sway bar is a spring and when twisted, or otherwise put into play, adds to the overall force needed to support the cars suspension at mid-turn.

Summation – Get this concept firmly in your grasp because we will be working with weight verses force as a major part of race car setup in the remainder of Level Three. We discussed some of this in Lesson 7, but we feel the need to drive this home to the students. It is so much a major part of working out the balance of the setup as we get into the rear world of race car setup and tuning.

Exam - In The Context Of This Lesson:

Weight Or Load Is The Measure Of What?

1) Spring Force

2) Gravitational pull on a body on earth

3) The combination of gravitational pull and centrifugal lateral force

4) 2 and 3

Force Is The Measure Of What?

1) Ride spring under compression

2) Bump device under compression

3) The sway bar under torsional loading

4) All of the above

Which Provides The Force Needed To Support Dynamic Loading?

1) The ride spring

2) The bump device

3) The sway bar

4) All of the above

Race Car Technology – Level Three

Lesson Fourteen – Where Does Force Come From?

Now we have defined weight verses force and now know that force is the mechanism that supports weight or loading. We need to now explain where this force comes from so we can know how to make changes to the forces and the contributors of those forces in order to setup the race car.

The knowledge of how the forces are administered and where they come from is relatively new. Individuals and companies who specialize in race car dynamics prediction and evaluation that have existed over the past five to ten years have combined their knowledge into what we are presenting in this school and in these lessons. It's not like this is old school knowledge, quite the contrary, this is extremely new school.

The force supporting the front suspension, and especially the right front, on this super late model asphalt race car is a combination of the ride spring force, the bump device force, and the sway bar force. We'll explain how all of that works in this Lesson.

Force Comes From Ride Springs – The first place the forces that hold and support the loads at the four tires comes from is the ride springs. This is fairly obvious, but important to point out and understand. Because there are different designs of race car, we need to define what the various ride springs are and how we might measure the force on those springs.

In a modern race car, the ride springs can be either coil-over springs, stock type "big" springs, torsion bar springs, and leaf springs. Each of these springs has a unique way they are mounted in the car. Let's examine each type of spring one at a time in relation to its role in creating force.

The Coil-over Spring – This is a very common type of modern spring that is found in not only pavement late models, but also dirt late models, dirt and asphalt modifieds, quarter midgets, prototype cars, and formula race cars. I might have left out a few types, but you get the picture.

The coil-over spring is mounted on a shock body most of the time. It can be mounted on a slider for some dirt car applications. It operates with a motion ratio meaning that it moves a lesser distance than the wheel, or in some cases a greater distance. It all depends on the motion ratio.

As covered before in previous lessons, the force is measured by dividing the motion ratio into the wheel weight. That is, whatever weight the wheel and tire supports, be it at static ride height or on the track at mid-turn, we can know the spring force by dividing the wheel load by the square of the motion ratio. This gives us the total force needed to support that weight or load, but it might not all be supported by the ride springs.

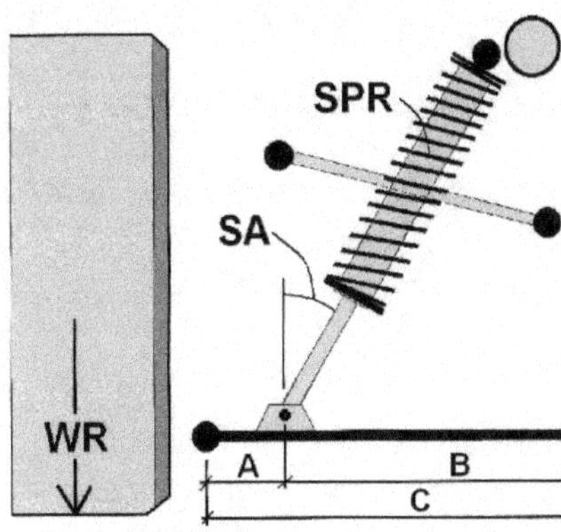

The force supplied by the ride spring is calculated using the motion ratio divided into the load the right front tire carries both statically on the scales and dynamically on the race track at mid-turn.

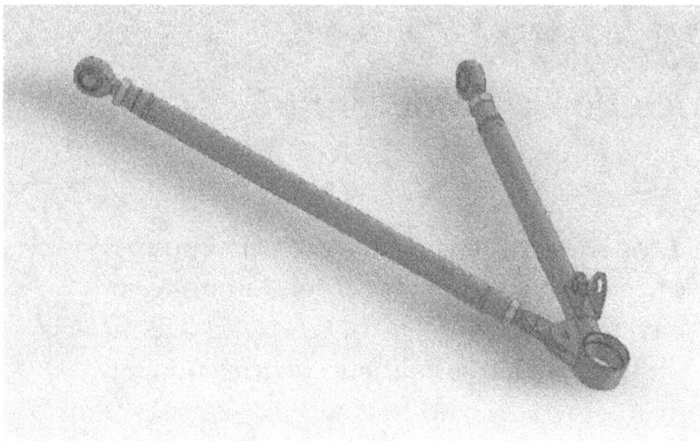

This sketch of a right side lower control arm on a super late model car shows the main link (right side), and the coil-over mount tabs. The tabs are about 2.50" from the center of the ball joint housing in red, and about 15.0" from the chassis mount for our example. The left link is the strut going back to the chassis.

Formula cars run coil-over shocks in most of the classes. The tire/wheel is connected to the rocker through a push rod (angled link shown in the right front going up to the upper part of the body). Then the rocker, inside the body, is connected to the coil-over. The motion ratio for this car is measured the same as for a late model, the movement of the coil-over divided by the vertical movement of the wheel.

The "Big" Spring – What we normally call a big spring car is one where the front, and rear in some cases, is supported by a larger spring that resembles the stock spring that comes in certain production automobiles. In stock car racing, this spring is mounted between the lower control arm and the chassis at the front of the car, and onto the rear axle or truck arms at the rear.

This is a much better depiction of a formula car and the coil-over design. The link coming from the lower control arm (center link on the right side of the photo) is connected to the rocker (black piece to the right of the right spring). Then this pushrod rotates the rocker to push the coil-over in compression. To find the motion ratio, we remove the coil-over, place a measured thickness of block under the tire and then measure the movement of the mounts for the coil-over. If we space the wheel up 1.00" and the coil-over compressed 0.75, then our motion ratio is 0.75 / 1.00 = 0.750. To find the force to support the wheel load, we divide the motion ration squared (0.5625) into the load amount. For a 600 pound load, we need a force of (600 / 0.5625) 1,066 pounds.

As for the front mount, the motion ratio is much different than we find in the coil-over application. The big spring is mounted much farther from the ball joint and therefore must be much stiffer in rate to support the same wheel weight verses the coil-over spring.

Whereas the coil-over spring might have a motion ratio of 0.850, a big spring car might see a motion ratio of only 0.645 or so. When we square those numbers, we get a value of 0.722 for the coil-over and only 0.416 for the big spring. It takes 1.7 times the coil-over spring rate for a big spring to support the same load. If we installed a 150 ppi (pound per inch) rated coil-over spring, to create the same force in a big apring, we would need a 265 pound spring. This is important for those who run the big spring cars and want to emulate the coil-over setups.

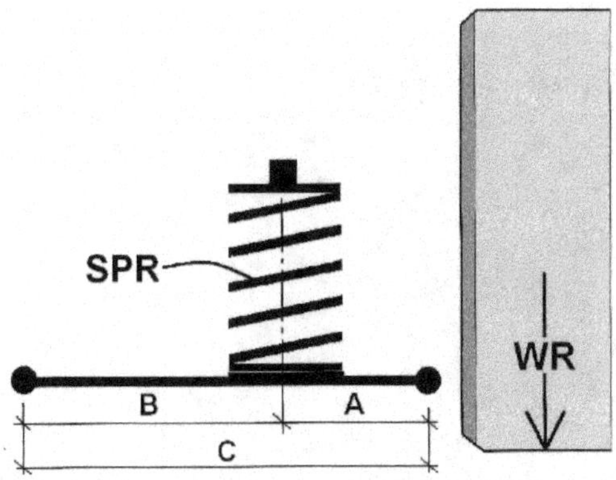

The motion ratio for a big spring car front suspension is calculated the same as a coil-over car, B divided by C, A being the distance from the ball joint to the center of the spring. Because the bump for these cars is on a much different motion ratio, we would translate this spring rate to a similar spring rate as if it were a coil-over mount. Since the bump is usually mounted on the shock mount, we would use the shock motion ratio for this calculation. Then we can just add the ride spring converted rate to the bump spring rate.

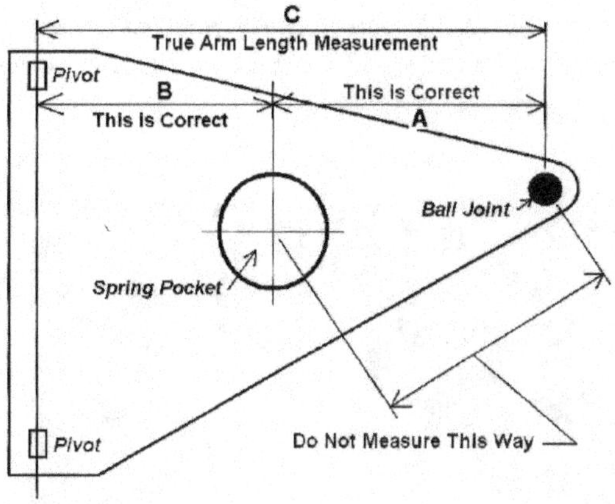

The correct way to measure the lower control arm length for a stock dimensioned lower control arm is this way. The measurement from the ball joint and center of spring to the chassis mounts is at a ninety degree angle to those mounts.

The Torsion Bar Springs – These types of springs are mostly used in sprint cars. Some older stock cars have used these springs because the stock automobiles used them. They are not popular in most modern race cars.

The rate is affected by the length of the bar, the diameter and wall thickness of the bar, and the arm length of the torsion bar arm. A sway bar is a type of torsion bar, but we'll get more into that later on. And, the torsion bar has a motion ratio depending on where it is mounted to the lower control arm.

Sprint cars usually use torsion bars as the suspension springs. Some use torsion bars in the rear and coil-overs in the front. It all depends on the design of the car. The torsion spring is much like a sway bar, but does actually act like a spring and not an anti-roll device.

Here we can see the arm mounted to the bar that runs through a tube under the rear body under the front of the number "8". The silver arm just to the right of the left rear tire runs forward to lay on top of the rear axle tube. Above that arm is the ride height/preload adjuster for the right side torsion arm.

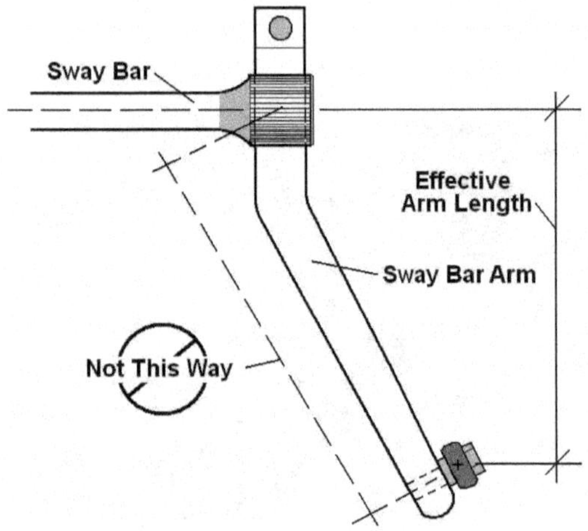

When we refer to the arm length for a sway bar or torsion bar, we car referring to the Effective Length and that is the distance at right angles to the bar of the center of the mount to the lower control arm to the centerline of the bar. The sketch shows the way to measure the arm length.

The Leaf Spring – The leaf spring is a type of ride spring normally used on the rear of a straight axle race car. Its rate depends on the number of leafs, the thickness of the leaf, and length of the leaf. The leaf springs are almost never used on the front of a race car, so for this discussion, we'll not include leaf springs for providing force.

Force Comes From Bump Springs - Every type of bump device is a spring. The different types react differently to the application of force. Whereas a coil bump spring and the bellows springs have linear spring rates, the bump stop type of spring has a progressive spring rate in compression, and digressive rate in rebound.

The bump spring only comes into play when the chassis has traveled far enough for the bump device to come in contact with a stop that represents the chassis. For a coil-over mounted bump, it is the shock body that contacts the bump device. When it comes into play, its rate is added to the rate of the ride spring in relation to the motion ratio of the bump to the motion ratio of the ride spring.

The wheel rate of a bump spring is dependent on the motion ratio, just like the coil-over and big springs are. If the bump is mounted on the shock shaft, then it has the same motion ratio and wheel rate ratio as the shock. If it is mounted separately from the shock, then it may have a different motion ratio and different wheel rate ratio. Remember this point.

This steel coil spring bump is now in play since the shock body has traveled far enough to make contact with the bump spring. The bump spring rate is added to the ride spring rate. When we measure force, we compress the shock to the mid-turn length and then read the force. This number represents the ride spring force and the bump spring force. If we remove the ride spring and compress into only the bump spring, we can know the force generated by the bump spring only.

This is a commonly used bump that is very soft and very non-linear in spring rate. It gains rate as a slow pace and then at a certain point begins to rapidly gain spring rate and force.

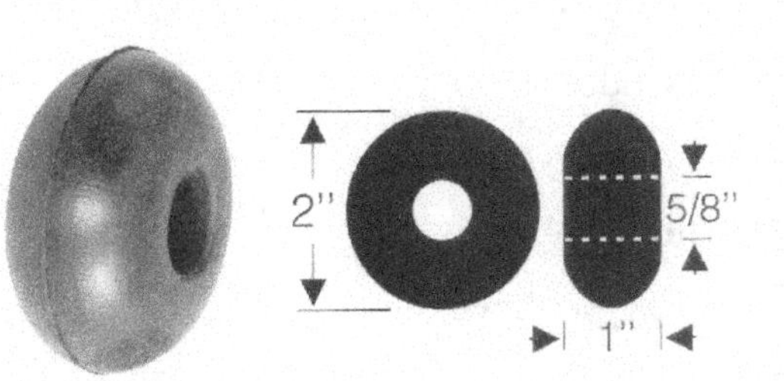

This type of bump stop is harder and provides more spring rate. These can be stacked for more travel and a softer bump spring rate. At some point of compression, these go basically solid.

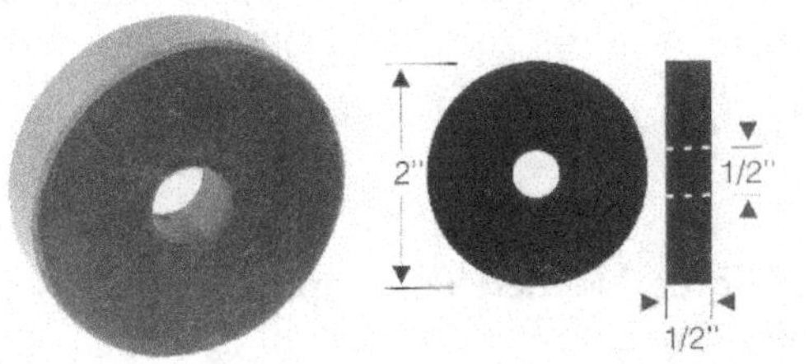

The "hockey puck" type of bump device has less travel before going solid, but provide a soft, yet stiff, spring rate. These too can be stacked for more travel and less spring rate.

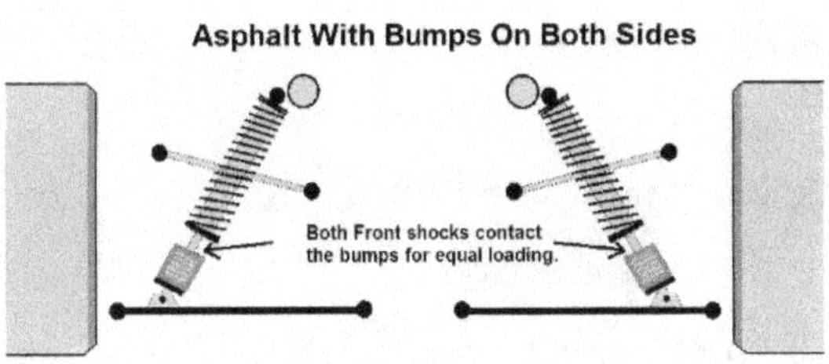

This illustration shows how bumps are mounted on the shock shaft to become inline with the ride spring so that we can add the forces and spring rates. If we mount the bumps on a fixture separate from the shock, mostly for convenience in making changes, then it would be a good idea to make sure that the motion ratio is the same for both the coil-over ride spring and the bump fixture.

Force Comes From The Sway Bar – The sway bar is a torsion spring, like we said previously. In chassis roll, it acts just like a ride spring. It also has a motion ratio for each side of the suspension. This motion ratio is calculated just like a coil-over or big spring, by its mounting distance from the ball joint verses the lower control arm length.

If the motion ratios are different side to side, then the sway bar can become like a ride spring. This is good to know and we'll delve into that deeper in the next Lesson. Just know that the larger the sway bar diameter, the stiffer the spring for roll resistance as well as ride spring uses.

In determining the total force needed to support the tire loads at mid-turn, the contribution of the sway bar spring rate and its force needs to be taken into account. Again, we will delve into that more later on, but at this point, know that the sway bar spring rate can contribute to the aggregate force needed to support the tire loads.

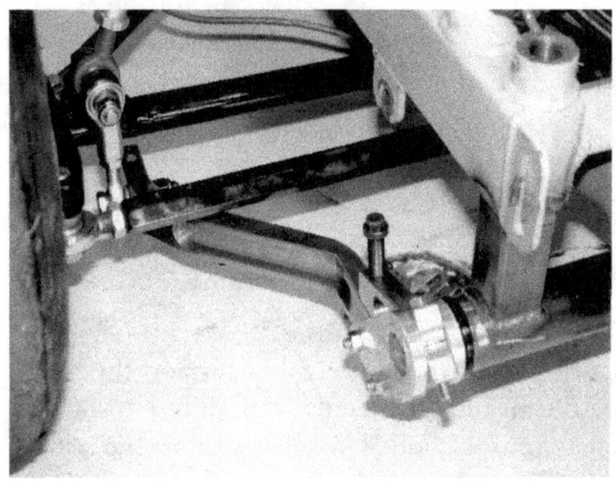

The sway bar is actually a torsion bar and provides an additional spring rate, mostly coming into play when the chassis rolls. For this installation, the sway bar has nearly the same motion ratio as the ride spring in the coil-over assembly because it is mounted to the bottom shock mount. There would be a slight difference in motions due to the shock mounting angle. We could find a point along the lower control arm where the motion ratios are then exactly the same. That point would be a short distance towards the chassis mount of the lower control arm from where it is now.

This road racing rear sway bar illustrates the idea that it is an added spring for the purpose of chassis roll. For these cars, with a solid axle rear suspension, the ride springs can be softer because the sway bar helps reduce chassis roll in the rear to match the front and rear roll angles for a balanced setup.

Summation – This is a very important Lesson. To review, the force needed to support the tire loads can, and does, come from different sources. These must be added together to come up with the force we will need to balance our setup.

The force can come from the Ride Spring, the Bump Device and the Sway bar as a spring. How we add those different springs together with their much different motion ratios will be explained soon. This is the art of modern day race car setup.

Exam - In The Context Of This Lesson:

The Concept Of Measuring Force Has Been Around For A Long Time?

1) True

2) False

Suspension Force Comes From Where?

1) The ride springs or torsion bars

2) The bump devices

3) The sway bar

4) All of the above

Motion Ratio As Discussed Is Defined As?

1) The wheel travel divided by the shock travel

2) The spring travel divided by the wheel travel

3) The sway bar arm travel divided by the wheel travel

4) The bump device travel divided by the wheel travel

5) 2, 3 and 4

The Force Needed To Support A Load Is Found How?

1) By multiplying the spring rate times the motion ratio

2) Dividing the spring rate by the motion ratio

3) By dividing the wheel load by the motion ratio

4) All of the above

A Sway Bar Can Be What?

1) An anti-roll device

2) Another ride spring

3) A torsion bar

4) All of the above

A Bump Device Can Be What?

1) Another ride spring

2) Linear in spring rate

3) Non-linear in spring rate

4) All of the above

Lesson Fifteen – The Sway Bar As A Spring

In Lesson Thirteen, we discussed how the force needed to support the chassis at both ride height and at mid-turn is supplied in different ways and the sway bar is one of those ways. In the context of that Lesson, the sway bar contributed as an anti-roll mechanism and had a spring rate that caused the stiffening of the suspension that reduced the roll angle. Now we will look at a different use for the sway bar.

The effect we will be studying is only used or to be applied to non-symmetrical AA-arm front suspensions that run on circle tracks and only turn one way. I know of no good use for this with symmetrical chassis like a road racing car.

If you are interested in road racing technology, this won't necessarily apply, but it's not a bad idea to understand the principle. That is because in some types of road racing designs such as a circle track car that was converted to road racing, the linkages or designs may lend themselves to be A-symmetrical by accident. This Lesson may help you identify that condition and make the needed corrections.

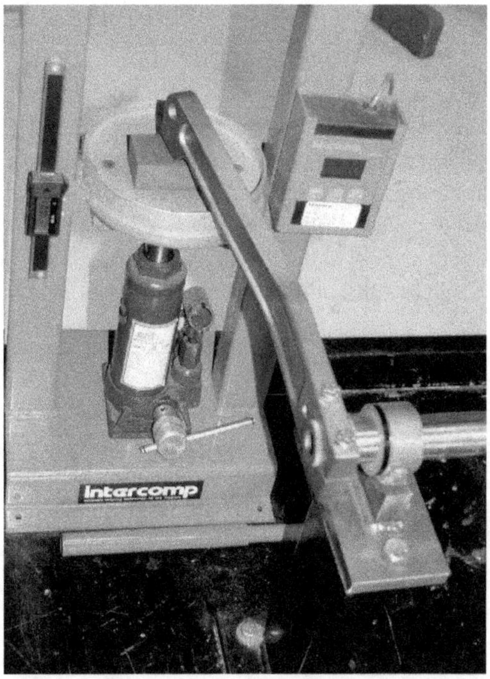

With the three-piece bar, we basically have a torsion bar and we can find the rate at the end of the bar using a fixture. We can also measure the force that bar produces on the car translated to the coil-over position with a Gale Force Sway bar tool.

We can rate a sway bar on a fixture, like this team did. That way we know exactly what the rate of the bar is at the end of the arm. This takes away all guess work. This one-piece bar will rate lower than an equal sized bar with equal effective length and equal length arms. That is because the hardness of the metal is less than with a three-piece torsion type of sway bar.

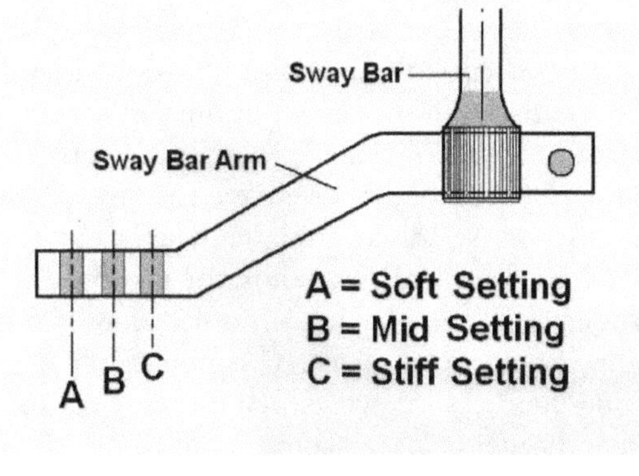

The rate of the bar is affected by the arm length. Some designs allow the team to change the length of the arm in the installation. A shorter arm provides more spring rate. If using this technique, be sure to move the control arm mounting holes too so that the link is parallel to the control arm and the sway bar arm.

Sway Bar In Roll – The sway bar, or anti-roll bar, was designed to reduce the roll of a AA-arm suspension by basically adding spring rate to each side of the suspension when the car rolled, and only when the car rolled. Well, that was the theory, and it works when the suspension is truly symmetrical.

So, if we have the same motion ratio on each side of the car, as the car dives, the sway bar has no anti-roll characteristics and the ends of the sway bar arms move the same amount vertically. There is no motion causing the bar to twist, so no anti-roll effect exists. That is the theory, but the application is a little different in most cases.

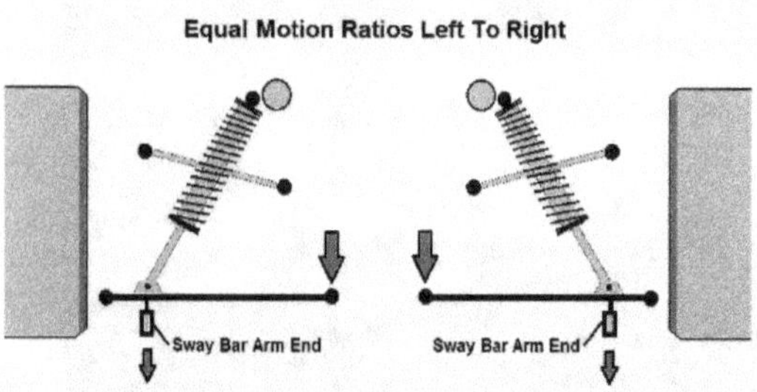

If the two sides are symmetrical, then the sway bar arms will move an equal distance as the chassis moves in dive. With this design, there will be no twisting of the sway bar caused by chassis dive, only chassis roll. This is common in road racing cars, but uncommon in circle track cars. If a circle track car has been converted to road racing, then the team must make sure that the motion ratios are the same side to side.

This is a concept little known or understood until very recently. With the advent of the soft spring setups where the car is made to run on bumps, and travel quite a bit, we discovered a phenomenon. We discovered that when we pushed the front end down onto the bumps, the sway bar, which was neutral at normal ride height, was now pre-loaded. The motion of moving the chassis down caused the sway bar to twist and become loaded some amount.

This is a major discovery and leads us to much more in the understanding of the sway bar and its contribution to the overall setup of the car and its contribution to the forces we need to support the cornering loads on the front tires.

The pre-load force of the sway bar can be adjusted using several different methods. This does not change the rate of the bar, only the pre-load and ultimate force of the system that includes the ride spring and the bump device.

For a race car using a one-piece sway bar, this is how it is adjusted for pre-load. The actual mounting point of the sway bar is moved up or down to twist the bar and add pre-load.

Sway Bar In An A-symmetrical System – For many circle track cars with offset chassis, and some that are called a perimeter, or symmetrical, the motion ratio of the sway bar mounts is not the same side to side. So, when the chassis moves vertically, and separate from the roll motion, the two ends of the bar do not move the same distance. In most cases, the left arm end moves less than the right arm end.

Relative to the chassis, the right arm end moves up more than the left arm end and this simulates what happens with chassis roll. Therefore, the sway bar supplies an anti-roll force with just chassis dive. Then when the chassis rolls, the bar is twisted to a greater extent.

Both of the actions, twist from chassis dive and twist from chassis roll, provide a force to the right front suspension that is added to the ride spring force and the bump force. If we only consider the ride spring and bump forces, we will be leaving out what could be a significant contributing force that helps support the mid-turn loads.

We have talked about recording the shock/spring travels so that we can simulate the force generated by the ride spring and the bump device. But what about the sway bar force, it needs to be measured too or we won't be able to properly adjust the ride spring and bump forces to what they need to be.

Suppose we need 2300 pounds of force to support 1500 pounds of loading on the right front tire. If the sway bar is supplying 400 pounds of force all by itself, then if we set the ride spring and bump combo to supply 2300 pounds of force, in reality we will be creating a force that is 400 pounds too high and our setup will not be balanced.

So, we need to know not only the force generated by the sway bar in chassis roll, but also in chassis dive. There are measuring devices that can provide the actual force the sway bar is generating when we put the chassis at the mid-turn attitude that simulates both chassis dive and roll. Then we can subtract that force from the total force needed to support the tire loading at mid-turn to then know how much force we will need from the ride spring/bump combination.

The Sway Bar Use As A Spring – So, far, we have talked about how the sway bar might act as both a spring in chassis dive, and as a spring resisting chassis roll. Let's look deeper into how that is happening and see if we can utilize this property to our advantage.

The motion ratios can be different in an offset chassis because the lower control arms are different lengths. If the end of the sway bar arm is mounted the same distance from the ball joint for the left and right lower control arms, then due to the difference in lengths, the ends will travel different amounts in chassis dive.

If the sway bar arm ends are mounted at different distances from the ball joints for the left and right lower control arms, then this difference in travel of the arm ends becomes greater. We can actually increase the dive force amount that the sway bar contributes by moving the mounting points of the sway bar arm ends.

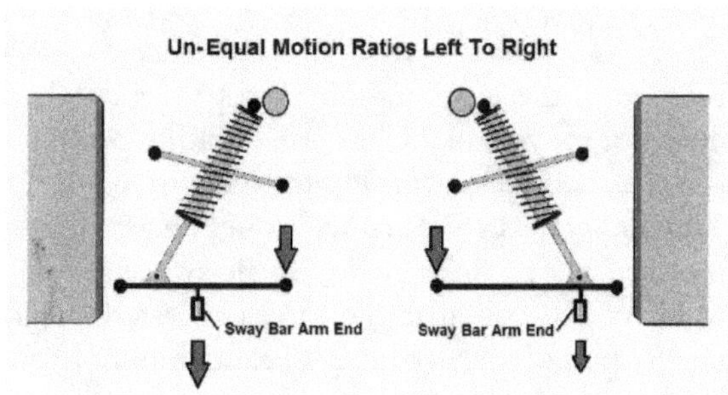

When we move the sway bar arm end towards the chassis mount, we change the motion ratio for that side. In chassis dive, the left arm end (left side of the sketch) will move more than the right side. This causes the sway bar to twist and puts force onto the right lower control arm. This force is added to the force of the ride spring and the bump at maximum mid-turn travel of the suspension. The total force needed to support the RF tire load at mid-turn is a product of sway bar force, ride spring force and bump force.

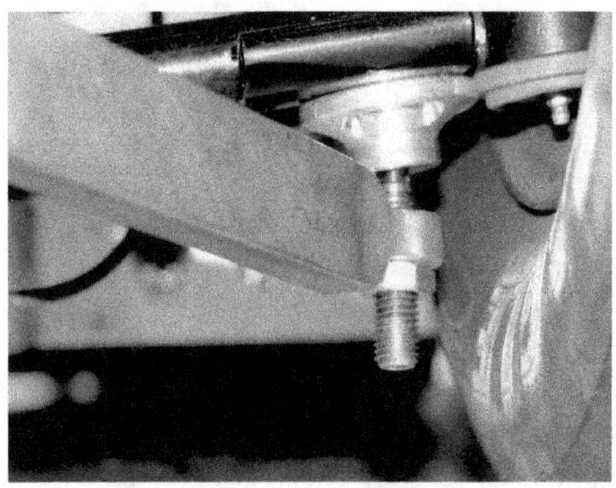

To change the motion ratio in this left side lower control arm, we need to install an arm with less top view angle that will move the contact point in towards the chassis mount. This decreases the motion ratio causing the left arm end to move a greater distance than the right arm end when the chassis dives.

If we move the left side arm end in towards the inner chassis mount, then the motion ratio becomes less. Let's exaggerate that so we can better understand this principle. As the chassis dives, the sway bar itself moves down with the chassis. The arm ends move much less in a normal system. So, let's take that a little farther.

Suppose we mount the right sway bar arm end right under the ball joint. Now this arm will not move vertically as the chassis dives because the ball joint doesn't move either. Now if we mount the left sway bar arm end under the left lower control arm chassis mount, it will move the same distance as the chassis and sway bar.

Now, with this exaggeration, the right arm end does not move and the left arm end moves with the chassis. Suppose the chassis were moving 3.00 inches in dive? Then we would be twisting the sway bar by three inches. If the sway bar rate were say 250 ppi, then the force generated would be 750 pounds. That is a significant amount of force. If the bar were a larger sway bar with a rate of 766 ppi, then the sway bar alone would generate the 2,300 pounds of force needed to support the tire loading at mid-turn.

This example is extreme for sure, and we would never mount the left side sway bar arm end in this manner. But what if moving the arm end in towards the chassis mount by an inch or two supplied the added force we need for some types of setups? Then we might just be on to something.

The Left Bump Only Setups – In a bump setup where the two sides of the front suspension had bumps installed, there would be plenty of force available to support the right and left front tire loads at mid-turn. Some race teams only run one bump and use it on the left side of the suspension.

Then the single bump team might run a stiffer RF spring than the LF spring, but never as much spring rate as the left side has when we combine the ride spring rate and the bump spring rate. So, what holds up the car? The sway bar does.

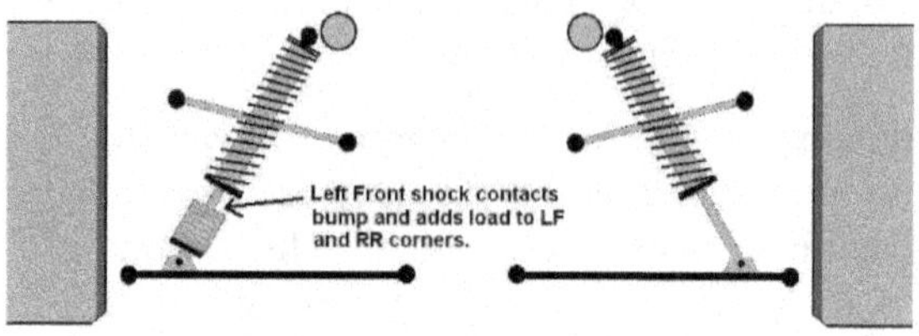

When a race car is designed to use only a left side bump (left side of sketch), there will be plenty of mechanism to provide force for mid-turn support. But at the right side, with no bump and a soft ride spring, there is nothing to provide the force needed to support that corner of the car at mid-turn. We need to install a larger sway bar and an unequal motion ratio design so that as the chassis travels, the sway bar will provide a force along with the ride spring that equals what is needed to support the tire loads at mid-turn. This is why single bump setups require a large sway bar, and double bump setups require a smaller sway bar.

From what we have learned, an A-symmetrical front suspension that has different motion ratios for the sway bar arm ends will produce a dive spring rate which can assist the RF ride spring in providing sufficient force to support the RF tire loading.

For our sample single bump car, we have a 150 ppi ride spring in the RF, a tire static loading of 650 pounds and a dynamic tire loading of 1500 pounds. For the ride spring to support the 650 pounds of static tire loading, it must travel 6.5 inches or so. We divided the 650 by our motion ratio squared of 0.660 to get that travel amount. This represents the pre-load on the ride spring at ride height.

The additional travel amount that spring would need for a 1500 pound mid-turn loading would be 8.5 inches more, and we just don't have that much distance available. Three inches of chassis travel is only 3.7 inches of shock travel. We'll bottom out long before that spring supplies enough force to support the car. So, we need more force.

If the sway bar twists a half inch at the arm ends in three inches of travel, then what size sway bar do we need to supply the added force to support the RF tire loading? The three inches of travel will give us 3 x 150 = 450 pounds of added force from the ride spring. We need a total of 2,300 pounds of overall force to support the 1500 pound so tire loading.

If the sway bar provided 985 pounds at ride height plus another 450 pounds from 3 inches of travel, then it is supplying a total of 1,435 pounds of force. We need a total of 2,300 pounds of force, or another 865 pounds.

A 1 15/16" sway bar is rated at around 1,730 ppi (pounds per inch). If the sway bar arm ends moved a half inch, then that bar would provide half the full inch rate, or 865 pounds of force in three inches of travel and exactly what we needed in additional force. Now we have the total force we need to support that tire loading at mid-turn.

Re-Cap - To recap all of this for you, the ride spring provided a total of 985 pounds (at ride height to support the tire load) plus 3 x 150 = 450 pounds (from chassis travel) for a sum of 1,435 pounds. We needed a total of 2,300 pounds of spring force to support the mid-turn tire loading, so we need 2,300 – 1,435 = 865 more pounds of force than we are getting from the ride spring alone.

Since a 1 15/16 inch diameter sway bar produces 1,730 pounds in one inch of arm movement, and the chassis dive produced a half inch of arm movement in 3 inches of dive, we divide the ppi rate of the bar by two to get 865 pounds of force, or exactly what we need added to the ride spring force to support the RF tire at mid-turn.

If our sway bar arm mount on the LF lower control arm was farther in from the ball joint towards the chassis mount, then the bar would twist more and we would then require a smaller sway bar to generate the same 865 pounds of force.

Suppose we had one inch of sway bar arm travel from the 3.0 inches of chassis dive, then we could use a sway bar rated at 865 ppi, or very close to a 1.50 inch diameter sway bar. Now we are beginning to understand why some teams run larger or smaller sway bars than others. It's all in the design and mounting points we have on our cars.

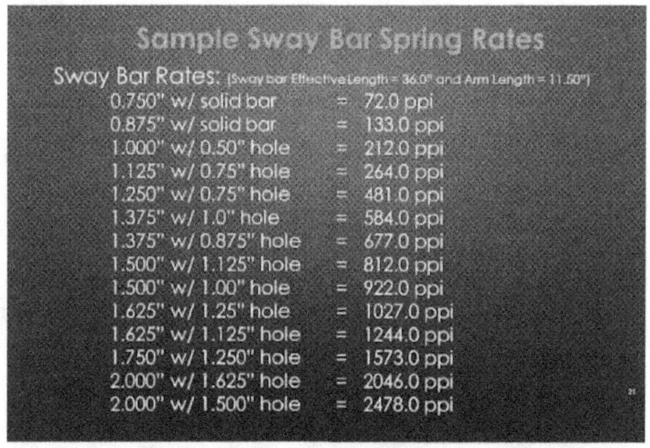
Sample Sway Bar Spring Rates

Sway Bar Rates: (Sway bar Effective Length = 36.0" and Arm Length = 11.50")

0.750" w/ solid bar	=	72.0 ppi
0.875" w/ solid bar	=	133.0 ppi
1.000" w/ 0.50" hole	=	212.0 ppi
1.125" w/ 0.75" hole	=	264.0 ppi
1.250" w/ 0.75" hole	=	481.0 ppi
1.375" w/ 1.0" hole	=	584.0 ppi
1.375" w/ 0.875" hole	=	677.0 ppi
1.500" w/ 1.125" hole	=	812.0 ppi
1.500" w/ 1.00" hole	=	922.0 ppi
1.625" w/ 1.25" hole	=	1027.0 ppi
1.625" w/ 1.125" hole	=	1244.0 ppi
1.750" w/ 1.250" hole	=	1573.0 ppi
2.000" w/ 1.625" hole	=	2046.0 ppi
2.000" w/ 1.500" hole	=	2478.0 ppi

This chart represents the spring rate of three piece sway bars, or torsion bars, for the dimensions shown. We can see how the rates increase as the sway bar diameter sizes increase. Double bumps setups can utilize much smaller diameter sway bars in the 0.750" to 1.250" sizes. For single bumps setups, much more force, or spring rate, is needed and we are talking about sizes above 1.500" up to nearly 2.00" diameters. The more un-equal the motion ratios, the less sway bar rate we will need for single bump setups.

Why We Use The Sway Bar As A Spring – This is exactly why we use a large sway bar for single bump setups. For a double bump setup, we already have plenty of force that is provided by the RF bump device, so we must use a much smaller diameter sway bar that has much less rate.

By the motion ratio rule, we can change the left side motion ratio to adjust the sway bar spring force at the RF corner. Professional race teams now experiment with different LF motion ratios to tune the forces at the RF corner rather than change the sway bar size.

Summary – Much of what we know comes from the application of new technology. One thing leads to another and at some point, different approaches and different results add up to a better understanding of the total package.

What you have just learned is less than a year old. It helps us understand how to setup our race cars. Now we know why we need a larger sway bar for the single bump setups, and why we must use a much smaller sway bar for double bump setups.

Exam - In The Context Of This Lesson:

Non-Symmetrical AA-arm Suspensions Are Common With?

1) Formula race cars

2) Circle track offset chassis

3) Circle track perimeter cars

4) Road racing cars

We Find The Working Rate Of The Bar By?

1) Measuring the diameter of the bar

2) Measuring the arm length

3) Rating it at the end of the arm

4) Finding the length of the sway bar

A Sway Bar Can Produce A Force In What?

1) Dive only

2) Roll only

3) Dive and roll

4) All of the above

A Symmetric AA-arm Suspensions Produces Force When?

1) The car rolls

2) The chassis dives

Non-Symmetrical AA-arm Suspensions Produces Force When?

1) The car rolls

2) The chassis dives

3) 1 and 2

For Single Bump Setups, We Need What?

1) A Non-symmetrical suspension

2) A large sway bar with a high rate

3) Additional Spring force at the right front

4) All of the above

For Double Bump Setups, We Need What?

1) A more symmetrical suspension

2) A smaller sway bar with a lower rate

3) Less sway bar force at the right front

4) All of the above

Race Car Technology – Level Three

Lesson Sixteen – Aero Downforce & Tire Loading

The goal of this school is to give you, the racer or want-to-be racer, real world knowledge about the subject matter that matters. We offer views on older technology just so you can get a perspective about whatever it is we are discussing. As far as aero technology is concerned, we feel that a complete picture of aero as it applies to race cars cannot be presented without addressing a few common, but erroneous assumptions.

In years past, some tracks allowed innovation in the body design and shapes. This late model that ran at New Smyrna Speedway in the mid-1970's utilized a high downforce nose using Flat Plate aero, and a low drag, low mounted roof. The tail was appropriately called a "whale tail" due to its shape and how large it was. We see utilization of flat plate aero as well as low pressure under the hood in this design.

I have formed my opinions by studying aerodynamics and talking to persons who live aerodynamics. And I have come to agree with them for some parts of aero and have offered my take on other parts of aero technology.

When you design an airplane or a race car that you will ultimately have to fly or drive, you must make sure you have a real world understanding of the subject of aerodynamic engineering. The classroom or cozy control room in a wind tunnel presents little risk and there is nothing to prove the results wrong, until you "fly" the airplane or race car.

Wind tunnels will give you tendencies and probabilities only, and that is why major airplane design companies have test pilots. Not one airplane ever designed and "flown" in a wind tunnel ever flew in test flights the exact same way. It's the same with race cars. Otherwise, every multi-million dollar F1 team would have the perfect aero properties in their cars design, and we know for a fact they don't.

So, the data we collect when using a wind tunnel has to be translated somewhat into information that will more closely resemble what happens when we "fly" the airplane or race car. There is various editions of software that will get that done and there is computational fluid dynamics software that will attempt to simulate what the wind tunnels, and real world flying, does.

These computer programs are not perfect and there is still room for practical knowledge of aerodynamics for the enthusiast and applications engineers in the discipline of aerodynamics.

Wind Tunnels Have Errors – The whole premise of a wind tunnel is to move air at a given speed into and over an object to measure lift, down-force, drag, etc. In a wind tunnel, the air is accelerated to some predetermined speed and the stationary object, the race car, is bombarded by the oncoming, energized air that has a lot of momentum. This is not what happens in nature with airplanes and race cars. They do the opposite by traveling at speed through relatively still air that has no energy or momentum. This is a major point to make. This is how you develop a practical knowledge of aerodynamics.

Some aero engineers will tell you, "Well, it's the same thing." No, it is not. Air moving at 60-100mph has a lot of energy and does not want to deviate from its path, kind of like a baseball traveling the same speed. When this energized air hits an object, the way it moves around that object is much different than if that object were moving through still air at the same speed. It has to.

Most veteran aero engineers, if they are honest, will tell you that the primary difficulty with wind tunnel data is converting it to reality. There exists some very expensive Computational Fluid Dynamics, or CFD for short, software programs that attempt to convert the wind tunnel data and make it portray what happens in the real world. The conversion is very complicated and obviously not perfectly accurate. And again, that's why we have race tracks and test pilots.

The other thing that causes errors in the data we get from a wind tunnel test is the close proximity of the walls. Most wind tunnels are too small because it is cost prohibitive to build it big enough so that the walls don't influence the results. Air is disturbed well beyond the distance to the walls in most wind tunnels when using full scale models. The presence of the walls restricts the movement of the surrounding air and causes a compression effect which alters the results.

The only "wind" tunnel that comes close to reality is the one Chip Ganassi is said to own called Laurel Hill in Pennsylvania. It is actually a tunnel through a mountain. There a test vehicle is run for a mile through the tunnel, through still air, and the pressure distribution that produces drag and down-force is measured. At least the test object is traveling through relatively still air just like an airplane or race car really does. While the tunnel still has walls that are too close, the results are much closer to reality.

The ideal aero testing for race cars would be done on a smooth flat surface outdoors at speeds that replicate the speeds the car would experience on a race track. Then the load and pressure sensors would record the aero influence on the car, much like test flights do with airplanes.

That Being Said – Now that we have that out of the way, let's get into how aero really works. The age-old depiction of an airplane wing, and one that I have used in the past, is not perfectly correct. Yes, the air traveling over the airplane wing travels farther and faster and thus may have less pressure than the air traveling under the wing.

This is often called the Bernoulli principle and it supposedly provides lift. But it really doesn't. The Bernoulli theory used a closed container and moving air. The air a race car flies through is not moving. It is being pushed out of the way. So, to say it flows is not exactly correct. It moves laterally and vertically as the race car speeds through it.

The Problem With Bernoulli – The Bernoulli principle states that, "As the velocity of fluid flow increases, the pressure exerted by that fluid decreases". This may have some validity in a wind tunnel where the "fluid" air is moving, but in the real world, our race cars and airplanes move through air that is not moving at all relative to the motion of the airplane.

What the air in the real world does is lay around until an airplane or race car moves through it at a high speed. The air is then pushed out of the way and displaced whereby it pushes against the surrounding air that is itself relatively still and not moving. Some of the displaced air is pulled along with the vehicle moving through it until it is released and allowed to flow back in a direction where it was before it was disturbed.

There is no way a wind tunnel can re-create the action we just described. It is so different from the scenario of moving air flowing over a still object that the two are completely different in the dynamics of the event.

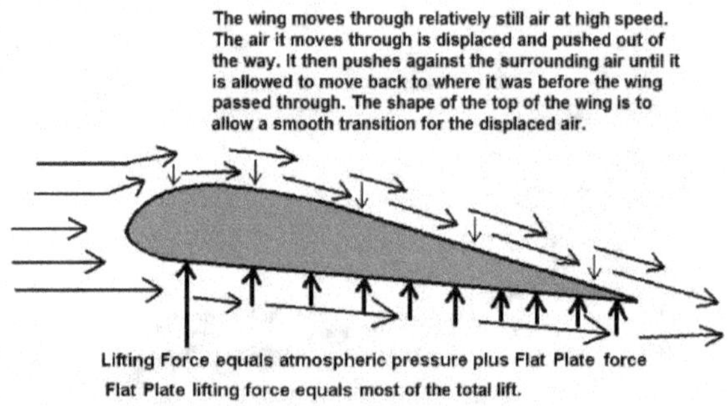

An airplane flies due to lift, but one seldom talked about component of lift is Flat Plate aero affect. It is the lift provided by the flatter underside of the wing moving through air at a slight angle that counts for a large percentage of the lifting component. The low pressure on the upper side, which is often erroneously explained using the Bernoulli's principle, reacts with the higher atmospheric pressure on the underside to provide the other part of lift. We can utilize the FP aero affect alone on our race cars where the conditions for Bernoulli's principle does not exist.

Most every airplane cruises with a wing attitude where the underside of the wing is at a small angle (the front higher than the back) to the air it is moving through. So, it acts much like a water ski to assist in holding up the airplane. This is called Flat Plate (FP) aero.

This early-2000's design of prototype Grand Am race car utilizes flat plate aero on the front of the car. These cars were very successful during their rein. They produced a lot of downforce.

Many aerodynamicists tend to discount this affect, but it can be a large percent of the total lift component. This is where we get into an important area of race car aero design.

In a race car, we may have flat plate areas where we can develop down-force to increase the grip of the tires. That is called FP aero down-force, but what about other areas of the car that are flat? What about the sides? We'll get into that in a minute.

In race car aero design, one often overlooked component of aero down-force is what is called Flat Plate Aero. On this outlaw late model, the front nose is angled down at about a 45 degree angle to create Flat Plate down-force. We'll explain how that works.

Down-force From Flat Plate – The sloped front of an Outlaw late model, the nose and hood on a '70's era super late model, the nose on a modern dirt late model and the wing on a sprint car all offer FP aero down-force possibilities.

The modern dirt late model race car utilizes flat plate aero not only at the front, but along the outer sides. These cars use rear steer in large amounts to get the rear of the car out and into the air. This provides flat plate aero sideforce that pushes the car towards the inside of the track, countering the lateral forces trying to push the car off the track.

This nose on an asphalt modified is relatively small in area, but effective. The raised parts on the outside edges help keep the air on the flat plate and helps prevent it from spilling off the sides.

One of the early proving grounds for short track racing was the annual February Speedweeks races at New Smyrna Speedway in Florida. Back in the late 1970's and early '80's, teams from all over the country would show up with "stock" bodied super late model cars that adhered to the strict rules of their perspective sanctioning bodies only to discover that there were no body rules at NSS. The local teams already had their car bodies tricked up.

Those who remember noticed that after about the second or third night of racing in the nine night series, most of the cars had been transformed into sleek, wedge shaped oddities with huge fantail rear spoilers attached. This is what we now understand to be an effective use of FP down-force.

With each of the examples, we may, or may not, have a low pressure on the other side of the FP, but the FP still generates its own force. The angle is critical and relative to the speed of the vehicle, just as it has proven to be with airplane design, but up to a certain angle of about 15-20 degrees, a lot of force can be developed.

As for the wing on a sprint car, most of the down-force from a sprint car wing is FP derived from air flowing over, and being deflected up, off the top flat plate. It acts much more like an outlaw nose. The shape of the bottom of a sprint car wing being curve like the top of an airplane wing, is there to transition the air flowing around the wing so that there is less drag.

And you can have too much angle in that wing. As the speed increases, the wing angle must be reduced in order to have maximum down-force while not creating too much drag. This is not unlike with an airplane where we trim the attitude of the plane and wings, and retract the flaps, as the speed increases.

A typical winged sprint car uses flat plate aero on its wings both as downforce as well as for side force from the side plates running at an angle to the direction of travel of the car. These wings can produce quite a bit of downforce helping these cars maintain very high speeds through the turns.

Flat Side Lateral Force – The concept of FP aero can be applied laterally in addition to vertically to assist in helping the car to turn. If a race car has relatively flat body work on the side of the car towards the outside of the turns, then if the car is run at an angle to the direction of travel, it can produce a force in a direction opposite to the centrifugal, or lateral forces trying to push the car off the track.

Cars where this can be a benefit are easy to pick out. Late model dirt cars, dirt modifieds, Outlaw late models are just a few. And, if we look at the large sideplates on a sprint car wing, we see where FP aero can be used if the car runs at just the right angle through the air.

In the case of the dirt late model, we know these teams rear-steer the car so that the rear tires are running outside the front tires. This puts the large flat side of the car at just the right angle to the air it is moving through to produce FP forces that are opposite of the lateral forces. Not equal to, but opposite in some amount. The tires still have to do the rest of the work, but the car goes faster with this FP advantage.

It's a similar situation with sprint cars. I have watched Outlaw sprint cars qualify at Volusia during Speedweeks at upwards of 135 MPH average. The weight of the car on the tires cannot hold the car in the turns at that speed. There must be FP aero down-force and side-force helping to keep the car on the track.

Spoiler Down-force Vs. Drag – When racers run tests with their cars in a wind tunnel, they experiment with spoiler angles. Sometimes the results can be miss-interpreted. Here is what I mean. A more vertical spoiler will produce much more drag than one angled at say, 55 degrees off of vertical.

The wind tunnel operator must compare changes in loading on each axle with the total loading of the car. So, let's say a vertical spoiler added 50 pounds to the rear axle, but removed 45 pounds from the front axle. The result is only 5 pounds of added down-force combined with a lot more drag, and represents more of a displacement of loads due to the leverage of drag. If the total loading on the car does not increase, you have no gain. You might as well take that 5 pounds of weight from the front of the car to the rear and remove the spoiler.

There has been much discussion about spoiler angles and their effectiveness. High angles (from horizontal) result in a lot of drag and very little actual downforce. This drag pulls on the rear of the car and does displace load from the front to the rear. Any increase in rear axle loading can be mostly attributed to load coming off the front axle and transferred onto the rear axle.

With the 55 degree spoiler angle, we might see a 40 pound increase in the rear loading and a 15 pound decrease in the front loading due to the same leverage affect from drag, but in this case the drag is much less. In this example, we see a total increase in the vehicle loading of 25 pounds from actual aero FP down-force.

Traditional Low Pressure Downforce – In a more traditional sense of looking at aero down-force, there is the phenomenon of high pressure opposite low pressure. The shape of the body on our race car can help produce low pressure inside the body panels to create even more down-force. What we need to do is design our bodies so that we can direct some of the air we are driving through around the car in order to vacuum air out from within the areas inside the body.

If we can use the movement of air that is being pushed out beyond the sides of the car to help vacuum air out of the engine compartment under the hood, we can lower the pressure along the underside of the hood. A higher pressure on one side of an object will push that object in the direction of the lower pressure, or high towards low.

To accomplish this, teams use wider, angled front noses that will direct the displaced air around the sides of the car beyond the wheel wells in such a way that a low pressure area is created just outside the wheels. Air then moves out of the engine compartment to fill this "void" and the pressure under the hood is reduced.

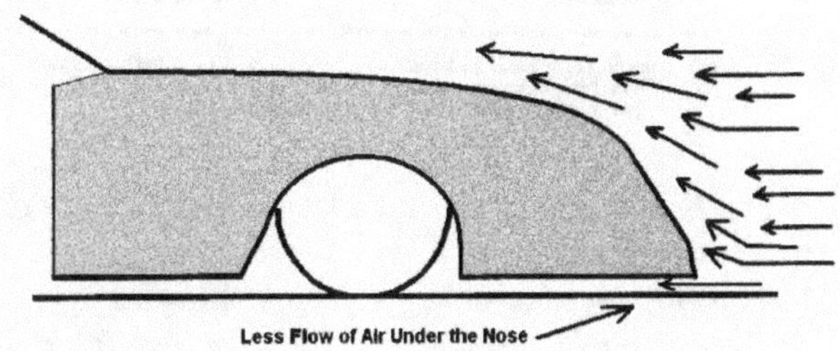

It is common in today's racing to position the race car low to the race track in the front. This accomplishes several beneficial things. First off, the Center of Gravity is lower, there is very little camber change as we previously discussed, and a low nose section cuts air off that would otherwise invade the under side of the hood where we are trying to create a low pressure area. If we can keep air from getting under the car, we can realize a greater low pressure effect under the hood.

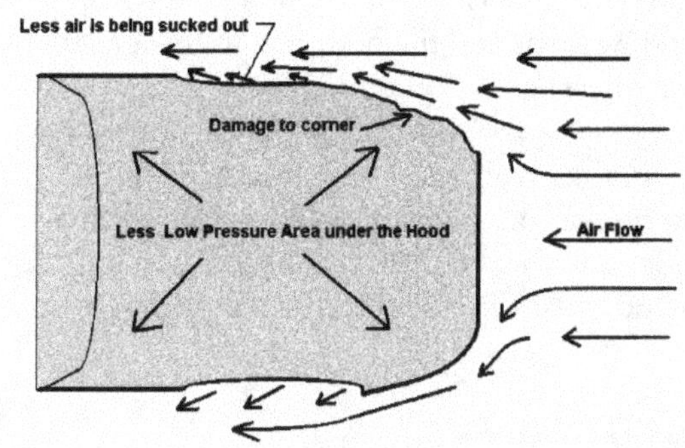

This illustrates how air being routed around the nose of a race car can cause air under the hood area to be pulled out into the low pressure area around the wheel wells. Any damage to the nose and fenders defeats this purpose.

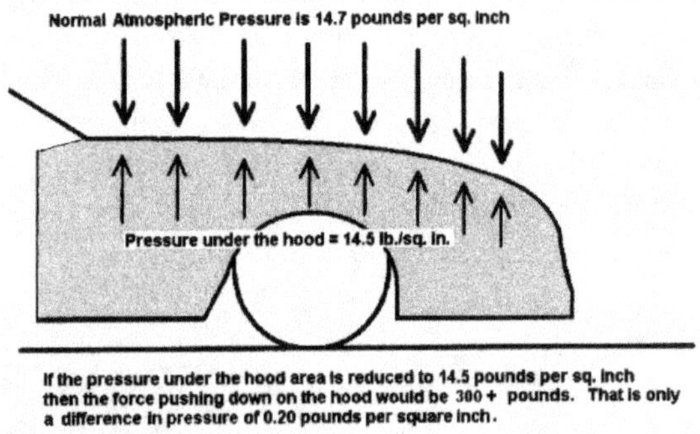

A simple calculation shows that for a small decrease in atmospheric pressure under the hood, we can see upwards of 300 pounds of downforce. This is distributed equally onto the front tires at about 150 pounds each. That is a lot of load gain and added traction.

The average atmospheric pressure at sea level is 14.7 pounds per square inch on all sides of an object, even on our bodies. If we reduce the pressure under the hood to 14.5 PSI, a drop of only 0.2 PSI, over an area of just a square yard we would generate about 260 pounds of down-force ($0.20 \text{ X } 36^2 = 259.2$ pounds).

Rear Aero - At the rear of the car, we can manipulate the shape of the spoiler, the rear window posts and the body just in front of the rear wheel wells. By routing the air that is flowing past the sides of the car out and to the sides of the rear wheel wells, a similar suction effect takes place to create a low pressure area under the rear deck. There are obvious limits as to body shape in this area, but a little reshaping can help.

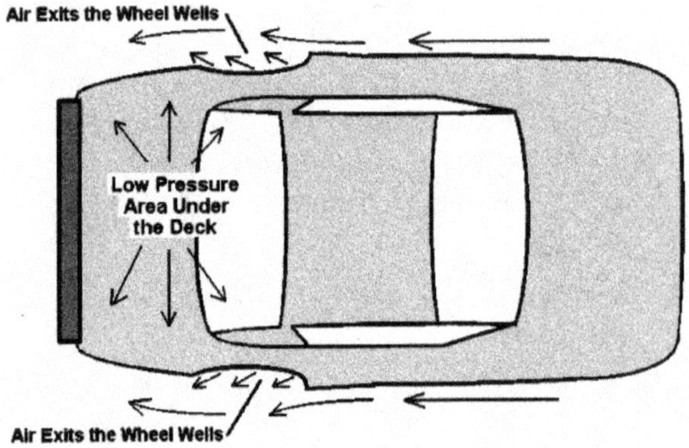

At the rear of certain bodied cars, we can create low pressure down-force by routing air out of the rear cavity under the rear deck. This gets tricky with current strict body rules, but most tech officials are looking elsewhere for "cheating".

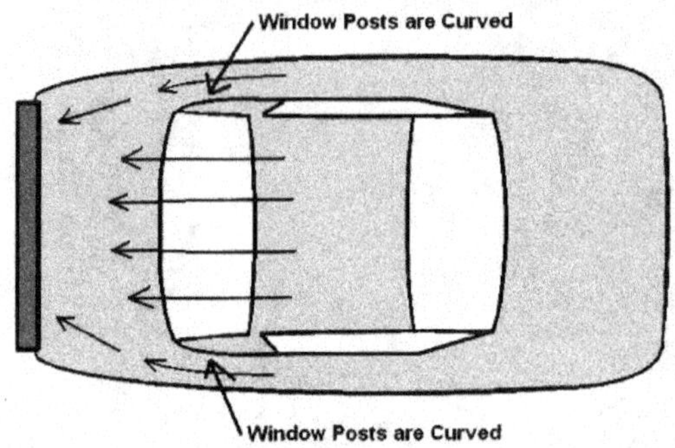

Another trick used by some professional race teams is to shape the window post areas so that they are curved allowing air to flow around the post and onto the rear spoiler. The greater amount of flowing air going onto the spoiler instead of over and around it causes more rear downforce.

Race tracks that are longer and have more banking require less down-force and would benefit from reduced drag. We can rethink how the air flows past the roof (green house area) and onto the rear spoiler. If we reshape the post that connects the rear window with the side window openings, we can direct air away from the spoiler and greatly reduce aero drag.

The modern day race car has a low attitude combined with a nose shape that provides low pressure under the hood, flat plate aero properties across the top of the hood, and rear air flow that enhances rear deck low pressure.

Summary - The key goals with stock car aero design and development is to create a body shape and running attitude that will provide more down-force and side-force to enhance turn speeds, produce less drag, and promote a more balanced race car. We can now see where and how we can utilize Flat Plate aero to increase down, as well as side, force.

Everything we do has to be done within certain limits. If we work hard to develop 600 pounds of down-force on the front end and the rear is not able to keep up with that high amount of grip, then the car will be loose and nobody can drive a loose car fast. So, there are limits to how far we can go.

Work towards a good balance of front to rear aero down-force to help produce more overall grip. Do not overdo your efforts to help the car aerodynamically at the expense of handling efficiency. Make sure the basic chassis setup is balanced, and then the combination of both aero down-force and handling will enhance the on-track performance.

Exam - In The Context Of This Lesson:

Modern Aerodynamics Is The Study Of?

1) Air flow across a wing or race car.

2) How air reacts when an object is moved through it

3) The Bernoulli principle of low pressure vessels

4) Wind tunnel data

How Is Moving Air Different Than Still Air?

1) Moving air resembles how a race car reacts

2) Moving air has less energy

3) Still air has no energy unlike moving air

4) It is how a wind tunnel works

Which Of The Following Are Deficiencies Of A Wind Tunnel?

1) The air is moving and has energy

2) The walls are too close to a full scale car

3) The data cannot be accurately converted to rear world numbers

4) All of the above

Flat Plate Aero Is Created When?

1) The air is moving and flows over a flat plate

2) The flat plate panel is moved through still air and is pushed in some direction

3) Low pressure air is on one side of a body panel

4) We have flat sides on the race car

Downforce From Low Pressure Is Caused By?

1) The air being evacuated out of a closed area

2) A higher pressure being created opposite a low pressure area

3) Air flowing around an opening outside a closed in area

4) All of the above

A Low Nose Helps Performance and Efficiency By?

1) Keeping air from invading the area under the hood

2) Providing a higher angle of the nose flat plate to produce more downforce

3) Creating a lower Center of Gravity

4) All of the above

Race Car Technology – Level Three

Lesson Seventeen – Practical Applications – Part One

All that we have done over the course of Levels One through Three has led up to this Lesson. This might be considered the holy grail of race car technology by some. It's not the final chapter, but one of the most important because here we will learn how to develop the actual setup for the race car.

Our approach to setup is designed to be the most efficient and quickest way to get fast and end up with a truly balanced setup, something we stressed all along. This approach may differ from some others in that we will tune for the Mid-turn portion of the track first, then tune the entry and then the exit. We do this for a reason.

The balance we seek in order to have the perfect setup will make our mid-turn performance the best it can be. Once we find that, our entry and exit balance will be much better also. Each one of these cars has made changes to find the mid-turn balance and each one probably got there a different way. There are many different ways to arrive at the ideal setup balance and we'll tell you what those are and how to make them.

The balance concept for dirt cars is arrived at and put together a little differently. Dirt cars need to be made to work on very slick surfaces at times. So, the setups vary as to the balance, whereas, on asphalt, the grip level of the racing surface remains fairly consistent. Here we see the 15 car has a rear roll angle that is much higher than the front roll angle. This condition will cause the left front wheel and tire to be lifted off the race track. This is a great illustration of the rear out-rolling the front.

The setup adjustments for a formula or prototype race car that has AA-arm front and rear suspensions is more limited than with a circle track car that only needs to turn one direction. The formula car must be symmetric in spring rates left to right. These car work mostly with the sway bar mechanism and front to rear aero balance to tune the setup balance.

The "tools" we will describe using are the tools we use in real life. And, these are the tools many race teams use successfully to setup their race cars, in both amateur and professional series. This is a disclaimer. Some of the tools we describe and use are tools we produce and sell.

Some of these tools can be obtained from other sources, some of them cannot be sourced anywhere else. So, we have to use these tools or we would be doing you a dis-service.

To clarify, one of the tools is a force measuring rig, or machine. We use the Gale Force Suspension rig because Bubba Gale is my partner in this school and that rig works very well. There are other rigs available that do basically the same thing, but when you understand how the GF rig works, you will understand the principle behind it and that is what counts.

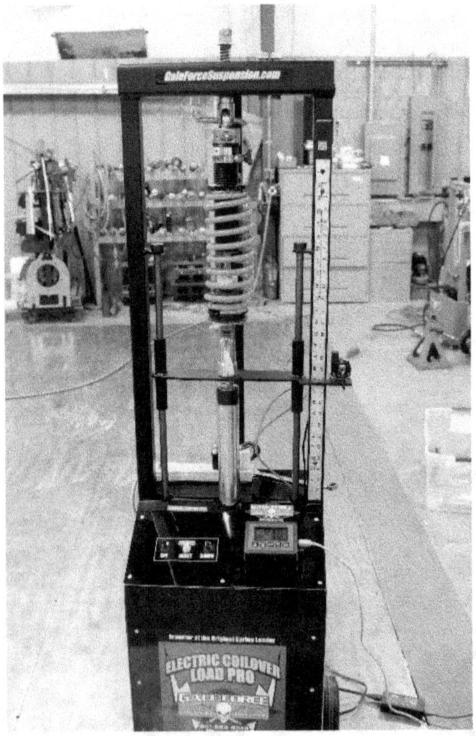

This is the Gale Force Suspension Load Pro spring loader. It measures the force produced by the spring and bump combination and uses the spring travel that is recorded on the race track. We will be using this tool to explain setup concepts because this type of tool is widely used in todays racing environment.

The other tool is the Chassis R&D Asphalt Setup software program that I developed in 1996. As far as I know, there is no other tool like this on the market. This tool has been used by me and thousands of racers over the past twenty years. If there were other tools like this that did what this tool did, I would tell you. There isn't, so we'll use this tool.

The setup tools help you to understand exactly what the goals are for ideal setup, how to get there, and really helps us to understand the significance of each setup change. As we go along with this Lesson, you'll understand more about that and come to appreciate that we use these tools in this school. This is where the racing industry is right now.

The Chassis R&D Setup software program was first offered commercially in 1996. Since that time, many thousands of programs have been used by racers all over the world. It calculates a predicted roll angle for the front and rear of a race car. On asphalt, we match the two roll angles to arrive at the perfect setup balance. We use this tool in the RCT courses because it is the only one of its kind available.

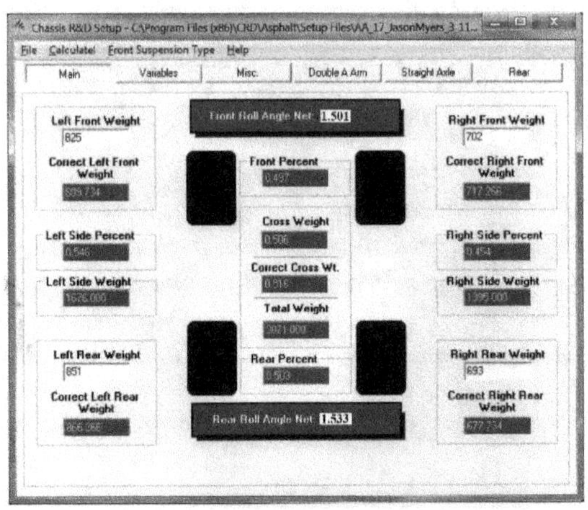

The CRD Asphalt setup software is very simple in concept, but very effective in the information it provides. The two answers it gives us are the Front and Rear Roll Angles. When we match these two angles, the two suspension systems are desiring to roll to the exact same angle and are working together. This is the balanced setup we are trying to achieve.

Where Do We Start Out? - Now that we got that out of the way, let's get started. The reason we tune the middle of the turns first is that when we achieve a balanced setup for the mid-turn, it often serves to balance the entry and exit. And, we always tune the spring and sway bar setups before we ever start tuning with the shocks. That's not to say that spring changes won't help entry and exit, it's just that those kinds of changes will be made while also keeping the mid-turn balance from changing.

The shock Lesson comes in the next Lesson and any shock expert worth their salt will tell you to first work out the basic setup before beginning to tune with the shocks. And that is what we are going to do. To explain, at mid-turn, the shocks and springs are in what is called a "steady state" condition. That means that for a short period of time, they are not moving. The car is not slowing and it is not accelerating and it is not moving vertically. That is the definition of steady state.

The middle part of the turn is also the slowest part and where the lateral G-forces, or centrifugal forces, are the greatest. It is here where the two suspension systems are rolled and compressed to the greatest extent. All of the way towards the mid-turn point, the G-forces are building and the radius is getting shorter.

The way we calculate the G-force is by using speed and turn radius, the more speed and the tighter the radius, the greater the G-force. Most overall turn layouts are elliptical in shape. That is, from the time the car first begins to turn until it gets to mid-turn, the radius is constantly decreasing. Then as the car begins to move away from the mid-turn, the radius constantly increases. If we plot this line, we will see that it is an ellipse. This is a very important concept to understand, especially for the drivers.

We will do some exercises in this Lesson where we apply gradually increasing G-force and see if our balanced setup stays balanced in the parts of the turn that is not mid-turn. The fastest driver will maintain a speed on entry that matches the turn radius providing the most performance possible. More on that later.

Mid-Turn Balance – The definition of mid-turn balance means that both ends of the car are trying to do the same thing. The "same thing" is defined here as roll to the same angle. If we take all of the effects placed on each suspension, like lateral force, gravitational forces, banking angle down force, etc. then each suspension system, front and rear, will want to roll to a specific angle.

Each suspension system is not free to roll to the exact angle it desires because in most cases, a stiff chassis connects the two suspension systems. If we read the chassis roll angle by doing a calculation using shock travels, it will not tell us what the desired roll angles are for each suspension, only the average of the two desires. Here, I will say that again.

If we read the chassis roll angle by doing a calculation using shock travels, it will not tell us what the desired roll angles are for each suspension, only the average of the two desires.

The setup software we told you about starting out in this Lesson will calculate the desired roll angle for each suspension system. This is the angle a suspension system would produce if that suspension system were not attached to the other suspension system. When we calculate both roll angles, we can then match the desires of the two systems by making changes to affect the roll angles, independently. We don't know of any other way to do this.

In the following examples and situations, we will be calculating the roll angles of the front and rear suspensions to demonstrate the magnitude of the effects of spring and sway bar changes, which is very valuable in working out our setups.

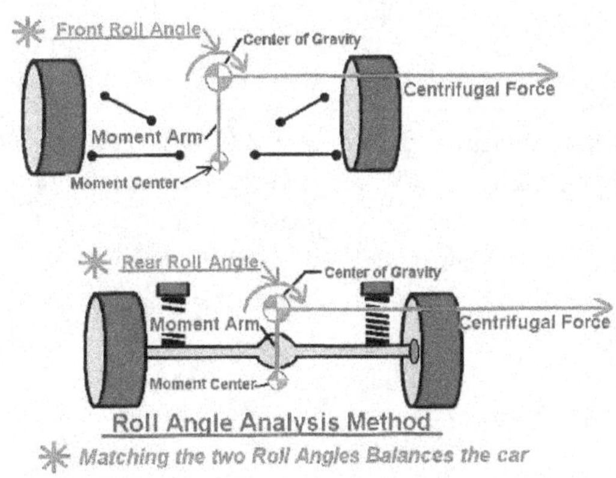

As shown many times in the past, our goal for most setups, especially on asphalt, is to match the desired roll angles of the front and rear suspensions. When we do this, the weight transfer becomes predictable and we can then have a balanced setup that will be fast and very consistent.

What Affects The Degree of Roll Angle? – The following are the effects that help determine what we can call roll stiffness, or the influence on the degree of roll angle for a suspension system. It is critical to know how these effects work before you ever begin to setup a race car.

Roll Center Height – The roll center for a suspension system is the bottom of the moment arm that tries to roll the car over when we are going through the turns. The length of the moment arm is a determining factor in the magnitude of the roll angle each suspension system desires. The longer the moment arm, the more the roll angle. This is a form of leverage where the longer the lever, the more power we have to move things, or roll the car.

So, if the roll center is higher, the moment arm is shorter, and the roll angle is less. If the roll center is lower, then the moment arm is longer and the roll angle is more. The roll center is a part of all suspension systems, AA-arm as well as the solid axle types and leaf spring types.

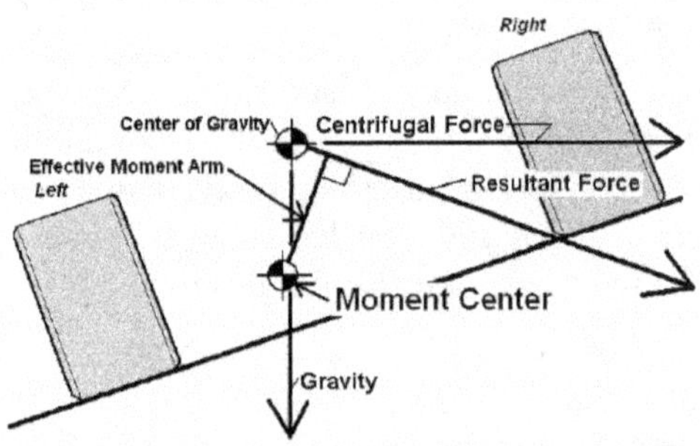

The Center of Gravity is the top of the Moment Arm and the Moment Center is the bottom of the moment arm. The height of the CG and the MC determine the length of the Moment arm and also the stiffness of the suspension in this AA-arm suspension. A longer MA causes a larger roll angle, and a shorter MA produces a smaller roll angle.

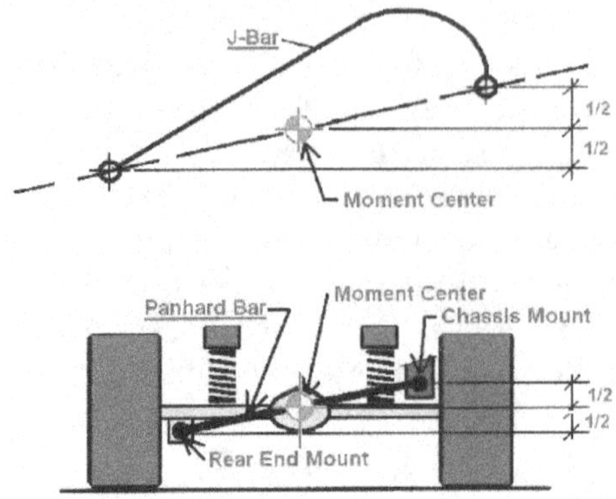

In a solid axle suspension, the Center of Gravity is the same, but the Moment Center is determined by the lateral restraint system, in this case a panhard bar. The MC height in this system is the average height of the center of the two end connecting bolts.

Center Of Gravity Height – The CG height of the sprung weight of the car, all of the components that are not included in the suspension, helps determine the magnitude of the roll angle. Like the roll center, the CG is a part of the moment arm and represents the top of the moment arm.

The higher the CG, the longer the moment arm and the greater the roll angle. The lower the CG, then the shorter the moment arm and the smaller the roll angle.

Spring Stiffness – Spring stiffness affects the roll angle. The stiffer the spring rates in a suspension, the less that system will roll. This is true if the spring rates are the same side to side (symmetric) or if they are not.

The spring stiffness is affected by the spring angle. In this installation, the rear spring being a coil-over design, is placed at a high angle off of vertical. This reduces the spring rate the car "feels" as well as shortening the spring base which is felt at the top of the springs in a straight axle suspension system.

This represents a swing arm spring installation. The coil-over assembly is mounted on the trailing arm and as such, creates a motion ratio. As the chassis moves vertically, the spring will move less than the chassis. The car feels less spring rate this way. The actual spring rate the chassis feels is a product of the motion ratio squared times the spring rate. In this car, that net rate is about half of the actual installed spring rate.

Spring Split – When there is a difference in spring rates side to side in a suspension system, that difference affects the roll angle. In a turn, if the inside (the turn) spring is softer than the outside spring rate, the roll angle is reduced. If the inside spring is stiffer than the outside spring rate, the roll angle is increased. These two statements consider that the overall average spring rate of the system is the same for both scenarios.

If we start out with two 200 ppi springs in a suspension system, and the roll angle is say 3.0 degrees, if we install a 175 ppi inside spring and a 225 ppi outside spring, then the overall average spring rate is the same at 200 ppi., but the roll angle will be less with the spring split, especially when the turn is banked with the inside lower than the outside.

If we again start out with two 200 ppi springs and then make a change to a 225 ppi inside spring and a 175 ppi outside spring, (the average spring rate of 200 ppi being maintained) then the roll angle will increase with the spring split.

Road racing cars must maintain a symmetry in spring rates, so they can never use this effect to tune the roll angles in the suspension systems on those cars. But for circle track cars, this is a major tuning tool.

Sway Bar Rate – For a suspension system that utilizes a sway bar type of mechanism, the spring rate, or stiffness, of the mechanism helps determine the roll angle. The stiffer the mechanism, the less the roll angle. The softer the mechanism, the greater the roll angle.

This shows a rear sway bar installation in a road racing car. Because this car is racing on a circuit where it must turn right and left, it cannot utilize spring split, or different spring rates side to side. Therefore, it must use a rear sway bar to help reduce the rear roll angle to match the front roll angle.

Lateral G-force – The lateral G-force helps determine the roll angle. The greater the G-force, the greater the roll angle. The lesser the G-force, the less the roll angle.

What is important to understand about the G-force influence is that for some suspension systems, when the G-force changes, the front and rear roll angles will change accordingly, but they don't necessarily change equally. If we put on new, faster tires, then the roll angle balance we had with old and slower tires can change. More on that later.

Track Banking Angle – The track banking angle influences the roll angle. The flatter the turn, the greater the roll angle becomes compared to a banked turn. So, the lower the banking angle, the more the roll angle. The greater the banking, the less the roll angle.

This is an easy concept to understand. For example, if we could drive around a turn with a 90 degree banking angle, the G-force would pull the car straight down into the track and there would be no rolling moment at all. So, the transition from a flat track to a banked one is taking us towards the 90 degree, zero roll angle, scenario.

The banking angle has an influence on the roll angles for cars who run spring split, or different spring rates side to side on both the front and back. The lateral force that creates the G-force is combined with gravity and the resultant vector is pointed somewhere between the tires. This serves to create a smaller roll force and a mechanical downforce pulling the chassis down.

Vehicle Track Width – The track width in a race car is known as the distance between the centers of the two tires at that end of the car. The wider this track is, the less the roll angle becomes for a AA-arm suspension system. A straight axle system is a little different and we'll talk about that next. In the AA-arm suspension, the spring rates are translated into wheel rates. The wheel rates become the suspension springs so to speak, and the wider the spring base, the less the roll angle. A narrow track is easy to roll whereas a wide track is harder to roll, just like any spring base.

Spring Base Width – As described above, the width of the spring base helps determine the magnitude of the roll angle. In a straight axle suspension system, the distance between the top of the springs represents the spring base. It is what the suspension feels as a spring base, similar to the AA-arm suspension feeling its spring base as the width between the tires.

And, in the same way as the AA-arm suspension, the wider the spring base, the less the roll angle and the narrower the spring base, the greater the roll angle.

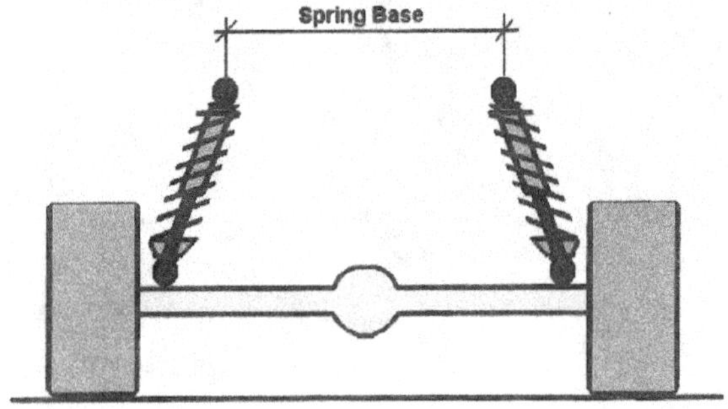

The rear springs on a straight axle rear suspension create the spring base. The car "feels" the spring base at the top of the springs. Some dirt cars have adjustable top mounts so that the upper mounts can be moved in or out. This creates an adjustment for the spring base width, which is a tuning tool used to change the rear roll angle.

This is an example of an adjustable right rear coil-over mount. It can be moved out to increase the spring base width for less roll angle, or in to shorten the spring base for more roll angle.

Why Match The Roll Angles? – We match the roll angles because we want the loads on the four tires to be ideal. That is, as we have discussed in earlier Lessons, at mid-turn, we want equally un-equal loading on the four tires. When this happens, the outside tires will carry the same load front to rear and the inside tires will carry the same load front to rear.

Remember in past Lessons where we explained that the most traction we get from a pair of tires on the same axle, or end of the car, is when they are equally loaded. Since in the real world, with weight transfer happening in the turns, there will never be equally loaded tires at mid-turn. The best we can hope for is to have as equally loaded tires as we can get and for the front and rear sets of tires to be equally un-equally loaded.

If the rear tires were more equally loaded than the front tires, then the rear would have more traction than the front causing a tight condition, or a car that understeers as the term implies. If the front tires were more equally loaded than the rear tires, then the car would be loose, or would oversteer as that term implies.

If the roll angles are not matched, then there will be un-even distribution of the loads, and the loading becomes very unpredictable and erratic. Un-equal loading and un-equal roll angles are the cause of slower mid-turn speeds and setups that don't perform the same as the laps increase on a continuous run.

Front With Higher Roll Angle – If the front suspension desired a greater roll angle than the rear, then it won't be able to achieve its desired roll angle and less weight will transfer at the front than would happen if the roll angles were balanced. Conversely, the rear would be pushed past its desired roll angle and then transfer more weight than would happen if the roll angles were matched.

In this case, the front would be more equally loaded and the rear more un-equally loaded. From what we have learned, the front would have more traction and the rear less traction. This car would be un-balanced and loose, or exhibit oversteer.

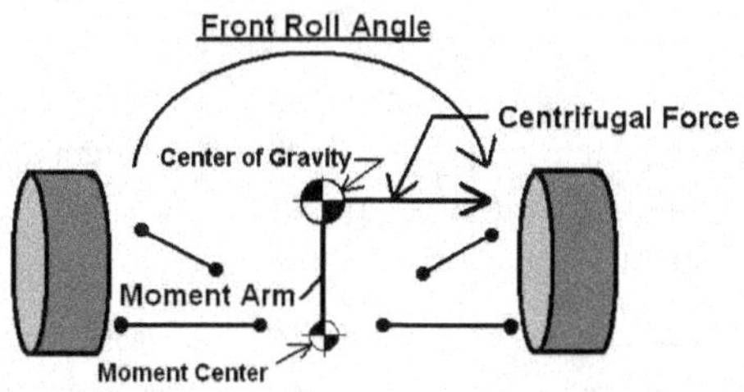

The Center of Gravity is the top of the Moment Arm for both the AA-arm suspension and the straight axle suspensions. The Moment Center is the bottom of the moment arm for both of these types of suspensions too. It is more difficult to change the height of the MC in a AA-arm suspension than it is to change it in a straight axle suspension. That is one reason why we usually tune the setup with the rear suspension spring rates and MC height for a stock car on circle tracks.

Rear With Higher Roll Angle - If the rear suspension desired a greater roll angle than the front, then it won't be able to achieve its desired roll angle and less weight will transfer at the rear than would happen if the roll angles were balanced. Conversely, the front would be pushed past its desired roll angle and then transfer more weight than would happen if the roll angles were matched.

In this case, the rear would be more equally loaded and the front more un-equally loaded. From what we have learned, the rear would have more traction and the front less traction. This car would be un-balanced and tight, or exhibit understeer.

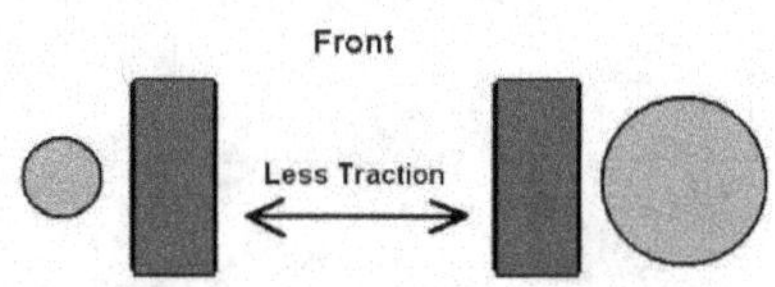

The result of an unbalanced setup where excess load transfers to the Right Front tire causing the rear tires to be more equally loaded. This would be a very tight handling car.

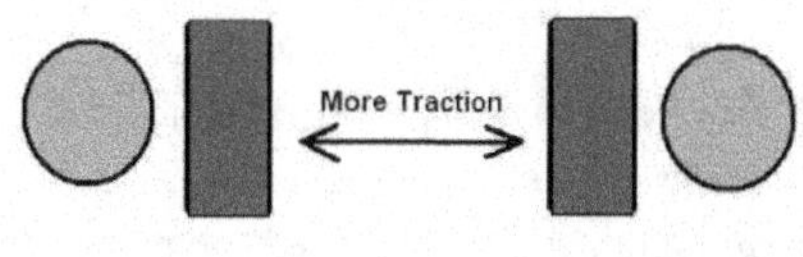

The sizes of the circles represent the amount of loading. If the car were setup tight as far as balance is concerned, then the rear tires will be more equally loaded and the front less equally loaded. From what we have learned about more equally loaded tires having more grip, this car would be tight. That is, the rear has more grip than the front.

Which End Do We Work On For Balance? – It is important to know which end to work on when we are trying to balance the setup. For a circle track car with a AA-arm front suspension and a straight axle rear suspension, we usually work with the rear suspension only.

That's not to say we cannot stiffen or soften the front spring rates, but if we have determined the correct stiffness of the front spring rates, it is much harder to effect a change at the front than at the rear. There are many more tools for adjustment at the rear in a straight axle suspension.

For a car with AA-arm suspensions front and rear, it becomes a matter of the platform attitude, or rake of the chassis when we decide which end to work with to achieve balance. For most applications of AA-arm front and rear systems, we adjust both ends if possible.

If the sway bar systems for one end is deficient, then we need to only work with that sway bar system to effect changes that will balance the setup. I have personally worked with a prototype car that from the manufacturer had a severely deficient rear sway bar system. The installed spring rates were fine.

The un-balanced condition this car suffered from was due to the fact that the rear sway bar was very weak and did not control the roll of the rear suspension. Through a series of tests, we determined that we needed more sway bar control and installed much stiffer components to solve the problem.

On some models of prototype cars, this one included, the sway bar is the primary tuning tool to create a balanced setup. We found a situation where the rear sway bar was not doing much work to help balance the car and the rear roll angle was much more than the front. When we redesigned the sway bar system and made it much stiffer in resistance to roll, the car became much more neutral in handling, and it came alive and picked up a lot of speed.

This is another car that when evaluated and balanced in roll angles became a multi-time winner at many different race tracks. This setup is more conventional and there are no tricks involved in setting it up. Once the balance is found, the setup is just maintained.

How Do We Work Towards Balanced Roll Angles? – If for example, we determine that the car is loose, or oversteering. This indication might come from the driver input, or tire temperatures. In all probability, the front is desiring to roll more than the rear. The basic way we would balance the roll angles and also tighten the car, or give more grip to the rear, is to do the following:

Decrease The Rear Spring Rates – If we decrease the average rear spring rates, we will then increase the rear roll angle. There is usually an optimum spring stiffness for a particular type of race car and we want to install spring rates that fall within that range of stiffness. So, this might not be the ideal solution.

For example, for a typical circle track car, the rear spring rates fall between 150 ppi and upwards of 300 ppi. We wouldn't think of installing a pair of 650 ppi springs to reduce our rear roll angle. It might work to do that, but the performance for the entry and exit portions of the track would be affected due to the extreme stiffness of those springs.

Introduce Rear Spring Split – Rear spring split has a pronounced effect on the roll angles. It is by far one of the most effective ways to change the balance on the race car. It is not a fine tuning tool, but a way to move quickly and in a major way towards a balance when the car way off. There are other ways to fine tune the balance that we will cover shortly.

Based on what we learned before in this Lesson, if we need to change the roll angles, we can introduce rear spring split. If the car is tight, with the rear out-rolling the front, we can soften the Left Rear (LR) spring rate, and the desired roll angle will become less. With this tight car, this change would serve to move more towards a balanced car. We can also stiffen the RR spring rate to increase the spring split to again reduce the roll angle and loosen the car.

If the car were loose, then if we increased the LR spring rate, or soften the RR spring rate, these two changes would increase the rear roll angle and tighten the car to bring it to a more balanced setup. Again, these are major changes and we will test our sample car in Part Two of this Lesson to see how much effect each of these changes produces.

For an example, for a circle track late model car running bump springs, we might split the rear springs and install a 150 ppi LR spring and a 275 ppi RR spring. That would produce a low roll angle. To equal that same roll angle with even rate rear springs, we would need to install two 650 ppi springs.

Rear Spring Rate Change – To a lesser extent, changing the average spring rate of the rear springs does change the rear roll angle. Stiffening the spring rates will reduce the roll angle and softening the spring rates will increase the roll angle.

As an example, if we stiffen the rear springs from a pair of 175 ppi springs to a pair of 200 ppi springs, we reduce the rear roll angle on a typical late model circle track race car by 0.2 degrees. That is not a big change. To make a comparison, that spring rate change is equals to a 0.25", or 1/4 inch change in the height of the panhard bar.

Changing The Height Of The Front or Rear Roll Center – Since the roll center is the bottom of the moment arm, we can make changes to raise or lower the roll center at each end of the car. Once again, for a AA-arm front and Straight axle rear combination of suspensions, the rear is much easier to make changes to than the front as far as roll center is concerned.

For a AA-arm front and rear type of race car, we probably wouldn't often make roll center height changes and would only work with spring stiffness and sway bar control.

For circle track cars, or for that matter, any car with the straight axle rear suspension, we can raise and lower the panhard bar. This device is called different things and goes by panhard bar, J-bar (because of its shape only), or track bar. Other devices that control the lateral movement of the rear end include a Watts link and the leaf spring.

Whatever system is used, and the panhard bar systems are by far the most popular and most used, we can more easily make changes to the rear roll center height by raising or lowering the bar, so that is what we do to fine tune the rear roll angle once we get close to the ideal angle.

An Example Of A Roll Center Height Change - The magnitude of the changes can be demonstrated easily. Using the setup software I described at the beginning of this Lesson, a typical late model circle track car might have a rear spring split of 150 LR and 275 ppi RR spring rates that are used to reduce the rear roll angle to match the front suspension stiffness. This car has a rear roll center height of 11.0 inches off the ground.

If we decrease the RR spring rate by 25 ppi to a 250 ppi spring, the reduced rate of the RR will increase the rear roll angle. To compensate for this change and maintain the same roll angle, we would need to raise the rear panhard bar (roll center) by 0.625 inches, or 5/8". As most veteran racers will tell you, that is a significant change in panhard bar height.

So, we can see where a seemingly insignificant 25 ppi change in spring rate at the rear makes a large change in the roll angle in the rear. In this case, the roll angle went up by 0.60 of a degree, or over a half degree. In Part Two of this Lesson, we will be installing springs and looking at the roll angles in typical types of race cars.

Spring Base Change – We can make changes to the rear roll angle by changing the width of the rear spring base. On some race cars mostly dirt cars, the top mounting point for the rear springs can be moved in or out. This changes the spring base width, and as we have learned, the spring base width is one determining factor in the final how large the roll angle is.

Again, for a typical late model race car, if we widen the rear spring base by moving the top of the two rear springs out by 2.0" each, this widens the spring base by a total of 4.0". This decreases the rear roll angle by a full half of a degree, or 0.50 degrees. This is also a significant change in the rear roll angle.

How Do Tire Loads And Forces Fit In? – In past lessons we talked about the importance of ideal tire loading and how to measure the spring forces to help achieve the ideal tire loading. When we can match the roll angles and balance the setup, we will then have ideal tire loading at mid-turn. One way to find the perfect balance, and matched roll angles, is to measure the existing spring force for a particular setup and then compare that to the spring force we would need to produce the ideal tire force.

So, not only do we have a tool to predict the roll angles and balance, we have a tool to measure the on-track spring forces that produce the tire loading. Once we know what is going on with the forces, we can then make adjustments to the setup to perfect the spring forces and ultimately the tire loading. We'll examine that process in Part Two of this Lesson.

Summary For Part One – This part one of Lesson Sixteen demonstrates how we can make changes to our setups to effect changes to the roll angles. We are working towards a balanced setup so that the wheel loads at mid-turn are ideal to produce the equal un-equally loaded tire loads that define a balanced setup.

For dirt racing, we don't necessarily match the roll angles in all cases. But knowing what changes will have what effect is important in any type of racing. We'll get more into the dirt setups in a separate section. For now, we come away with a better understanding about what we can change, and how much of a difference each change can make. There are major setup changes to get us close quickly, and more minor setup changes that serve to fine tune the setup.

Exam - In The Context Of This Lesson:

Which Part Of The Turn Do We Tune For First?

1) The entry
2) The mid-turn
3) The exit
4) All of the above

Mid-turn Steady State Is Defined As?

1) No acceleration
2) No braking
3) Maximum lateral loading
4) All of the above

Mid-turn Balance Means?

1) The car is not tight
2) The car is not loose
3) The front and rear roll angles are the same
4) The wheel loads are all the same

A Car That Is Setup Un-balanced And Tight Has?

1) Too much front roll angle
2) Too much rear roll angle
3) Equal roll angles
4) A low cross weight percent

We Can Measure Each Chassis Suspensions Roll Angles By?

1) Calculations using shock travels
2) Knowing the roll resistance of each end
3) Using a dedicated computer program
4) Using a pull down rig

For A Car With A Solid Axle Rear Suspension, We Tune Balance By?

1) Adjusting the front components
2) Adjusting the rear components
3) Changing the front geometry
4) Raising the rear of the car

We Can Adjust The Rear Roll Angle By?

1) Changing the spring rate split
2) Adjusting the rear roll center height
3) Changing the rear spring base width
4) All of the above

Which Rear Change Causes The Most Change In The Roll Angle?

1) Changing the average spring rate by 25 pounds
2) Changing the rear roll center height by ¼ inch
3) Introducing a 25 pound rear spring split
4) Changing the rear spring base by one inch

Race Car Technology – Level Three

Lesson Eighteen – Practical Applications – Part Two

This final Part Two of Practical Applications began in Lesson Sixteen will involve actually setting up race cars for the purpose of creating a dynamic balance. We won't be tuning with shocks, or for entry and exit performance here. This is all about mid-turn balance. We'll tune the other parts of the turns in Lessons to come. We need to get this process presented and understood by you before we go on.

Granted, there are a lot of race car types out there in the racing world and the students who enroll might have different interests and different race cars. The overall goal of Online Racing School with the Race Car Technology courses is to try our best to include information about all types of race cars.

The fact remains that circle track racing makes up the bulk of racing in the US. In other parts of the world, that might not be so true. We truly believe that the methods we present can be used, with some obvious tweaking, for every type of race car.

In this part Two, we will place a heavy emphasis on circle track race cars, only because there are a lot more of those here in the US where ORS is produced. But, it will be important to follow along as we go even if your racing is road racing stock cars, or formula and prototype racing. There is an interconnection between what we do for circle track cars and what would be done with other types of race cars.

The real difference between circle track cars and non-circle track ones is that it becomes easier to setup the circle track cars because they only turn one way. Race cars that turn both ways are somewhat limited in what can be done to achieve the balance we desire.

No matter what the type of race car, the latest technology in setup information works to get the car lower and more level to the race track. Even in the stock classes, teams are going ever softer on the ride springs to achieve a low attitude. This provides better aero efficiency as well as a much lower Center of Gravity.

How It's Always Been Done – Throughout racing history, the initial setup for a new car, or one we purchase used, but without information, is to engage in trial and error. If you can find the original manufacturer, or if your car is new and you are in contact with the car builder, they can provide a "ball park" setup to start you out.

If that is not an option, then most racer teams will seek out information from fellow racers who race in that division with that type of car. Or, they can ask consultants who may work alone or in partnership with some of the parts manufacturers like shock companies. After getting that first setup and trying it out, you're off and running into the setup world of racing.

This is the way it has always been done. We install a setup that might and might not include all of the unique characteristics of your car. Each car is different, and our sample car we use to develop the setups here is no doubt different than your car in many different ways.

To name a few, the CG is probably a little different due to different weight drivers, etc., the weight distribution, the motion ratios, the front and rear geometry, the bumps used if that applies, and more. No two cars are exactly alike even if they come off the assembly line together. Once you get the car, what you do with it makes it unique.

We're going to use very little trial and error to setup our sample car, and only then to fine tune the setup. We'll know all of the critical information about the car going in. When we get finished, we'll have the correct geometry front and rear, the correct weight distribution, and the correct spring rates. So, let's setup a race car shall we.

The stock classes as well as the Pro Late models are being setup in special ways. Experimentation by top teams throughout the last five years has yielded big gains in performance.

Choosing Spring Rates – Our choice of spring rates is dependent on the type of race car we will be racing and the actual type of setup we will run. In every class of race car, there are certain norms as far as spring rates.

For example, in road racing, both stock car and prototype, the spring stiffness is high and the front spring rates are usually 10-20% higher than the rear spring rates. These cars utilize front and rear sway bars to assist in controlling the roll of each suspension system. The overall control of the roll angles for road racing cars is tuned using the sway bars in most cases.

For asphalt circle track cars, there are presently two basic setups, conventional and bump setups. The rules for each division might allow or dis-allow the use of bump devices. The overall goal for most circle track race cars, dirt and asphalt, is to run lower in the front and particularly with asphalt cars, to reduce the overall average roll angle.

The combination of spring rates and sway bar rates serve to produce a lower attitude and one that has less roll angle. We've learned a lot about these various combinations over the past few years or so and we will pass that knowledge on to you. Now let's setup a real race car.

Choosing spring rates is an art that is backed up with science. We have tools available to us now that can predict and measure the chassis setup balance and the dynamic tire loading. There is less and less trial and error involved in creating modern day setups.

Conventional Metric Stock Car – The metric stock division race car based on a production car must use the stock control arms and stock rear suspension parts. The rear moment center is very high in these cars and can run from 13.5 up to 15.0 inches or more off the ground. This makes the rear of the car very stiff in roll resistance. So, we need to compensate for that high rear moment center by softening the spring setup in the rear. The roll angles will be higher than we see with the late models because the CG is much higher and the springs are softer overall.

A typical conventional setup for a metric four link stock car running a 1/3 mile track with 12 degree banking on 8" grooved tires might look like this:

CG @ 18.5" height.
G-Force = 1.25
Spring Rates: LF = 600 ppi, RF = 600 ppi, LR = 200 ppi, RR = 175 ppi, (this is a big spring front suspension)
Sway bar size = 7/8" solid,
Cross weight = 47.4%, (these cars have a high front percent, 52.5% in this case)
Rear Moment Center Height: 14.00".
Roll angles are 3.50 front and 3.60 rear.

This stock class car is built from a Metric Monte Carlo production car. These cars are very popular in the stock classes, but must be setup a little differently than other types of stock class cars that have.

Metric Stock Class Asphalt Running Big Springs

Corner Weights		Spring Rates		Rear Moment Center
738	648	600	600	14.0"
830	584	200	175	

Cross Wt % = 47.4%

Center of Gravity = 18.50

Sway Bar Rate = 0.875 w/ 0.187" wall

Roll Angles = 3.50 / 3.60

Rear Percent = 46.5%

When we refer to a "big spring" car, we car saying it uses stock dimensioned and designed lower control arms in the front with the springs mounted in the stock position. And, the rules might also require using the bigger 5" diameter springs at the rear also. Here we see the yellow springs in the front end of this stock chassis. The spring is resting in a pocket in the lower control arm and the top is controlled by a placement inside the frame rail. This spring can be adjusted for height by use of a screw jack bolt seen above the spring. When we are setting up the car, we need to keep in mind that the motion ratio of this big spring car is much different than if it were mounted like a coil-over in a late model.

Conventional Late Model – In days past, and again it is always good to know where we came from, the setups were what we now call, conventional. Even the conventional setups changed over the period from the late 1990's to the mid 2000's. A typical conventional setup for a late model stock car running a 1/3 mile track with 12 degree banking on 10" slicks might look like this:

CG @ 16.5.
G-Force = 1.50
Spring Rates at LF = 300 ppi, RF = 300 ppi, LR = 175 ppi, RR = 175 ppi,
Sway bar size = 7/8" solid,
Cross weight = 52.4%, (the low range of cross weight – it is common to use left side percent numbers)
Panhard bar was set at 9.00 left and 11.0 right.
Roll angles are 4.20 front and 4.20 rear.

Conventional Late Model Asphalt

Corner Weights		Spring Rates		Panhard Bar
758	616	300	300	9.00" / 11.00"
832	558	175	150	

Cross Wt % = 52.3%

Sway Bar Rate = 0.875" Solid Bar

Roll Angles = 4.20" / 4.20"

Rear Percent = 50.3

Center of Gravity = 16.5"

This is an actual setup I ran on a car in 1996 that won a lot of races in the mid-Atlantic region. It is balanced with the front and rear roll angles at 4.50 degrees.

In more conventional times just a few years ago, the race cars looked like this with the front valence riding several inches above the race surface. Now these cars ride much lower and flatter to the track. There is a difference of several tenths of a second in lap times between the two types of setups. We know because we ran head to head against top bump teams using conventional setups that were balanced.

In the mid-west, the cars ran on much flatter tracks for the most part, and the setup would change for the G-force, RR spring rate and panhard height. This car running on a 6-8 degree track using the same tires might make the following changes to the setup: RR spring to 150 ppi, G-force at 1.35 and panhard bar at 11.0" left and 12.0" right. Note that we had to move the panhard bar up 2.0" just to compensate for the softer RR spring.

Soft Conventional Late Model – Th pro, or limited, late model type of car is very similar to the super late model car in our next example. These are by far some of the most popular types of race cars in today's racing and in some cases these cars cannot run bump devices. So, we install soft front springs, larger sway bars and then tune the rear roll to match the front. A typical soft conventional setup for a late model car running a 1/3 mile track with 12 degree banking on 10" slicks might look like this:

Our sample car NOT running bumps will have the following setup:

CG @ 15.5".
G-Force = 1.75
Spring Rates at LF = 125 ppi, RF = 150 ppi, LR = 150 ppi, RR = 200 ppi,
Sway bar size = 1.500" with 0.25" walls, (these soft front spring setups need a much larger sway bar)
Cross weight = 52.8%,
Panhard bar set at 9.50" left and 10.25" right,
Roll angles are 2.50 front and 2.50 rear.

Soft Conventional Late Model Asphalt

Corner Weights		Spring Rates		Panhard Bar
738	648	125	150	9.50 / 10.25
830	584	150	200	

Cross Wt % = 52.8
Sway Bar Rate = 1.50" / 962 ppi
Roll Angles = 2.50 / 2.50
Rear Percent = 50.5

This chart represents a soft conventional setup where the car is designed to use coil-over springs in the front.

Soft Conventional Late Model Asphalt
Big Springs

Corner Weights		Spring Rates		Panhard Bar
734	652	225	250	9.00" / 11.00"
834	580	150	200	

Cross Wt % = 53.1
Rear Percent = 50.5
CG = 15.5"
G-force = 1.75
Sway Bar Rate = 1.50" / 962 ppi
Track Banking = 12.0
Roll Angles = 2.30 / 2.30

Here we have a soft conventional setup where the car uses stock big springs in the front mounted in the stock location.

Bump Spring Late Model Setup – For the bump spring setup late model, we now get into the very stiff front spring rates creating much less roll angle. These setups are designed to get the car low in the front and to reduce the roll angle significantly. For the front spring rates, we add the ride spring rate to the bump spring rate. Also, these cars are usually allowed to run softer 10 inch slicks.

CG @ 14.0". (the nose is at least 2.0 inches lower than with a soft conventional setup)
G-Force = 2.00 (the lower CG and softer tires generate more G-force)
Spring Rates at LF = 1625 ppi, RF = 1625 ppi, LR = 150 ppi, RR = 275 ppi,
Sway bar size = 0.875" solid (we went back to a smaller bar)
Cross weight = 53.0% (there is a high range that some teams use by adding 6-7 percent)
Panhard bar set at 9.00" left and 8.00" right (a reverse split)
Roll angles are 1.10 front and 1.10 rear.

This is a Pro Late model, and part time super late model, that runs on bump springs. The setup sample car we listed was based on this very car that we setup several years ago.

NOTE: Bump stop setups are very similar to the bump spring setups. We only need to determine the spring rate of the bump at the compression travel it is operating at. If we record a gain of 750 pounds of force in a half inch, then we can assume the bump spring rate is double that in one inch, or 1500 ppi. We then add that rate to the ride spring rate.

Bump Late Model Zero Roll Setup (Experimental) – For the bump spring setups used today in the super late model division, we are seeing some different setups. This design was first spoken about in a Circle Track article I wrote some years ago. Everyone at that time was trying to reduce the roll of the cars. So, I spoke about creating a zero roll angle setup and how to do it. Now we see some teams, at some race tracks, trying that concept.

The design involves creating a positive roll angle at the front and a negative roll angle at the back with the same roll angle number, but in reverse. The front wants to roll right, and the rear wants to roll left. These two roll angle desires cancel each other out leaving us with a net zero average chassis roll angle.

This school is about teaching and that is why we are making this known. This is a concept that is used in racing today. We don't necessarily promote it, just acknowledge it exists. Here is how it works.

For our bump super late model sample car, we generated a front roll angle of 1.10 degrees. If we arrange the rear spring rates (using rear spring split) and raise the rear moment center height (panhard bar height), we can realize a negative 1.10 degrees of roll angle. That negative 1.10 degrees of roll will cancel out the front positive 1.10 degrees of roll angle.

Here are the numbers:

CG @ 14.0".
G-Force = 2.00
Spring Rates at LF = 1625 ppi, RF = 1625 ppi, LR = 150 ppi, RR = 300 ppi, (increased rear spring split)
Sway bar size = 0.875" solid
Cross weight = 59.0%, (the high cross weight puts load on the LR tire compensating for a loose setup)
Panhard bar set at 11.50" left and 9.00" right, (raised the panhard bar to reduce the rear roll angle)
Roll angles are 1.10 front and (-) 1.10 rear. (now we have a net zero average chassis roll angle)

If you look at the front valence on this car, it appears that the setup is based on the zero roll angle concept that we explained. We call this the equal and opposite roll angle setup. This is becoming popular and testing will tell if it is better than a setup that has equal positive roll angles.

Late Model Asphalt Bump Setup
Zero Roll Angle

Corner Weights		Spring Rates		Panhard Bar
750	624	125/1500 1625	150/1500 1650	11.0" / 9.0"
841	550	150	300	

Cross Wt % = 53.0
CG Height = 14.0"
Rear Percent = 50.5%
Sway Bar Rate = 1.50" / 0.25" wall
Roll Angles = 0.92 / (-) 0.99
G-force = 2.00
Track Banking Angle = 12.0

This is the basic zero roll angle setup. This setup uses a stiff RR spring to help achieve the negative rear roll angle. There are other ways to setup this zero roll concept.

Late Model Asphalt Bump Setup
Loose In With RR Pre-load

Corner Weights		Spring Rates		Panhard Bar
732	654	125/1800 1925	150/1800 1950	13.0" / 12.0"
836	578	150	150 / 225	

Cross Wt % = 53.2
Rear Percent = 50.5%
Sway Bar Rate = 0.875" / 135#
Roll Angles = 0.85 / (-) 1.20
Bump Rate = 1800 / 1800

If we install a RR spring rate of 150 and pre-load it with a force equal to the previous stiffer 300 ppi spring to support the RR during entry and at steady state mid-turn, then we can create a setup that is loose into the corner. Because the rear roll is a higher negative number than the positive front roll angle, the front will have more grip. This makes the car loose in, or turn in much better. At mid-turn, as the car begins to accelerate, the added load being transferred from front to rear loads the RR spring beyond the pre-load and now engages the softer 150 ppi spring. As we will see, this softer spring rate tightens the car off the corner.

Late Model Asphalt Bump Setup
Tight Off With RR 150 Spring

Corner Weights		Spring Rates		Panhard Bar
732	654	125/1800 1925	150/1800 1950	13.0" / 12.0"
836	578	150	150 / 225	

Cross Wt % = 53.2

Rear Percent = 50.5%

Sway Bar Rate = 0.875" / 135#

Roll Angles = 0.85 / + 1.20

Bump Rate = 1800 / 1800

Once the load on the RR spring surpasses what is needed at mid-turn, the 150 ppi spring rte comes into play and we can see where the rear roll angle goes from a negative (-) 1.20 to a positive 1.20. Since the rear roll angle is now more than the front roll angle, the setup becomes a tight setup helping to drive the car off the corner.

We aren't sure this is the best setup you can run, but we acknowledge it is being used. In theory, it might work very well. What it also does is try to twist the chassis. A balanced setup is where both ends of the car are trying to roll to the same angle. In this setup, the two ends are not in sync and are working against each other.

Road Racing The Late Model – For road racing using a late model type of chassis, we are turning right and left, so this setup must be different than what we develop for a circle track car. We cannot use spring split in the rear to help reduce the rear roll angle. Therefore, we will need to use a rear sway bar.

There are several other differences we see. The overall spring stiffness is going to be much more for the road racing car. And, the weight distribution must be such that we end up with a 50/50 cross weight percent. The cross weight must be the same for right and left turns.

The weight distribution is a very critical element in these cars. If possible, we need to find a front to rear percent that works to create an ideal weight distribution at mid-turn when we start out with a 50% cross weight. We think we can. Here is a typical late model car that has been converted to road racing. We use the same basic car as in the previous circle track examples, only we make the suspensions symmetrical side to side.

CG @ 15.5".
G-Force = 1.30
Spring Rates at LF = 600 ppi, RF = 600 ppi, LR = 400 ppi, RR = 400 ppi
Front Sway Bar Size = 1.375" w/ 0.125" wall
Rear Sway Bar Size = 0.75" solid bar
Cross weight = 50.0%, (the cross weight percent must be 50% for a road racing car)
Panhard bar set at 10.00" left and 10.00" right,
Roll angles are 1.54 front and 1.63 rear.

This basic setup was run on an AGT class Grand Am road racing car that won three championships in the early 2000's.

Road Racing A Late Model Asphalt

Corner Weights		Spring Rates		Panhard Bar
666	666	600	600	10.0 / 10.0
634	634	400	400	

Cross Wt % = 50.00

Front Sway Bar Rate = 1.375" / 454#

Rear Sway Bar Rate = 0.750" / 60.0#

Roll Angles = 1.55 / 1.55

Rear Percent = 48.80

This is what a circle track car that has been converted to a road racing car might look like. The cars have more roll angle than a circle track car due to the fact that they must turn both ways, so we cannot utilize spring split or bump setups in these cars. Roll angle is not a bad thing if it is controlled.

In a circle track car that has been converted to a road racing car, or any AA-arm front and solid axle rear road racing car, it is a necessity to run a rear sway bar to help control the rear roll. Most of these sway bars are small in diameter being 0.75" or smaller. This one is adjustable for arm length.

Summary - How we finalize the setups we run beyond spring rates and roll center heights all depends on the goals for rear steer and the shock package that gets us into and off of the corners. In the coming Lessons we will explain how to select shocks and how to tune your setup at the race track. For now, just know that we all need a starting point that represents the basic, baseline balanced setup.

Keep in mind that any setup you get from someone else will not be what they win with. Nobody gives out the winning setup. It will be generic. Even coming from the chassis manufacturer, the setups you might get will be generic and the best setups might be saved for the house car, or a favorite race team that shows the manufacturers product well and that has already has some success.

The best setups are developed inhouse and are best kept a secret. As we stated, every race car is a little different than the next, even if they come off the assembly line together. Concentrate on creating an ideal setup for the chassis you run that provides performance based on the dimensions and weights of that chassis, not someone else's.

Next, we will discuss how shocks can enhance the entry and exit performance now that we know how to develop the setup that created a balanced mid-turn condition. A few Lessons later, we will tell you how to fine tune the setup at the race track.

Exam - In The Context Of This Lesson:

The Primary Part Of The Turns We First Setup For Is?

1) The entry.

2) The mid-turn

3) The exit

4) All of the above

The Primary Setup Goal For All Types Of Race Cars Is?

1) A balanced setup with equal roll angles

2) Soft front springs

3) Higher rear moment centers

4) Very little roll angle

The Metric Stock Class Car Has What Unique Characteristic?

1) They run big springs in the stock location

2) The CG is higher

3) The rear moment center is very high

4) All of the above

The Transition From Conventional To Modern Setups Involve Which?

1) The use of soft front ride springs

2) Use of high rear spring split

3) The use of large sway bars or stiff bump devices

4) All of the above

Road Racing Cars Differ From Circle Track Cars How?

1) They run much stiffer spring rates

2) They run rear sway bars

3) The cross weight must be 50%

4) All of the above

Race Car Technology – Level Three

Lesson Nineteen – Advanced Racing Shock Tuning In Transitions

In keeping with the theme of the setup part of the Online Racing School, Race Car Technology courses, we tune the Mid-turn portion of the turns first. Then we tune the Entry and then Exit in that order. One of the major tools we use to tune entry and exit are the shocks. With just a basic understanding of how the dynamics of the race car works, we can choose our shock rates correctly to complement our setups.

In the modern world of racing, for both dirt and asphalt competition in circle track racing, and also for road racing, shocks have become one of the most important fine tuning tools we have. They can complement the current setups, especially the bump setups, if applied correctly.

The following information is useful whether you are running more conventional setups, bump setups or road racing. Shocks affect the speed of movement and the load distribution at the four corners of the car as we transition from high speed to minimum steady state mid-turn speed, and then to up high speed again.

One of the most basic and important things to remember is this. Shocks do not affect the handling of the car if they are not in motion. To say that you can tune your mid-turn, steady state, handling with shock changes is wrong. Any competent shock expert will tell you that up front.

Even though we cannot affect the mid-turn handling with shocks, conditions that begin with corner entry that negatively affect the mid-turn handling can be solved with shocks. This is where the thought that shocks affect mid-turn handling began. If there are no entry problems and you need to solve mid-turn handling problems, shocks are not the answer.

Shocks Work With Springs - Controlling wheel movement would be much easier if all we had to work with was the shocks. But in reality, our race cars are supported by a set of springs. Basically, we always want to match our shock

We can influence the load distribution on the four tires through the use of different rates of shock resistance to movement. For some types of race car, the entry and exit performance can be tuned by the use of shocks. We must remember that shocks do not affect loading on the tires when they are not in motion. Shocks only influence loads when they, and the suspension, are in motion.

For certain types of race cars, it becomes important to allow the corners of the car that are compressed to rebound and stay in contact with the track. The car on the right is allowing the left front tire to remain in contact with the track surface. The car on the left has the left front tire riding off the track, maybe as a result of too much rebound in the LF shock.

rates to the spring rates, and/or the bump device you might be using.

Springs resist compression and promote rebound. The shock doesn't need to do much compression control because the spring resists compression. It is hard to compress a spring, and the more rate the spring has, the harder it is to compress.

Springs promote rebound. When a spring is compressed, it wants to return to its normal free height. The more rate the spring has, the faster it wants to return to its free height. Less than adequate shock rebound control causes the suspension to move too quickly.

So, based on the above, in all situations, the shocks rebound rate will always be greater than the compression rate for all suspension shocks because the spring helps resist compression and promotes rebound.

As we install stiffer springs, we would naturally need to increase the rebound resistance and decrease the compression resistance. Stiffer springs would include adding bumps to one or more corners of the car. Each bump device has a spring rate that is measured in the range of motion it is operating in. Bump springs have a more consistent rate than bump stops and the rates of both are added to the equivalent ride spring rate.

With bump setups, we may say we have a soft spring setup because we install 150 ppi ride springs at the front in our coil-over car, but when we add the stiff bump rates to the ride spring rate, we end up with 1500-2000 ppi or more for the front spring rates.

Therefore, with those setups, keeping with the idea that the shocks must control the installed spring rate, we must run shocks with a rebound rate upwards of 1500-2000 pounds at 3-5 inches per second of speed.

The shock must control both the ride spring and if so equipped, the bumps spring or other bump device. The shocks control the overall spring rate and that includes all of the springs.

"Tie-Down" Shock Terminology – This brings us to an important discussion. The use of the term "tie-down" has been around for some time to describe shocks that are high in rebound resistance. The idea initially was that if we install these high rebound shocks, we can tie one or more corners of the car down and keep the tire "attached" to the track. This way of thinking is completely wrong.

First of all, none of the tires are connected to the track surface, and they are free to rise and fall according to what the other three corners of the car are doing at that instant on the track. If our setup causes that corner to move vertically because load has shifted off of it, then the load will come off regardless of how much "tie down" we design into the shocks. You just won't see much suspension movement if the rebound is too high.

We may be led into thinking that the "tie down" worked. No, it did not. If enough load comes off that corner, or the setup is unbalanced enough, the tire might well come off the track surface even without suspension movement. The existence of lack of movement doesn't mean the loads on the tire stayed the same.

This unloading of the tire will occur on the inside front or inside rear tire normally in a race car while turning. So, we cannot tie the inside front tire down and we cannot tie the inside rear corner down, although I've seen some teams try. Lack of suspension movement does not mean the load remains on the tire.

I repeat: Lack of suspension movement does not mean the load remains on the tire.

What we can do is control the spring rate of the ride spring plus any bumps we have installed. If your left rear spring is a 175 ppi spring, then you need to use a normal shock rebound rate associated with that spring or risk taking much of the load off that tire on entry into the corners. On some tracks when using too much rebound, not only will you be loose in, but also loose off due to the way the track banking transitions.

For race cars that ride on very stiff springs, the motion of the chassis is minimal. If the shocks offer control only when they are moving, then with little motion, there will be very little control offered by the shocks. Instead of moving these shocks through inches of motion on the shock dyno, they should be cycled through increments of one inch at the most in many cases.

Entry Tuning With Front Shocks – Let us understand one thing up front in our discussion about shock tuning. We have made it very clear that if the shock is not moving, it has no control over load distribution. When using bump setups where the front suspension moves onto, and rides on, the bumps on initial corner entry, there will be very little movement of the shocks throughout the turn until the car exits the corner and load shifts off the front causing the shocks to come off the bumps.

So, in relation to bump setups, the entry discussion presented here is mostly irrelevant. This discussion is for setups and race cars using more conventional springs to support the front suspension.

Moving on: If we split the front shock compression rates with a RF shock using a stiffer compression rate than the LF shock, then, while the suspension is in motion due to load being transferred to the front on entry, the RF suspension will move slower than the LF suspension. Additional load will be transferred onto the RF and LR tires causing a momentary increase in the cross weight percent in the car.

It is important to note here that the load transfers almost immediately when a force is presented to enact that transfer, such as applying the brakes. If on entry we transfer 300 pounds from the rear to the front, the 300 pounds goes to the front in an instant.

The distribution of that 300 pounds between the two front tires (while the suspension is in motion, and while in the process of assuming a new attitude that will support the additional load) will depend entirely on differences in stiffness of the suspension systems at all four corners. Stiffness is defined as the resistance to movement influenced by the shocks and springs.

The result of all of the above is this. The slower moving (or stiffer) corner in compression will momentarily retain more of the transferred load while the suspension is in motion.

Cross weight, as we commonly know it, is defined as the percent of the combined RF and LR weight divided by the total vehicle weight. If the cross weight percent increases, then a car turning left will be tighter on entry and the car might be faster if that is the desired effect. This is exactly why it has been said that a stiffer RF shock will speed up load transfer to that corner, although the reality of that statement is not entirely true.

Later in the entry event, some of the load that has been transferred onto the RF due to that corner being stiffer in compression will transfer over to the LF tire as the shocks stop moving and the car reaches a steady state at mid-turn. Then the normal cross weight you set the car at will apply.

If the car is already tight on entry, after having eliminated common causes of tight entry such as rear miss-alignment, rear steer or brake bias issues, then an opposite effect can be utilized. If we install a LF shock that is stiffer than the RF shock, and/or run a stiffer LF spring than on the RF, then we can effectively reduce the cross weight in the car on entry while the suspension is in transition by loading the opposite diagonal, the LF and RR. As one diagonal goes up in percentage of supported weight the other goes down.

The front shocks on bump type of setups result in very little movement from the point of initial entry until the car exits the corner and begins acceleration. This lack of movement means that the front shocks have very little influence on entry characteristics.

Entry Tuning With Rear Shocks – In a typical circle track car turning left, the left rear corner is usually sprung soft to enable the high spring split across the rear to reduce chassis roll. To keep from running a very stiff right rear spring rate, we use a lighter left rear spring in the range of 125 to 150 pound per inch.

Now we need to setup our shocks to work with that light spring. Since our spring rate is low, our shock rebound only needs to be sufficient to control that softer spring rate.

Most left rear shocks are valve'd to a higher rebound setting than what is needed because back in earlier days using conventional setups, the LR spring rate was higher. If we use a shock that worked for the older setup and is too high in rebound, then on entry to the corner, tire loading will be reduced on that tire and the car will become loose.

This is a very common problem and easily solved by just reducing the rebound in the LR shock. The LR shock does not need very much rebound, only enough to control the spring that is installed, or less.

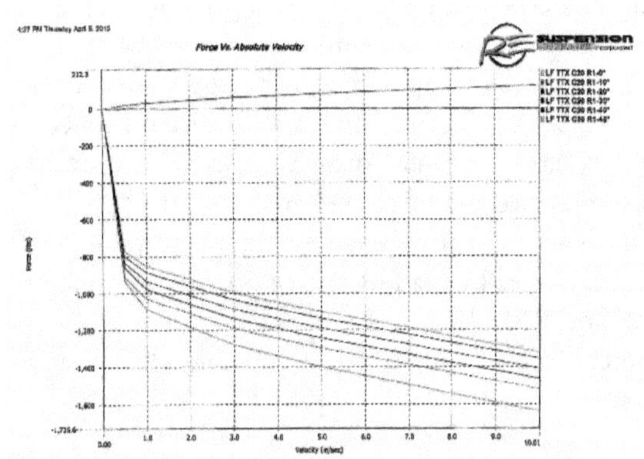

This represents a shock dyno readout sheet showing the various shock rates through different bleed settings. The bottom readings are for rebound. We can see that this shock is adjustable for rebound from 1,000 pounds of resistance to almost 1,300 pounds of resistance at 3.0 inches per second speed of movement. This is also a digressive shock in that the resistance initially ramps up when approaching 1.0 ips speed and then tapers off as the speed increases up to 10.0 ips of speed.

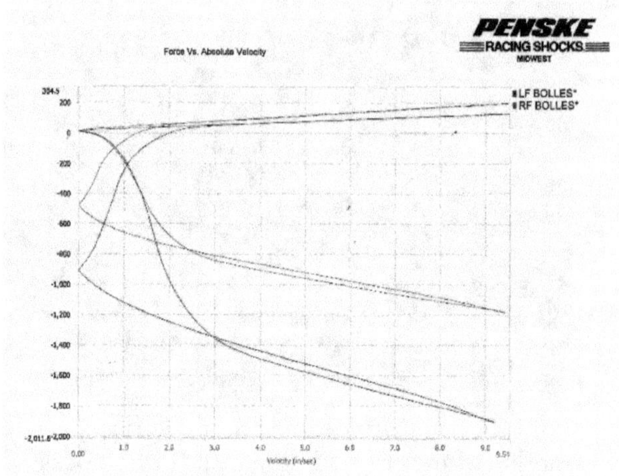

Here we see a more typical pair of front shocks built for bump setups. The LF shock (shown as a blue line) has a "nose" rate (initial force at the start of movement) of around 900 pounds, and around 500 pounds for the RF shock (red line). The rate at 3.0 ips then climbs to 800 for the RF shock and around 1,350 pounds for the LF shock. These relatively high rates are necessary to control the combined spring rates of the ride spring plus the bump device.

It is useful to use shocks that are adjustable for rebound and compression so that when you change springs or bumps, you can then adjust the shocks to match the new rates. This shock has a rebound adjuster that we can see just above the eye. A tool is used to "sweep" the adjuster to restrict or open up a needle valve in the shock shaft.

Exit Tuning Using Split Valve Shocks – Corner exit performance can be improved by utilizing the shocks. This is done by either working with the compression settings in the rear shocks and/or working with the rebound settings in the front shocks. In a car with equal rear spring rates, a stiffer compression setting in the LR shock than in the RR shock will load the LR and RF corners while the shocks are in motion.

Load is transferred to the rear under acceleration and while the rear suspension is in motion, this split will tighten the car by increasing the cross weight percent. Keep in mind that this motion and event happens very quickly, so the effect is short termed.A shock with a stiffer rebound rate at the LF corner can help accomplish the same effect by causing a slower movement of that suspension and a more rapid transfer of load off of that corner which in turn increases the percentage of load supported by the RF and LR tires.

If the RR spring is somewhat stiffer than the LR spring, there will be a loosening effect on acceleration when load transfers to the rear. The stiffer RR spring causes more of the load to be placed on the RR and LF corners reducing the cross weight percent. This event last much longer than the event caused by the shock being in motion.

By installing a LR shock with a much stiffer compression rate, and a RR shock with a much softer compression rate, you can momentarily nullify the negative effect of the stiffer RR spring. This serves to equalize the unequal resistance to compression due to the dissimilar spring rates and helps keep the car tight on exit. Again, keep in mind that this motion and event happens very quickly, so the effect is short termed.

A more permanent cure for this problem is to run a softer RR spring that has been preloaded to a force equal to what is needed to support the RR corner at, or near, the mid-turn portion. Then when the load has been transferred to the rear on acceleration, the RR will compress at the softer spring rate. In this way the load is more evenly distributed among the LR and RR tires. If you understand this statement, you have found a nugget of gold to use in setting up your race car.

This car has very little rebound in the front shocks as evidenced by the high attitude of the front valence when accelerating off the corner. With this little control, the springs in this bump setup are not being controlled properly through the entry and mid-turn portions and that can cause problems on a track that is rough.

Putting All Of This to Good Use – In order to utilize the configurations we have discussed here, we must be able to use a range of different rates of shocks in order to find the right combination for our car at a particular race track for a particular setup. For a team that races at only one track, the process is fairly simple.

You would experiment to find the fastest set of shocks and ones that suit the driver's style and stick to those while also staying within the boundaries of physics. For teams that travel to different tracks, some changes might be necessary if the setup (read as spring rates) needs to change and/or the track layout is different from the track you are used to.

Most shock experts agree with these basic statement:

1) The shock package should be softer overall when racing on dirt and when the track is flatter when on asphalt for the conventional setups.

2) Get your basic setup close to being balanced by tuning for mid-turn before trying to tune with shocks. Shocks cannot solve basic handling balance problems that mostly occur at mid-turn.

3) Higher banked tracks require a higher overall rate of shocks and springs as opposed to flat tracks. This is because of the higher speeds and the extreme amount of downforce.

4) Shocks that are mounted farther from the ball joint should have more overall compression and rebound control than if they were mounted closer to the ball joint. That is because with each

inch of travel of the wheel, the shocks that are mounted farther away will move at a slower speed which means less resistance in both rebound and compression. (See the shock graphs) This is also true of shocks mounted at high angles to the direction of motion.

5) Of the two transitions, tune entry performance first. If there are no entry problems, make small changes if you want to experiment to see if entry speeds can be improved. Entry problems include a tight or a loose car. By far the worst problem would be the loose-in condition. This can involve an alignment problem, but far too many times, the problem involves a LR shock that is too stiff in rebound.

6) Tune exit performance last. Exit problems can include a car that pushes under acceleration or one that goes loose under power. Be sure that you do not have a Tight / Loose condition where the car is basically tight in the middle and goes loose just past mid-turn. This is fixed with spring rate and/or panhard bar adjustment, etc. and is the first thing we are supposed to work on.

7) On dirt race tracks, reduce rebound settings on the left side and decrease the compression rates on the right side for dry slick surfaces to promote more chassis movement. This helps to maintain grip as the car goes through the transitional phases of entry and exit.

8) For the bump setups on asphalt, the whole shock package must be much different than when running conventional or soft conventional setups. The bump spring rates (either bump rubbers, bump stops, or bump springs) will be very high and so the shock rebound rates must match those high rates in order to control the ride spring and bump device.

Using a spring/bump combination that is rated in the 1500-2000 ppi range will need a shock that is rated at around near that range in the 3 - 5 inches per second range of movement. Usually the ultra-low speed rate of the shock will be comparatively high too and it is helpful to use a "nose" rate of between 500 and 800 ppi or more at zero inches per second speed of movement.

Nose means that it takes that much force to cause the shock to move initially. The shock will not move until that amount of force is applied to the shock.

The angle of the shocks as well as the motion ratio to wheel movement both affect the rate of the shock that influences wheel movement and chassis movement. A high angle like this one causes a reduction in efficiency of almost 20%. If this shock were rated at 150 pounds at 3.0 ips, then when installed on the car would only provide around 120 pounds of rate at the same vertical wheel speed. With this installation, 3.0 ips of wheel movement would equal 2.4 ips of shock movement. The same is true for motion ratio at the front end.

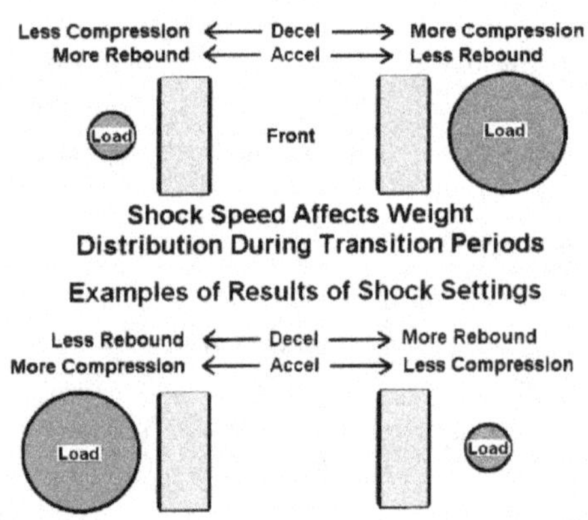

This chart represents how a shock can provide added loading on the RF and LR tires during the transitions listed. This is true for setups where the shocks actually move during these transitions and the loading influence only occurs for a short period of time, while the shock is actually moving.

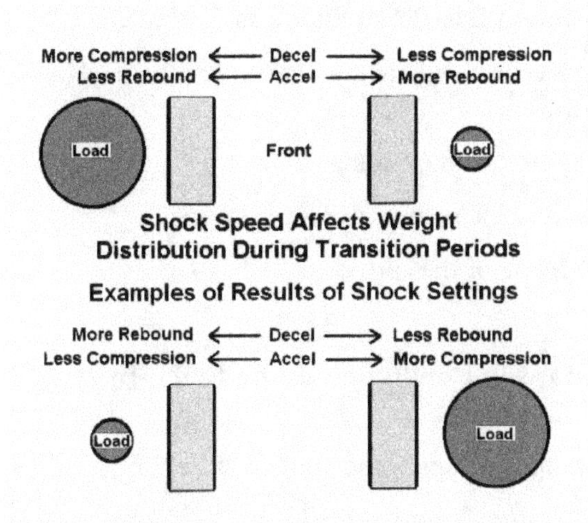

This chart represents how shock tuning can influence the added loading on the LF and RR tires during the transitions listed. Again, this effect is only present for a short period of time and only while the shock is in motion.

It is recommended that most amateur race teams consult with shock retailers and/or consultants that specialize in shock technology. If you don't have you own shock dyno to evaluate your shock rates, then seeking outside help is a must. Then the team can make intelligent decisions on how to setup the shock package.

Summary – The shock rates and changes provided here are representative of trends that can enhance your handling package. Before any of this can work, the setup must be balanced, the steering characteristics must be ideal and the car must be aligned properly. If not, nothing you will do by tuning your shock package will fix those problems.

Shock tuning is the last thing to experiment with in order to try to increase the race cars performance. It is a necessary step in finding the ideal total handling package. That said, before setting up your car and choosing your shocks, evaluate what you will need by matching the shock rates to the spring rates you will run.

Exam - In The Context Of This Lesson:

Shocks Affect Which?

1) Mid-turn handliing

2) The speed of chassis movement

3) How fast weight transfers

4) The amount of weight transfer

Shocks Do Their Work When?

1) They are mounted closest to the ball joint

2) The car is at mid-turn

3) They are in motion

4) They are mounted near the springs

Shocks Control What?

1) The compression of the springs

2) The rebound of the springs

3) The chassis attitude during transitions

4) All of the above

In Most Cases, Shocks Are Designed With?

1) More rebound than compression control

2) More compression than rebound control

Using A Tie-down Shock Is A Way Of?

1) Keeping the tire in contact with the track surface

2) Maintaining loading on the tires

3) Controlling chassis dive

4) None of the above

Loose-In Condition Can Be Cured By Which?

1) More compression control in the LF

2) More rebound in the LR

3) More compression in the RF

4) More compression in the RR

Loose-Off Condition Can Be Helped By Which?

1) More compression control in the LR

2) More rebound in the RF

3) More compression in the RR

4) More compression in the LF

Bump Setups Require More What?

1) Compression in the front shocks

2) Rebound in the rear shocks

3) Front rebound to control the high spring rate

4) Front compression to control the high spring rate

Which Do We Tune First?

1) Entry handlling

2) Exit handling

Race Car Technology – Level Three

Lesson Twenty – Testing and Tuning The Setup At The Track

To this point in RCT Level Three we have studied the important concept of balance, forces, roll angles, weight transfer, dynamic loading, angle of attack, what makes traction and force verses weight. We learned how a sway bar can be used as a spring. Then we ran through various setups for the mid-turn performance and the use of shocks for the transitions. All of that leads us to this all-important lesson.

Now we go to the track and try all of this out. One of the primary attractions of racing is the fact that everything you do with your race car will be evident once it hits the track. If it is fast, you did your job well. If not, there is more work to do, period. Every setup has its day on the track.

Here we will explain how to do a test session and how to tune the setup if you think it needs a bit of improvement. Hopefully this takes place before the season starts and there is enough time to properly examine how your setup is working.

Once we get the setup established, we need to go test at the race track. Not only do we need to evaluate the setup for balance and speed, we need to compare our performance to other cars we will be racing against. We never need to seek the fastest lap times without regard for creating the dynamic balance we have talked about continually in this school. Some setups are fast for a few laps, but in the long run, the fastest lap times at the end of the race are what wins races.

Goals - The overall goal of testing is to develop the setup combination that will be both initially fast and also stay fast for a long time. It should be good on the tires, comfortable for the driver, and out run the competition all the way to the last lap.

On asphalt, the setup we end up with is probably the one we will qualify and race with given small changes between the two if the rules allow. For dirt, we will make setup changes that may be required for the different track conditions. That does not mean we cannot test on a track that is consistent.

A primary goal for dirt cars might be to just learn the process of making changes to meet the track conditions. There is an order and logic to adapting to changing track surface grip levels. Becoming comfortable with making those changes can be a huge performance gain.

The following is a description for testing a circle track late model type of car. No matter what type of race car you will be using, many of the important points will pertain to your car. Obviously, for road racing, there will be no spring split, weight distribution tuning, or other circle track specific changes. We will address road racing with stock cars and formula cars because we have experience with those types of cars.

Road racing cars are more limited in what can be changed to create the balanced setup. The springs and suspension must remain symmetric and the weight distribution must stay at 50/50% cross weight. We tune these types of cars with spring rates and sway bar stiffness mostly. If the car has a solid axle rear suspension, we may be able to change the rear roll center height if it is equipped with an adjustable panhard bar.

Road Racing Weight Distribution

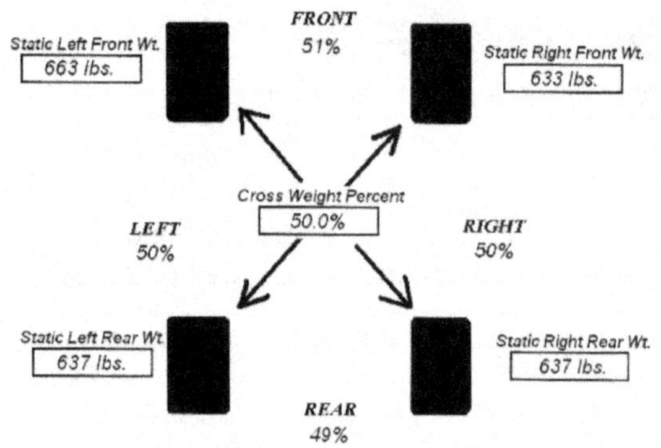

The road racing cars cannot adjust the weight distribution as far as cross weight is concerned. They may be able to move weight fore and aft to arrive at a weight distribution that works with a 50-50% cross weight. This is one of the secrets to creating the perfect road racing setup.

Cross Weight Related to Front Percent

Front Percent	46%	48%	50%	52%	54%
Cross Weight	59.1% X	55.6% X	52.1% X	48.6% X	45.0% X
Rear Percent	54%	52%	50%	48%	46%

This chart shows the relationship of cross weight to front-to-rear percent. In a typical circle track race car, or even a circle track car that has been converted to road racing, there is an ideal cross weight that will work to provide the ideal wheel weights at mid-turn, based on the front to rear percent of weight distribution. Based on this, a road racing car would need to design the car with around 48.8% rear weight distribution. That way, the car can have a 50% cross weight distribution that will work for turning left and right.

Pre-testing Preparation / Planning – It is most important to know your car before you go to the race track for practice or testing. All of the things we talked about in Levels One through Three are what leads up to the setup. If those items are not in order, there is little reason to go test.

If you are going to a new track, take into consideration the banking and transitions. If it has a different banking angle than you are used to, a different spring setup might be in order. High banked tracks need higher spring rates overall and have little need for traction enhancing technology. If the track is flatter, include methods of creating bite off the corners into your planning.

Collect information before you go to your test. Check Ackermann, check rear alignment, eliminate bump steer and make rear steer settings that are appropriate. Weigh the car with different spring combinations and note what changes need to be made to bring the original weight distribution back to the baseline.

You should make notes detailing what needs to be done to the car before you go out onto the track. Also, list your test session priorities and post them on the car. This team was preparing the car before a big test and wanted to make sure all of the items had been taken care of and checked before departure. Once at the track, we tend to forget the order of testing we discussed at the shop.

Arrival – On arrival at the track, establish a pitting position for the car that is relatively level. You should have easy access to the tool cart as well as the trailer and other track facilities that may be needed. Mark the spots around the tires so you can always park the car in the same position after each run.

Weigh the car before testing and after all of the testing is done at the end of the day, re-weigh the car to see how the weight distribution might have changed from the various adjustments. If the track scales are different from yours as to cross weight, note the difference and adjust what you read each time to what the shop scales read.

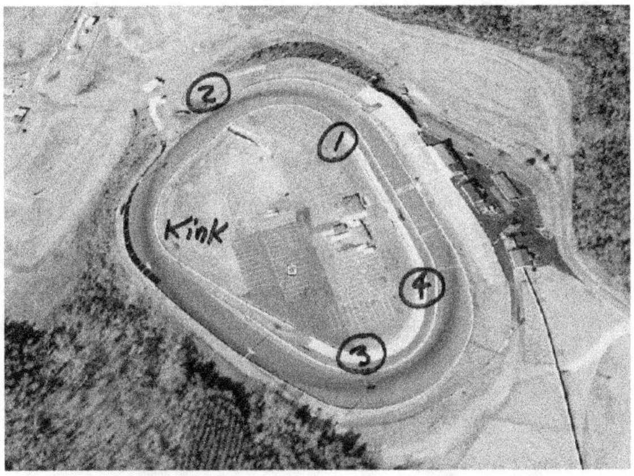

A well-prepared race team might print out an aerial photo of the race track they will be racing at. This view gives a lot of information about the turns and the line drivers have been using (the dark line around the track is where every one drives). This is Concord Speedway outside of Charlotte, NC. The configuration is different than most tracks. The part labeled "kink" is an extension of Turn Two because there is no real straight between Turn Two and the Kink. The length of the turning portions of this track is very long.

Find a spot at the track that is mostly level and park the car in the same spot every time it comes in. That way you can take measurements and set the sway bar the same each time without deviation in the numbers.

How to Measure Track Performance –We need to measure the on-track performance. There are two components to speed, the motor/gear, and the chassis setup combination. Since we work on these separately, we need to measure them separately.

If we have lap times that include turn segment times, we can then compare our times with our competition. Turn segment times tell us all we need to know about how good the chassis setup is verses the other cars we are racing against. Remember what we talked about, if we can improve the mid-turn speeds, we will usually also improve the straightaway speeds.

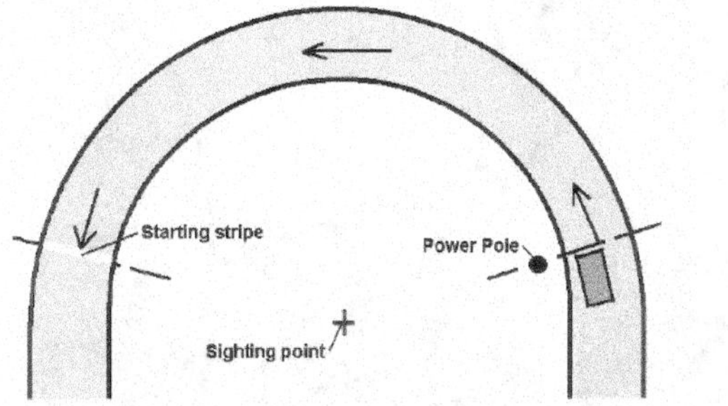

Be sure to take segment times that represent the speed through the turns. If you are several tenths off the fast whole lap times of your competitors, but just as fast through the turn segments, then the engine might need looking into. When you are the fastest car through the turn segments, you can stop making changes to your setup, or if you do, be prepared to go back right away if no further gain is seen.

The First Set of Runs - The driver should initially make several slower circuits and then a few faster laps to "shake down" the car the first time out. This establishes that the brakes work as expected, the wheels are on tight, the air will stay in the tires, there are no water or oil leaks and the transmission and rear end lubricants will be brought up to temperature. We should do two more longer runs following the initial outing before we can expect to get meaningful tire temperatures.

Have the driver initially run the turns at a speed lower than normal and note the position of the hands. Then once the car is up to speed, the driver should again note where the hands are and if the steering is significantly different, the car is either tight or loose.

After each run, record the tire pressures and temperatures, tire sizes, and engine water and oil temperatures. Keep hard copy records of the data in addition to digital records that may be stored in the tire temperature device or a computer.

Once the driver is confident that the car is sound, longer and faster runs can be done. As you make your next series of runs, try to have the driver stay out at least 10 laps so that the tire temperatures will be sufficient to show how they are working. The exception to this is when the car is obviously not right. Then the driver should come in and explain the problem.

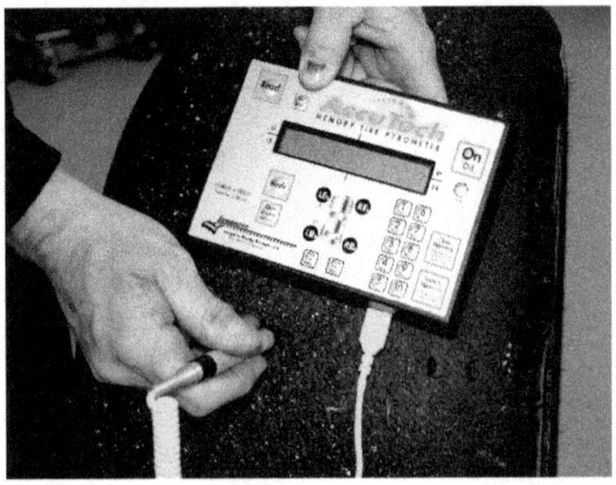

Record your tire temperatures and refer to them for tire pressure adjustments and camber changes. Bring the tires up to race temperatures and analyze your tire data. Correct any camber or tire pressure problems before making major setup changes. The priority starting out is handling balance (i.e. neutral handling), then tire temperatures related to pressures and cambers, then on to setup balance.

When we first go out and run hot laps, we need to quickly react to the tire temperature and make changes to the tire pressures and cambers before we make any major setup changes. The pressures and cambers help to create the largest tire contact patch possible. This adds a lot of grip to a tire without ever making a spring or other setup change.

Initial Evaluation - Evaluate the tire cambers, pressures and overall handling balance. Make quick adjustments to the front tire cambers and all four tire pressures if the temperatures dictate. Do not make chassis adjustments until the tire issues have been corrected.

Record the driver comments as well as crew comments as to the handling and engine performance. If the car is not neutral, now is the time to make changes to improve the handling while working to maintain a balanced setup.

There is a difference between Handling Balance and Dynamic Balance. The car is neutral in handling balance when it is neither tight nor loose. We can easily adjust most cars so that they will be neutral. This may make the car faster, but it is not our primary goal. We need the car to be both neutral in handling and balanced in how the front suspension and rear suspension are working.

The first runs should be made to shake down the car. We want to make sure there are no fluid leaks, and that the brakes and engine are working correctly. The driver can relay comments about the general handling once the car gets up to speed.

Evaluate the data from every run. If possible, try to use a data acquisition system to record shock travels, steering, throttle and more. Keep good notes as to your progress and how each setup change affected the car and the way the driver feels. Keeping detailed notes helps to refresh your memory so that you can look over the progress of the test later on in the quiet of your shop. Convert written notes to computer notes so you won't need to worry about losing them.

Mid-Turn Performance First – We need to always evaluate and correct the mid-turn performance first. To balance the car at this Steady State point on the track will also help to balance it on entry and exit.

We can interpret the balance of the car by evaluating the tire temperatures. These tell us how much work each tire is doing in relation to the other tires. We are looking for more equal temperatures on pairs of tires on each side of the car.

Changes to the panhard bar height and/or spring rates can help add temperature to a tire that is too cool. More uneven front tire temperatures indicate a tight car. High RR tire temps indicate a loose car. Once the tire temps come to be more equal on each side, the handling balance must be tuned with cross weight.

Remember that a tight car will also often be loose off the corner. This is called "tight-loose" because the initial condition, being tight, leads to the other condition, becoming loose off, and fixing the tight condition will often solve the loose condition.

If the car were tight with the original setup, to make it neutral in handling, the team might have found a cross weight percent that worked to make the car more neutral. As we make changes to the setup to heat up the LF tire, the car will then begin to have more front grip. This can make the car loose, so we then need to increase the cross weight to bring it to a more neutral handling condition.

Remember that spring split in the rear and panhard bar height changes are the most effective ways to re-balance a circle track car. As you make those changes, you will also need to tune the handling balance with changes in cross weight.

Tire wear can tell us a similar story when racing on dirt. Dirt teams rarely take tire temperatures. They do feel the tires for temperature, so we know that they consider the temperatures to be important. But tire wear can also tell us how hard a tire is working.

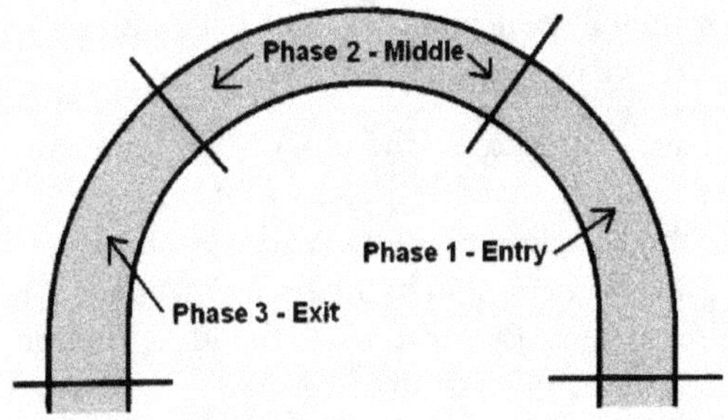

For tuning the setup, there are three turn segments that we work on. The place to start is in the mid-turn segment where poor handling and dynamic balance will affect the other two segments. We perfect the mid-turn portion first, and then the entry and exit portions in a way so as to not affect the mid-turn performance we have gained.

Mid-turn Changes - To change mid-turn balance, we can do one or more of the following:

1) Raise or lower the rear Moment Center by moving the panhard bar or J-bar up or down. For leaf spring cars, we can raise or lower the actual spring, but that is not an easy thing to do. Metric four-link cars also have a tough time changing the rear moment center height and must rely on other methods for changing the balance.

NOTE: For road racing cars with AA-arm front and rear suspensions, the roll centers can be changed to effect a change in handling. In the authors experience, changes to the roll centers in AA-arm suspensions have significantly improved the grip for that pair of tires. Lower roll centers help create more roll angle. Changes to the upper control arm angles can improve the camber change characteristics to help improve the size of the contact patch.

2) You can change the rear spring rates. Softening the right rear springs, and/or stiffening the left rear spring will increase the rear roll and will tighten the car, as will softening both rear springs. The inverse is true, stiffening the RR spring and/or softening the LR spring will loosen the car.

3) Softening the front springs will help the car turn, but to a lesser degree than making rear spring changes. Spring split at the front also has less affect and has more influence on entry characteristics than on mid-turn.

4) Installing larger or smaller sway bars will have an effect on handling. The stiffer the bar, the less the front will want to turn. So, to help cure a tight car, we can go to a softer sway bar.

5) Increase or decrease the cross weight percent. As we make changes to the cross weight, we affect the handling of the car and we can easily make the car neutral in handling by making cross weight changes. But, this is not the ideal method by any means, it is just the easiest.

6) We can increase or decrease the stagger. This is never an acceptable way to tune the handling of your race car. For a defined radius of turn, there is an ideal stagger that will allow the car's rear wheels to roll around that radius and not influence the direction the car travels from following that radius. Learn what stagger your track needs and use that.

Summary – In this Lesson, we talked about our goals for testing, now to plan out a test and how to tune for mid-turn, the first consideration. In the next Part Two of this testing theme, we will tune for Entry and Exit performance and explain how to evaluate the results of your test.

Exam - In The Context Of This Lesson:

The Overriding Goal For The Test Is To?

1) Find the fastest lap times

2) Create a dynamically balanced setup

3) Tune the setup for a perfect handling balance

4) Compare our lap times to our competitors

The Order We Need To Solve Handling Problems Is?

1) Entry first, then middle and then exit

2) Exit first, then entry and then middle

3) Middle first, then exit and then entry

4) Middle first, then entry and then exit

Road Racing Cars Easily Cannot Change What?

1) Weight distribution front to rear

2) AA-arm Roll Centers

3) Cross weight percent

4) All of the above

The Very First Changes We Make Are?

1) To the spring rates

2) To the panhard bar heights

3) To the tire pressures and cambers

4) To the shocks

To Correct Dynamic Balance Issues, We Change The?

1) Spring rates

2) Panhard bar heights

3) Sway bar sizes

4) All of the above

To Correct Handing Balance Issues, We Change The?

1) Spring rates

2) Panhard bar heights

3) Sway bar sizes

4) Cross weight

Race Car Technology – Level Three

Lesson Twenty-One – Tuning At The Track, Part Two

Entry Tuning – When evaluating and making changes to the car to correct entry problems, do not make a change that will affect the mid-turn portion of the track. You should have already tuned the mid-turn handling and balance. If you must change a spring to help entry, then you need to re-tune the mid-turn balance.

Entry problems can be caused by rear alignment issues or incorrect shock rates, mostly in the LR corners of the car. Over-driving the entry can make a car push. Make absolutely sure that the rear end is aligned properly and square to the centerline of the car. Do not install a high rebound left rear shock. And don't let your driver "dive bomb" the entry to the corners.

Excess LR shock rebound may cause the car to be loose on entry as load is transferring to the front while braking. The LR shock should allow the LR tire to move in rebound to help it maintain contact with the racing surface as the car pitches forward and to the right on entry.

Spring split has some effect on entry performance too. At flatter tracks, a stiffer LF spring over the RF spring helps entry stability in most cases. Remember that spring changes also affect the Dynamic Balance of the car and you will need to re-evaluate the tire temperatures and make changes to the panhard bar height to re-balance the setup after a spring change.

On entry, the car is turning in, braking, and transferring weight that is mostly going to the right front corner. We need to provide more compression control with the RF travel and less rebound control for the LR as weight comes off that corner.

Entry Changes - To change entry performance, we can do one or more of the following:

1) Rear alignment is the number one cause of entry problems. The cause relates to either the miss-alignment of the rear tires or by the rear steering of the rear end. A car can become tight or loose on entry and that can translate to mid-turn problems. You should have checked and corrected any rear alignment problems long before you came to the track. Rear alignment and rear steer are not mid-turn tuning tools.

2) Shocks affect entry. Shock rates that restrict movement of one or more corners of the car can negatively affect entry. A LR tie-down shock, or one with excessive rebound control will help cure a tight-in condition by loosening the rear, but this is considered a crutch and not a valid tuning tool.

The two corners most affected by the dynamics of corner entry are the LR corner and the RF corner. A RF shock that is too stiff on compression can cause a tight condition on entry and a LR shock that is too stiff in rebound can cause a loose condition on entry.

3) Brake bias changes affect corner entry. There is an ideal brake bias that will allow adequate braking of the front and rear sets of tires based on the loads those tires carry. Different cars with different Center of Gravity heights will require different brake bias.

Tune your brakes so that wheel lockup occurs simultaneously at the two ends of the car under heavy braking. We do not want the brake bias to influence entry handling characteristics. Never try to correct a car that is tight into the corner by increasing the rear brake bias or fix a loose-in car by increasing front brake bias.

4) Setup Changes to solve corner entry problems? We never want to make changes to our spring rates, sway bars, weight distribution or moment centers to solve entry problems without making sure we then bring back the mid-turn balance. If you do that, we will certainly change your mid-turn handling in a negative way. You should have

already tuned the car so that the mid-turn handling was balanced correctly.

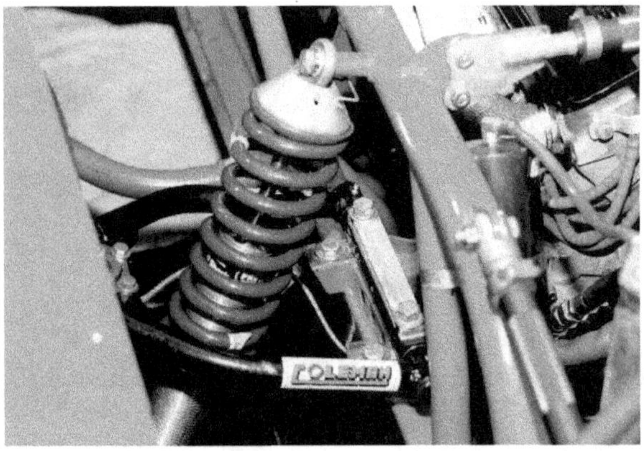

Be sure to match your shock rates to your spring rates. It's OK to go a little higher on the rebound rate verses your spring rate, but you don't want to install a 'tie-down", or very high rebound setting where the spring doesn't require that. Use a normally valved shock on a corner with a more conventional spring rate. High rebound shocks are designed to be used with bump stops and bump springs mostly, and certain soft spring conventional setups that we described in Lesson 17.

Exit Tuning – Problems associated with corner exit involve either a tight-off or loose-off condition. If we use the wrong methods to improve either of those, we might then end up with a car that no longer handles in the middle. So, the changes we make to improve exit performance should never change the mid-turn balance. Changes to spring rates, spring split, panhard bar height and cross weight will all affect, and probably ruin our mid-turn balance. So, just how do we tune exit performance?

The tracks where we usually see exit issues are mostly the flatter tracks with associated lack of grip or at tracks where the transition in the banking cause problems. The combination of lateral forces that come from turning the car and the torque associated with power application tend to overload the grip capability of the rear tires. So, we need to develop ways to increase the amount of grip the rear tires have available on exit while being careful not to affect the mid-turn balance we have established.

We can experiment with various designs of Pull Bars, Push Rods, Lift Arms, and rear steer that happens only on acceleration. The goals are to reduce the "shock" to the rear tires upon initial application of power and increase the total rear grip by introducing rear steer (to the left) into the rear geometry. The more the rear tires are steered, the more traction they will develop, just like the front tires when they are steered. This is the "angle of attack" syndrome we learned about earlier.

There is a limit to how much rear steer we can use before the car becomes too tight and we cannot steer the front wheels enough to overcome it. A few ten-thousandths of an inch of fore and aft rear wheel movement can be felt by the driver on asphalt. Most movements are in the range of 0.050" to 0.125".

One design element uses about 0.063" of RR wheel movement to the rear on entry, and then 0.125" RR wheel movement to the front on exit. This steers the car to the right by 0.063" on entry and a net of 0.063" to the left on exit.

Loose Off Condition – We can use Rear Steer to solve loose off problems. If we know we are good through the middle, then a loose off condition can be solved with the application of rear steer that happens only upon the application of power.

The most common way to create rear steer in cars equipped with the three link suspension is to increase the angle of the left trailing link, the front being higher than the rear. If we also decrease the anti-squat by decreasing the angle of the third link, when the car sits down on turn exit under acceleration, the LR wheel will be moved back creating rear steer.

We can tune the amount of rear steer by changing the angle of the LR link. This works very well and compensates for the rear tire drift caused by adding power. The rear tires will attain a new angle of attack to gain traction and if we cause rear steer, we immediately give the rear tires that angle of attack it needs without needing to step out.

Shock rates can temporarily increase the cross weight percent on exit to tighten your car off the corners. If you run shocks with a stiffer compression rating on the LR corner than on the RR corner, then when the shocks move as the car squats coming off the corner under acceleration and while the loads transfer to the rear, then the LR corner will momentarily carry more load and the LR and RF will then share that increased load.

Throttle Control - This is a learned art and will allow the rear tires to maintain their grip on the track surface and help to provide better acceleration. Once we lose grip in the rear, we must back off the throttle until we regain grip before we can continue to accelerate. By exercising throttle control, we may feel like we are giving up performance, but in reality, we are providing the most acceleration possible.

Throttle control is defined as the modulation of the gas pedal through a range of motion, never moving quickly from one position to another, in order to keep the tires in contact with the track surface. The rate of change in throttle position must be altered depending on your position on the track and through the corner, so the driver must develop an educated foot.

Many dirt drivers report that they never got past half throttle over the course of an entire race in which they won. This means that they were working from off throttle to half and many points in between. It is the development of efficient throttle modulation that is one of the most effective tools you can use to promote bite off the corners.

We can utilize rear steer to gain entry performance and to create more rear tire angle of attack. We have described ways to create rear steer and how much to use in previous Lessons. Make sure you test these configurations before trying to race with them. A little rear steer goes a long way.

Balance Evaluation Using Tire Temperatures – The question becomes, how do I know when the race car setup is balanced dynamically? There are several ways to evaluate the setup for dynamic balance. One of the first and most simple ways is to compare the average tire temperatures for each of the four tires. By now, we probably won't have any one tire that is way hotter or cooler than the others.

We mostly look at the left side tire temperatures for circle track cars, and front to rear average temperatures for road racing cars. In today's circle track world, the LF tire may be either cooler or hotter than the LR tire. Let's clarify that.

LF Cooler – If the LF tire is cooler than the LR tire, and the setup is more conventional in spring layout and having two positive roll angles, then the setup is tight. That is, the rear is rolling more than the front. This takes more load off of the LF than would otherwise transfer due to lateral load transfer. The front pair of tires end up with less grip than the rear pair of tires.

LF Hotter - If the LF tire is hotter than the LR tire, then the front is out-rolling the rear. This means that we have arranged the spring rates such that the rear wants to roll less than the front. This will transfer more load off the LR tire than would otherwise transfer due to lateral load transfer.

The situation with the LR being cooler than the LF may be due to a setup where the front and rear roll angles are equal and opposite, as we have discussed. The front wants to roll to a positive roll angle and the rear wants to roll to an equal value, but negative roll angle. These two will then cancel out each other, but this leaves the LR less loaded than if the two roll angles were the same and both positive.

Chart 1

Left Front

Loc 12	Loc 11	Loc 10
146	154	150
Avereage Temp 150.00		
Pr Cold	Pr Hot	Pr Gain
20	23	3

Right Front

Loc 1	Loc 2	Loc 3
211	204	207
Average Temp 207.33		
Pr Cold	Pr Hot	Pr Gain
24	29	5

Front Average = 178.67

Session #3

Left Rear

Loc 9	Loc 8	Loc 7
162	168	158
Average Temp 162.67		
Pr Cold	Pr Hot	Pr Gain
20	25	5

Right Rear

Loc 4	Loc 5	Loc 6
216	214	210
Average Temp 213.33		
Pr Cold	Pr Hot	Pr Gain
24	32	8

Rear Average = 188.00

Tire temperatures can tell a lot about how the chassis is working and if the setup is balanced or not. In this example, we see where the LR average tire temperatures are almost thirteen degrees hotter than the LF temperatures. And, the RR tire temp. is six degrees hotter than the RF temps. This car is setup tight based on the cooler LF tire temps. and is probably loose off the corner. This is commonly called, "tight/loose" because the tight condition through the mid-turn phase actually causes a loose off condition. This also can make the rear average temperatures hotter than the front temps.

Balance Evaluation Using Force Evaluation – Another way to determine the dynamic balance is to physically measure the right front force that the spring and sway bar are together putting on the RF tire. For every type of race car running a certain type of race track, there is a combination of the four tire forces that is ideal. This situation means that the perfect equal-unequal tire loading is taking place at the front and rear.

This has been covered in past Lessons, but to review, ideally, we need for the front and rear pairs of tires to be equally un-equally loaded. A pair of tires on any race car in todays racing will never be equally loaded when going through mid-turn. Because we have load transfer from the inside tires to the outside tires, the inside tires will always be less loaded than the outside tires.

We call this un-equal loading. We want the front and rear to be equal in the un-equal loading so that the outside pair of tires carry the same tire load, and the inside pair of tires carry the same tire load.

Under this situation, if we can measure what the RF tire load is at mid-turn, then we can adjust the setup so that the force supporting that tire provides the perfect tire loading. But, we have to know what force we are looking for.

In a typical super late model circle track car, the RF tire load at mid-turn might be around 1,500 pounds. If the spring motion ratio were say 0.8125, then the MR squared is 0.6602. If we divide that number into 1,500, we get 2,272 pounds of spring force needed to support that mid-turn tire load.

There are many different products that will measure the spring force if we know the exact shock travel at mid-turn. We support the Gale Force Load Pro because it was the first on the market and offers excellent tech support to the user.

Remember too that the force needed to support the RF tire load is often a combination of the ride spring, the bump device and the sway bar spring rates. Measuring the ride spring and bump spring rate combination is fairly simple. It takes a special tool to measure the sway bar spring force. Gale Force also has a product that will measure the sway bar force at the shock/spring position. This is important because we need to be able to add these two forces together.

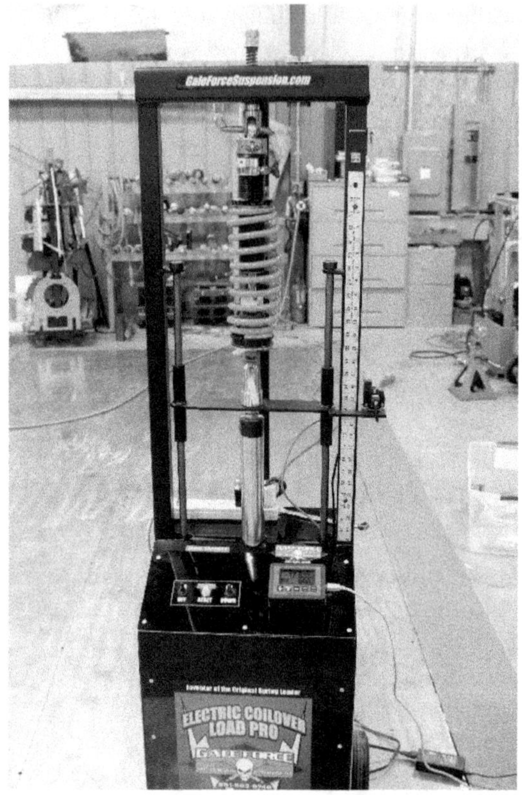

The Gale Force Load Pro rig allows a team to re-enact the forces that occur on the race track. If we know the shock/spring travel at mid-turn, then we can then move the coil-over to that same length on the Load Pro and then record the spring force. There is an ideal spring force for the RF that will provide the ideal tire loading for that car on a particular track.

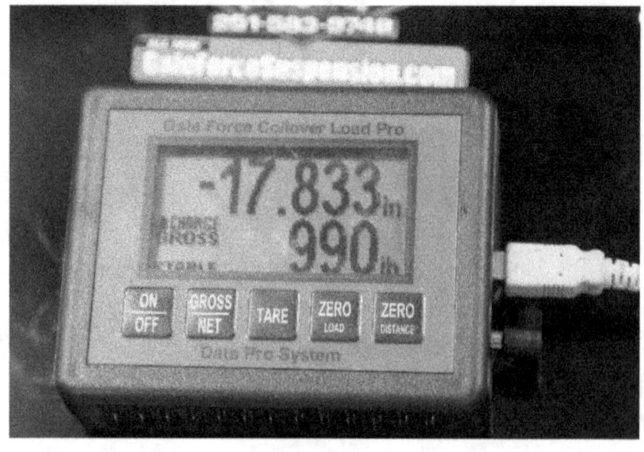

Here we see the readout from the Gale Force Load Pro. The shock length is 17.833 at ride height and the force reading for the ride spring is 990 pounds. When we compress the shock/spring and bump combination to the length it is at mid-turn, then we can read the force that the spring combination provides to help hold the RF tire load. The sway bar is a part of the combined force that supports the tire load, and it can be added to the spring and bump forces.

The ride spring and bump provide much of the force needed to support the RF at mid-turn. The sway bar also contributes to that total force. Therefore, we need to know how much force is added by way of the sway bar. Gale Force also makes a sway bar tool that measures the sway bar force at the coil-over position so that the ride spring/bump force can be added to the sway bar force. This gives us the total spring force supporting the RF tire.

Once we make adjustments to the setup so that the RF spring force is correct, then the other tire forces must also be correct because each tire force is interdependent on the others. If one changes, then the other three also change. If 2,272 pounds of combined spring force is what we need, then we adjust the setup until we read the shock travel that will supply that force.

Errors In Shock Travel Readings – It is possible to read errors into the shock travels. If you are reading shock travels using a travel indicator mounted on the shock body, then you need to know that the travel represents only the mid-turn shock travel and not travel from any other influence.

Another influence might be due to braking into the corners. If the RF shock travel we see on braking exceeds the shock travel that happens at mid-turn, then our reading is in error. We might think there is more force on the tire than there actually is at mid-turn.

The way to avoid this is to either use data acquisition and read only the mid-turn shock travel, or run laps that avoid heavy braking into the corners, but provide maximum speed through the mid-turn portion. The driver needs to work with the team to accomplish this.

Another way to reduce the braking influence is to use Anti-dive geometry in the RF and LF upper control arms. This helps prevent over-travel of the shock while entry braking. The Anti-dive will not interfere with any other aspect of the setup if applied in measured amounts.

Normally a team will use a shock travel indicator such as the one seen here. This is adequate if the user knows how to eliminate errors in the reading. On some tracks, entry braking that causes the nose of the car to dive, can put more load and shock travel on the RF suspension than what we would see at mid-turn. So, care must be taken to run a set of laps where braking is diminished and the maximum shock travel is due to mid-turn suspension travel.

Here is an example of entry braking where the RF has traveled probably more than what it will at mid-turn. Anti-dive geometry can help solve this problem that will cause errors in our shock travel data.

One of the solutions to errors in shock travel readings is to use data acquisition on the car during testing. Then the team can look at the data and isolate the shock travels through the mid-turn portion. Any additional travel that might come from entry braking, or even transitioning up onto the banking when going out for a run, can be isolated.

End Of The Test - Always save your sticker tires for the last run of the day after the car is all dialed in. If the setup is good, make a qualifying run on fresh tires. After that run, do a 30 or 40 lap run on those newer tires and see if the lap times stay consistent. A truly balanced setup will provide lap times that fall off less than your competition as more and more laps are run on a single set of tires.

Review your notes when you get back to the shop and learn from both the gains and losses. All of the results are valuable and the more we learn about the effects of changes, the better we can make quick adjustments during a racing event. The top teams make a point of knowing how each chassis adjustment affects all of the other parameters involved with their setups.

Incorrect tire stagger, bent shocks, suspension binding and poor alignment are some of the peculiarities that can ruin a test session. If radical setup changes do not seem to provide the expected result, look for a mechanical problem and fix it. Keep your test notes available for review. Test as often as you can afford and whenever the track is available. If you can develop a comprehensive plan for your testing, your performance will get better and you will enjoy your racing experience that much more.

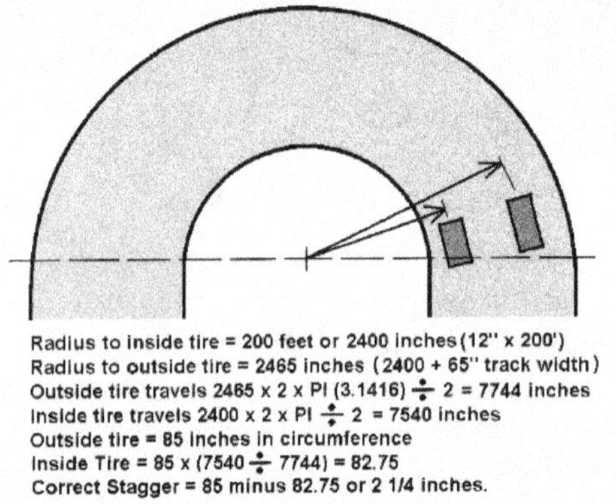

The correct stagger you will need for a particular track is fixed and not to be used as a tuning tool. We never want to adjust stagger to mask a handling problem. Using incorrect stagger is a crutch and will only provide a temporary solution. Running the wrong stagger can cause a multitude of handling problems. The example shows a typical calculation for stagger on a track with a 200 foot radius turn.

The data showing the corner the tire was mounted on, the track it was run on, the date and the set numbers

are important things to keep track of. Some teams will write that information on the tire. Keeping the information printed on the tire saves time and confusion when deciding which tires you need to put on the car next.

Conclusion - The test procedures are based on the assumption that you have already solved the most critical issues facing your race car. Back at the shop, you have aligned it, checked and corrected the moment center design, checked for binding in the suspension, rebuilt the shocks, set the weight distribution and done all of the other maintenance things we know we should do.

The last thing to do is run the car. Teams that have learned how the various changes will affect the handling of the car will go through the process much more quickly. If you're just now learning these things, take good notes and concentrate on what is happening with each change. Ask your driver a lot of questions so that you will know exactly what changes to make and how far to go with each change. Once you complete the test and have fine tuned the setup, it will be time to go racing.

Exam - In The Context Of This Lesson:

Loose-In Problems Might Be Caused By?

1) Brake balance issues
2) Shock valving that is too high in rebound
3) Rear steer issues
4) All of the above

Tight-In Problems Might Be Caused By?

1) Brake balance issues
2) Too much right front shock compression
3) Overdriving the corner
4) All of the above

Loose-Off Problems Might Be Caused By?

1) Rear steer issues
2) A stiff RR spring
3) A car that is tight in the middle
4) All of the above

Tight-Off Problems Might Be Caused By?

1) Rear steer issues
2) A soft RR spring
3) LF shock rebound too high
4) All of the above

Tire Temperatures Tell Us What?

1) If our spring rates are too high
2) If our spring rates are too low
3) How the tires are working
4) When the weight distribution is wrong

Reading The Right Front Total Spring Force Tells Us What?

1) How the tire temperatures will look
2) The loading on the RF tire
3) If the RF spring rate is too high
4) If the RF spring rate is too low

The Sway Bar Does What?

1) Resists chassis roll
2) Adds RF spring rate
3) Adds to the RF suspension force
4) All of the above

Errors In Reading Shock Travel Come From?

1) Misreading the tape measure
2) Going up onto the banking
3) Chassis dive from braking into the corners
4) All of the above

Race Car Technology – Level Three

Lesson Twenty-Two – Post Race Analysis

We now have one race under our belt with the changes we have made. The true measure of success, or of making the right decisions, is how the race car performs in an actual race. The proof is in the pudding so they say and that might be one of the primary attractions people have to going racing and being a part of a race team. The results of your efforts are immediately known, to everyone.

That part, the knowing, can be a scary process to go through. Every "setup guy" knows exactly what I'm talking about. But, remember this one thing, every bit of information is valuable. Even the so called "failures" are not really that. They are, for the smart racer, opportunities to grow and learn. I've had my share of those.

It may take several races and an accumulation of information before your car is where you want it to be setup wise. And then there is the part that is the hardest to work on, the driver. In most cases, unless you have a proven, winning, driver, there is a learning process that every driver has to go through. Even for experienced drivers, the setup that is balanced and tuned correctly might feel different than anything they have experienced before. Just driving the balanced setup might take time to adjust to.

When you go through the process and work out a successful setup, there's nothing like it. To see all of your hard work end up in victory lane is priceless. We created ORS to help put racers on the right path to success. We can't do it for you, but with the right information and the right tools, you can surely do it yourself.

And, if the car was close, but not all the way there, the driver can have a better opportunity during the race to fine tune their thoughts on how to make it better, if not for anything else, his benefit.

Timing During The Race – The first point I will make about Post Race Evaluation is this. Make sure during the race that you have someone taking and recording lap times for your car as well as the leaders. This tells us a lot about our possibilities.

If your car is leading, then the lap times will tell us how much performance gains we have over our competition. If we are chasing the leaders, the lap times will tell us how much work we have to do to be competitive enough to lead and possibly win races.

If your car is off in lap times to the leaders, then it becomes necessary to know where the loss is being produced. You will need to have someone time your turn segment times against those of the leaders. This tells us a lot.

In a personal experience several years ago, I setup a Pro Late model car and felt we had a good balanced setup. In the first race we ran, the dad of the driver went to turns one and two and timed our car and the leaders. I went to turns three and four and did the same. He ended up telling me that we were about a tenth faster in one and two and I recorded laps that were the same times as the leaders at my end of the track.

We were off in total lap times by 0.5 second. The car was losing at least 0.25 seconds on each straightaway. This told us that we needed to work on the motor. If we didn't take the segment times, we could have easily assumed we needed to work on the setup, just like many teams would do.

If you are off on segment times, then obviously you need to find a setup that works to produce a faster mid-turn speed. Or, the driver needs to work on their driving style.

This is a tough one to call. We'll be providing a driving school at some point in time and we'll address this issue in more detail then. But suffice it to say, in our experience, throttle and brake modulation is the difference between a good driver and a not so good driver. This has been proven over and over going back many years in racing. A lot of time can be lost between the lifting of the throttle and the point of mid-turn, due to the way the driver operates their foot.

That said, let's look at some things that might not have been perfect for your first race. Much of this is a review of previous information we have presented, but in the context of after the race evaluation.

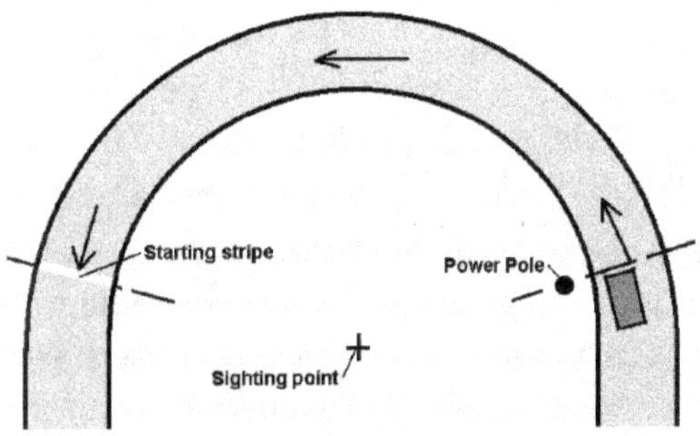

A major step in understanding where your car is performing, or not, is to take segment times for that part of the race track that is from entry lifting and the beginning of braking to where the car starts to straighten out onto the straights. The timing in this portion tells us how well our setup is working against the competition. It is a much better measure than total lap times.

Entry To The Corner – If the car had entry problems, think about when this problem occurred. Was it at the very start and got better as the race went on? If so, the drivers driving style might be something we can look at. Over-driving the corner on entry can cause a push from over-braking. This comes from the excitement of, well, the start of the race. If this might be the case, don't work in the car, work on the driver.

If the entry problem showed up later in the run, then maybe the RF tire is being used too much. This can be caused by too much front brake bias or maybe a tight setup. We can still have a tight setup while being neutral in handling. If other adjustments crutched the tight setup, then the tight part will eventually show up. Here is why.

When the race pace slows later on in the race, the actual lateral G-forces become less. When the G-forces go down, typically the car goes loose. This is a little advanced in our thinking process, but proven to be true. A good balanced setup will in most cases be balanced through a wider range of G-forces than one that is not balanced. I personally worked out a setup for a NE Tour modified team this year where the roll angles stayed the same through G-forces from 2.0 to 2.5 G's. The actual roll angle number went up as the G's increased, but the setup remained balanced. That is a wide range, but fairly typical of a truly balanced setup.

If the car on entry was good for one line, but not so good on another line, like good low but not so good high, then the shock package might need to be examined. There are usually different transitions in the change of the banking on entry and exit for different lines. The low line might have less abrupt transition than the high line does. Shocks can help the car transition through these changes in banking angle.

Entry tuning is the second phase of the on-track setup routine. When we get the middle worked out to create a balanced setup, many of the entry and exit problems will be solved. If your driver goes out and runs laps and has entry or exit problems, look first at the mid-turn performance before starting to change the setup to solve the other two segment problems.

Tight Or Loose Middle – The most obvious reasons for being tight or loose through the middle segment of the turn involves the basic spring/rear roll center parts of the setup, but then again, we were supposed to have worked that out in testing, right? If you didn't have a chance to test and just had short practice sessions before the race, then maybe when you put on those new tires at the start of practice, they masked the problem.

Definitely take tire temperatures immediately after the race if at all possible. These tire temps. tell more about the setup than the ones you collected from shorter practice runs.

If the tight or loose condition was present the entire race, then work on creating a more balanced setup by making setup changes. These longer runs will definitely provide more information than the shorter practice runs.

Look at the tire wear to determine if you are over-using one or more tires. If the car is setup tight, then the RF and RR tires might be abused. The RR tire will show more heat and wear than any of the others from either a loose setup, or a tight setup that produces the "tight/loose" syndrome, which we have talked about.

An Important Point - In yesteryear, the primary cornering problem usually involved a tight car because traditionally, most race cars didn't turn very well. What we know now about front geometry was not so widely known then. Now, in today's racing, we have learned how to produce larger contact patches for the front tires, and more than anything else, we have learned how to put a lot of loading onto the LF tire by using the bump setups and other methods.

Here is where the problem comes in. And this information comes as a result of a more recent experience that the owners of this school came across. With a dynamically loose setup, or one where the front has more grip than the rear, the rear tires will slip out and find an Angle Of Attack (remember that Lesson?) that will provide the grip they need to get through the turns.

This added AOA stresses the rear tires, especially the RR tire, and they cannot continue to provide this level of grip for a long time. In a race where we discovered the setup was very loose, the car ran fairly neutral for up to 30 laps before it began to go to handling loose. It never had great drive off the corners, but that could have been a natural result of the race track we were running on.

Only later did we discover that this car had a much lower CG than originally estimated. So, the setup we installed was loose for that reason. We learned something from this. A dynamically tight setup becomes evident much sooner than a dynamically loose setup.

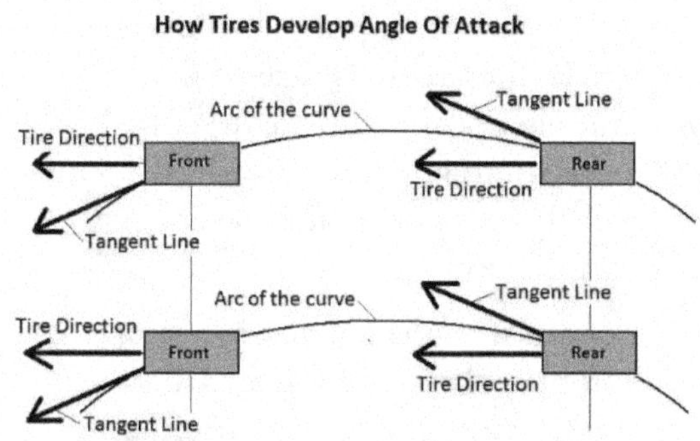

We have learned that a car will be more forgiving when the setup is loose verses when it is tight. Some of that is dependent on the driver and what they like. Some drivers can drive a loose car, some cannot. For those who can, a setup that is more towards the loose side can be faster. That is because as the rear slips out, it finds more Angle of Attack and gains grip. The car then becomes more neutral and not so loose.

This photo is taken near mid-turn and we can see where the rear tires have slipped out and are now at a higher angle of attack to the direction the car is traveling. This is exactly what we are talking about in this section. By running at this angle, the rear tires have more grip and the loose condition goes away.

Exit Performance – If your car is good on entry, good through the middle, but loose or tight off the corner, again, when did this happen? In the previous case of the Dynamically Loose setup we described, as the car began to go loose through mid-turn after the 30 lap point, it also became very loose off the corners.

So, in contrast to the tight/loose setup that is very common throughout the past twenty years or so, we now have a Loose/Loose syndrome happening. The car is setup loose but doesn't show it until later in the run.

If in a race, because of many caution periods, the car happens to run short sprints, the loose condition from the loose setup may never show up. But, the drive off the corner will never be as good as it could be. That might just be a clue and something to consider when evaluating the actual race performance.

Going back to the case of a tight/loose setup, the loose off condition will show up fairly early in the run, or race, because the rear tires don't have as much time to seek the added AOA that will help mediate the problem. Just as the rear tires are slipping out, the driver gets into the throttle and the rear tires cannot adapt that quickly. The driver will report a loose car.

When a race car is exiting the turns, there is a lot to consider that is different than what we do for entry and mid-turn. We still don't want to disturb the mid-turn performance by making wholesale spring and roll center changes that will disturb the mid-turn performance.

How To Tell If The Car Is Balanced - The key to realizing and recognizing a truly balanced setup is how the car performs over a longer number of laps run. In one of my first experiences with a race team, we had developed a truly balanced setup. We ran a full 50 laps on sticker tires at a test before the season started. The lap times fell of about 2.0-2.5 tenths of a second by lap 20 in the run. The next 30 laps, the times did not fall of at all. They stayed very consistent within a few thousandths per lap. That car won a lot of races.

The goals for a team are up to the team to determine. If the races are short, 35 laps or less, and there will inevitably be a few cautions thrown in, then maybe a jack rabbit setup that won't be good in the long run might just be the best setup to run. For those conditions at that kind of track, maybe this kind of setup will work better. Only testing and racing will tell.

If your racing involves longer races beyond 35 laps, and in the later laps there are usually less cautions to break up the runs, then by all means, the balanced and more consistent setups will usually beat all others.

It is true of most racers, even in today's racing, that they test and practice to find a setup that will produce the fastest lap times. These setups will often be un-balanced in either the tight or loose direction, depending on the type of race track. The team can definitely get away with this if the runs and the race duration are shorter. If your type of racing involves longer duration and longer runs, then the totally balanced setups will be superior in the long run, literally.

One of the benefits of a balanced setup is the ability for the driver to run different lines, high or low, through the turns. In a test we did, a modified Tour car was balanced through a range of 2.0 to 2.5 G's. In many cases, as the tires wear, a group of cars will slow from a half second to a full second or more in the course of a race. Since G-forces are dependent on speed through the turns, as the tires wear, the turn speeds decrease and the G-forces go down, So, a more consistent setup that stays balanced through a wide range of G-forces will stay balanced the entire race.

There will be situations where you will be forced to run different lines, so a balanced setup is critical in being able to compete and be fast no matter what groove you end up in.

Summary Lesson Twenty-Two – The actual race event is where we can learn a lot about the way our car is setup. The setup is only part of the equation, the driving being the other part. We can evaluate the tire temperatures and tire pressures and we can determine that we have a good and dynamically balanced setup. Once we know that, we can move on to evaluating our motor and gearing combination, and also to driver considerations.

Gather the data and record the data. A lot of time is spent looking over all of the information gained in testing and the race by experienced and competent crew chiefs. Make information collection a priority.

Most of all, use the data to make a decision and then go with it. Don't go back and forth with your thoughts. Too many teams do this. The race will tell you what you need to do next. Trust what the race tells you and make adjustments for the next race. It's a process that is fun if you do it right, because success is its own reward. And once you find that perfect setup, don't change a thing, just maintain it.

Exam - In The Context Of This Lesson:

Which Information Is The Most Valuable?

1) What works to make the car better

2) What works to make the car worse

3) Neither, both are valuable

Lap Times Tell Us Where Our Performance Is Lacking?

1) True

2) False

Entry Problems Can Be Caused By Which?

1) Brake bias errors

2) Rear alignment

3) Overdriving the corners

4) All of the above

Which Of These Might Show Up Earlier In The Race?

1) A tight setup

2) A loose setup

A Balanced Setup Will Be Good Where?

1) Through a wide range of G-forces

2) On the longer runs

3) Through the mid-turn portion of the turns

4) All of the above

How Can We Tell If The Setup Is Balanced?

1) By examining tire wear and tire temperatures

2) Less loss of lap times and performance

3) Consistent performance

4) All of the above

Race Car Technology – Level Three

Lesson Twenty-Three – Level Three Summation

Congratulations! You have just completed the Race Car Technology Level Three course. We hope you have gained a better understanding of your race car and what it takes to set it up. Whether you are coming from Level One all of the way to this finishing point, or just the Level Three, we have made every effort to provide the most current and useful information that exists in today's racing community.

For those who have taken the complete journey from Level One, you have learned about all of the parts and pieces that make up a modern-day race car. Then when you worked through Level Two, you learned how to work with those parts to assemble and make correct all of the chassis components in preparation to setup the race car for competition.

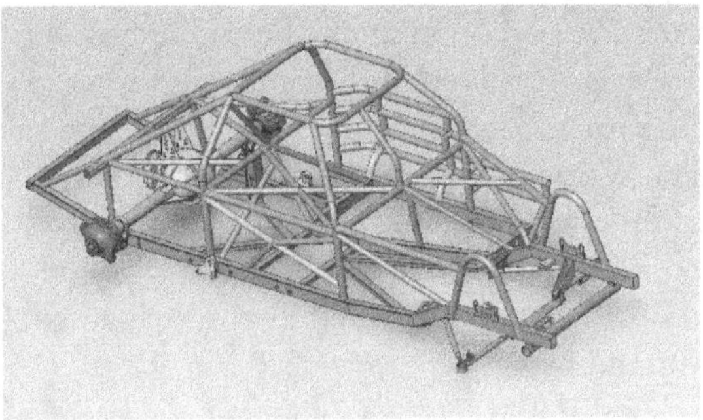

We started out with our Level One courses that taught the students all about how a race car was constructed and named all of the parts and pieces that go into a modern day race car.

Now in Level Three, we learned about the engineering that goes into the decisions we will make about how to setup a race car. We now know important information about both the car and the race track we will be racing on that will make us successful.

We taught the students, in Level Two, how to adjust all of those parts and pieces so that they would work together to improve the performance of the race car. We discussed bump steer, alignment, Ackermann, rear steer and much more.

This all helps us to make intelligent decisions about both the initial setup we will use, but also how to tune that setup once we hit the race track. We now know how to evaluate the setup during testing or practice and what to do after we have raced the car in actual competition.

The racing environment is ever changing and the way we design and setup the race cars is also changing as time goes on. Just when you think you've got it all figured out, someone, or some group comes up with another plan and takes all of this to a new level.

The level we are at right now, in April of 2018 as I write this Summation, is very advanced from when I started my career some twenty years ago. But, that being said, the basics are still the same as to how to best make the four tires work. IN todays racing, we are now causing the cars to run at a much different, and more advantageous, attitude than ever before.

This thing called Race Car Setup is not magic. It is science and the overall theme is one that everyone can understand because it is simple in concept and fairly easy to apply.

You just have to have the keys to the puzzle, and that in just a few words is what this school is all about. We provide the necessary keys to the puzzle.

Now, in this Level Three, we learned how to work with the forces and mechanical motions of the race car so that we can make our four tires gain the most grip possible for a much longer period of time. Understanding the forces and how to design the car to take advantage of those forces is what advanced race car setup is all about.

We started out with the intent of giving you the student the knowledge and the tools to be able to understand these basic principles and to teach you how to apply them to any race car. We hope we have done our job.

There will be no more school quizzes. The final exam is when you get to work on a real race car, hands on, and experience the thrill of going through the process of preparing that car for competition and then seeing it compete. It is both scary and exciting at the same time.

But, once you have learned the car and worked out all of the bugs, the first time you stand in the winner's circle, you'll look back with pride that you had the courage, fortitude and commitment to see this journey through to the end. That's the way it has been for the founders and all of our instructors in the Online Racing School. And I can tell you from personal experience, there's nothing like it.

You too can be a member of a race team, if you're not already involved with one, you just have to ask. Now that you have knowledge of how a race car works, and how to set one up, any team would be happy to have you join them and to be part of their success.

Thank you so much for trusting us to provide your advanced education in Race Car Technology. As to the future of RCT, we will be adding more course books that will be more focused on individual routines and processes involved in the preparation, setup and process of racing different types of race cars. We will enlist the help of industry professionals to act as the Instructors for those courses. These will be names that are well known and respected in our racing community.

Now, go out there and play with a race car. If you don't already have a team to work with, go to the race track and choose one. That is what I did and it worked out very well. Most teams are looking for educated and motivated team members. The fact that you have completed this school and showed an interest proves you are worthy.

Our school will continue to grow and we will be adding more and more books as time goes on. Please keep checking back to look for additions to the school, as well as new content in those courses. Plus, we will be adding courses that deal specifically with routines and processes involved in setting up and racing in competition. These new courses will be presented by authors that are very experienced and well respected names in the industry.

Good luck in all of your racing endeavors and always stay safe.

Signed, Bob Bolles, Race Car Technology Courses

Exam - In The Context Of This Lesson:

Which Information Is The Most Valuable?

1) What works to make the car better

2) What works to make the car worse

3) Neither, both are valuable

About the Author

Bob Bolles has been a hands-on motorsports engineer for over twenty-five years. Although he holds a degree in the Mechanical Engineering, his skill and experience with racecars comes from working directly on the chassis and with many hundreds of race teams. He likes to think that he comes from much the same mold as a few others before him such as his friend, the late Dr. Smokey Yunick, who also wasn't afraid to "get his hands dirty" to effect change. Like Smokey, Bob loves this sport and enjoys the interaction with others who also love it.

The nice thing about racing is that you cannot BS what you know or what you develop. It either works or it doesn't. Proof is just a few laps away. Before Bob started his research into racecar dynamics and engineering, he struggled with the very same problems that most race teams continue to struggle with today. He just knew there was a better way and if he just looked hard enough, the answers might come. Well, they did. That information is shared in the pages of the RCT series.

Over the years, Bob has worked with a high degree of success on virtually every type of stock car raced in the United States, as well as sophisticated formula type race cars. Whether it is an asphalt stock car, modified, or a dirt late model, his techniques and methods have improved handling and opened the door for winning. He has engineered cars that have won major asphalt late model championships, touring championship, modified championships, road racing championships, and dirt late model races including the Dream and the World 100 at Eldora Speedway.

It is this broad level of experience and his development of new technologies that qualifies him to write about these important subjects. Not a day goes by that Bob isn't speaking with or helping racers. With contacts all across America, he is truly in tune with the pulse of auto racing on a technological level that very few individuals enjoy.

The software Bob developed in the mid-1990's is still being used by championship winning teams throughout the U.S., Canada, Australia, New Zealand and Europe. His dream of being able to help all racers in their pursuit of success and enjoyment is fast becoming a reality.

Bob has been a professional motorsports technical writer serving as the Senior Technical Editor for Circle Track magazine for over fifteen years and currently as a technical contributor to Speedway Illustrated magazine.

Race Car Technology- Full Course represents Bob's fourth book in this series. Bob is also the author of RCT - Levels One thru Three. These books are being used by college instructors to educate future motorsports engineers.

www.ingramcontent.com/pod-product-compliance
Lightning Source LLC
LaVergne TN
LVHW080008210926
848481LV00011B/2971

Race Car Technology

Full Course – RCT L1, L2, L3

CRD Publications presents the Race Car Technology Full Course book. By reading this book, you will receive all of the information contained in RCT Level One, Level Two and Level Three. These three combined books teach you all about the parts and pieces of a race car, how to design and setup those cars for maximum performance and all about current engineering principles and applications.

You have taken the first step in furthering your knowledge and understanding of the systems and parts of a dedicated race car, how to work with those parts, and advanced engineering concepts and applications. This book will help you:

- Become familiar with the all of the components on a race car,
- Learn how the various parts of the car interact,
- Learn how to setup and tune the different chassis components, and
- Become educated in the physics and dynamics all race teams use to develop winning setups.

The RCT Full Course book will explain how all of the relevant systems that race teams and engineers work with on a daily basis can be put together into a winning package. The information is State of the Art, and up-to-date in every aspect. Learn what winning teams already know and how they produce performance.

Most of the information is relevant to all race cars, and some of the information is specific to certain types of racing and specific race cars. Every race team will benefit from the extensive range and detail contained in the RCT series of books. The basic technology is common among a wide range of race cars from simple street stock circle track cars to sophisticated prototype road racing and formula cars.

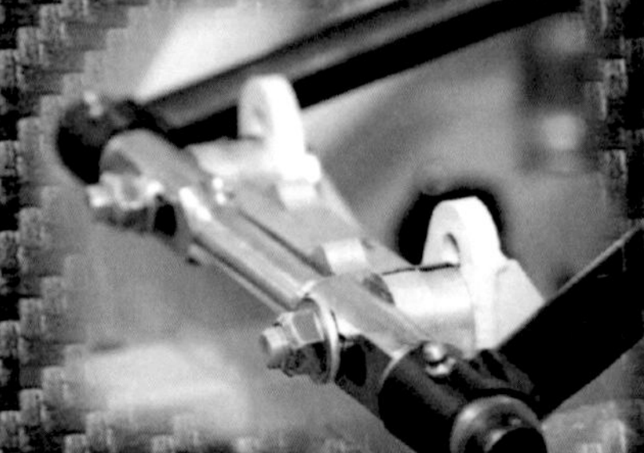

$135.00
ISBN 978-1-7324884-6-5
13500>
9 781732 488465